Water Security in a New World

Series editor
Zafar Adeel, Pacific Water Research Centre, Simon Fraser University
Burnaby, BC, Canada

More information about this series at http://www.springer.com/series/13406

Regina M. Buono • Elena López Gunn
Jennifer McKay • Chad Staddon

Editors

Regulating Water Security in Unconventional Oil and Gas

 Springer

Editors
Regina M. Buono
Center for Energy Studies
Baker Institute for Public Policy,
Rice University
Houston, TX, USA

Lyndon B. Johnson School for
Public Affairs
University of Texas
Austin, TX, USA

Jennifer McKay
School of Law, Division of Business & Law
University of South Australia
Adelaide, SA, Australia

Elena López Gunn
ICATALIST
Madrid, Spain

Chad Staddon
Department of Geography and
Environmental Management
University of the West of England
Bristol, UK

ISSN 2367-4008   ISSN 2367-4016   (electronic)
Water Security in a New World
ISBN 978-3-030-18341-7   ISBN 978-3-030-18342-4   (eBook)
https://doi.org/10.1007/978-3-030-18342-4

This Springer imprint is published by the registered company Springer Nature Switzerland AG.
The registered company address is: Gewerbestrasse 11, 6330 Cham, Switzerland

Foreword

Water is integral to the production, transportation, and use of most forms of energy, and more so for shale oil and gas. This book, *Regulating Water Security in Unconventional Gas and Oil*, enhances our understanding of regulatory and policy regimes around the world in relation to the use of water for the unconventional production of oil and gas, namely, through hydraulic fracturing. Legal, policy, and regulatory issues surrounding the use of water for hydraulic fracturing are present at every stage of operations: operators must understand the legal, political, and hydrological context of their surroundings, procure water for use in the fracturing and extraction processes, gain community cooperation or confront social resistance around water, collect flowback and produced water, and dispose of these waters safely. It is expected that climate change will impact rainfall and water availability, intensifying the water deficit in many regions. This fact is most evident in oil- and gas-producing dry areas.

The book offers an interdisciplinary approach that includes chapters looking at water security in energy production in general (the "water-energy nexus"—a research area to which I have deep ties and a chapter to which my students contributed); public opinion of hydraulic fracturing (especially with regard to water supplies); implications for groundwater management and protection; the impacts of hydraulic fracturing on agriculture, municipalities, and other stakeholders competing for water supplies; potential conflicts between hydraulic fracturing and water as a human right; induced seismicity; and more. The book includes 18 issue-focused and geographically distinct studies written by scholars from around the world. One of the strengths of this volume is the inclusion of the work of scholars local to the countries discussed, in particular to those countries in which shale oil and gas are most prominent (the United States, Australia, Argentina, China, Canada, and the United Kingdom) and to those countries that have yet to be examined extensively (Ukraine, Russia, and Poland). The book also considers countries in which unconventional production may develop in the near future (South Africa, Mexico).

My research team's experiences in San Antonio, Texas, align with much of the findings in the book. The Eagle Ford Shale play, just south of Bexar County, Texas, has substantial fracturing activity. Significant expansion is expected to lead to a large increase in the production of oil and gas from shale. Between 2008 and 2014, the number of new drilling permits issued leapt from only 26 to 5613 permits.[1] Groundwater use has been found to be 2.5 times greater than the rate of recharge of the aquifers from which water is drawn for oil and gas production (Arnett et al. 2014). The region already faces growing concerns about the future of its water resources: a 40% deficit is projected, even in the absence of hydraulic fracturing demands. The primary concerns associated with water demanded for hydraulic fracturing are the decline in quantity available for other uses (municipal and agricultural) and the necessary disposal, treatment, or recycling of the reclaimed water. Thus, the region has substantial water issues related to (1) the growing fracturing-based energy sector, (2) the rapidly growing municipal sector, (3) the active irrigated food production sector, and (4) the important regional environmental issues, such as the emerging water deficit. This book examines many of these issues and considers how various jurisdictions have attempted to address them in law and policy.

By analyzing and comparing various approaches to these issues from around the globe, the book provides insights into how policy, best practices, and regulation may be developed to advance the interests of all stakeholders. While culture, geography, and local factors potentially make transferring "good practice" from one place to another challenging, there is value in examining and understanding the components of different legal and regulatory regimes: these may assist in the development of better regulatory law and policy moving forward. This book presents a valuable step toward that objective.

Professor and Dean, Faculty of Agricultural   Rabi H. Mohtar
and Food Sciences   mohtar@tamu.edu
American University of Beirut   mohtar@aub.edu.lb
Beirut, Lebanon   http://www.wefnexus.tamu.edu

TEES Professor, Biological & Agricultural
Engineering Department and Zachry Department
of Civil Engineering
Texas A&M University
College Station, TX, USA

Reference

Arnett B, Healy K, Jiang Z, LeClere D, McLaughlin L, Roberts J, Steadman M (2014) Water use in the Eagle Ford Shale: an economic and policy analysis of water supply and demand. http://hdl.handle.net/1969.1/151989

---

[1] http://www.rrc.state.tx.us/media/8675/eaglefordproduction_drillingpermits_issued.pdf

Acknowledgments

Regina M. Buono

Many thanks to the Center for Energy Studies at Rice University's Baker Institute for Public Policy for supporting my efforts on this volume and to my family for their unending patience and support in everything I do.

Elena López Gunn

To my husband, Dave, and my beautiful, special kids, Niko, Alex, and Anna Wendy, for their endless patience and support despite hours stolen from family time. Also to my wonderful team of co-editors, a real challenge to bring this book to port while residing on three different continents, but we did it!

Jennifer McKay

I would like to acknowledge my wonderful children, James and Monica, for their love and support. In addition, colleagues from the Law School at the University of South Australia, especially Professor Wendy Lacey and the university in general, for resources devoted to this research. The terrific editorial team who negotiated many changes and with good humor worked on this project. Finally, law students in two courses who were encouraged to debate the many legal issues pertaining to the processes involved.

Chad Staddon

My wife, Mary Beth, and daughters, Sophia and Isobel, deserve recognition for their forbearance while this complex project was completed—I promise that I will now start making Sunday waffles again! I am grateful to my colleagues, Enda Hayes, Jenna Brown and Sarah Ward, for their good humored criticism during our many discussions of the book and its broader themes, to Calum Foster and Harry West for producing the maps, and to Wayne Powell for patient and meticulous copy editing. Alas, no waffles for you! I would also like to acknowledge the on-going support of the Lloyd's Register Foundation, a charitable foundation helping to protect life and property by supporting engineering-related education, public engagement, and the application of research. For more information, see: www.lrfoundation.org.uk.

Contents

Part I
Framework and Context

Chapter 1
Regulating Water Security in Unconventional Oil and Gas: An Introduction

Regina M. Buono, Elena López Gunn, Chad Staddon, and Jennifer McKay

Abstract The last 20 years have seen dramatic growth in the production of oil and gas from shale, as production techniques developed in the latter half of the twentieth century have advanced under largely favorable economic conditions. Hydraulic fracturing is a well stimulation technique in which sand and other proppants suspended in fluids are forced at high pressure through cracks in shale rock to free hydrocarbons to flow to the surface. This requires significant volumes of water and presents challenges for protecting nearby humans and the environment from water, air, and noise pollution, as well as other effects of the activity. Legal, policy, and regulatory issues related to the use of water for hydraulic fracturing are present at every stage of such "unconventional" operations. Operators must understand the legal, political and hydrological context of their surroundings, gain community cooperation or confront social resistance, procure water for use in the fracturing and extraction processes, collect flowback and produced water, and dispose of these waters safely. A recent study found significant increases in water use for hydraulic fracturing and wastewater production in major shale gas and oil production regions,

R. M. Buono
Center for Energy Studies (CES), Baker Institute for Public Policy,
Rice University, Houston, TX, USA

Lyndon B. Johnson School for Public Affairs, University of Texas, Austin, TX, USA

E. López Gunn
iCatalist, Madrid, Spain

University of Leeds, Leeds, UK

C. Staddon (✉)
Department of Geography and Environmental Management, University of the West of England, Bristol, UK
e-mail: chad.staddon@uwe.ac.uk

J. McKay
Professor of Business Law, Law School, University of South Australia Business School, Adelaide, SA, Australia

with attendant increases in water-use intensity over time (i.e., water use normalized to the energy production) (Kondash et al. 2018). The water volumes required for hydraulic fracturing are only likely to grow over time, as will the challenges of meeting that demand, and of disposing of wastewater associated with that production. This book considers how regulators and other decision makers have addressed many of these issues, considering varying legal frameworks, political systems, social acceptance, and geologies around the world (Fig. 1.1).

Keywords Hydraulic fracturing · Framework · Water-energy nexus · Acquisition · Waste disposal · Geology · Geography · Politics

1.1 Introduction

The last 20 or so years have seen dramatic growth in the so-called "unconventional production" of oil and gas from shale, as production techniques developed in the latter half of the twentieth century have advanced under largely favorable economic conditions. Hydraulic fracturing is a well stimulation technique in which sand and other proppants suspended in fluids are forced at high pressure through rock cracks to free hydrocarbons to flow to the surface. This requires the use of significant volumes of water and presents challenges for protecting nearby humans and the environment from water, air, and noise pollution, as well as other effects of the activity. Legal, policy, and regulatory issues related to the use of water for hydraulic fracturing are present at every stage of unconventional operations. Operators must understand the legal, political and hydrological context of their surroundings, gain community cooperation or confront social resistance, procure water for use in the fracturing and extraction processes, collect flowback and produced water, and dispose of these waters safely. A recent study found significant increases in water use for hydraulic fracturing and wastewater production in major shale gas and oil production regions, with attendant increases in water-use intensity over time (i.e., water use normalized to the energy production) (Kondash et al. 2018). The water volumes required for hydraulic fracturing are only likely to grow over time, as will the challenges of meeting that demand, and of disposing of the resulting wastewater associated with that production. This book considers how regulators and other decision makers have addressed many of these issues, considering varying legal frameworks, political systems, social acceptance, and geologies around the world (Fig. 1.1). It also highlights common challenges faced by operators and regulators attempting to normalise hydraulic fracturing within the modern energy industry.

1.2 Some Conceptual Reflections and Objectives

Advancements in unconventional oil and gas production have been heralded as industry-changing developments that have had—and will continue to have—important implications for energy production scenarios around the world. New

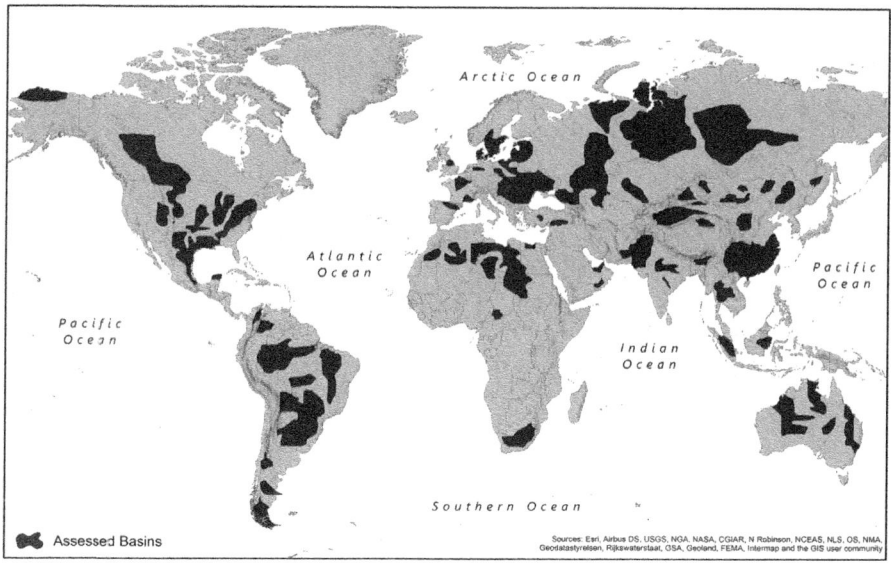

Fig. 1.1 Major Shale Gas Basins around the World (Callum Foster, UWE, Bristol Cartography)

technology has now made vast expanses of new resources economically accessible and profitable, presenting huge money-making opportunities. The enormous supply of natural gas made available by these advances has been hailed as the transition fuel to a low-carbon future, though some have also expressed concerns that society will become mired in the transition if it diverts investment from renewable energy sources. This revolution, however, arrives as freshwater supplies around the world grow scarcer, a problem compounded by climate change. This tension brings operators and other interests who seek to exploit these resources into real or potential conflict with others over water.

The objectives of this book are to offer different perspectives, data, and experiences regarding the issues that unconventional energy development—hydraulic fracturing—presents for water security around the world and to examine how these challenges are being addressed by regulators, operators, and other stakeholders in various locations. By analyzing and comparing approaches to these issues from around the globe, we hope to glean insights into how policy, best practices, and regulation may be developed to advance the interests of all stakeholders. While it is not possible to simply "cut and paste" aspects of successful regulation and governance from one place to another, there is great value in examining and understanding the components of legal, governance and regulatory regimes that work relatively well and those that perform relatively poorly, as lessons drawn from both may assist in the development of better law and policy. The laws and regulations herein were current when the chapters were completed in 2017 and 2018, but as with all analyses of current and evolving policy issues, we are trying to hit a moving target. Technological, regulatory and political changes may move faster than the publishing process, but it is our hope that the lessons learned will remain useful and pertinent for future policy development.

This topical book has had a long gestation, reaching back to initial discussions between some of the co-editors and authors in 2016. We met as a group for the first time at the IWRA World Water Congress in Cancun, Mexico, in June 2017. Two of the editors are legal scholars (Buono and McKay) and two are social scientists (López Gunn and Staddon), which led to some discussion about a common framing for the book around what might be called "critical socio-legal studies". This approach attempts to balance the "doctrinal" approaches favoured by legal scholars with social scientific concerns about broader social, economic and political context. In commissioning chapters, attention was paid to ensure regional and thematic coverage. An example was the specific commissioning of a chapter on indigenous communities and the community consultation processes in Canada (Chap. 15). The resulting collection is somewhat eclectic and multidisciplinary, drawing on research methods from several disciplines, including law, politics, and economics.

We asked the authors to approach the issues in a practical manner with policy in mind—grounded in data and theory but directed at generating useful recommendations for stakeholders evaluating the opportunity or threat of future hydraulic fracturing operations or attempting to understand the current and future implications for water security of operations in play now. We sent all potential contributors these framing questions:

- (i).   What are some of the key features/issues of water security (and for whom?) in the context of hydraulic fracturing?
- (ii). What is the interaction between the political economies of water and of energy in your case study?
- (iii). How do we understand multilevel policy in the jurisdiction?
- (iv). How does science interact with policy regarding hydraulic fracturing?
- (v).   What is the role of conflicts and resolution measures to date, e.g., potential conflicts between hydraulic fracturing and water as a human right, in your case study? What are the key factors in community cooperation or confrontation and the current social resistance around shale gas extraction?

All authors addressed these questions and added their own particular reflections and research on the complex issues relevant to their particular domain. Many were also able to learn from, and refer to, other draft chapters, creating a stronger sense of cohesion amongst chapters. As the book developed, we commissioned a few additional chapters (e.g. Chaps. 9 and 10) and asked some authors to include aspects not in their original drafts, for example on community-based protests against unconventional methods, as these emerged as common themes.

The book takes an interdisciplinary approach. The authors in this volume come from a variety of institutions and perspectives—many in academia and policy think tanks, but some from business, the consulting community, non-governmental organizations, and the practice of law. The authors' disciplinary backgrounds are varied, encompassing everything from agriculture to zoology, and from law and economics and geography. The approaches taken were, accordingly, methodologically diverse, wherein some authors have made explicit their research approach and others describe the issues, relying on their deep knowledge of the topic. For example Chap. 8 uses political ecology and critical discourse analysis and presents data from interviews with two Argentinian Parliamentarians and two industry representatives. Other chapters use human rights-based approaches (Chap. 3), legal methods such as doctrinal analysis (Chaps. 5, 6 and 13), or are more regional and descriptive (Chaps. 10 and 18).

The book includes chapters looking at water security in energy production in general (i.e., the "water-energy nexus" or how unconventional oil and gas production interacts with the human right to access to water), as well as issue-specific and geographically (or jurisdictionally) distinct case studies written by scholars from around the world. One of the strengths of this volume is the inclusion of the work of scholars local to the countries being discussed and, in particular, those countries where shale oil and gas are most prominent, either because resources are already under exploitation or because existing reserves are in exploration (namely the United States, Australia, Argentina, Canada, and the UK). We have also attempted to include consideration of those countries where shale oil and gas production could soon develop (e.g. Poland, Mexico, South Africa and China).

1.3   Organization of the Book

With the objective of contributing to the goals outlined above, this book is divided into four sections. After this introductory chapter, Part I is devoted to broad frameworks and overviews of social issues related to the propagation of hydraulic fracturing. Parts II and III turn their attention to the production processes themselves, considering how operators acquire water for use in fracturing processes and manage the waste products, especially the wastewaters, generated. Specifically, Part II addresses water use and availability for hydraulic fracturing operations, examining how operators—or would-be operators—in various parts of the world (including some regions already suffering from water stress) obtain water for use in their operations, and how regulators seek to balance the allocation of water with prior existing uses such as agricultural, industrial, or municipal demand. Part III addresses the management of water at the other end of the hydraulic fracturing process, considering how operators and regulators manage both the water used in fracturing and stimulating the source rock, and the water naturally produced by the formation that is brought to the surface by operations. Part IV takes a deeper dive into specific cases, examining how jurisdictions as varied as Poland, South Africa, and Brazil address often similar questions related to the use of water in unconventional energy production. In the concluding chapter, the editorial team offers thoughts on the emergent issues and themes, putting forth recommendations for stakeholders working in the field, and suggesting a tentative research agenda to further enhance understanding of this important topic.

1.3.1   Part I: Frameworks and Context

The first section of the book lays out a range of broad frameworks and issues that overlay the unconventional production of oil and gas and its effect on water supplies. Ranging from the water-energy nexus to the human right to water, topics here consider hydraulic fracturing from the broader perspective—i.e. the conceptual issues and challenges the technique presents no matter where in the world operations are being conducted.

A holistic understanding of water security as related to hydraulic fracturing would be incomplete without considering the intricacies and trade-offs inherent in the water-energy nexus. In Chap. 2, Ahmed M. Mroue, Gabrielle Obkirchner, Jennifer Dargin, and Jordan Muell address this topic, considering issues related to hydraulic fracturing from the nexus perspective. Hydraulic fracturing is a critical point of tension in the water-energy nexus due to both its water-intensive nature and the potential water pollution risks it poses. Decision makers managing elements of the water-energy nexus often have conflicting preferences or face opposing pressures and scenarios associated with environmental, economic, and social tradeoffs.

The authors discuss a number of potential solutions, as well as tools developed by their research, to conduct rigorous tradeoff analyses and modeling of scenarios to assist decision makers and thus facilitate better, more sustainable, resource planning. These tools consider energy and electricity production, as well as the social, economic, and environmental impacts of these activities by quantifying relationships and trade-offs between water, energy, and transportation. Some of these tools have been applied to understanding energy production in the Eagle Ford shale region of Texas, where water use for hydraulic fracturing "is in direct competition with water use for agriculture and municipal purposes" (Mroue et al., Chap. 2). In areas experiencing this kind of demand by energy interests, population growth—especially in areas with sizeable municipal areas—is predicted to cause potential water shortages for competing sectors. The authors also review existing and developing technology and policy solutions that address key water-energy nexus challenges and call for the creation and application of policy that better addresses the nexus of water and energy, as well as mechanisms to assist institutions in transitioning from standard sectoral policy towards a holistic nexus approach that views the water and energy sectors simultaneously—including water and energy security interlinkages.

Robert Palmer, Damien Short, and Ted Auch in Chap. 3 chart the international development of the human right to water and contextualize its relevance to discussions around unconventional energy, with a case study focus on the United States. Serious concerns have been raised over the effect of a range of polluting activities associated with the hydraulic fracturing process. The authors argue that water impacts of unconventional energy extraction are one of the most contentious and widely publicized environmental and human rights issues today. The scientific literature, NGO policy reports and other grey literature, and the testimony of many local residents, indicate the likely impairment of the human right to water for residents living near the site of hydraulic fracturing operations. Perhaps the major issue regarding water use is the shifting of the resource from society to industry and the demonstrable lack of supply-side price signalling that would prompt the oil and gas industry to reduce or stabilize their water demand per unit of energy produced.

Jim Bradbury and Courtney Cox Smith, practicing attorneys working in the heart of Texas's Barnett shale—ground zero for attempts to impose local control over fracturing operations in hydrocarbon-loving Texas—examine in Chap. 4 the widespread controversy in the United States over hydraulic fracturing and its potential implications for surrounding areas, as well as selected other areas of the globe. Public opposition to hydraulic fracturing is often closely linked to water risks. Bradbury and Cox Smith pinpoint clear areas of conflict and public concern over hydraulic fracturing. Specifically, they examine issues related to the use of water in arid environments, mandatory disclosure of water volumes and chemical components, baseline testing of local water supplies, induced seismicity, contamination risks related to hydraulic fracturing and the disposal of waste materials, noise pollution and residential proximity to operations, and methods of resolving or minimizing conflict and opposition. They also consider instances in which efforts have been made to prohibit hydraulic fracturing in its entirety within circumscribed geographic areas, including the state of New York, portions of Texas, and a few European countries. Controversies are briefly examined and discussed for Texas, Colorado, New York, the United Kingdom, and Spain.

1.3.2 Part II: Acquiring Water for Hydraulic Fracturing: Conflicts and Regulatory Issues

As noted above, water is essential to unconventional energy production in a myriad of ways. Part II of this book considers how and where operators acquire water for use. The volumes of water needed to support fracking operations are a key concern and may act as a limiting factor, particularly in regions that already suffer from water scarcity. A 2014 report by Ceres found that vast numbers of hydraulically fractured wells have been installed in places with high or extremely high water stress, and that more than 55% of wells were located in areas experiencing drought (Freyman 2014). A more recent study by Kondash et al. (2018) found that water use per well for hydraulic fracturing in major shale gas and oil production regions of the U.S. increased by up to 770% from 2011 to 2016. Regulatory frameworks governing water and energy are complex: they are often not aligned with one another or are blind to each other, and jurisdiction over a resource or issue may be fragmented between multiple agencies. Operators must compete with agricultural interests, municipalities, industry, and other would-be water users, many of whom can be powerful adversaries with long-entrenched water-usage rights. This part of the book provides a close examination of water allocation and acquisition regimes, as well as related issues, in Argentina, China, Russia, the United Kingdom, and the United States.

Given that the United States was the first nation to actively pursue widespread hydraulic fracturing of shale reserves, it is not surprising that there is a larger literature on water and shale gas deriving from the American experience than from elsewhere. Yet, as Gabriel Collins and Julie Rosen demonstrate in Chap. 5, these issues need to be considered on a state-by-state basis. What happens in one state (for example, Texas) may be quite different to what happens in another (say, Pennsylvania). Even more complex is the position of shale plays and water sources that *cross* state boundaries. Despite jurisdictional slicing and dicing, one trend that remains clear across the board is that water demand for shale exploitation is significant and likely to continue to grow well into the future. Collins and Rosen (Chap. 5) tell us that:

> at its near-term peak in the fall of 2014, daily frac water demand across the U.S. unconventional plays was 8.3 million barrels per day, as calculated from FracFocus disclosure data. To put that number in perspective, it is approximately 3.5 times the daily average water use of Washington D.C., or enough to fill NRG Stadium (home of the Houston Texans[1]).

The authors note that Texas accounts for approximately half of total water abstractions for hydraulic fracturing in the United States, with four other states (North Dakota, Pennsylvania, Colorado and Oklahoma) accounting for most of the other half. Moreover, their research has identified a new trend, a rise in the number of "super fracs," where water consumption is significantly higher per well due to either supercharging existing wells or drilling more densely from existing well pads. Both strategies are designed to boost capital efficiency and are unlikely to be scaled back without stronger regulation.

Collins and Rosen argue that there are three ways that operators can obtain water necessary for shale gas production: (1) operators can source it themselves (via surface water or ground water abstraction), (2) operators can buy it from existing water services providers or rights holders, or (3) operators can recycle flowback or produced water. In the United States, the option an operator chooses is conditional on the local/state structure of water rights. In Texas, for example, groundwater rights are held by the property owner, creating a situation where operators may acquire water rights as part of the same private deal that gave access to the land necessary for well pad development. As we will see, in other jurisdictions (e.g., the UK or Spain), it is more common for subsurface rights to be vested in a central or regional government. And even in other areas of the U.S., rights acquisition can become much more complex. In Colorado, the axis of power between private interest and public regulation is mediated by a system of "water courts," which can and do exercise significant allocative authority, especially in over-allocated catchments and during drought periods (such as the last several years). There are cases, reported by Collins and Rosen, of water services "middlemen" arising in specific basins, buying rights from rights holders and selling them on to operators as "bundled offers."

Other jurisdictions where unconventional energy production is occurring have approached issues of allocation and use differently. In the UK, where active commercial hydraulic fracturing is only just getting under way, government-issued

---

[1] NRG Stadium has the capacity to seat approximately 72,000 people.

licences are required for water abstractors even if they own the land they are operating on. Jenna Brown notes in Chap. 7 that estimations of water demand for fracking are quite variable, but that even high-end estimates potentially aggregate up to total water volumes of less than 1% of total abstraction for industry and agriculture when compared to annual licensed water abstraction in England and Wales. Staddon et al. (2016), however, noted that the national picture is less relevant than the localized impact at catchment scale, where it is possible for the required volumes to impact the local water balance negatively. Brown recognizes the need for a regulatory approach that is sufficiently flexible to accommodate interannual variation in water availability. In the U.S, this is greatly complicated by the "rigid" nature of water rights allocation, something the UK is attempting to address through a restructuring of the abstraction management regime (Staddon 2014). In the U.K., as Brown (Chap. 7) notes:

> … from 2020 all existing licenses will shift to a new system of water abstraction permits. Under the new permit regime, the volume of water that can be abstracted (surface water or ground water) will depend on the source availability. A new charging system will mean that water taken from high risk/low resilience sources will cost more, water abstracted from abundant sources will be less expensive.

As in the United States, it may be that a middleman market for third party water suppliers opens, especially given the commercial water market "opening" undertaken in England and Wales from April 2017.

Another potentially large market for unconventional operations is China, explored in Chap. 6 by Libin Zhang, Sheng Shao, Fang Dong, and Jiameng Zheng. Though the shale gas sector in China is still in its infancy, the Chinese government has set ambitious goals for shale gas development, driven by the growing energy demand and challenging environmental conditions—including the need to reduce the use of coal, which China has in abundance. The Sichuan Basin in southwestern China is the largest shale gas region, wherein perhaps over 55% of the country's technically recoverable shale gas reserves are located (US EIA 2015). The Fuling field, which is China's first large-scale shale gas commercial production project, is located in the Sichuan Basin (Sinopec 2016). Water shortages present an unresolved challenge for many basins in China. As we see in other chapters, hydraulic fracturing requires access to large amounts of water, which can be in short supply, as in the arid and endorheic Tarim Basin in northwestern China (the second largest shale gas region in China).

As in the UK, ownership of Chinese water resources vests in the state acting on behalf of the common good. The authors of Chap. 6, on China, note that:

> The State has also adopted a water use planning and control regime including, among others, the water resources allocation plans (the "WRAPs"), the industry-specific quotas, and the region-based total quantity control (Water Law, Art. 45 & 47). Therefore, a shale gas developer, like other industrial water users in China, may only have the rights to acquire and use water subject to certain allocation plans, but cannot own the resource.

Also like the UK, China is experimenting with more market-based water allocation mechanisms. Though it is still early days, the China Water Exchange (CWEX) is the mandated exchange platform where shale gas developers may seek to purchase the right to abstract water from transferors. By 2018, only ten deals have been closed on the CWEX since its opening in mid-2016, and none of them is related to hydraulic fracturing.

Bernaldez and Rocio consider in Chap. 8 the advent of unconventional production in Argentina. Argentina possesses what the US EIA has described as "world-class shale gas and shale oil potential—possibly the most prospective outside of North America" (US EIA 2015). The US EIA estimates Argentina has 802 Tcf (trillion cubic feet) of shale gas in-place out of 3244 Tcf of technically recoverable shale gas resources and 27 billion barrels of technically recoverable shale oil (US EIA 2015). In July 2011, Jorge Sapag, the governor of the Argentine province of Neuquén, inaugurated "the first multi-fractured horizontal well aiming at shale gas in Latin America" (Bernaldez and Rocio, Chap. 8). This was provocative because fracking is quite controversial in Argentina, with more than 50 cities and districts having banned it outright. Moreover, energy exploitation in parts of Argentina has an indigenous rights dimension to it, as indigenous communities have long claimed territorial rights and denounced the environmental contamination caused by oil and gas extraction. Now these communities also struggle against problems deriving from unconventional exploitation, as discussed in Chap. 3 of this volume by Palmer et al. on the human right to water. In Bernaldez and Rocio's view, which may also apply to other case studies in this volume, debates about sustainable water management in unconventional gas exploitation typically fail to challenge the modernist discourse of environmental management, which sees resources as put on earth for human use. In other words, there is no inherent problem in draining rivers dry because the alternative, as one Argentinian industry representative put it is that "95% of the water … flows into the Atlantic Ocean" and is lost (Bernaldez and Rocio, Chap. 8). This may lead to general disregard for other meanings of water, decoupled as it is from its hydro-social context.

As conventional oil production declines in Russia's Western Siberian Basin, the Russian government is encouraging the development of 'tight oil' reserves through multi-stage horizontal hydraulic fracturing operations. Owen King, in Chap. 9, reviews the existing literature on the hydrological, environmental and social impacts of this technique, which are limited by the lack of historical data. The further absence of understanding of the implications for specific hydrological contexts such as subarctic Siberia, and the unpredictable nature of these systems due to climate change, means that consideration of environmental and social consequences of the expansion of fracking is severely restricted. This is exacerbated by an absence of effective environmental regulation by the Russian state. King's contribution identifies the potential implications for the 'hydro-social' security of communities 'downstream' of these new extractive industries and identifies the critical gaps in knowledge. He also reflects on insights into the experience of the indigenous Khanty people, relating the changing materiality of the western Siberian waterscape to the global exploitation of the oil that lies beneath it.

In Ukraine, shale gas development intersects with highly charged geopolitics as one of two promising shale gas basins is located in eastern Ukraine, site of an ongoing civil war between pro-Russian separatists and the Ukrainian government.

As Mitryasova, Pohrebennyk and Staddon's chapter explains, initial attempts at development, starting in 2013, had to be abandoned due to the escalating civil war. Some Ukrainians even contend that one of the separatists' war aims has been to prevent development of these potentially rich shale gas deposits. In the other major shale gas deposit, Oleska, located in the western part of the country, it is not war, but lack of a clear regulatory structure and poor initial test well results that is hampering shale gas development. Ukrainian environmental regulation appears still to lack a clear ecosystems focus, meaning that it is ill-suited to managing the complex trade-offs attendant on shale gas exploitation. There are also frictions between the central government, which supports shale gas (partly on an energy sovereignty argument related to the civil war and its tense relations with Russia) and local councils that do not see the benefits to them deriving from permitting shale gas development in their jurisdictions. As elsewhere there is also considerable civil opposition to shale gas development.

1.3.3   Part III: What Goes Down Must Come Up: Disposing of Water From Hydraulic Fracturing

The chapters in Part III consider the options that exist in the jurisdictions of study for disposing of flowback and produced water from unconventional production, as well as related issues such as disclosure requirements, liability for spills or seepage of wastewater, induced seismicity, and the regulation of wastewater recycling. In both conventional and unconventional production, the primary fluid by volume exiting a well is, in fact, water, which can present operators with an expensive logistical challenge—one that, as volumes increase, will become an ever more pressing challenge. In the United States, Kondash et al. (2018) found that volumes of flowback and produced water generated within the first year of production increased up to 1440% from 2011 to 2016. Water produced from wells during the hydraulic fracturing process may also vary in characteristics compared to water produced from conventional wells—especially at the beginning of production—by carrying additives (corrosion or scale inhibitors, disinfectants, friction reducers, acids, or surfactants), as well as substances (such as radionucliedes) present in underlying geology.

Like the water acquisition regimes explored by Collins and Rosen, the law and policies governing the disposal of wastewater related to hydraulic fracturing in the United States are also largely a matter of state law that varies across the jurisdictions with active shale plays. Webb and Zodrow in Chap. 11 explore wastewater challenges and disposal practices in six U.S. states with jurisdiction over five shale plays: the Permian, Eagle Ford, Bakken, Marcellus, and Niobrara. Their analysis highlights the complexity of produced water regulation, treatment, and disposal within the United States, which varies dramatically in accordance with the geologic and cultural contexts of each state. Notable differences in management practices in the states exist, but most jurisdictions examined manage their produced water primarily by injecting it deep underground, either as permanent disposal or as part of enhanced oil recovery (EOR) efforts. Primary regulatory authority over underground injection rests with the U.S. Environmental Protection Agency (EPA) under the Safe Drinking Water Act (42 U.S.C. § 300f et seq.). That law established the Underground Injection Control (UIC) Program, through which the EPA (or a state, if the state has assumed primary responsibility) regulates the injection of produced water into non-producing formations with the objective of preventing the contamination of underground sources of drinking water. Injection for EOR is considered part of the production process and, as such, is regulated by the state in which it occurs.

Webb and Zodrow focus in particular on Pennsylvania and Texas, both of which have experienced large shale booms, but which have regulated the practice and disposal of wastewater very differently. Pennsylvania, with geology ill-suited for deep well injection, has recycled and reused more produced water than other states where disposal wells are more common and economical. Texas has leaned more heavily on injection wells for disposal, but the loosening of regulations and streamlining of the permitting process for recycling have resulted in a small increase in uptake of the practice. The authors suggest that states that desire to increase the recycling of produced water should simplify their regulatory frameworks on the issue, but caution that lawmakers must ensure the simplification does not compromise environmental protections. Drawing on Pennsylvania's experience—in which geology has limited the feasibility of widespread disposal by injection—the authors also point out that restricting the injection of wastewater by regulatory means may encourage operators to recycle, though they acknowledge the likelihood of this is small, given political action by producers.

In Chap. 12, Tina Hunter and David Campin examine the regulation of produced water and flow back fluid in Australia. Australia faces issues related to the regulation of produced water from coal seam gas production in Queensland, as well as shale gas exploration and production in Western Australia and South Australia. The issues associated with produced water from coal seam gas vary in content and regulation from those associated with shale gas extraction. Hunter and Campin focus, in particular, on the management of produced water from coal seam gas, which is currently the sole unconventional petroleum resource under commercial development in Australia. Dewatering—the process of depressurising coal seams by removing water from the coal measure and allowing the gas to desorb from the coal cleats and flow to the wellhead—results in large volumes of produced water. These volumes can be a challenge for project operators to process and dispose of. To manage the use of produced water from coal seam gas, the Queensland Government has created a hierarchy of options for operators, encouraging beneficial use where possible, and then allowing for treatment and disposal in ways that minimize and mitigate harmful environmental impacts.

Brett Miller's chapter, Chap. 13, focuses particularly on the qualitative aspects of water use and the regulatory challenges linked to the complexity of scientific debates around pathways for the potential contamination of groundwater. Miller reviews the many potential implications of hydraulic fracturing operations for groundwater, exploring the direct, indirect, and natural contamination pathways that may threaten drinking water supplies. These risks include the subsurface migration of methane, accidental surface spills, leak-off implicating fracturing fluids, well-casing integrity, and water table interactions with produced water. He also examines the disconnect between legal and scientific standards of causation and the uncertainty involved in assessing liability where groundwater has been contaminated.

Miller reviews research on hydraulic fracturing and groundwater to date and finds that studies have tended to "support the proposition that, properly conducted, hydraulic fracturing, in and of itself, likely does not elevate the risk of groundwater contamination" (Chap. 13). But, like Collins and Rosen, Miller worries that the *aggregate* impact on groundwater of numerous independently considered operations seems sometimes to be lost (or perhaps are not considered) in regulatory considerations of impact. Increased production inevitably provides more chances for operator error, thus increasing the risk to groundwater supplies. Moreover, decisions by the U.S. government to exempt shale operations from provisions within the *Safe Drinking Water Act (1974)* have made it particularly difficult to establish "the heightened standard of causation as it relates to establishing legal liability for potential groundwater contamination." However, by the time of publication of a 2016 U.S. EPA report, it was more readily acknowledged that groundwater contamination could take place in certain circumstances, including "injection of fracture fluids into wells with inadequate mechanical integrity (thereby allowing contaminants to move to groundwater resources), injection of fracture fluids directly into groundwater resources, and disposal of wastewater into unlined pits resulting in contamination of groundwater resources" (Miller, Chap. 13).

Miller also considers the potential for groundwater contamination and the challenge that the management of flowback and produced water presents for operators, an issue complicated in the United States by the imposition of commercial secrecy on the specific chemical "recipes" used by operators. This means that, unlike in Europe where operators must disclose the chemical composition of frac fluids in compliance with the REACH Directive, in the U.S. it is often the case that regulators and environmental action groups must undertake forensic examination after incidents have occurred (which also adds a burden to the public budget). Another particularly well-known issue involves the potential contamination of groundwater with shale gas (methane) itself, most spectacularly publicized in the documentary film *Gasland*. Here, as Miller shows, different state courts have come to different conclusions in cases brought against operators, though "the fact that 16% (n = 6,896) of … wells were hydraulically fractured at depths shallower than one mile may increase the potential for contamination events considering the limited vertical separation in particular instances". (Miller, Chap. 13). A problem arises when newer horizontal wells are drilled from older vertical wells originally drilled to different structural standards. Such "frac hits" seem to be regionally specific, with Oklahoma, for example, registering more than 400.

Researchers have called for both best-practice regulations and legal liability regimes to protect groundwater resources (Merrill and Schizer 2013). Miller (Chap. 13) highlights this dual layer system, emphasizing that "regulatory oversight must be backstopped by liability and enforcement regimes, or the regulations will be ineffective". He also notes that "best-practice regulations may be ineffective for novel risks associated with unconventional oil and gas activities with respect to the underlying scientific and legal uncertainty." Miller argues for (1) aquifer-specific restrictions that limit the depth at which horizontal wells can be hydraulically fractured to ensure adequate separation between the shale interval and the groundwater table; and (2) the administration of groundwater sampling surveys to set baselines for water quality prior to operations.

Monika Ehrman in Chap. 14 considers the issue of induced seismicity, which is earthquake activity caused by anthropogenic activities. Induced seismicity is often defined as increased seismic activity relative to historical levels. Ehrman examines the effect on geologic stability of injecting wastewater from oil and gas production in a variety of jurisdictions, with a focus on Oklahoma and Texas. Oklahoma has experienced more induced seismic activity than any other jurisdiction, suffering 585 magnitude 3-plus earthquakes in 2014 alone. Texas has experienced fewer and less strong earthquakes. Regulatory agencies and scientists in both states are working to understand the complex geologic structures and predict the effect of volumetric and pressure differentials on seismic stability. Oklahoma, perhaps because of the relative severity of the series of earthquakes, has taken a more aggressive approach to regulating disposal wells, while Texas is more skeptical regarding oil and gas wastewater induced seismicity. Texas authorities have denied the existence of a clear link to oil and gas activity despite studies by academic experts purporting to establish such a connection. Ehrman also considers induced seismicity in Canada and the United Kingdom, where the hydraulic fracturing process itself appears to be the cause of induced seismic events, and in the Netherlands, where concern over seismicity has been a main factor for public opposition.

1.3.4   Part IV: Regulatory Regimes and Issues: A Regional Perspective

Part IV of the book considers several jurisdictions around the world in which hydraulic fracturing has been undertaken or in which resource deposits are such that decision makers have sought to implement the production technique. Running through these chapters are themes of conflict between water uses for the environment, cultural heritage, or agriculture and the new demands due to unconventional oil. These countries all have poor hydrological data, as well as poor support for the administration of laws regulating unconventional sources, even where the state owns the resource. Public outcry can be intense against unconventional exploration, but geopolitical factors can lessen resistance. In Poland—where energy independence is important—the social license to operate afforded to operators is greater. Communities are more accepting of the risks or costs of hydraulic fracturing when they are balanced with the perceived benefits of greater energy independence, and when they themselves have a role in deciding development applications.

Deborah Curran in Chap. 15 considers the regulation by moratorium associated with aboriginal and treaty rights in Canada. This important chapter examines the constitutional issues and conflicts inherent in a system where the provinces both regulate water use and seek to expand (in both the private and public sectors) the production of oil. Conflict is often resolved with the oil interests taking precedence over the long-term water stewardship ambition. In Canada and around the world, indigenous communities are lobbying for a watershed-based approach to water management and for proper collaborative structures to be embedded in this regulation to achieve long-term water stewardship. The Canadian legal system has overturned water licenses for hydraulic fracturing where the hydrological information was considered to be incomplete and inadequate. Overall, there are huge oil reserves in Canada, and Curran sees this problem continuing in the future. The other problem is the cumulative effects of horizontal wells, and the science and data here are also considered to be inadequate.

Loretta Feris and Bill Harding in Chap. 17 look at the Karoo region, a fragile, semi-desert ecosystem in South Africa, and the regulation proposed to manage shale gas development therein. South Africa is a federation but has more power in the center to regulate. Two relevant acts—the National Water Act and the Minerals and Petroleum Development Act—include the sustainable development principle. This principle incorporates the three pillars of economics, environmental value, and social values; it requires consideration of the ways in which water and potential pollution will impact on the Karoo, its people, and economic activities currently pursued there. South Africa actively embraces the precautionary principle[2] as well. Unconventional gas resources cover 20% of South Africa, and the government enacted a moratorium for a few years, which ended in 2012. The moratorium was primarily due to public outcry over the proponent's poorly drafted environmental management plan. The South African legislation lacks enforcement and the scientific data is also often not available.

In Chap. 18, Anna Mikulska examines the geopolitics of Poland, whose reliance on Russian sources for power creates a community that is willing to embrace unconventional oil production. The geologic, economic, legal, and bureaucratic environments have failed to meet the initial expectations and facilitate production of new unconventional wells, though there is some use of the technology to provide gas by enhancing recovery from old reservoirs. One issue that has proved to be a barrier in Poland is that the state owns all mineral rights, even those on private properties. Thus, to engage in shale exploration, an entity needs to apply for a concession from the state. This process has been particularly cumbersome and costly, and initially too risky for overseas investors, especially as the concession period was only five years. In order to facilitate future exploration, this changed in 2014 to a concession valid for between 10 and 30 years. Poland is already a water-stressed country. The environmental and water regulations have been correspondingly strengthened to

---

[2] The precautionary principle is designed to facilitate decisionmaking under uncertainty when there is a threat of serious or irreversible damage. It enables decision makers to adopt precautionary measures when scientific evidence about an environmental or human health hazard is uncertain and the stakes are high. The principle is enshrined in a number of international treaties.

meet requirements under the European Union Water Framework Directive. Regarding future hydrocarbon production, Mikulska argues that the investment permit must detail the ways the development would consider the water environment. The application should contain a description of a balanced approach to exploration and production and should ensure that potential negative environmental effects of the activity are minimized.

Bárbara Bittencourt and David Meiler in Chap. 19 examine Brazil, which is looking to provide energy security by supplementing dwindling conventional supplies with other sources of energy. Brazil is one of the largest oil producers in the world, yet the regulations to control this have many shortcomings. The relatively young constitution (1988) provides that the federal state is the owner of mineral deposits and these are separate from the soil. The constitution also has a sustainable development objective. In 1995, the federal union was authorized to hire private state-owned companies to explore and produce energy under concession agreements. With regard to gas, the states have a role in distribution to the customer. Shale gas was regulated in 2014 with laws requiring operators to have an environmental management system and to perform a risk assessment prior to gaining approval. However, many are critical of this approach and the environmental authorities do not have processes in place. State governments are legally responsible for issuing the licenses, but where the resource covers more than one state, the national government takes over responsibility through the Brazilian national environmental authority. There are several uncertainties in the operation of this body and these have resulted in injunctions to stop the exploration of shale at the present time.

Andrés Sánchez in Chap. 16 considers the incentives for Latin American countries to achieve energy security, and the subsequent decisions by Argentina, Colombia and Mexico to explore hydraulic fracturing. The U.S. Department of Energy has assisted many Latin American companies with technical advice. The Latin American countries in this chapter, however, have state ownership of the companies exploiting the resources, and this is very different from the American system. Argentina, which re-nationalized its oil companies in 2012, gets 90% of its energy from fossil fuels, leaving it vulnerable to fluctuations in world prices. The country has abundant shale resources and is hence incentivized to use these. Argentina's General Environmental Law devolved upon both the Ministry of the Environment and Sustainable Development and the provincial governments the power to protect biological diversity; however, the country lacks policies to achieve these ends. This has resulted in protests in some parts of the country and moratoriums. Mexico exports oil but imports 81% of the natural gas it uses. The country has extensive resources of shale gas and 1000 wells. Until 2014, the State of Mexico or its companies had exclusive rights to exploit the country's minerals, but this was reformed to decrease barriers to investment and facilitate more private sector participation. In addition, new laws were created to oversee and regulate production standards and account for indigenous communities. There have been extensive protests against hydraulic fracturing in Mexico, mainly with regard to water pollution. Colombia, too, has felt pressure on its resource base, leading the state-owned company to attempt to increase production. The Colombian government has defined technical requirements for exploration of wells, as well as the social and economic parameters for fracking projects. The very powerful Attorneys General Department has insisted that all governments use the precautionary principle.

1.4 Concluding Remarks

The concluding chapter of this volume offers the reader an overview and synthesis of hydraulic fracturing around the world. There are several important common and differentiating elements that emerge from the book. First, there are clear and well-defined common issues that regulators in all jurisdictions have to consider in the management and oversight of hydraulic fracturing operations. Chief among these are how access to water for hydraulic fracturing and the wastewater produced by the practice are regulated. Here, previous path dependencies and existing legal frameworks (e.g., on property rights) will be key. Second, two fundamental correlates of the successful uptake of unconventional production of gas and oil are the need for energy security and the acceptance of the population. Hydraulic fracturing presents a trade-off between geopolitics and local potential impacts of production activities. Third, hydraulic fracturing sits in an evolving landscape, where many factors at play will determine whether shale is indeed a transition fuel towards a low carbon future, or whether it is a means to extend the life of the current energy system. In either case, strong messages come out from these chapters on the importance of having a robust regulatory regime for this new socio-technical challenge of hydraulic fracturing and the opportunities it presents. We hope that you enjoy reading the book as much as we have enjoyed bringing it together.

References

Freyman M (2014) Hydraulic fracturing and water stress: water demand by the numbers. https://www.ceres.org/sites/default/files/reports/201703/Ceres_FrackWaterByNumbers_021014_R.pdf.

Kondash A, Lauer NE, Vengosh A (2018) The intensification of the water footprint of hydraulic fracturing. Science Advances 4:eaar5982

Merrill TW, Schizer DM (2013) The shale and gas revolution, hydraulic fracturing, and water contamination: A regulatory strategy. Minnesota Law Review 98:145–264

Sinopec (2016) Sinopec corp 2015 communication on progress for sustainable development. http://english.sinopec.com/download_center/reports/2015/20160329/download/20160329001en.pdf. Accessed 13 Oct 2016

Staddon C (2014) Abstraction reform and water security: the view from England and wales, The Environmentalist. Special issue on Water Security, October 2014, pp 70–73

Staddon C, Hayes ET, Brown J (2016) Potential environmental impacts of 'fracking' in the UK. Geography 101(2):60–69. ISSN 0016-7487 Available from http://eprints.uwe.ac.uk/28892

US EIA (2015) Technically recoverable shale oil and shale gas resources: Argentina. Available at https://www.eia.gov/analysis/studies/worldshalegas/pdf/Argentina_2013.pdf

Regina M. Buono is a nonresident scholar at the Center for Energy Studies (CES) at Rice University's Baker Institute for Public Policy and a doctoral student in public policy at the Lyndon B. Johnson School of Public Affairs at The University of Texas at Austin. She previously served as the Baker Botts Fellow in Energy and Environmental Regulatory Affairs at CES. Prior to that, she was an associate with McGinnis, Lochridge & Kilgore, LLP, in Austin, Texas, focusing her practice in the areas of water, administrative, and endangered species law. Buono has also worked in various roles with the Texas legislature and as a consultant to oil and gas companies, designing a habitat credit exchange to achieve compliance with the U.S. Endangered Species Act. She holds bachelor's degrees in international relations, political science, and Spanish from the University of Arkansas, a J.D. from The University of Texas School of Law, and an M.Sc. in water science and governance from King's College London.

Elena López Gunn is Founder and Manager of ICATALIST, an associate researcher in the water group at the University of Leeds, UK, and an associate professor at IE Business School. She has more than 15 years' experience in projects and publications on a wide range of subjects, mainly related to innovation, water governance, agriculture, adaptation to climate change, hydrological planning, partnership models, public policies, sociological analysis on environmental issues and knowledge management. López Gunn has led the elaboration, management, and coordination of various applied research projects in the Horizon 2020 and LIFE programs of the European Commission. She holds degrees in economics and social studies from the University of Wales, a Masters in environment and development from the University of Cambridge, and a PhD in geography from the University of London.

Chad Staddon is Professor of Resource Economics and Policy in the Department for Geography & Environmental Management at the University of the West of England, Bristol and Global Director of the International Water Security Network. Through more than 100 published outputs, Chad's research revolves around the social, political and economic issues related to sustainable resource management. Current projects focus on the historical geography of urban water systems around the world, water-energy trade-offs in unconventional oil and gas operations, and economic policy for resilient urban water services. He received his PhD in geography from the University of Kentucky in 1996 for research on the political economy of water (mis)management in post-communist Bulgaria.

Professor Jennifer McKay is Professor of Business Law at the Law School at the University of South Australia Business School. She has conducted research supported by government and the private sector on sustainable development of freshwater and the development of governance models. This work has been undertaken in Australia, India and the US. The governance models have been developed for urban and rural water and she has made law reform suggestions, appeared before State and Commonwealth Committees and toured and gave presentations at Congress in Sacramento and Utah whilst on a Fulbright senior Scholarship to Berkeley Boalt School of Law. Prof McKay has a BA Hons and PhD from the University of Melbourne where she used social science methods to research water and environmental management, a LLB from Adelaide, GDLP UniSA and Diploma in Human Rights law from American University in Washington DC.

She has received local recognition for her work and is proud to serve as a sessional Commissioner on the Environment, Resources and Development Court of SA. Her TedX talk in 2019 was entitled Duty to Cooperate: Making Messy Mosaic Laws into Jigsaw Laws to manage the environment in a sustainable way. https://www.youtube.com/watch?v=qeGsWmu0RPY

Chapter 2
Water-Energy Nexus: The Role of Hydraulic Fracturing

Ahmed M. Mroue, Gabrielle Obkirchner, Jennifer Dargin, and Jordan Muell

Abstract This chapter considers some challenges attendant on optimising water-energy trade-offs in hydraulic fracturing, focusing on the interplays between constantly evolving technologies (e.g. use of treated effluent, brackish water or even waterless methods) and regulatory systems, using the Eagle Ford shale play in Texas as a case study. Regulators and higher level policy-makers often have conflicting preferences associated with the specific trade-offs (environmental, economic and social) that come within their purview. Therefore, it is very important to understand the basic trade-offs of the water-energy nexus when addressing nexus issues such as energy resources mining and production, water production, treatment and allocation, power plant construction and environmental impacts.

Keywords Water-energy nexus · Water security · Trade-offs · Decoupling · Technology · Eagle Ford

2.1 The Water-Energy Nexus

Consumers living in developed societies expect an immediate supply of water and energy through opening a faucet and flipping a switch. However, these consumers may be unaware of the significant interconnections between water and energy. This lack of awareness could potentially lead to abuse natural resource allocation at both the regional and national levels.

A. M. Mroue (✉)
Water-Energy-Food Nexus Initiative, Texas A&M University, Houston, TX, USA
e-mail: ahmed.mroue@cheniere.com

G. Obkirchner
Department of Geography, Texas A&M University, College Station, TX, USA

J. Dargin · J. Muell
Zachry Department of Civil Engineering, Texas A&M University, College Station, TX, USA

© Springer International Publishing AG 2020
R. M. Buono et al. (eds.), *Regulating Water Security in Unconventional Oil and Gas*, Water Security in a New World,
https://doi.org/10.1007/978-3-030-18342-4_2

Water and energy are drivers for economic and social growth, yet both energy and water securities are exposed due to the deep interdependency of water and energy systems on one another. Water is an input to almost all phases of energy production: fossil-fuel production, transport and refining; electricity generation; biofuel irrigation and processing; and even emission controls. Water security is crucial for energy security. At the same time, energy is needed to extract, desalinate, treat and transport water. Energy security is crucial for water security. Given the projected increase in demand for water and energy, understanding the water-energy link is key to addressing future potential resource sustainability challenges (which we call "hotspots"), developing well-rounded policies, and implementing technologies that mitigate risks (Fig. 2.1).

The drilling process for both conventional and unconventional oil and gas is a major user of local water resources. Production of fossil fuels, such as shale oil and gas, unconventional drilling techniques is rapidly increasing around the world (Mroue et al. 2018). While conventional production techniques for oil and natural gas are also water intensive (especially secondary and tertiary oil recovery processes), unconventional production processes are perceived as the main concern (Rahm 2011). Oil and gas in shale plays are produced by hydraulic fracturing, a technique that uses extensive amounts of water in drilling and fracturing the formation. Hydraulic fracturing includes horizontal drilling and multistage fracturing using water jets to reach shale gas reserves. Moreover, the large volumes of water used in hydraulic fracturing are mixed with chemicals, which is a primary reason for environmental concern, as it is associated with water reservoir contamination as well as with use of considerable quantities of land. According to the U.S. Environmental Protection Agency (EPA), hydraulic fracturing uses two to five million gallons of water per well (Rahm 2011). The wastewater produced by the fracturing process is comprised mainly of the fluid (water and chemical additives) used to drill and fracture the shale plays. Several methods of wastewater disposal

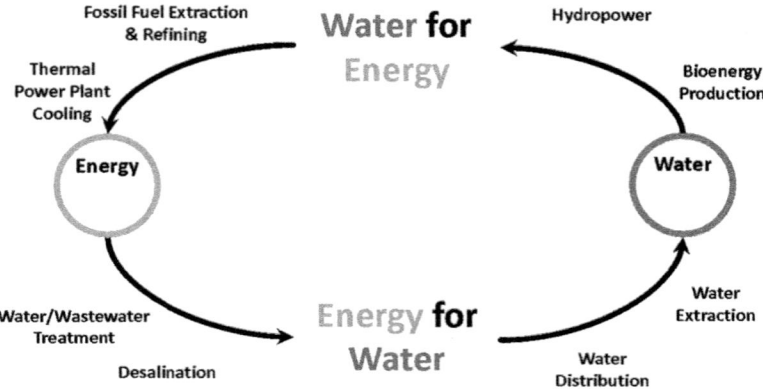

Fig. 2.1 Schematic of the water–energy nexus and the interconnected parameters

are currently used, including underground deep injection and discharge to surface water after treatment (Environmental Protection Agency 2010).

The issues around hydraulic fracturing lie squarely at the center of the water-energy nexus. The push for hydraulic fracturing is framed as a matter of energy security (increased production of oil and natural gas) yet comes with potentially significant costs to the security of water resources (Office of Research and Development 2010). Global demand for energy is rising, and consequently energy's demand for water is also rising. According to the International Energy Agency (IEA), energy accounted for 10% of global water withdrawals in 2016. Most of these withdrawals were for electricity generation, as well as for the production of fossil fuels and biofuels. The IEA projects that water demand for energy will increase over the period to the year 2040. Water withdrawn is expected to rise by 2%, while water consumed to rise by almost 60% (IEA 2016).

Yet, up to this date and despite this undebatable interdependence, water and energy are regulated independently. Policy makers often disregard the interconnectedness of water and energy, which results in contradictory water and energy policies (Hanlon et al. 2013). Both water and energy policymakers seek optimal sustainable solutions, but from an optimization point of view, neither provide optimal solutions for their sectors because the systems are decoupled in the approach to their respective policies. The nexus of these two systems is beginning to gain attention on multiple (national and international) levels (Poumadere et al. 2005). Working separately exposes the water and the energy systems to risk, introducing vulnerabilities to both: droughts, heat waves, water contamination, grid outages, and unfair competition for water.

This mutual dependency makes water a limiting factor or a weakness in the global energy system in both access and security. At the same time, energy is a vulnerability in the global water system in both supply and security. The supply, demand, management, and security of these systems are impacted by many variables, such as climate change, population growth, technology advancements, and practiced policies. Some of these variables are looked at as being out of the direct control of policymakers, such as climate change and population growth. But other variables such as technology and policy—are mainly under the control of policymakers.

Research has shown that there are ways to mitigate the risks of vulnerabilities of the water-energy nexus: policy and technology. Policies and technologies are not only capable of alleviating allocation stress points in the water-energy nexus, but also reducing water and energy demand such that an increase in energy demand would not be accompanied by an increase in water demand and vice versa. Such policies and technologies already exist; however, the implementation is accompanied by trade-offs that should be carefully considered through a robust nexus focus by policymakers, industry stakeholders, researchers and consumers. This chapter considers the importance of the water-energy nexus in the context of hydraulic fracturing, focusing on the tradeoffs inherent in the nexus and the role of policy and technology, using the Eagle Ford shale play as a case study.

Sustainable development in both energy and water requires a new, integrative approach based on the water-energy nexus. More importantly, with an integrative system, successful realization of water-energy policies and technologies can be much more effective to ensure sustainable development and avoid contradictory and unintended consequences of resource mismanagement (IEA 2016).

2.2 Unconventional Production: Hydraulic Fracturing in the Water-Energy Nexus

Hydraulic fracturing has been a technique used for over a century to increase oil and gas production, taking its first form as an exploding nitroglycerin "torpedo" to create fractures in oil-bearing rock and facilitating hydrocarbon flow to a well. Modern day hydraulic fracturing was developed in the 1990s by George P. Mitchell, who combined the use of hydraulic fracturing with horizontal drilling. This method of hydraulic fracturing combined with horizontal drilling is a form of unconventional oil and gas recovery because it produces from a different source than conventional recovery. Conventional and unconventional oil and gas come from the same geologic formation, but unconventional produces from the source formation while conventional produces from a reservoir, or cap formation. A permeable reservoir accumulates migrated oil and gas from the source rock and becomes trapped below a low permeable cap formation. This creates a pocket of fossil fuels that can be extracted through a vertical well. What makes unconventional recovery different, is that oil and gas is being extracted from the low permeable source rock, usually shale or tight sand. Using horizontal drilling, a well bore will increase its contact with the source rock as it travels through a formation, increasing the area it will produce oil and gas from. With an increased surface area, when the source rock is hydraulically fractured, large amounts of water and proppants are pressurized in a well structure to create fissures that drastically increase its permeability to allow oil and gas to flow to the well bore. This technique is valuable in that operators can artificially create flow to a well instead of relying on the natural migration and capture of the oil and gas. With the conception of these techniques, operations have been able to produce fossil fuels from resources that were once thought of as inaccessible and put countries such as the United States on a track towards energy independence.

Hydraulic fracturing is a critical point of tension in the water-energy nexus due to both its water-intensive nature and also its potential risks to water pollution. While the practice has made huge innovations to increase oil and gas production leading to increased prosperity, the many environmental issues that follow need to be weighted accordingly. The agriculture and municipal sectors are some of the main competitors for water, sustaining both food production and a modern way of life. While the total land used for oil and gas production makes up only 7% of the energy-land footprint—less than biofuel production and coal mining—oil and gas production requires a constant supply of new land to continue production, whereas agricultural sources remain in one place (Manfreda 2017). This adds potential competition for land in the future as oil and gas production grows and its need for new resources increases.

Regarding water pollution, the EPA has recorded a number of cases of water contamination due to hydraulic fracturing. These incidents, although not extensive, have tainted ground and surface water on a local scale, from flowback and produced water surface spills, breaches in well integrity of both oil/gas wells and disposal wells, and discharges of inadequately treated wastewaters into freshwater sources. An error of this magnitude has consequences in the other sectors of a community and can halt production of other life-sustaining resources. While hydraulic fracturing is a large resource in fueling the world's energy demand, it also has the potential to negatively affect other life-sustaining resources while lowering the environmental quality of communities in highly active regions.

A single hydraulic fracturing operation can use between three and eight million gallons of water, depending on the length and geology of the well. Although concerns arise over water consumption, water used in hydraulic fracturing makes up less than 1% of the total industrial water usage in the U.S. (Manfreda 2017). While this number as a whole is not alarming, hydraulic fracturing can pose threats on a regional scale where certain areas face droughts. A global study found that approximately 40% of shale plays occur in areas of high to extremely high water stress, calling for a need to assess a play's regional water resources in order to withdraw water responsibly and in a way that will not hinder the functionality of society on a micro scale (Hanson 2017).

In addition to oil and gas, oil and gas wells produce large amounts of water—called produced water—often in greater volume than the oil and gas actually produced. These production wells also provide a market for those willing to treat and sell produced water. Produced water is different than flowback water in that it is water produced from the hydrocarbon formation and is often highly saturated in dissolved solids, heavy metals, and naturally occurring radioactive materials, requiring expensive treatment to bring to reusable conditions. This has the potential to benefit a community where hydraulic fracturing is occurring but it is not usually economically feasible, leading oil and gas companies to opt for deep injection disposal. While commercializing produced water may still be uncommon, research is being done to improve costs of treatment to one day make its reuse more universal. Flowback water is the water injected into a well during the hydraulic fracturing process. Up to 40% of the total volume injected returns to the surface as flowback containing the added proppants and chemicals, along with additional dissolved solids from the formation. Treating and reusing flowback water for hydraulic fracturing is a growing practice as on-site treatment stations improve their capabilities to process more water, but once again are limited due to high operational costs.

Unconventional oil and gas production requires more tools, management, and technological applications than conventional production. In a conventional well, basic costs include the vertical well, drill stem, and casing. Unconventional wells have additional costs, including thousands of feet of extra piping for the horizontal well section, water, chemicals, and the management required to access, transport, and treat source/flowback water. Due to the added cost components of hydraulic fracturing and horizontal drilling, it is much more expensive than building a vertical well. Due to this higher production cost of unconventional wells, operators are limited to producing only when market prices are profitable, usually when oil is around $60 a barrel (Trainor et al. 2016). This causes unconventional operations to halt when prices become too low. The advantage compared with conventional production is that it is more resilient to changes in market price and can still operate when oil price is as low as $30 barrel, creating a synergy and reliance between the two forms of production.

The issue with conventional production is that many of the conventional deposits have already been tapped, creating the need for exploring new oil and gas resources. Currently in the U.S., conventional gas accounts for approximately 30% of total production and is expected to decrease to 24% by 2035 (EIA 2016). Aside from having limited source regions, conventional well counts in the U.S in 2015 show less than 1500 wells were producing more than 400 barrels of oil a day, versus over 4000 for unconventional wells (Jolly 2013). These counts reflect the fact that horizontal wells greatly outnumber vertical wells, but show just how profitable horizontal wells are. Out of the lowest rate of production, less than 15 barrels of oil per day, only 2% of the wells were horizontal (Jolly 2013). While horizontal wells produce at a much higher rate, they produce a majority of its lifetime recoverable oil and gas in its first few years of operation. Given this steep production decline in horizontal wells, there is high pressure on companies to constantly develop new sites to remain at their steady production levels. Depending on the geology of a location, production after just a couple years may drop to the point when continuing production becomes uneconomical, resulting in the plugging of the well and moving to a new location that has not been tapped yet. What this cycle results in are many abandoned wellbores not being currently maintained, serving as potential conduits of left over fracture fluid, formation water, and natural gas to seep into surrounding aquifers. While this is an issue with all wellbores, the vast number of horizontal wells could pose a more widespread risk to underground sources of drinking water.

2.3   Tradeoffs of the Water-Energy Nexus

In the water-energy nexus, complex interconnections such as cooling power plants, fracking shale, and powering desalination demand robust management solutions. Indeed, an optimization of such interconnections on various scales may be unreachable. However, rigorous tradeoff analysis and the modeling of scenarios can provide a pathway for decision makers to navigate and influence water-energy nexus synergy. The decision makers managing the water-energy nexus often have various conflicting preferences and scenarios which are always associated with tradeoffs: environmentally, economically and socially. Therefore, it is very important to understand the basic tradeoffs of the water-energy nexus when addressing nexus issues such as energy resources mining and production, water production, treatment and allocation, power plant construction and environmental impacts.

The extraction of water from deep wells, treatment, desalination and long-haul transportation are all energy-intensive activities. With sufficient energy, these water security challenges can be addressed and solved enduringly. However, questions remain as to how we can reach such a scenario, and what are the associated tradeoffs? Going toward a water-secure scenario will require vast added amounts of energy.

As mentioned earlier in the chapter, almost all energy activities require water as a major input, especially the processes of mining for energy resources and generating electricity, and that is when energy constraints become water constraints. Stillwell et al. (2017) detailed the water consumption for a variety of fuel sources. Consumption, which depends widely on technology and materials, is shown in Table 2.1. The life cycle water input of biofuel production can vary greatly by region, from rain-fed to irrigated crops. There can be no industrial fuel production, or electric power generation, without water. Power plants require water for cooling, and depending on the technology and fuel used, withdraw and consume various amounts of water. A once-through cooling system, for example, withdraws a significant amount of surface water but returns it with minimal consumption. A closed-loop cooling system reuses the same water, so has a much lower withdrawal but results in more water consumption from evaporation from cooling towers and other processes. The water footprint, or net impact on water supply, of electricity generation was also illustrated by Stillwell et al. and given in Table 2.2. These footprints are partitioned into withdrawal and consumption; most power plants remove and return a large quantities of surface water for cooling, while some is evaporated or lost through other means. This loss is considered water consumption; it is lost locally or becomes unavailable.

Water cannot be treated and transported over a great distance or from great depths without significant electricity and fuel use. In another study, Stillwell et al.

Table 2.1 Water consumption for different fuels produced

Fuel Source		Consumption (gal/GJ)
Natural gas		
	Conventional	0.19
	Unconventional	1.7–6.4
Petroleum		
	Gasoline	7.4–104
	Diesel	7.0–114
Biofuels		
	Corn Ethanol	459–1040
	Soy Biodiesel	423–2890
Hydrogen		86–131

Source: Stillwell et al. (2017)

Table 2.2 Water withdrawal and consumption for different electricity sources and cooling technology

Electricity Source		Withdrawal (gal/MWh)	Consumption (gal/MWh)
Coal			
	Open-loop	20,000–50,000	100–317
	Recirculating reservoir	300–24,000	300–700
	Cooling tower	500–1200	480–1100
Natural Gas Steam Turbine			
	Open-loop	10,000–60,000	95–291
	Cooling tower	950–1460	662–1170
Natural Gas Combined-Cycle			
	Open-loop	7500–20,000	20–100
	Recirculating reservoir	5950	240
	Cooling tower	150–300	130–300
Nuclear			
	Open-loop	25,000–60,000	100–400
	Recirculating reservoir	600–13,000	560–720
	Cooling tower	800–2600	581–845
Concentrated Solar Power			
	Cooling tower	725–1100	725–1100

Source: Stillwell et al. (2017)

(2009) examined the potential for long-haul seawater desalination. Treatment and distribution of conventional surface water requires 4.4 and 24.1 MWh/d respectively. Treating brackish groundwater or seawater increases the energy footprint of water supply to 78–195 MWh/d and 196–330 MWh/d respectively, which does not include any conveyance (Stillwell et al. 2009).

Energy production and generation has an impact on water and air quality, producing emissions and increasing the temperature in surface water where cooling water is returned, and has the potential to impact ecology. Chemicals from hydraulic fracturing have the potential to contaminate surface or groundwater through leaching or runoff if stored in lagoons. The injection of produced water from hydraulic fracturing into deep formations for disposal removes potential water supply from the hydrologic cycle. Chemicals can enter surface and groundwater through spills from oil pipelines and mines, or from refineries in disasters like floods. Water supply and quality impacts the efficiency of thermoelectric power generation and recovery in oil and gas production. Thus, water has a significant impact on energy industries that are a major economic component worldwide, and which are essential to maintaining standards of living.

These tradeoffs weave a complex web of interactions at different scales. As discussed previously, they exist within siloes of different decision makers in both private and public institutions. Energy and water have various important tradeoffs acting in both directions with clear financial, environmental, and social implications.

2.4   The Eagle Ford Case Study

The Eagle Ford shale play, located in South-Central Texas, is one of the most economic and prolific shale-oil producers in the nation. Spanning over 30 counties, the Eagle Ford contains approximately 3.4 billion barrels of technically recoverable oil and 20.8 trillion cubic feet of technically recoverable natural gas (EIA 2011, 2012). This play has seen exponential growth since the advent of hydraulic fracturing and the drilling of its first well in 2008. The region accounted for 85% of the total increase in Texas' production from 2010 to 2011. Using hydraulic fracturing technology, operators in the Eagle Ford more than doubled its natural gas production and increased its oil production six-fold (EIA 2016). Production rose drastically until 2014 when oil prices began to decline from $100/barrel to $30/barrel, causing a harsh decline in new production rigs. Production reached a minimum of less than 50 rigs in 2015 and has since been slowly increasing with only some minor setbacks. The play is currently producing from approximately 75 production rigs, compared to 2014 during its peak at approximately 1400 rigs (EIA 2018). Despite economic hindrances, the Eagle Ford currently produces 20% of the nation's oil and 10% of its natural gas (EIA 2018).

Water use for hydraulic fracturing, taken into account as water for mining in the Texas Water Development Board's State Water Plan, is in direct competition with water use for agriculture and municipal purposes in the Eagle Ford region. While mining water use is expected to peak in 2030, the region's population is expected to grow from over three million in 2020 to over five million by 2070 (TWDB 2016). This increase in population predicts the largest of potential shortages in the municipal sector where San Antonio is the largest metropolitan area of the region. Following this, are water shortages for steam electric power and manufacturing. Although demand for water in the mining sector is expected to decrease after 2030, the sector is expected to undergo potential shortages of 22% of what it demands, compared to 15% and 31% for the municipal and irrigation sectors (TWDB 2016).

Mining water use for the entire region constitutes approximately 4% of the total water demand, but at a county scale mining can account for a large majority of the water usage. McMullen County, located in the southwestern portion of the play, has a population of just over 700 people. Mining water accounts for 90% of total water use in the county (TWDB 2016). Taking its population into account this number seems reasonable, but irrigation in the county is predicted to experience shortages of 100% of what it demands, with mining at 67%. These statistics show that on a regional scale water use for hydraulic fracturing may appear minimal, but locally it can pose threats to other sectors of individual counties and may potentially harm the performance of communities that rely on these other resources for food and income.

Examples of nexus tools developed to support energy production in Texas and the Eagle Ford region to facilitate in infrastructure planning are the WET (Water-Energy-Transportation) 2.0 tool and EPAT (Energy Portfolio Assessment Tool)

(Mohtar et al. 2015; Mroue et al. 2018). These tools focus on energy and electricity production and the social, economic, and environmental impacts of these activities.

Under varying production or market price scenarios, WET 2.0 quantifies the interrelations and trade-offs between water, energy, and transportation. This tool is dynamic in selecting variables that characterize the Eagle Ford as it undergoes technical advances, such as increased lateral length of wells and increased water reuse. This tool offers a decision support system to operations in developing sustainable road integrity, controlling emissions, minimizing water use, and decreasing the energy footprint for the entire region of the Eagle Ford. A user can also create and compare varying scenarios under a social-environmental-economic index to see in what areas a plan may work more favorably.

More recently developed than WET 2.0, EPAT offers a platform designed to quantify the environmental needs as well as the environmental and economic outcomes from energy portfolio scenarios for Texas. Using this tool, a user can define current or desired energy production plans through different sectors, such as coal, oil, natural gas, bioenergy, solar, wind, nuclear, and hydropower. For each energy portfolios, the tool quantifies water consumption, land use, carbon emissions, and revenue for Texas. Both tools offer valuable insight into planning energy development impacts, and while their current focus is specific to Texas, their framework may be expanded to other entities with added data. These tools have the potential to reform policy with their holistic approaches and can help aid officials in making more informed decisions when they are able to view a snapshot of projected outcomes from a given energy scenario.

2.5   Potential Transformative Solutions in the Water-Energy Nexus

The analysis and presentation of synergies and tradeoffs between the water and energy resource systems is an important component of nexus studies because it provides stakeholders with guiding principles for reducing water loss and carbon emissions, while also meeting context-specific economic and sociocultural expectations. The inclusion of non-technical factors of a policy or technology solution are important since they set forth the system capacity of local constraints. Both a challenge and benefit of the nexus resource management approach is the array of stakeholders involved in the decision-making process, including the food and agriculture, industrial, economic, public health, financial, energy, water, and environmental sectors. Each interest helps determine the direction of policy with regards to water-energy issues.

Policies form the core of regulatory standards and have a significant role in resource allocation and use. However, the application of policy that addresses the nexus of water and energy is lacking. Mechanisms are therefore needed to assist institutions in transitioning from standard sectoral policy towards a holistic nexus approach that views the water and energy sectors simultaneously. Successful policy development can be accredited to the transparency and sharing of perceived risks and anticipated benefits across the spectrum of involved stakeholders (Dijk et al. 2015). Expert stakeholders from academia and industry have a role to play in shaping the public perception of emerging practices and technology (Eyck 2005) thus their potential for reaching the interests of policymakers. Ensuring policymakers are informed on the latest technological advancements and scientific understanding pertinent to the water-energy nexus is essential for the shaping of policy that optimize trade-offs and reduce direct and indirect negative impacts. However, this calls for the refinement of policy structures; going forward, it is fundamental for policies to be configured based on scientific evidence with direct contribution from cross-sector stakeholder collaboration.

Insufficient research and development of alternative technology applications for the water-energy nexus, in addition to low sociocultural awareness of their potential benefits, consequently results in a low priority for policymakers and low incentive for industries. A significant factor contributing to the weak investment and understanding can be related back to the knowledge gap between the scientific research community and decision-makers. Moreover, the persisting gap hinders development in terms of national and global resource management and climate objectives. Examples of existing and novel policy solutions that address key water-energy nexus challenges are presented and described below.

2.5.1 Increase Incentives for Water Reuse and Recycling in Fracking and Water Cooling

Local oil and gas producers and power plants are witnessing firsthand the impact of water availability on the success of their operations. Likewise, policymakers are taxed with figuring out the best strategies for managing scarce water resources. 27% of all US shale resources are found within areas of high water-stress (Webb 2017). It does not come as a surprise that drilling permits and opportunities for hydraulic fracturing have been denied in several states where the shale gas industry is active due to cases of low water-availability (Middleton et al. 2015). Several approaches can be applied to reduce withdrawals from freshwater resources and increase water-use efficiency:

In conventional practices, wastewater from fracking activity is disposed into injection wells, being removed from the hydrological cycle virtually forever (Webb 2017). The Fasken Oil & Ranch, Ltd. has been recognized in Texas and in the shale gas industry for its implementation of water recycling policy into its daily operations. In 2014, the small oil and gas producer discontinued all use of freshwater from the Ogallala Aquifer for its drilling and well completion operations (RRC 2017). Freshwater was replaced by recycled wastewater from fracking activity and brackish water from the Santa Rosa aquifer formation substituted the water.

Recycling and filtration facilities allow the water to be reused up to 80 times, while more than three million barrels had been processed since 2013 (Muscat 2015). To generate incentive for producers to cut-back on freshwater use, wastewater recycling needs to be a cost-competitive alternative to existing practices. Imposing fees on freshwater use may encourage industries to consider wastewater recycling and alternative water resource supplies, provided that the amount of fees meets or exceeds the cost of recycling. Similarly, disposal of wastewater from fracking activity can be discouraged by increasing injection fees, which currently go for an average rate of $2 USD per barrel of water (Webb 2017). This would force producers to seek alternative sources and explore water-less fracking techniques, or dry-cooling systems for power plants.

2.5.2 Encourage Use of Alternative Water Resources Including Brackish Water and Municipal Effluent, in Order to Preserve Fresh Water Supplies

Oil and gas producers in the Eagle Ford basin are already taking advantage of municipal effluent as an alternative to freshwater for fracking. The Apache Corporation in the area purchases three million gallons of treated effluent per day from College Station, Texas. As noted above, the Fasken Oil & Ranch, Ltd. is renowned in Texas for its use of brackish water for fracking operations. While this eliminates the dependency on freshwater resources, brackish water and municipal effluent will also face challenges of competition in the future, as both are currently being used for agriculture and municipal sectors.

2.5.3 Decoupling Water and Energy Sectors

Decoupling of the water and energy sectors means reducing the dependency of energy on water, and vice versa, in ways that are economically viable and will have low environmental impact. The decoupling begins with developing transparency of nexus tradeoffs in order to reduce sectoral dependency. Waterless fracking is one technological approach that decouples the fracking industry from water use but still requires more studies before wide-scale applications. The most significant challenge facing the energy sector is the availability of water resources. Freshwater, brackish, and municipal effluent each having competing users, and demand for them will only continue to increase into the future. Agricultural production and municipalities both are using brackish and effluent waters to meet growing demands.

2.5.4 Make Treatment Technologies more Economical

Despite the potential water savings and environmental benefits resulting from wastewater recycling policy, use remains limited due to costs of recycling treatments. Without financial incentives and regulation, industries will more likely continue use of freshwater and dispose of wastewater into injection wells. Policymakers in Texas realized this and developed a new permitting process for new recycling facilities that offers producers tax incentives to recycle (Webb 2017). Alternative fracking using non-aqueous fluids nearly eliminates the use of water in fracking. However, it is unlikely that industry will switch to non-aqueous working fluids unless there is a demonstrable and reliable increase in production that justifies the increased costs of alternative fracturing methods (Middleton et al. 2015).

2.5.5 Reduce Pollution Risks Associated with the Disposal of Fracking Water

The biggest potential source of environmental contamination is flowback and produced water, which is highly contaminated with hydrocarbons, bacteria and particulates, meaning that traditional membranes are readily fouled flowback water as well as post-well completion water (production or produced water) are contaminated with hydrocarbons many of which are classified as hazardous, which along with significant bacteriological content means that this water cannot be reused without significant treatment (Maguire-Boyle et al. 2017). Much of the concern around fracking surrounds the potential for chemicals to contaminate surface and groundwater (Webb 2017). In the case of the Marcellus Shale formation in Pennsylvania, insufficiently treated water from shale gas operations had been discharged into rivers creating a major public health concern.

2.5.6 Transformation through Technology

At the center of the nexus are the technologies producing, transporting, and channeling water and energy to fuel our livelihoods. Understanding these technologies within a nexus environment is a crucial step forward in identifying sustainable and resilient management and policy strategies that target multiple sectors simultaneously and address multi-scale resource challenges. In the context of this chapter, technology is discussed as a physical mechanism or process developed in order to gain a specified outcome (Rao et al. 2018). In the case of the water-energy nexus, technologies are physical mechanisms that have the general purposes of (a) production and provision of energy and water sources, and (b) water filtration.

Transformative solutions through the application of emerging technologies in the water and energy sectors specific to the area of hydraulic fracturing are introduced and their role in policy are described.

2.5.7   Waterless Fracking

Waterless fracturing technologies emerged due to concerns of formation damage, water consumption, and contamination risks associated with conventional fracking methods (Wu 2016). Liquid carbon-dioxide is one example of a non-aqueous fluid showing to be a promising alternative to water-based fracking. Results of a Department of Energy (DOE) sponsored experimental study on hydraulic fracturing showed that the use of CO_2 resulted in up to five times more gas production in comparison aqueous fluids and significant cutback on water use (Moridis 2017). Widespread application and acceptance of waterless fracturing in the shale gas industry is limited due to several technical factors noted by Middleton et al. (2015): "[the added expense] of capturing, pressurizing, and transporting carbon-dioxide, [the need for] robust accounting of CO_2 emissions and storage, pressure safety at the site, separation of hydrocarbons and brine from the flowback CO_2, and re-pressurization of flow-back CO_2." More development on waterless fracking methods is needed before they can be a viable alternative to traditional water-based methods. With successful applications of waterless fracking, the technology offers a significant step towards the decoupling of the water and energy sectors.

2.5.8   Renewable Energy Water Integration and Zero-Liquid Discharge Desalination

Direct reuse of wastewater from shale gas production is generally not feasible due to its high contamination which could have detrimental impacts on the health of shale formations if reused without proper pretreatment. Emerging technologies for zero-liquid discharge (ZLD) desalination provide promising applications in shale gas wastewater management. ZLD desalination uses thermal and membrane-based processes, while the selection of the most appropriate desalination method depends on physiochemical composition of the wastewater being treated (Onishi et al. 2017b). ZLD is an appealing technology for improving the overall sustainability of shale gas industry by increasing water-use efficiency (as much as 75–90% of wastewater can be reclaimed for reuse) and eliminating the environmental health risks of discharging of highly saline and contaminated wastewater

(Onishi et al. 2017a). Furthermore, integration of renewable energy resources with desalination have potential to reduce costs associated with desalination while creating opportunities to cut greenhouse gas emissions and reduce dependencies on fossil fuels (Rao et al. 2018).

2.5.9 Nanotechnology

In recent years, nanotechnology is becoming increasingly popular in the oil and gas industry, finding applications in drilling, drill-in, completion, stimulation, and exploration and exploitation of oil and gas (Fakoya and Shah 2017). In order to access oil and/or gas reservoirs beneath subterranean rock formations, high pressure pumping of fluids into the wellbores is needed to stimulate and breakdown the rock formations (Al-Muntasheri et al. 2017). Nanoparticles are of particular interest due to their small size (1–100 nm), enabling them to travel smoothly through the porous rock formations without blockage and damage (Franco et al. 2017). Other noted benefits of nanotechnologies in hydraulic fracturing include wellbore stability during drilling operations and reservoir sensing (Al-Muntasheri et al. 2017). Field trials conducted in Colombia show that applications of nanotechnology can increase the productivity and reserves of oil and gas (Franco et al. 2017).

2.6 Conclusion

In order for policymakers to make informed decisions about certain policy goals, it is critical that they have a fundamental understanding of the interplay between the technical components of the water-energy nexus, including emerging technologies and water and energy sector challenges. Policymakers must also have a broader vision and thorough understanding as to how technologies may be used as vehicles for achieving certain policy goals. Further research and development is needed to develop cost-effective policies that will discourage freshwater withdrawals for consumptive activities and encourage more efficient water use practices (recycling, brackish water, municipal wastewater). Furthermore, policy development needs to incorporate cross-sectoral dialogue in order to facilitate knowledge transfer from field experts to policymakers. From this discussion, the link between policy and technology becomes clear: while policy defines a specified goal, the application of technologies can serve as the means for achieving the policy goal. Lastly, coupling policy and technology will be essential in preparation for future challenges related to competition over water resources as more sectors resort to the use of alternative grey water resources to meet growing demand.

References

Al-Muntasheri GA, Liang F, Hull KL (2017) Nanoparticle-enhanced hydraulic fracturing fluids: a review. SPE Prod Oper 32(2):186–195

Dijk HV, Fischer A, Marvin H, Trijp HV (2015) Determinants of stakeholders' attitudes towards a new technology: nanotechnology applications for food, water, energy, and medicine. J Risk Res 20(2):277–298

Environmental Protection Agency (2010) Scoping materials for initial design of EPA research study on potential relationships between hydraulic fracturing and drinking water resources. Environmental Protection Agency, Washington, D.C. Available at https://yosemite.epa.gov/sab/sabproduct.nsf/0/3B745430D624ED3B852576D400514B76/$File/Hydraulic+Frac+Scoping+Doc+for+SAB-3-22-10+Final.pdf

Eyck TAT (2005) The media and public opinion on genetics and biotechnology: mirrors, windows, or walls? Public Underst Sci 14(3):305–316

Fakoya MF, Shah SN (2017) Emergence of nanotechnology in the oil and gas industry: emphasis on the application of silica nanoparticles. Petroleum 3(4):391–405

Franco C, Cortés F, Zabala R (2017) Nanotechnology applied to the enhancement of oil and gas productivity and recovery of Colombian fields: a review. J Pet Sci Eng 157:39–55

Hanlon P, Madel R, Olson-Sawyer K, Rabin K, Rose J (2013) Food, water and energy: know the nexus. GRACE Communications Foundation

Hanson VD (2017) The fracking industry deserves our gratitude. Natl Rev. July 6, 2017) available at https://www.nationalreview.com/2017/07/fracking-industry-united-states-energy-independence-oil-middle-east-venezuela/. Accessed 24 May 2018

Jolly D (2013) France upholds ban on hydraulic fracturing, The New York Times, October 11, 2013

Maguire-Boyle SJ, Huseman JE, Ainscough TJ, Oatley-Radcliffe DL, Alabdulkarem AA, Al-Mojil SF, Barron AR (2017) Superhydrophilic functionalization of microfiltration ceramic membranes enables separation of hydrocarbons from Frac and produced water. Sci Rep 7:12267

Manfreda J (2017) The real history of fracking. OilPricecom, 24 Feb 2017

Middleton RS, Carey JW, Currier RP, Hyman JD, Kang Q, Karra S, Jiménez-Martínez J, Porter ML, Viswanathan HS (2015) Shale gas and non-aqueous fracturing fluids: opportunities and challenges for supercritical CO_2. Appl Energy 147:500–509

Mohtar R, Blake J, Shafiezadeh H (2015) WET Tool (Water-Energy-Transportation application to assess impacts related to hydraulic fracturing in the Eagle Ford Shale). http://wefnexustool.org/wef/

Moridis, G. (2017) Literature review and analysis of waterless fracturing methods. Energy Geosciences Division. Lawrence Berkeley National Laboratory. Available at http://www.ourenergypolicy.org/wp-content/uploads/2017/04/CERC_WaterlessFracturing.pdf. Accessed 11 April 2018

Mroue AM, Mohtar RH, Pistikopoulos EN, Holtzapple MT (2018) Energy portfolio assessment tool (EPAT): sustainable energy planning using the WEF nexus approach – Texas case. Sci Total Environ 648:1649–1664. https://doi.org/10.1016/j.scitotenv.2018.08.135. Elsevier, (2019)

Muscat S (2015) US pioneers water saving in fracking, but China's nascent shale sector lags. China Dialogue. https://www.chinadialogue.net/article/show/single/en/7987-US-pioneers-water-saving-in-fracking-but-China-s-nascent-shale-sector-lags. Accessed 12 April 2018

OECD/IEA (2016) Water energy nexus. Excerpt from the World Energy Outlook 2016. Available at https://www.iea.org/publications/freepublications/publication/WorldEnergyOutlook2016ExcerptWaterEnergyNexus.pdf. Accessed 11 April 2018

Office of Research and Development (2010) Science in action: building a scientific foundation for sound environmental decisions. Environmental Protection Agency, Washington, D.C

Onishi, VC, Reyes-Labarta JA, Caballero JA, Antunes CH (2017a) Wastewater management in shale gas industry: alternatives for water reuse and recycling, challenges and perspectives. EMCEI 2017 – Euro-Mediterranean Conference for Environmental Integration. Available at https://rua.ua.es/dspace/bitstream/10045/71671/1/Poster_18EMEC_II_UA_2017pdf. Accessed 12 April 2018

Onishi VC, Reyes-Labarta JA, Caballero JA, Antunes CH (2017b) Zero-liquid discharge desalination of hypersaline shale gas wastewater: challenges and future directions. EMCEI 2017 – Euro-Mediterranean Conference for Environmental Integration. Available at https://rua.ua.es/dspace/bitstream/10045/71669/2/Poster_1EMCEI_Concept_UA_2017.pdf. Accessed 12 April 2018

Poumadère M, Mays C, Le Mer S, Blong R (2005) The 2003 heat wave in France: dangerous climate change here and now. Risk Anal 25(6):1483–1494

Rahm D (2011) Regulating hydraulic fracturing in shale gas plays: the case of Texas. Energy Policy 39(5):2974–2981

Rao P, Kostecki R, Dale L, Gadgil A (2018) Technology and engineering of the water- energy Nexus. Annu Rev Environ Resour 42:407–437

Stillwell AS, King CW, Webber ME (2009) Desalination and long-haul water transfer: a case study of the energy-water Nexus in Texas. In: Proceedings of the ASME 3rd international conference on energy sustainability. American Society of Mechanical Engineers, San Francisco, CA

Stillwell AS, Mroue AM, Rhodes JD, Cook MA, Sperling JB, Hussey T, Burnett D, Webber ME (2017) Water for energy: systems integration and analysis to address resource challenges. Curr Sustain Renew Energy Rep 4:90–98

Texas Railroad Comission (RRC) (2017) Texas energy companies win top national environmental Stewardship Award –Pioneer and Fasken recognized for water conservation efforts. Railroad Commission of Texas. Accessed on 4 Dec 18. http://www.rrc.state.tx.us/all-news/100217b/

Texas Water Development Board (TWDB) (2016) 2017 Texas state water plan. https://2017.texas-statewaterplan.org/region/L

Trainor AM, McDonald RI, Fargione J (2016) Energy sprawl is the largest driver of land use change in United States. PLoS One 11(9)

U.S. Energy Information Administration (EIA) (2011) Review of emerging resources: U.S. shale gas and shale oil plays. U.S. Department of Energy, Washington, DC. https://www.eia.gov/analysis/studies/usshalegas/pdf/usshaleplays.pdf

U.S. Energy Information Administration (EIA) (2012). Eagle Ford oil and natural gas well starts rose sharply in first quarter 2012 – on April 23, 2012. U.S. Department of Energy, Washington, DC. https://www.eia.gov/todayinenergy/

U.S. Energy Information Administration (EIA) (2016) State profile and energy estimates – Texas on September 24, 2015. U.S. Department of Energy, Washington, DC. https://www.eia.gov/state/?sid=TX#tabs-4

U.S. Energy Information Administration (EIA) (2018). Crude oil production – Texas on May 13, 2018. U.S. Department of Energy, Washington, DC. https://www.eia.gov/dnav/pet/PET_CRD_CRPDN_ADC_MBBLPD_M.htm

Webb RM (2017) Changing tides in water management: policy options to encourage greater recycling of fracking wastewater. William Mary Environ Law Policy Rev 42(1):85–144

Wu K, Paranjothi G, Milford JB, Kreith F (2016) Transition to sustainability with natural gas from fracking. Sustainable Energy Technol Assess 14:26–34

Ahmed M. Mroue is a Market Analyst with Cheniere Energy, conducting quantitative and qualitative research on the global liquefied natural gas (LNG) industry and market to support Cheniere's commercial group with market insight and analysis. Ahmed is also a Research Associate at the Water-Energy-Food Nexus Initiative at Texas A&M University. His research interests focus on strategic sustainable energy development and energy and environmental policy analysis, and the Water-Energy-Food Nexus. As a Research Associate, his research projects included: (1) NSF San Antonio Case Studies (SACS) - Water-for-Energy subgroup; (2) developing Energy Portfolio Assessment Tool (EPAT), a nexus energy portfolio tradeoff analysis tool in support of policymakers with sustainable development goals (published). Mroue co-authored two journal publications on the Water-Energy and Energy-Water Nexus and is contributing to additional scientific papers on topics of Water-Energy-Food-Climate Nexus modelling and socio-economic assessment of hydraulic fracturing. He holds a B.E. in Mechanical Engineering from Lebanese American University and a M.Sc. in Energy from Texas A&M University.

Gabrielle Obkirchner is a member of the Water-for-Energy subgroup of the WEF Nexus Group, researching groundwater availability for hydraulic fracturing in the Eagle Ford shale of Texas and its role in an area with diverse water demands. Gabrielle's research interests extend to groundwater modelling, management and policy, and improving water resource sustainability. Gabrielle holds a B.S. in geology from Texas A&M University and is pursuing a M.S. degree in water management and hydrological science at Texas A&M University.

Jennifer Dargin is a research assistant at the Water-Energy-Food Nexus Research Group and a doctoral student in civil engineering at Texas A&M University. She specializes in nexus modeling tools and their application in policy and decision-making. Her research focuses on the potential of emerging technologies, including machine learning and remote sensing, in improving data analysis and predictive modeling of nexus resource systems. Dargin holds a B.S in public health science from the University of Massachusetts-Amherst. Throughout her academic and professional career, she has dedicated her work towards developing solutions for and creating awareness of water resource scarcity challenges in water-stressed regions of the world including Jordan and Pakistan.

Jordan Muell is an engineer with Provost & Pritchard Consulting Group, working with farmers and water planning coalitions in the Central Valley of California to navigate new agricultural and water resources regulatory programs and engage in public processes to shape water policy. From 2016 to 2018, he was a member of the Water-Energy-Food Nexus Initiative at Texas A&M University, working in the Water-for-Energy subgroup in the San Antonio Case Study. Muell completed his thesis, entitled "Developing a Farm-Scale Water-Energy-Food-Waste Nexus Framework for a Closed-Loop Dairy Concept." Muell holds a B.S. in biological agriculture from the University of Iowa and a M.Sc. in civil engineering from Texas A&M University.

Chapter 3
The Human Right to Water and Unconventional Energy

Robert Palmer, Damien Short, and Ted Auch

Abstract Hydraulic fracturing for oil and gas is an emotive subject, generating passionate arguments both pro and con. Some scholars argue that a 'human right to water' (HRW) approach could usefully enshrine in law the priority of human needs over industrial uses, in hydraulic fracturing and other sectors. This chapter explores the existing status of the HRW in international law and in the constitutions and statutes of some nations around the world. It appears that attempts to link struggles over HF's impact on water resources with the HRW have so far foundered on a lack of clear unambiguous HRW declarations that can be tried in courts of law.

Keywords Human right to water · Extreme energy · United Nations · Fracking

3.1 'Unconventional' Extreme Energy and Water

This chapter will chart the international development of the right to water and contextualise its relevance to discussions around unconventional energy via data from the country with arguably the most mature and extensive industry, the USA. To begin, it is important that we define our terms. How do we define 'unconventional' energy? To answer that question it is perhaps pertinent to explain first 'conventional' mineral extraction. In simple terms, it is the extraction of readily available and relatively easy to develop oil and gas from reservoirs trapped in natural geological structures, generally sandstone and carbonate rocks. In the not too distant past,

R. Palmer
The Open University, Milton Keynes, UK

D. Short (✉)
Human Rights Consortium and the School of Advanced Study, University of London, London, UK
e-mail: Damien.Short@sas.ac.uk

T. Auch
FracTracker Alliance, Camp Hill, PA, USA

© Springer International Publishing AG 2020
R. M. Buono et al. (eds.), *Regulating Water Security in Unconventional Oil and Gas*, Water Security in a New World,
https://doi.org/10.1007/978-3-030-18342-4_3

natural geological processes that took place over hundreds of thousands of years provided plentiful hydrocarbon resources. Conventional gas uses traditional methods to extract primary deposits held in underground reservoirs created by the geological processes. Unconventional natural gas resources, such as coal bed methane, tight gas and shale gas are termed 'unconventional' minerals because the porosity, permeability, fluid trapping mechanism, or other characteristics of the strata from which the gas is extracted differ greatly from conventional sandstone, siltstone and carbonate (limestone) sources.

Hydrocarbon resources were relatively easy to exploit using the conventional gas development methods. In essence conventional mineral resources 'pool' into convenient reservoirs ready to be exploited, whereas unconventional minerals are present in the entire rock strata. The problem lies therein. In order to extract enough of that unconventional gas, tightly trapped in vast shale plays and coal seams, an amount of extraordinary scientific and engineering developments are going to be required: certainly it is not a straightforward undertaking. The technique, widely coined as fracking, has been the subject of controversy because of the potential effects that hydraulic fracturing and related oil and gas production activities may have on human health and the environment. The advent of unconventional oil and gas development (UOGD) poses threats to the natural support systems that are necessary for life, all life, specifically air and water. Here we are concerned with perceived threats to the planet's water resources.

As the easier to extract 'conventional' reserves are gradually depleted and global demand for energy rises, there is increasing pressure to exploit so-called 'unconventional' energy sources. It is a fact that oil and gas reserves are running out and whilst a small number of oil fields are discovered each year, thus adding to the global reserves, at the same time, ever larger volumes are being extracted that cancel out any new discoveries: we are rapidly reaching what is termed as 'Peak Oil'. Peak Oil is a hypothetical event in time, based on M. King Hubbert's theory, when global oil production reaches its maximum rate of extraction, after which production will terminally decline. The exact timing of Peak Oil is complex and outside the remit of this chapter, but outside estimates places the event between 2020 and 2035, although dates after 2030 have been considered as implausible by some experts (Mobbs 2004). In simplest terms, Peak Oil equates to the world's oil producers being unable to sustain historical increases in their production: resources will be in freefall. However, demand for oil will remain deep-seated; not just for energy but throughout the entire commodity supply chain of the developed world. UOGD represents a last-ditch attempt to bridge the gap of dwindling oil and gas resources using methods beyond an artificial lift or tradition methods to increase production.

In the simplest sense, 'unconventional gas,' for example, is natural gas obtained from secondary deposits via techniques and methods of production that are, in a given era and location, considered to be new and different. Often these new, unconventional techniques involve 'stimulation' processes such as hydraulic fracturing, the focus of this book. The International Energy Agency contrasts 'conventional and 'unconventional oil' thus:

'Conventional oil is a category that includes crude oil – and natural gas and its condensates. Crude oil production in 2011 stood at approximately 70 million barrels per day. Unconventional oil consists of a wider variety of liquid sources including oil sands, extra heavy oil, gas to liquids and other liquids. In general, conventional oil is easier and cheaper to produce than unconventional oil. However, the categories "conventional" and "unconventional" do not remain fixed, and over time, as economic and technological conditions evolve, resources hitherto considered unconventional can migrate into the conventional category.'

A distinction between conventional and unconventional oil and gas extraction needs to be made. Essentially the equivalent amount of valuable hydrocarbons acquired from conventional wells dwarfs unconventional wells when compared on a well to inventory ratio. A simple comparison can be made: a handful of conventional wells are needed to produce the same amount of hydrocarbons as hundreds of fracking wells situated on scores of well pads.

One could say that as hydrocarbon resources become ever more difficult to exploit, what was once unconventional becomes the conventional. There are a number of problems with using such constructed categories as key descriptions in important policy discussions, but they are even more problematic when analyzed conceptually. Michael Klare sought to capture the element of 'risk' that was missing with benign terms such as 'unconventional' when he first coined the term 'extreme energy' to describe the range of relatively new, higher-risk, non-renewable resource extraction processes that have become more attractive to the energy industry as the more easily accessible supplies dwindle and we reach 'Peak Oil'. However, similar to the term 'unconventional', this definition of extreme energy—as a category—is highly problematic as it is dependent upon specific examples; it lacks explanatory or predictive power and leaves open the question of who decides which extractive techniques qualify (Short et al. 2015).

A preferable conceptual understanding is extreme energy as a 'process', whereby extraction methods grow more intense over time, as easier to extract resources are depleted. The foundation of this conception is the simple fact that those energy sources which require the least amount of effort to extract will be used first, and only once those are dwindling will more effort be exerted to gain similar resources. Extreme energy, in this sense, is evident in the history of energy extraction in the change from gathering 'sea coal' from British beaches and exploiting 'natural oil seeps', to opencast mining and deep-water oil drilling. Viewed in this light, the concept of extreme energy becomes a lens through which current energy extraction efforts can be explained and the future of the energy industry predicted. Using this extreme energy lens necessitates an understanding of 'the amount of energy which is needed to obtain energy', as in this process it is that value which is continually rising. This value may be calculated as either 'net energy' or 'energy return on investment' (EROI), whereby net energy is the available energy for use after subtracting the energy required for extraction, and EROI is the percentage of energy produced divided by the amount required for extraction. When charted together, the net energy resource available to society is seen to decrease along with EROI in a curved mathematical relationship, which forms the 'energy cliff' – i.e., the point at which EROI becomes increasingly low and net energy drops to zero (Short et al. 2015). In the extreme energy process the economic system can be conceptualised as consisting of two distinct segments, the part which is extracting, refining and producing energy (the energy industry) and everything else, which just consumes energy.

What needs to be clearly understood is that the energy industry is in the rare position where the commodity which it produces is also the main resource it consumes. Therefore, as energy extraction becomes more extreme, while the rest of the economy will be squeezed by decreasing energy availability and rising prices, the energy industry's rising costs could be offset by the rising revenues it receives. The net result will be a reallocation (through the market or otherwise) of resources from the rest of society to the energy industry, to allow the energy industry to target ever more difficult to extract resources (Short et al. 2015). This process is ongoing as easier-to-extract resources are depleted and we loom ever closer towards 'Peak Oil', with some arguing we have already reached peak 'conventional' oil and gas but are yet to reach total peak – which includes potential 'unconventional' reserves. What's more, data from unconventional extraction methods, such as hydraulic fracturing and tar sands extraction, show that industry is increasingly lurching towards the net energy cliff. Such action on the part of some of the largest and most commercially successful trans-national corporations may only be understood as the logical result of the extreme energy process – there simply are no longer enough easier-to-extract 'conventional' resources available to meet the needs of growth driven capitalism.

Another issue surrounding the extraction of oil and gas goes beyond the issue of energy usage. Despite coal and gas's ability to supply a significant amount of our contemporary energy needs, oil quite literally makes the developed world go around. Oil is not simply restricted to providing most of our energy needs for domestic and commercial sectors it, among others, oils the bearings of industry, binds our roads together and is the principle ingredient in a vast array of materials in society, particularly plastics and chemicals. Accordingly, dwindling supplies will have a detrimental knock-on effect all the way through the global commodity supply chain. Peak Oil and Gas will not only mean an inevitable price hike in energy, commodities, such as plastics, will also become extremely expensive. UOGD – arguably – offers some respite to valuable oil resources that would be better allocated to the production of commodities rather than producing energy. What is problematic is that this possible and, inevitably, temporary solution to offsetting the realities of oil and gas reserves in terminal decline could also bring unacceptable risks to public health and the environment, and even that is leaving aside arguably most important issue of the UOGD discussion – the contribution to greenhouse gasses and climate change of a much less efficient form of energy.

Even so, under the process of extreme energy, one of the most precious resources that is being reallocated away from society to industry is *water*.

Water impacts are one of the most contentious and widely publicized environmental, and as we argue herein, human rights issues, connected with unconventional energy extraction, including but not limited to: groundwater contamination, water use, and contaminated water waste disposal. For example, unconventional gas production is a highly water-intensive process, with a typical single well requiring around 5–11 million gallons of water, and an average well-pad cluster up to 60 million gallons, to drill and fracture, depending on the basin and geological formation.[1] The vast majority of this water is used during the fracturing process, with large volumes of water pumped into the well with 3300–5000 thousand tons of sand (i.e., proppant) and chemicals to facilitate the extraction of the gas; the remainder is used in the drilling stage, with water being the major component of the drilling fluids. Once that water is used by the industry it is no longer a useful resource for society and must be disposed of in what are called Class II Salt Water or Brine Disposal wells. While increasing quantities of water are being recycled and reused in the US, freshwater is still used in high quantities for the drilling operations as 'produced' water is more likely to damage the equipment and reduce the chance of a 'successful well'. The industry's requirements (Stark et al. 2012) for such quantities of freshwater is clearly a serious concern in water-scarce regions of the world and in places with high cumulative demand for water. Furthermore, the relationship between water demand and 'produced' water production is highly correlated across many of the US Shale Plays with waste production a linear function of water used during the hydraulic fracturing process (See US Case Study Below).

The large quantities of water used by the fracking industry is but one of many serious concerns. The contamination of groundwater sources (Fischetti 2013) from failure in the well casing over time (Ingraffea et al. 2014)—what industry refers to as 'zonal isolation' failure—is a very serious issue across regions that have seen considerable fracking development to date and has duly featured as a central public relations battleground for industry and pro-fracking governments (Short et al. 2015). Even so, arguably the most concerning issue with the use of water for hydraulic fracturing is the issue of produced/waste water treatment and disposal often simply referred to as 'waste water management'. And yet, the risks in this regard go well beyond the concerns of corporate risk minimisation. Indeed, the whole process of dealing with fracking's waste water is a highly risky business for local populations and the environment with considerable risks of water or soil contamination from surface leaks and spills (Olmstead et al. 2013), but perhaps the most concerning issue with waste water is that it can contain significant amounts of radioactive material (Mobbs 2014) due to the "naturally occurring hypersaline brines associated with the formations targeted for natural gas production" (Warner et al. 2013). For instance radium has been found to be building up in rivers downstream of shale gas waste discharge points in Pennsylvania (Mobbs 2014). Vengosh et al. (2014) summarised the overall risks posed by fracking development for water as being four-fold:

- Contamination of shallow aquifers in areas adjacent to shale gas development through stray gas leaking from improperly constructed or failing gas wells.
- Contamination of water resources in areas of shale gas development and/or waste management by spills, leaks, or disposal of hydraulic fracturing fluids and inadequately treated wastewaters.

---

[1] Average figures obtained from the US based www.fracfocus.org website. FracFocus is the national hydraulic fracturing chemical registry. FracFocus is managed by the Ground Water Protection Council and Interstate Oil and Gas Compact Commission, two organizations whose missions both revolve around conservation and environmental protection.

- Accumulation of metals and radioactive elements on stream, river and lake sediments in wastewater disposal or spill sites, posing an additional long-term impact by slowly releasing toxic elements and radiation to the environment in the impacted areas.
- The water footprint through withdrawals of valuable fresh water from dry areas and overexploitation of limited or diminished water resources for shale gas development.

In addition to concerns over groundwater and surface water pollution, other primary concerns include air pollution, greenhouse gas emissions and radiation. Whilst the issue of aesthetics to the environment is often overlooked as a primary concern in the UOGD debate, vast tracks of areas of outstanding natural beauty have been scarred across the United States of America. All these concerns have been exacerbated in the US by the Energy Policy Act 2005 (infamously known as the Halliburton Loophole), which, for example, exempts unconventional oil and gas production and delivery from the Safe Drinking Water Act (SDWA) (Vann et al. 2014).[2] In terms of water, the Halliburton Loophole essentially addresses the issue of water quality; however, it is not trite to say that there is a succinct link between water quality and quantity. Palpably, if groundwater and shallow aquifers are polluted/contaminated by hydraulic fracturing and related oil and gas production activities – and no longer viable as a freshwater resource – then water quantity is also diminished. The issue of the right to water encompasses both water quality and quantity: both are essential facets to the right which, in turn, is essential to the 'minimally good life' and the realisation of all human rights.

3.2 Status of the Right to Water in International Law

The contemporary legal basis for a right to water at an international level is imprecise and uncertain. This chapter, in part, examines the development of the right to water in terms of international law. In examining the development of the right to water, and its contemporary legal status quo, we seek to explore the potential impact of the human right to water through the 'fracking' dialogue and the impact of UOGD on people's ability to realise the right. The connection between UOGD and the right to water requires consideration because the contrast offers a dramatic example of public decision-making diminishing perceived and guaranteed human rights. Here we are concerned with effects of the UOGD industry on water quality and the inequitable situation (for those who live in close proximity of well pads) that arises with regards to not just accessing the vital resource but also having a sustainable and potable supply.

In both human and environmental terms, biology dictates that water – after air – is the most important resource to safeguard human and ecological survival (Cullet

---

[2] Also the Clean Water Act, the Clean Air Act and the Comprehensive Environmental Response, Compensation, and Liability Act (CERCLA).

2013; McCaffrey 2001). Freshwater is a vital resource for natural ecosystems, human physical and mental health, and various human socioeconomic needs. Many of the contemporary water issues have historical foundations. Indeed, Malcolm Langford refers to 'millennia-old struggles over the ownership of water, the pollution and depletion of water sources, and conflicting water uses' (Langford 2005). The importance of a global sustainable supply of freshwater is ubiquitous but 'water resources are under pressure to meet future demands due to population growth and climate change' (Richey et al. 2015; Kundzewicz and Döll 2009). Further, it has been argued that the threat posed by worldwide groundwater depletion to global water security is far greater than is currently accepted (see in general Famiglietti 2014). According to the United Nations around 1.2 billion people live in regions of water scarcity, and a further 1.6 billion people live in regions of economic water shortages (United Nations 2014).

The international community has been faced with old threats and new challenges since the onset of the twenty-first century. In view of this, what is novel is the scale of old problems and impending environmental threats from global warming and, due to advances in technology, far more invasive new industrial practices than have been witnessed in the past, of which UOGD is a prime example. In recent times water scarcity has been the issue that has driven the debate with regards to the human right to water. The hypothesis Curry makes is that a global effort is required to coordinate endeavours to combat freshwater scarcity. In his opinion, the recognition of the right to water (by the UN) 'would be a building block to initiate the chain of decisions necessary to prevent the dire effects of water scarcity' (Curry 2010). However, in a setting where the introduction of invasive industrial practices threatens both equitable accesses to water and to devastate water quality, UOGD has added a different dimension to the right to water, which is emerging ever greater as a human right that needs to be recognised at a universal level (Brown Weiss 2012; Giupponi and Paz 2015).

Whilst there is an impending global crisis with regards to water scarcity, the western world, owing to a perceived abundance of the resource, has maintained an artificial stance and ignores, what can be described as, an impending crisis (see generally Lal Panjabi 2014; Gusikit and Lar 2014). This is a precarious situation as even coordinated efforts to combat future global water scarcity are foreseen to fall short of addressing looming human health, environmental and financial crises. Palpably, water scarcity is linked to food insecurity which, in turn, leads to human migration as populations seek to settle in other – more resource rich – geographic locations.[3] With perceived threats regarding conflict over resources, and the real possibility of 'Water Wars' as a consequence of food insecurity and water scarcity, a UN covenant recognising the human right to water 'will not solve water scarcity by itself, but it will establish the framework necessary for implementing any solution' (Barlow 2006; Curry 2010).

---

[3] Human migration in this context is taken to mean the movement of human populations to settle in other geographic locations, due to any other cause other than intended and voluntary movement. In that respect, movement associated with disaster, conflict, or forced migration is perceived to constitute human migration.

Whilst it is beyond the scope of this chapter to give a detailed commentary on human migration and resource conflicts, it is important to highlight associated difficulties that are likely to exacerbate the water scarcity crisis if UOGD continues on proposed sites around the world. Palpably, from the available scientific evidence, the vast quantities of freshwater used in the process will put an unacceptable strain on vital water resources. Unfortunately, the current policy on addressing human migration and resource conflicts is inconsistent with guaranteeing fundamental human rights and freedoms. The Oxford Research Group (ORG) recognised that:

> The current security paradigm adopted by most governments and their defence forces is based on the flawed premise that insecurity can be controlled through military force or containment, thus maintaining the status quo. This has been termed the 'control paradigm' (Brock 2011)

In human rights terms, the 'control paradigm' strikes at the very heart of human dignity, life and health. Within the narrative of the 'wide range' of human rights documents, particularly since the 1970s, elements of the right to water are required to be adequate for human dignity, life and health. Indeed, as will be seen below, in accordance with Article 11(1) and 12 (ICESCR) these three key elements are all-pervading and embody, perhaps personify, the right to water's normative content in international soft law documents. Curry acknowledged that the right to water, like all other human rights, is 'derived from a basic acknowledgment of the dignity of all human beings.' First declared in the UDHR, human dignity stands as the 'minimum definition of what it means to be human in any morally tolerable form of society' (UNHCR 2009). He argues that a lack or a denial to clean freshwater fails to meet this minimum standard of dignity and that a right to water would ensure that industrial technologies will not take priority over domestic uses. At present it would also ensure that those who disrupt access to clean freshwater are held accountable (Curry 2010).

It is argued by the ORG that a new approach to security is required that addresses the drivers of conflict: 'curing the disease' rather than 'fighting the symptoms'. The concept of 'sustainable security' is one viable alternative and addresses human rights concerns protected by the UDHR (Curry 2010). Perhaps, the Bolivian alternative to the 'control paradigm' is the precursor to sensible resource security: Bolivia's army already have a role in the 'protection of mother earth' (Escobar 2010). Whereas there remains a military element to the Bolivian solution, the precursor to any such action is safeguarding vital natural resources rather than vested industrial interests. Famous for its 'Law for the Rights of Mother Earth' (2010) (Higgins 2012), where 'protection of water from contamination' is a right, Bolivia's Constitution (2009) also established a precedent by formally recognising water as a 'basic human right'. In fact, Article 124 of the Constitution states that citizens who violate the 'constitutional regime of natural resources' is recognised as committing an act of treason against the country.

What is noteworthy is that the Bolivian constitution protects human dignity, life and health through giving the environment standing in the courts through its constitution; thus, highlighting the inalienable nexus between human and environmental rights. It can be asserted that the expansion of UOGD will impair any attempts at addressing water scarcity and curtail those rights. Particularly as the technology will both deplete and contaminate available water resources. Despite Bolivia's recognition of water as a human right, there have been recent concerns over UOGD and its compatibility with their laws. The issue of whether the technology can be introduced in the country has met with fierce resistance including the matter being discussed in the International Tribunal for the Rights of Nature (December 2014). It can be asserted that, even where progressive developments towards recognising and codifying the right to water exist, UOGD poses a significant threat to human health and the environment as companies such as Halliburton move to exploit Bolivia's 48 trillion cubic feet of shale gas, which could permanently contaminate 242 billion litres of water and emit 2.6 billion tonnes of carbon dioxide. The question is: will the Law of Mother Nature allow the country to frack rare water resources? (Hill 2015).

Another major facet to the global water scarcity crisis is an exponential increase in population in the future; thus, owing to the vast quantities of water used in unconventional oil and gas production, the introduction of UOGD will place an insurmountable strain on freshwater resources globally. According to the Population Reference Bureau (2017) the world's population will reach 9.8 billion in 2050, up 31 percent from the current estimation of 7.5 billion (PRB 2017). A major concern is that some of the largest potential oil and gas reserves are in countries that already have a water scarcity issue, for instance, China. However, China's population (like Europe) is predicted to decline by 2050, albeit there will still be around 1.3 billion people. By 2050, India will be the most populous country in the world (1.7 billion) and has huge potential unconventional oil and gas reserves, particularly in the Cambay basin. With a population of 1.3 billion today, its groundwater resource is already being depleted at an alarming rate. The introduction of UOGD (and 323 million people) will palpably overwhelm the fragile resource. In addition, there is a large UOGD potential in Africa; again, freshwater is at a premium on the continent. However, thirty African nations are expected to at least double in population by 2050. Nigeria, for instance is expected to have the world's second largest population increase, by 220 million people; and Nigeria has a large UOGD potential with the same threat to freshwater reserves.

Contemporary media coverage and the political backdrop to water scarcity is overshadowed by 'the impending energy crisis and the search for sustainable solutions' (Curry 2010). UOGD is at the forefront of political aspirations to address the energy crisis and instead of specific steps being taken to produce a sustainable water policy or any relevant technological changes to promote such policies the invasive technology of hydraulic fracturing threatens to be the antithesis to combating water scarcity. In the meantime, weak laws are being watered down by intense corporate lobbying where human rights and environmental concerns are secondary. The balance of lobbying between the wider shareholders and corporate lobbyists has become so disparate that those who pursue human rights and environmental issues from a non-industry perspective are drowned out by the financial might of corporations. Corporations can launch massive offensives to influence issues which put profit above all other issues; hence, the type of regulation that makes them more competitive. UOGD has been a prime example (Higgins 2010). The 'perpetuation of ignorance' surrounding water scarcity will seemingly create human rights abuses owing to the disastrous circumstances associated with it; such as, water-related diseases, famine, drought and inevitable fatalities to humans, flora and fauna (Higgins 2010).

This chapter advocates the growing consensus that if UOGD became widespread in the western world and beyond then the escalating water scarcity issues associated with the technology could be catastrophic. Human rights protagonists have sought to ground the right to water in international law in the decades following the International Covenant on Economic, Social and Cultural Rights (ICESCR) but renewed attempts to establish the human right to water have been plagued by a 'lack of legal articulation' ever since (Bulto 2011). Whilst some interpret the right as lacking 'an explicit and comprehensive expression in international human rights law' and therefore does not exist in that context (Bulto 2011), it can still be derived – albeit within a limited scope – from the explicitly protected rights recognised within the ICESCR. Nevertheless, such a derivation is not without criticism, as will be seen in the next section (see generally Tully 2005; Dennis and Stewart 2004).

According to its Charter, the United Nations has succinct purposes including; maintaining international peace and security; developing amiable relations among nations; and to cooperate in solving international economic, social, cultural and humanitarian problems. In addition, it seeks to promote respect for human rights and fundamental freedoms and to be central to coordinating the actions of Members in reaching such ends. These aims were reaffirmed and clarified in the 2000 United Nations Millennium Development Goals (MDGs) which were expected to be achieved by the year 2015; one of those goals was to eradicate extreme poverty and hunger. Although the issue of whether MDGs were achieved by 2015 is outside of the scope of this chapter, realising a human right to water is fundamental to the eradication of poverty and hunger. Accordingly, it is questionable whether the introduction of UOGD and the control paradigm safeguard the raison d'être of the UN and strive to attain MDGs in the future.

On the surface, the right to water 'has been recognised in a wide range of international documents, including treaties, declarations and other standards' (UN OHCHR 2010) but the reality is such that water remains unrecognised explicitly as an autonomous human right in international treaties, although international human rights law involves specific obligations relating to access to safe drinking water and sanitation.[4] Whilst one could say that the right ought to be derived from numerous human rights documents as they are intended, an explicit right to water has failed to be adopted by States universally (Huang 2008). This is perhaps why the overall theme within recent literature suggests that the right to water has been a 'latent component' of other socioeconomic rights contained within the ICESCR, other international human rights treaties and other water related treaties (Bulto 2011). For instance, in 2001 the European Committee of Ministers recognised that the rights to

---

[4] The Convention on the Elimination of All Forms of Discrimination against Women, adopted in 1979 (art. 14 (2)); International Labour Organization (ILO) Convention No. 161 concerning Occupational Health Services, adopted in 1985 (art. 5); The Convention on the Rights of the Child, adopted in 1989 (arts. 24 and 27 (3)); and The Convention on the Rights of Persons with Disabilities, adopted in 2006 (art. 28).

be free from hunger and the right to an adequate standard of living that are contained within international human rights instruments include the right to a minimum quantity of water of satisfactory quality.

The summary of the development of the right to water that follows will demonstrate that the intentions of the Committee on Economic, Social and Cultural Rights (CESCR) has been to 'articulate a pre-existing right' (Bulto 2011). Moreover, despite the existence of a prior or contemporary autonomous right being disputed, it can be demonstrated that the right has a firm legal standing, particularly when supported by environmental law, international water law and State related jurisprudence. This is evident in a wide range of national legal instruments which contain State duties/obligations and entitlements of citizens with regards to, among others, access to water and sanitation.

Fundamentally, the right to water—at an international level—resides in Articles 11 and 12 of the International Covenant on Economic, Social and Cultural Rights (ICESCR) 1966. Thus far, since no intergovernmental organisation enjoys exclusive responsibility for water resources, the choice of the appropriate law-making forum falls to governments (Tully 2005). Constitutional provisions can be put into operation in three principal ways: development of legislation, enforcement in courts, and in political discourse. There are, palpably, no guarantees – legal or otherwise – that the insertion of access to water in a constitution will lead to its inevitable implementation. However, establishing the right to water and sanitation within constitutions and a series of State obligations that create the necessary legal, social and economic conditions represent a step towards ensuring the realisation of the right. Importantly, many national constitutions impose specific duties upon the State to ensure availability, quality, and accessibility for their citizens (Centre on Housing Rights & Evictions,2008). Kenya, for example, has included in its constitution that every person has 'the right to water in adequate quantities and of reasonable quality' and that 'every person has the right to a reasonable standard of sanitation'; thus, both the right to water and sanitation have been enshrined in law.[5]

3.3 The Development of the Right to Water

At the national level, despite the absence of a ubiquitous right, the right to water and sanitation has been progressively more recognised in constitutions, legislation and courts globally (Centre on Housing Rights & Evictions 2008). Some countries have broad provisions addressing not just the quantity of drinking water, but the quality of water and sanitation services holistically; however, universality is far from being achieved. It is noteworthy that most water laws that been adopted since General Comment No. 15 (below) and that are currently being drafted (or under revision)

---

[5] For a comprehensive list of States that have adopted the right to water and sanitation within constitutions and law (and the Articles etc. in which they are contained) see Centre on Housing Rights & Evictions, 2008.

contain provisions based on the human rights dimension of access to water. For instance, the Uruguayan 2004 referendum based on a constitutional amendment regarding public ownership of water supply and water and sanitation provisions was supported by two-thirds of the population. Subsequently, the Constitution of Uruguay has been amended and now stipulates that '[a]ccess to drinking water and access to sanitation constitute fundamental human rights'.

The origins of the 'right to water' can be traced to 1946 when, whilst adopting its constitution, the World Health Organization declared that 'the enjoyment of the highest attainable standard of health is one of the fundamental rights of every human being' (Curry 2010). In 1948, with the adoption of the UDHR, Article 25 used similar language as the WHO two years earlier and bound the right to an adequate standard of living with the right to health. The right to water was not addressed in either document. Reminiscent of the ICESCR two decades later, which also omitted water, we can deduce that many of the interpretive issues that have curtailed the development of the right to water not merely originate from these early omissions but have continued in the same vein (Curry 2010). However, a continual stream of declarations and treaties has followed the ICESCR; it is palpable that many of them have attempted to realise the right over the decades.

In 1977, the United Nations Water Conference in Mar del Plata, Argentina first established the concept of basic water requirements to meet fundamental human needs. Its Action Plan (United Nations 1977) asserted that everyone has the right to have access to drinking water in a quantity and quality that meet their basic needs. There are three human rights treaties that unequivocally declare 'water' as a right: The Convention on the Elimination of All Forms of Discrimination against Women (CEDAW), the Convention on the Rights of the Child (CRC) and the Convention on the Rights of Persons with Disabilities (CRPD). The CRPD will be dealt with below. In 1979, the Convention on the Elimination of All Forms of Discrimination against Women (CEDAW) was signed (adopted 1981). Article 14(2)(h) (CEDAW) specified that States shall ensure women the right to 'enjoy adequate living conditions, particularly in relation to … water supply'.[6] A decade later the CRC was adopted; again, the CRC explicitly referred to water within its text.[7] Whilst the provisions of the ICESCR negate to mention water, both the CEDAW and the CRC mention the human right to water, as part of the right to development (CEDAW) and in relation to a universal right to health (CRC). In the context of opposing UOGD, the CRC is of vital importance because it imparts children with the right to clean drinking water free from the 'dangers and risks of environmental pollution' (United Nations 1989).

The Mar del Plata Conference was further established when Agenda 21 was adopted at the United Nations Conference on Environment and Development in

---

[6]Article 14 (2)(h): 'To enjoy adequate living conditions, particularly in relation to housing, sanitation, electricity and water supply'.

[7]Article 24 (2)(c): 'To combat disease and malnutrition, including within the framework of primary health care, through, inter alia, the application of readily available technology and through the provision of adequate nutritious foods and clean drinking-water, taking into consideration the dangers and risks of environmental pollution'.

June 1992 (Rio). Agenda 21 denoted as the 'Programme of Action for Sustainable Development', included a separate chapter on freshwater resources (Chap. 18). Chap. 18 endorsed the Resolution of the Mar del Plata Water Conference that all peoples have the right to drinking water whilst naming this principle 'the commonly agreed premise' (United Nations General Assembly 1993; Bulto 2011). Again in 1992, The Protocol on Water and Health to the United Nations Economic Commission for Europe's 'Convention on the Protection and Use of Transboundary Watercourses and International Lakes' sought to protect water resources (used as sources for drinking water) by ensuring States take appropriate measures to prevent them from being polluted (UN OHCHR 2010).

Another important development in 1992 came in the guise of the Dublin Statement (United Nations 1992; Tully 2005).[8] Of note, the Statement advocated – in Principle One – a holistic approach linking 'socio-economic development with environmental protection' when addressing effective resource management (Tully 2005). Dublin therefore brings to the fore the inalienable nexus between human rights and environmental rights, which is essential to recognise within the UOGD debate, particularly because provisions on the right to enjoy a healthy environment (formally adopted in most constitutions in 1992) may afford a legal basis for the improvement of water quality, for instance, through the prevention of pollution and the provision of adequate sanitation. Integrated water resource management (IWRM) has been developed to allocate water with reference to efficiency and sustainability and is defined as 'the co-ordinated management of water, land and related resources with a view to maximising socio-economic welfare in an equitable manner without compromising ecosystem sustainability' (Tully 2005; United Nations Commission on Sustainable Development 1998; Global Water Partnership 2000).

Whilst it is unclear how a human rights approach to water as a vital 'socio-economic' resource and an environmental rights approach to water as an 'environmental' resource can be reconciled, the UOGD debate necessitates the adoption of human rights and environmental law because the technology poses significant risks to both human health and the environment. Accordingly, fundamental environmental principles and fundamental human rights and freedoms can be utilised in tandem to secure the right to water. Tully correctly recognised an example of the possible affiliation/homogenisation stating: 'that environmental legislation which employs pollution abatement schemes and economic incentives for water conservation or recycling could usefully complement a human rights framework'(Tully 2005). Dublin was, in effect, a preparatory meeting for the Rio Summit later in 1992 (United Nations Conference on Environment and Development 1993). At that summit Agenda 21 referred to as the 'Programme of Action for Sustainable

---

[8]Whilst it is outside the remit of this chapter, it is important to note that the Dublin Statement introduced the issue of 'affordable price' into the development of the right to water. Principle 4 of the statement stated that 'it is vital to recognise first the basic right of all human beings to have access to clean water and sanitation at an affordable price'. In terms of that development, pricing water has both its merits and disadvantages e.g. culturally: it would be contrary to some cultures to pay for water whereas other capitalist cultures would demand a price for the resource (but arguably capitalism is the antithesis to an autonomous right to water).

Development'. In particular, Chap. 18 of the Agenda (water resources) not only endorsed the principle of UN Water Conference in Mar del Plata (that all peoples have the right to water) but affirmed the principle as a 'commonly agreed premise' (United Nations Conference on Environment and Development 1993; Bulto 2011; Salman and McInerney-Lankford 2004).

In 1994, within the Programme of Action of the International Conference on Population and Development, States affirmed the right to an adequate standard of living for citizens and their families, including adequate food, clothing, housing, water and sanitation. In 1995 the General Assembly adopted the United Nations Principles for Older Persons. In paragraph 5 the Committee referred to the 'basic rights' of 'access to adequate food, water, shelter, clothing and health care'. In Paragraph 32 the Committee affords those basic rights whilst attaching 'great importance' to the principle of 'Independence', which 'demands for older persons the rights contained in article 11 of [ICESCR 1966]'. In 1996, the United Nations Conference on Human Settlements (Habitat II) adopted the Habitat Agenda; again, water and sanitation were recognised as part of the right to an adequate standard of living. Paragraph 11 of the Agenda is noteworthy because it comprises Article 11(1) verbatim, except for adding the words 'water and sanitation'.

In 1997 the 'Water Convention' (United Nations 1997) coined the concept of 'vital human needs', which the International Law Association defined as 'waters used for immediate human survival, including drinking, cooking, and sanitary needs, as well as water needed for the immediate sustenance of a household' (International Law Association 2004). Peter Beaumont denotes drinking water as the 'most' vital of human needs (Beaumont 2000), which is important in this context, particularly as the phrase is a shorthand expression for the 'minimum core of the human right to water' (Bulto 2011). The Convention entered into force in 2014 and according to the 'statements of understanding' concerning the Convention and contained within the *Report of the Sixth Committee convening as a Working Group of the Whole* are comments stating that '[I]n determining "vital human needs", special attention is to be paid to providing sufficient water to sustain human life, including both drinking water and water required for production of food in order to prevent starvation' (Beaumont 2000; International Law Commission Working Group 1997). According to McCaffrey (2001), the Convention's provision on vital human needs 'is consistent with the human right to water'. The Convention, unlike the soft law provisions that erstwhile provide for the right to water, is binding on ratifying States and consequently can be used directly as a normative source of the human right to water and thus provide 'supportive legal authority for General Comment No 15' (see below) (Bulto 2011).

In 2000, the UN General Assembly resolution on the Right to Development, saw the 'strongest and most unambiguous' (Salman and McInerney-Lankford 2004) statement recognising a human right to water to date. The resolution affirmed that 'the rights to food and clean water are fundamental human rights and their promotion constitutes a moral imperative both for national governments and for the international community' (UN OHCHR 2000). Also in 2000, General Comment No 14 (CESCR) on the right to the highest attainable standard of health stressed that the drafting history of the ICESCR (particularly the wording of Article 12(2) ICESCR) accepted that the right to health extended to the underlying elements of health, including access to safe drinking water and sanitation (UN CESCR 2005).

In 2002, General Comment No 15 (GC15) of the UN Committee on Economic, Social and Cultural Rights (CESCR) gave the most complete and persuasive interpretation of the human right to water thus far and represents the 'first instance' of a UN body explicitly signifying that the right to water is contained within the ICESCR (Curry 2010; Hiskes 2010). GC15 declared that the 'human right to water is indispensable for leading a life in human dignity' and 'entitles everyone to sufficient, safe, acceptable, physically accessible and affordable water for personal and domestic uses'. The scientific evidence, that is the basis for objections to UOGD, raises clear-cut questions whether the intentions of the Committee are to be realised.

The Committee set out that States must 'adopt effective measures to realize, without discrimination, the right to water'. The notion of discrimination in the context of GC15 is important in the UOGD debate as the technique is not possible in all locations, owing to geological conditions. Only those who live in areas where unconventional oil and gas extraction methods can be adopted are under threat; hence there is a succinct element of discrimination surrounding the issue based on location. Within the normative content of the right to water and the entitlements and freedoms it contains, one such entitlement includes the right 'to a system of water supply and management that provides equality of opportunity for people to enjoy the right to water' (Curry 2010; Hiskes 2010). Again, the geological nature of UOGD prevents equality of opportunity under the umbrella of the right. The right's normative content ensures that water (and water facilities/services) must be accessible to all without discrimination; therefore, if a legal industrial practice prevents equitable access to water it is in contravention of the right to water. Alternatively, a freedom of the normative content of right is 'the right to maintain access to existing water supplies necessary for the right to water, and the right to be free from interference, such as the right to be free from…contamination of water supplies' (UN CESCR 1995). Accordingly, the normative content of the right to water should protect people and the environment from a 'highly risky and contaminating technique' that uses huge amounts of water and highly toxic chemicals (Hill 2015).

It is noteworthy that GC15, which addresses interpretive issues arising from ICESCR, has become the source of debate that denies ascending the human right of water into a universally binding right. Nonetheless, the language across the various documents of international human rights law is, in the main, consistent. Prior to the adoption of General Comment 15, international human rights documents that either implied or explicitly referred to the human right to water were inclined to vary slightly in their description, but the language was along the same lines (Hiskes 2010). Whilst the Comment should not be understated, it has proved problematic because it has been criticised for not having a succinct definition (see generally Tully 2005). Langford observed (2005), 'it is one thing to recognise a human right, it is another to define what it means'; but, on the other hand, the Comment elaborated the normative content of the right under the ICESCR.

It has been reasoned that weaknesses in General Comment 15 (based upon Articles 11 and 12 of the ICESCR) 'indicates deficiencies within the content of the standard' of the right to water (Cahill 2005). Without a succinct definition, concerning various facets of the right to water, the perceived 'interpretive creativity' (Tully 2005; Bulto 2011) of the CESCR has raised questions concerning whether the right exists. In a damning article by Dennis and Stewart (2004), the Committee is accused of being 'aggressive' in its effort to 'deconstruct' the right to an adequate standard of living, as laid down in Article 11 (food, water, clothing, and housing). By this means, they assert that the Committee 'has overridden the decisions of the negotiators and taken positions inconsistent with the views of states'. However, Gleick observed (in notes from the original debate through the construction of the UDHR) that the provisions for food, clothing and housing were not intended to be all-inclusive, but representative or indicative of the 'component elements of an adequate standard of living'; such standards can, and indeed have, change over time (Gleick 1999).

What is, perhaps, most damning to the recognition to the right to water is Dennis and Stewart's accusations that the Committee rewrote Article 11 'by resurrecting and adopting alternatives that were considered and rejected by the negotiators' (Dennis and Stewart 2004). In Tully's opinion, the Committee took an approach which 'undermines the principle of legal security by reading into a legal text a content which simply is not there'. Dennis and Stewart go further and proclaim that the source of a *separate* right to water is 'virtually without precedent'. In order to justify that assertion, they refer to the CRC as the 'only international human rights instrument that even mentions water' and that they are unaware of any 'mention of water in the negotiating histories of the UDHR or the ICESCR' (Dennis and Stewart 2004). Palpably, such an insight is erroneous in light of the documents highlighted above, particularly when the Legal Resources for the Right to Water and Sanitation report is considered (Centre on Housing Rights & Evictions 2008).

Dennis and Stewart's and – subsequently – Tully's denial of the long-standing existence of the right was challenged again in the Convention on the Rights of Persons with Disabilities (CRPD[9]). One of the human rights treaties that explicitly refers to water as part of a protected right, the CRPD was later than their articles and refers to ensuring equitable access to clean, affordable water. Accordingly, despite the objections of such academic commentary, key actors within the various UN Committees outwardly disagree when both drafting new Conventions and interpreting past treaties; the right to water, for all intents and purposes, is enshrined in human rights discourse and continues to be incorporated in new treaties.

It is evident that the development of the right to water is chronicled within the documents mentioned herein and culminated in the adoption of the 2010 UN General Assembly Resolution on the Human Right to Water and Sanitation (United Nations General Assembly 2010). The problem is that that Resolution has muddied the waters somewhat, as the plenary sessions suggest. We can only speculate

---

[9]Article 28 (2)(a)'To ensure equal access by persons with disabilities to clean water services, and to ensure access to appropriate and affordable services, devices and other assistance for disability-related needs'.

regarding why certain vested economic interests continue to oppose a universal recognition of the right. Despite any judicious read-through of the treaties and documents suggesting that the right does exist and should be recognised, those vested interests impinge the development of the right to water. In the face of such objections, the question as to whether the right will become autonomous is set to continue.[10] Cullet reasons that, generally speaking, the different developments that have emerged over the past couple of decades have elevated the right to water 'to a principle of international custom' (Cullet 2013; Bates 2010). What may be confusing, therefore, to the layman, activist or, perhaps, policymaker in a search for redress against UOGD, is that the human right to water is seemingly recognised in international human rights treaties but, at the same time, it is rarely recognised as a legal tenet or enforced: when individuals or groups seek protection they find that access to justice is often fraught with insurmountable difficulties.

The reality is that attempts to establish the right to water, based on the ICESCR alone, have proven slow and controversial (Bulto 2011). Nonetheless this should come as no great surprise considering the original negotiations are, overall, beyond incarnate recognition. Negotiations that took place in the 1950s and 1960s are incongruent with contemporary scientific knowledge and modern societal needs: the right to water has developed in line with human development (including population increase), scientific advancements and the advent of globalisation. It can be seen as peculiar, owing to the immediate link between water and life, that legal instruments have only been giving increasing importance to the right to water over the past two to three decades. Alternatively, this development is a response to new knowledge and better awareness concerning a set of environmental threats with relatively recent foundations.

The impending – potentially catastrophic – threats with regards to water scarcity, particularly from heavy industry, require international action. UOGD is a matter in question here; certainly, the industry is an environmental threat on numerous levels and (despite industry misdirection and media misinformation regarding its actual age) the way the various methods and scale combine effectively make it 'a new technology'. In consideration that this type of oil and gas production only became operational in the twenty-first century it is accurate to assert that it is still being developed and the 'manual' is still being written. UOGD has not, and never will be,

---

[10] The United States, for example, denied the existence of the 'right to water and sanitation', as described in the resolution, as being reflected in an international legal sense. This denial relates back to the omission of the term 'water' and the right to water in the ICESCR. In addition, the US criticised the resolution for undermining the work done, regarding HRs, in Geneva. In essence, the US reflected other countries sentiments that the right to water (and sanitation) equated to creating a new human right, stating that '[the resolution] attempted to take a short cut around the serious work of formulating, articulating and upholding universal rights'. In contrast, Egypt's stance is based on the understanding that 'it did not create new rights, or sub-categories of rights, other than those contained in internationally agreed human rights instruments'. The Resolution was passed with 122 recorded votes, but there were 41 abstentions. The plenary sessions to the Resolution provide a narrative regarding why those 41 States abstained from the vote. See General Assembly Adopts Resolution Recognizing Access to Clean Water, Sanitation as Human Right, by Recorded Vote of 122 in Favour, None against, 41 Abstentions' (64th General Assembly, Plenary, 108th meeting). Available at: www.un.org/press/en/2010/ga10967.doc.htm

a steady state operating procedure; it will never be perfected. This is largely why the regulatory framework cannot respond to the new science: comprehensive, bespoke guidelines are required to regulate the technology.

To deny the development of the right to water owing to the objections of historic negotiators is, at best, short sighted, at worst, ignoring the raison d'être of the United Nations and the fact that laws themselves develop. An important but modest question is lost within the relevant literature that needs to be addressed: why was the right to water omitted from the ICESCR? The omission can be one of three things: a drafting oversight; a deliberate omission; or it was considered and rejected by the original negotiators. Most of the contemporary literature is concerned with trying to find 'novel' ways in which to find the right from within the wide range of documents (above) or alternatively to deny the right existed in the first place (Tully 2005). There is a viewpoint that it was a deliberate omission, which suggests the original drafters (ICESCR) took it for granted that the right would be recognised in the provisions. Indeed, Tully revealed that water was *deliberately* omitted by the drafters 'as an explicit right on account of its nature: like air, it was considered so fundamental that its formal inclusion was unnecessary' (Tully 2005; Gleick 1999).

It has been argued that the 'intrinsic link with life' means that water is included as a right 'in any human right instrument or bill of rights', regardless of whether or not it is formally incorporated within the list of recognised rights (Cullet 2013; Gleick 1999). There is a degree of rationalisation when water's vital role is considered; unquestionably, academic commentary and the CESCR have been frustrated by attempts to deny the right. Bulto (2011) observed that 'the absence of a comprehensive guarantee for the human right to water in the universal human rights treaties has variously been dubbed "odd," and "startling"' and, in light of prior General Comments and international instruments adopted in earlier decades containing references to right (the CESCR endeavoured to break new ground by unequivocally affirming that the ICESCR contains provisions that implicitly contain an autonomous human right to water (Bulto 2011)) the CESCR were responding to being 'confronted continually with the widespread denial' of it by States. The Committee's role must be born in mind; they must consider reports submitted by UN member States on their compliance with the ICESCR, thus perceived difficulties must have been central when the Committee convened for GC15.

Undeniably, one is immediately struck by the fact that both Articles 11 and 12 (and the Covenant) are devoid of a separate right to water or the term 'water'; this omission has in turn created legal and political uncertainty, thus limiting the development and legal status of the right to water. When read in context, however, such uncertainty can be viewed as being overly doctrinaire because good sense and sound judgment would advocate that including the term 'water' would be superfluous to requirements in the Covenant at that time.

Certainly, the content of Article 11 demands a right to water in order for the right to 'adequate living standards' to be achievable. Further, in recognising a 'fundamental right of everyone to be free from hunger' (UN OHCHR 1976), water would be essential for that right because crops need water, water is needed for cooking and livestock need to drink. Furthermore, States would be useless in their obligations to disseminate knowledge 'of the principles of nutrition' or 'developing or reforming agrarian systems in such a way as to achieve the most efficient development and utilisation of natural resources' without safeguarding water resources (UN OHCHR 1976).

The wording of Article 11(1) is pivotal here. It states:

> The States Parties to the present Covenant recognize the right of everyone to an adequate standard of living for himself and his family, including adequate food, clothing and housing, and to the continuous improvement of living conditions.

The development of the right to water is lost in an interpretive minefield that first began, over seventy years ago. It is clear from the onset of any UN movement towards recognising the right to water that an etymological absurdity has developed alongside it, which, in turn, has weakened its legitimacy as a self-standing human right. This linguistic paradox can be seen to revolve around the term 'food'. The Concise Oxford Dictionary defines food as 'any nutritious substance that people or animals *eat or drink* or that plants absorb in order to maintain life and growth'. Certainly, water is defined as a nutrient and is an indispensable ingredient in many mainstay foods – essential to human survival – across the globe. The biological and chemical descriptions that define water as a foodstuff are beyond our remit but as a 'nutritious substance that people eat or drink … in order to maintain life and growth' water can be defined as food for the purposes of Article 11. The right to food is recognised in the 1948 Universal Declaration of Human Rights (UDHR) as part of the right to an adequate standard of living, and is protected, as a right, by the ICESCR 1966. Palpably, there is an anomaly that requires reconsideration if all human beings have the right to adequate food and the right to be free from hunger but have no right to adequate water or the right to be free from dehydration (UN OHCHR 2010 – Factsheet 34).

The interpretation of the ICESCR has become central to the development of the right to water within the literature. Bulto (2011), in particular, pays specific attention to the manner in which the CESCR 'read in' the right from the implicit terms of Articles 11 and 12 of the ICESCR. This interpretive approach – the purposive approach – that the Committee undertook during GC15 has been unceremoniously criticised by the likes of Tully and Dennis and Stewart; despite both methods being universally accepted methods of interpretation (Tully 2005; Dennis and Stewart 2004). In an attempt the prevent interpreting the ICESCR in a narrow and restricted sense, the Committee sought to elucidate the purpose of the legislation to reduce ambiguities in a manner that best serves the object and purpose of the Covenant. The most visible criticism for using this approach is the way the right to water has been derived from succinct guaranteed human rights. However, those criticisms seemingly originate from the Committee's (again perfectly acceptable legal method) interpretation of the word 'including' in Article 11(1). Preceding the list of rights mentioned in the ICESCR, the Committee reading of the Article was such that those rights were 'not intended to be exhaustive' (Hiskes 2010). Using this rational (again accepted) interpretative method, Curry recognised that 'since the right to water is a natural extension of those rights listed in the ICESCR, recognition of the right to water is as essential as all others mentioned' (Curry 2010).

An additional factor with regards to the content of the right to water is 'quality'. This factor has rarely been approached in the literature but is fundamental to public grievances concerning UOGD. Considering the type issues highlighted by Vengosh and the language of GC15, which states that 'the water required for each personal or domestic use must be safe, therefore free from micro-organisms, chemical substances and radiological hazards that constitute a threat to a person's health', the technique comes heavily under fire because it injects chemical substances during numerous phases of the process and leaves radiological contaminants above and below the surface free to pollute. Therefore, UOGD can contaminate sub-surface and surface water resources; both constitute 'a threat to a person's health' and to the environment that supports human health.

3.4   The Right to Water and 'Fracking'

Referring States to WHO guidelines, GC15 states standards should ensure 'the safety of drinking water supplies through the elimination of, or reduction to a minimum concentration, of constituents of water that are known to be hazardous to health'. The General Comment also requires water to be of an acceptable colour, odour and taste for each personal or domestic use'. Taking Pennsylvania as an example, owing to the incidents and complaints that have been reported to the Pennsylvania Department of Environmental Protection (DEP), it is clear that actual incidents have occurred where quality issues undermined GC15 and, ultimately, the ICESCR in this context. According to Inglis and Rumpler (2015):

> Drilling poses major risks for our water supplies, including potential underground leaks of toxic chemicals and contamination of groundwater. There are at least 243 documented cases of contaminated drinking water supplies across Pennsylvania between December 2007 and August 2014 due to fracking activities, according to the Pennsylvania Department of Environmental Protection (DEP).

In fact, it is commonplace for O & G companies to be required bring in 'replacement water supplies for residents, construct new drinking water wells, or otherwise modify their existing water wells' in order to make the water potable following 'fracking' operations (Inglis and Rumpler 2015). Some companies have been 'cited'. For instance, in November 2012, Carrizo (Marcellus) LLC was cited (violation 653,937) for failing to properly restore a contaminated drinking water supply following drilling operation in Forest Lake Township, Susquehanna County (Pennsylvania Department of Environmental Protection 2014).

In cases like Forest Lake (and there are many), it is difficult to dispute that UOGD affects both the quantity and quality of water resources. Whereas legislation setting technical standards for drinking water quality is commonplace in countries around the world we are reminded of the Halliburton Loophole that exempts fracking operations from the SDWA in the USA. The human rights aspect to fracking operations has arisen in response to what can be described as a belief that governments have failed to adequately protect citizens and their environment. Cooperation between the industry and its regulatory agencies - in ways that fail to respect human rights standards - has been perceived. The Halliburton Loophole is an illustration of a government and regulations failing to protect its citizens owing to corporate lobbying.

The nexus between human health, human rights and the environment and environmental rights has been succinctly forged over the last decade or so; in Europe (thus the UK) the establishment of the right to live in a healthy environment, born out of the Article 8 ECHR's right to respect for one's 'private and family life and home', is the embodiment of that profound connection between humans and their environment. Consequently, it should come as no surprise that human rights norms articulated in international declarations, conventions and treaties have come into question in the UOGD debate. Traditionally human rights advocacy is sought when government agencies, legislatures and courts regularly fail to adequately protect the process, health, safety and other rights of their citizens. Therefore, the status of the right to water has come to the fore since the advent of the technology. The recognition of the right has become instrumental to protecting other inalienable rights.

It can be asserted that when human rights standards are applied to the issue a ban (rather than regulation) is supported; especially as the emerging science suggests that the risks to public health are widespread and costly. Concerned citizens often see very good science - that suggests the risks are too high - brought into the public domain and ignored by decision-makers and a regulatory framework that is not geared to respond to new science, or in other words, in this case, an advanced industry in its infancy. In essence, it is evident that our laws are not prepared for either the new science that facilitates unconventional gas extraction or the scientific evidence that questions the safety and economic viability of UOGD. Arguably, existing regulation does not adequately regulate UOGD (for example, in general Environment Impact Assessments do not have a category that covers fracking, and in some countries (e.g. in the UK there is no fracking specific regulation when the associated risks are quite simply not the same as conventional extraction.) and existing technology is often found wanting when it comes to mitigating harm – nonetheless many governments are aggressively promoting the technology. Accordingly, the right to water (within the human rights discourse), in conjunction with environmental law and water law, is poised to take centre stage in the 'fracking debate'.

At the outset of the US shale revolution there was quite a bit of discussion about "energy independence", jobs, tax revenue, and short shrift was paid to concerns about issues such as air quality, community cohesion, ecological effects, and water quality and/or quantity. However, 5–15 years out from the beginning of the UOGD revolution we are seeing that realized tax revenue is often an order-of-magnitude less than projections, job creation has been replaced by job migration from shale play to shale play, and directional wells are proving short-lived (i.e., 85% declines in productivity from year 1 to year 2). The latter, along with expansion into less productive plays like the Antrim in Michigan, has resulted in the UOGD industry requiring more and more resources in the form of chemicals, sand (Carroll and Wethe 2016), and water to stimulate production in the name of shareholder return.

Furthermore, the exponential increase in water demand, as well as ever longer laterals, has resulted in parallel rises in liquid and solid waste production. This waste increase has created tremendous stressors in states like Oklahoma and Ohio where "induced seismicity" or anthropogenic earthquake activity has increased from an average of 21 M3+ quakes between 1973 and 2008 to 659 of M3+ in 2014 alone according to the United States Geological Survey (USGS). These seismic events are not limited to smaller events; magnitudes exceeding 5.8 have been witnessed in Pawnee and Cushing, Oklahoma. Incidentally Cushing happens to be the site of the US's largest strategic commercial crude oil storage terminal prompting questions about whether the UGD's excessive production of waste and demand for water may be compromising their own infrastructure and undercutting the notion that UGD will lead to "energy independence".

Examples of increasing resource demand include Chesapeake Energy's 'Propageddon' lateral in Louisiana, which required more than 25 tons of frac sand (5–6 times the average amount of sand needed in a typical fracked lateral). From a water demand perspective names like "Purple Hayes", "Outlaw", and "Walleye" have been given to "Super-Laterals" in Ohio and West Virginia where they exceed 17–20,000 feet in length, or 2.3–2.6 times average lateral length in these two states. These laterals require more than 85 million gallons each, which translates into 4470 Gallons Per Lateral Foot (GPLF). To put this demand into some perspective, and assuming the average American uses 33,000 (USGS Water Science School 2016) gallons of water per year, this is roughly equivalent to the annual water demand of 2587 Americans.

Traditionally Appalachian West Virginia and Ohio laterals require 970–1080 GPLF with demand growing at a rate of 11–22% per year. As an example of how much liquid – and potentially radioactive – waste is produced we estimate that 11–12% of the freshwater used in the fracking process comes back to the surface as "brine" and must be disposed of in Class II Salt Water Disposal Injection Wells. Put another way an 85-million-gallon lateral would likely produce 9.8 million gallons of liquid waste, which is equivalent to the total amount of water in 15 Olympic sized swimming pools.

These recent developments call into question our existing resource demand models, which estimate resource demand is increasing by 7–30% per year, while oil and gas production is declining by 85% per year per well. Yet, more importantly these trends raise concerns about watershed ecological security and/or resilience, public water supply robustness, and the increasing importance of the UOGD's water demand in modelling any given watershed's redundancies in the face of less frequent and more intense precipitation events resulting from climate change.

At the present time UOGD's water demand amounts to roughly 14% of residential water demand but exceeds 65% in counties as geographically disparate as Carroll County, Ohio, Richland County, Montana, and Wetzel County, West Virginia. The most extreme example of how residential water demand is in conflict with UOGD's demand is Doddridge County, West Virginia where UGD demand is nearly double that of residential demand. These demands greatly exceed the "precautionary principle setting 20% of the natural runoff in a region as the upper limit of... consumptive use" by any one industry described by Sposito (2013). The crux of the matter is that the industry is only charged \$4.25–6.25 per thousand gallons, which only amounts to 0.25–0.28% of well-pad costs and is less than half of what residential users pay here in the United States (Porter 2014). Furthermore, when the resulting liquid wastes are produced the UOGD industry is charged \$00.05–00.20[11] per gallon disposed (e.g., 0.0044% of well-pad costs) across the country's thousands of Class II Injection Wells with examples of such wells from Eastern Ohio below.

Data from Colorado on how they price water for the fracking industry is even more concerning. Our research has found that Windsor Town Council sold Great Western Oil & Gas (GWOG) up to 65 acre ft. (1 acre foot of water = 325,851 gallons) per year for the next 10 years. That contract can be renegotiated at that time. The Town Council reports that the cost that GWOG paid is \$400/acre ft. That brings into the Town of Windsor a meagre \$26,000 annually should the entire 65 acre ft. be taken, and it seems likely that GWOG will use the water rather quickly, based on past consumption and the requirements of the industry. Therefore, that brings the cost per gallon to \$0.00122 or one tenth of one penny per gallon, not \$00.01 but \$00.00122.

Such data highlights a crucial problem, there is no supply-side price signal demanding the UOGD industry reduce or stabilize their water demand per unit of energy produced. An additional issue concerns anecdotal evidence pointing towards the UOGD industry relying on highly fragile and ecologically critical 1st- and 2nd-order streams throughout Appalachia, when their demand cannot be met by documented water withdrawals agreements with conservancy districts. At the present time, research points to a 22–25% gap in our understanding of where this industry's water demand is coming from; thus, leaving frontline communities and policy makers in the dark regarding how this known 'unknown' environmental externality will manifest in the coming years and decades.

Resource demand in the UOGD industry is directly related to the global price of oil and gas, with water demand increasing exponentially as the price of oil and gas declines. This forces the industry to rely on resources known to generate a disproportionate Return-On-Investment (ROI) relative to the price paid for the resources. As an example, the water demand inflection points we have documented in the Marcellus and Utica Plays of Southern Appalachia happened to coincide with a 50% decline in the global price of Brent Crude and West Texas Intermediate Oil between Q1–2014 and the end of 2016.

3.5 Conclusion

Despite its widespread use in the United States for over a decade, hydraulic fracturing has only recently been scrutinized to determine the industry's effects on human rights. Under the special procedures of the HRC, the Special Rapporteur on the human right to safe drinking water and sanitation, Catarina de Albuquerque,

---

[11] This range is from Ohio alone and does not speak to a nationwide fee structure for UGD liquid waste disposal.

concluded her 2011 mission to the United States by outlining serious concerns over the effect of a range of polluting activities associated with the hydraulic fracturing process. She observed a distinct:

> … policy disconnect…between polluting activities and their ultimate impact on the safety of drinking water sources. The absence of integrated thinking has generated enormous burdens, including increased costs to public water systems to monitor and treat water to remove regulated contaminants and detrimental health outcomes for individuals and communities. (United Nations Human Rights Council 2011)

There have been scores of scientific studies that have revealed water contamination due to fracking processes. Ingraffea's review (2014) of the compliance reports from conventional and unconventional oil and gas wells drilled in Pennsylvania between 2000 and 2012 revealed that casing/cement impairment is six times more likely to occur in unconventional wells than in conventional wells. Such flaws may result in cases of subsurface gas migration into the water supply, as has already occurred in the state. Indeed, published data demonstrates evidence of:

> Contamination of shallow aquifers with hydrocarbon gases… contamination of surface water and shallow groundwater from spills, leaks, and/or the disposal of inadequately treated shale gas wastewater… [and] accumulation of toxic and radioactive elements in soil or stream sediments near disposal or spill sites…from hydraulic fracturing throughout the United States (Vengosh et al. 2014).

Qualitative data from Colorado has further revealed complaints of water contamination from residents living near fracking sites that are often intentionally misunderstood, assigned a different cause, or diluted by state regulatory bodies (Opsal and O'Connor Shelley 2014). Recently, the Pennsylvania Department of Environmental Protection disclosed details of 243 cases in which fracking companies were found by state regulators to have contaminated private drinking water wells in the last four years (Wall Street Journal 2014). In a much delayed survey of existing scientific literature on this topic (not a new data set), the US Environmental Protection Agency found 'scientific evidence that hydraulic fracturing activities can impact drinking water resources under some circumstances. The report identifies certain conditions under which impacts from hydraulic fracturing activities can be more frequent or severe:

- Water withdrawals for hydraulic fracturing in times or areas of low water availability, particularly in areas with limited or declining groundwater resources;
- Spills during the handling of hydraulic fracturing fluids and chemicals or produced water that result in large volumes or high concentrations of chemicals reaching groundwater resources;
- Injection of hydraulic fracturing fluids into wells with inadequate mechanical integrity, allowing gases or liquids to move to groundwater resources;
- Injection of hydraulic fracturing fluids directly into groundwater resources;
- Discharge of inadequately treated hydraulic fracturing wastewater to surface water; and
- Disposal or storage of hydraulic fracturing wastewater in unlined pits resulting in contamination of groundwater resources. (U.S. EPA 2016)

Cumulatively, the scientific literature, NGO and other policy reports and the vital testimony of local people indicate the likely impairment of the right to water for residents living near fracking sites. Even so, from the data presented here, we can see that perhaps *the* major issue regarding water use is the shifting of the resource from society to industry and the demonstrable lack of supply-side price signals that would demand the UOGD industry reduce or stabilize their water demand per unit of energy produced. Thus, in the US context alone, there is considerable evidence that the human right to water is seriously undermined by the UOGD industry and given its spread around the globe, this could soon become a global human rights issue.

References

Barlow M (2006, November) A UN Convention on the right to water: an idea whose time has come, Blue Planet

Bates R (2010) The road to the well: an evaluation of the customary right to water. Rev Eur Community Int Environ Law 19(3):282, 289

Beaumont P (2000) The 1997 UN convention on the law of non-navigational uses of international watercourses: its strengths and weaknesses from a water management perspective and the need for new workable guidelines. Int J Water Res Dev 16:475, 483

Brock H (2011) Competition over resources: drivers of insecurity and the global south, Oxford Research Group. http://www.oxfordresearchgroup.org.uk/publications/briefing_papers_and_reports/competition_over_resources_drivers_insecurity_and_global_so

Brown Weiss E (2012) The coming water crisis: a common concern of humankind. Transnat Environ Law 1(1):153–168

Bulto TS (2011) The emergence of the human right to water in international human rights law: invention or discovery? Melbourne J Int Law 1:12

Cahill A (2005) 'The human right to water – a right of unique status': the legal status and normative content of the right to water. Int J Human Rights 9(3):389–410, 389

Carroll J, Wethe D (2016, October 20) Chesapeake energy declares 'Propageddon' with record frack, Bloomberg Markets

CESCR (1995) General comment no. 6: the economic, social and cultural rights of older persons, E/1996/22. Available at: https://www.refworld.org/docid/4538838f11.html

Centre on Housing Rights & Evictions (2008) Legal resources for the right to water and sanitation - international and National Standards, 2nd edn. Available at: www.worldwatercouncil.org/fileadmin/wwc/Programs/Right_to_Water/Pdf_doct/RWP-Legal_Res_1st_Draft_web.pdf. Accessed 13 Mar 2019

Cullet P (2013) Right to water in India—plugging conceptual and practical gaps. Int J Hum Rights 17:56–78

Curry E (2010) Water scarcity and the recognition of the human right to sage freshwater. Northwest J Int Hum Rights 9:103–121

Dennis MJ, Stewart DP (2004) Justiciability of economic, social, and cultural rights: should there be an international complaints mechanism to adjudicate the rights to food, water, housing, and health? Am J Int Law 98(3):462–515

Escobar A (2010) Latin America at a crossroads: alternative modernizations, post-liberalism, or post-development? Cult Stud 24(1):1–65

Famiglietti JS (2014) The global groundwater crisis. Nat Clim Chang 4(11):945–948

Fischetti M (2013, August 20) Groundwater contamination may end the gas-fracking boom, Sci Am, http://www.scientificamerican.com/article/groundwater-contamination-may-end-the-gas-fracking-boom/

FracFocus. FracFocus 2.0: hundreds of companies. Thousands of wells, FracFocus, www.fracfocus.org

General assembly adopts resolution recognizing access to clean water, Sanitation as human right, by recorded vote of 122 in favour, none against, 41 abstentions (64th General Assembly, Plenary, 108th meeting). Available at https://www.un.org/press/en/2010/ga10967.doc.htm

Gleick PH (1999) The human right to water. Water Policy 1(5)

Global water partnership, towards water security: a framework for action, Stockholm, 2000

Giupponi MBO, Paz MC (2015) The implementation of the human right to water in Argentina and Colombia. Anuario Mexicano de Derecho Internacional 15(1):323–352

Gusikit RB, Lar UA (2014) Water scarcity and the impending water-related conflicts in Nigeria: a reappraisal. J Environ Sci Toxicol Food Technol 8(1):20

Higgins P (2012) Earth is our business. Shepheard-Walwyn, London

Higgins P (2010) Eradicating ecocide. Shepheard-Welwyn, London

Hill D (2015, February 24) Is Bolivia going to frack "Mother Earth"? The Guardian https://www.theguardian.com/environment/andes-to-the-amazon/2015/feb/23/bolivia-frack-mother-earth

Hiskes RP (2010) Missing the green: golf course ecology, environmental justice, and local "fulfilment" of the human right to water. Hum Rights Q 32:326–341

Huang, L.-Y. Not just another drop in the human rights bucket: the legal significance of a codified human right to water (2008), 20 Fla J Int Law

Inglis J, Rumpler J (2015) Fracking failures: oil and gas industry environmental violations in Pennsylvania and what they mean for the US. Environment America Research & Policy Center.

Ingraffea A et al (2014) Assessment and risk analysis of casing and cement impairment in oil and gas Wells in Pennsylvania, 2000-2012. Proc Natl Acad Sci 111(30):10955–10960

International and National Standards – 2nd edition (Right to Water and Sanitation Programme CENTRE ON HOUSING RIGHTS AND EVICTIONS, 2008) http://www.worldwatercouncil.org/fileadmin/wwc/Programs/Right_to_Water/Pdf_doct/RWP-Legal_Res_1st_Draft_web.pdf

International Labour Organization (ILO) Convention No. 161 concerning Occupational Health Services, adopted in 1985 (art. 5)

International Law Association, 'Water resources law: fourth report' (Paper presented at the Berlin Conference, Berlin, Germany, 2004) 12

International Law Commission Working Group, Report on the Convention on the Law of the Non-Navigational Uses of International Watercourses, UN GAOR, 51st sess, 6th comm, Agenda Item 144, UN Doc A/51/869 (11 April 1997) 5 [8]

Kundzewicz ZW, Döll P (2009) Will groundwater ease freshwater stress under climate change? Hydrol Sci J 54(4):665. https://doi.org/10.1623/hysj.54.4.665

Lal Panjabi RK (2014) Not a drop to spare: the global water crisis of the twenty-first century. Georgia J Int Comparat Law 42(2)

Langford M (2005) The United Nations concept of water as a human right: a new paradigm for old problems? Water Resour Dev 21(2)

McCaffrey SC (2001) The law of international watercourses: non-navigational uses. Oxford University Press, Oxford

Mobbs P (2004) Turning the world upside down, the world today. J R Inst Int Aff 60(12):16

Mobbs P (2014, July 8) An abuse of science – concealing Fracking's radioactive footprint, Ecologist, www.theecologist.org/News/news_analysis/2469495/an_abuse_of_science_concealing_frackings_radioactive_footprint.html

Olmstead SM et al (2013) Shale gas development impacts on surface water quality in Pennsylvania. Proc Natl Acad Sci 110(13):4962–4967

Opsal T, O'Connor Shelley T (2014) Energy crime, harm, and problematic state response in Colorado: a case of the fox guarding the hen house? Crit Crim. https://doi.org/10.1007/s10612-014-9255-2

Pennsylvania Department of Environmental Protection, Office of Oil and Gas Management, Oil and Gas Reports. www.portal.state.pa.us/portal/server.pt/community/oil_and_gas_reports/20297. Accessed 18 Sept 2014

Population Reference Bureau (PRB) (2017) World population data sheet (August 2017) www.prb. org/Publications/Datasheets/2017/2017-world-population-data-sheet.aspx

Porter E (2014, October 14) The risks of cheap water, The New York Times

Richey AS, Thomas BF, Lo M-H, Reager JT, Famiglietti JS, Voss K, Swenson S, Rodell M (2015) Quantifying renewable groundwater stress with GRACE. Water Resour Res 51:5217–5238. https://doi.org/10.1002/2015WR017349

Salman SMA, McInerney-Lankford S (2004) The human right to water: legal and policy dimensions. World Bank, Washington, DC

Short D, Elliot J, Norder K, Lloyd-Davies E, Morley J (2015) Extreme energy, 'fracking' and human rights: a new field for human rights impact assessments? Int J Hum Rights 19:697–736

Sposito G (2013, November) Green water and global food security. Vadose Zone J 12(4)

Stark M, et al. (2012) Water and shale gas development: leveraging the US experience in new shale developments, Accenture December

Tully S (2005) A human right to access water? A critique of general comment no. 15. Netherlands Q Hum Rights 23(7). https://doi.org/10.1177/016934410502300103

U.N. Human Rights Council [HRC], Promotion and Protection of All Human Rights, Civil, Political, Econ., Social and Cultural Rights, Including the Right to Development: Report of the Independent Expert on the Issue of Human Rights Obligations Related to Access to Safe Drinking Water and Sanitation. U.N. Doc. A/HRC/12/24 (July 1, 2009), para 57

U.S. EPA Hydraulic fracturing for oil and gas: impacts from the hydraulic fracturing water cycle on drinking water resources in the United States (final report). U.S. Environmental Protection Agency, Washington, DC, EPA/600/R-16/236F, 2016

United Nations (1977) Mar Del Plata Action Plan of the United Nations Water Conference. Available at: www.internationalwaterlaw.org/bibliography/UN/UN_Mar%20del%20Plata%20Action%20Plan_1977.pdf. Accessed 19 Mar 2019

United Nations (1979). Convention on the Elimination of All Forms of Discrimination Against Women, adopted 18 Dec. 1979, G.A. Res. 34/180, U.N. GAOR 34th Sess., art. 14, U.N. Doc. A/34/46 (1980), 1249 U.N.T.S. 13 (entered into force 3 Sept. 1981)

United Nations (1992) Dublin statement on water and sustainable development, international conference on water and the environment: development issues for the 21st century, UN Doc. A/CONF.151/PC/112

United Nations (1997) Convention on the law of the non-navigational uses of international watercourses adopted by the general assembly of the United Nations on 21 May 1997 (entered into force on 17 August 2014).

United Nations (1989) Convention on the Rights of the Child, adopted 20 Nov. 1989, G.A. Res. 44/25, U.N. GAOR, 44th Session, art. 24, U.N. Doc A/44/49 (1989), 1577 U.N.T.S. 3 (entered into force 2 Sept. 1990)

United Nations (2006) Convention on the rights of persons with disabilities and its optional protocol (A/RES/61/106) was adopted on 13 December 2006 at the United Nations headquarters in New York (effective 3 May 2008)

United Nations Commission on Sustainable Development, Report of the Expert Group Meeting on Strategic Approaches to Freshwater Management, Harare, 6th Session, New York, UN Doc. E/CN.17/1998/2/Add.1

UN Committee on Economic, Social and Cultural Rights (CESCR), General Comment No. 6: The Economic, Social and Cultural Rights of Older Persons, 8 December 1995, E/1996/22, available at: https://www.refworld.org/docid/4538838f11.html. Accessed 13 Mar 2019

UN Committee on Economic, Social and Cultural Rights (CESCR), General Comment No. 15: The Right to Water, 11–29 November 1995, E/C.12/2002/11

United Nations Conference on Environment and Development, Report, UN Doc A/CONF.151/26/Rev.1 (Vol 1) (1 January 1993)

United Nations General Assembly (1993) Report of the United Nations conference on environment and development, UN Doc A/CONF.151/26/Rev.1 (Vol 1) (1 January 1993) annex II, [18.47]

United Nations General Assembly (2010) Resolution 64/292, the human right to water and sanitation, UN Doc. A/RES/64/292

United Nations Human Rights Council (2011) Report of the Special Rapporteur on the Human Right to Safe Drinking Water and Sanitation, Catarina de Albuquerque: Mission to the United States of America, A/HRC/18/33/Add.4

UN Office of the High Commissioner for Human Rights (OHCHR), Fact sheet no. 34. The right to food. GE.10–11463–April 2010–13,735. Available at: www.ohchr.org/Documents/Publications/FactSheet34en.pdf. Accessed 13 Mar 2019

UN Office of the High Commissioner for Human Rights (OHCHR), Fact Sheet No. 35, The Right to Water, August 2010, No. 35, available at: www.refworld.org/docid/4ca45fed2.html. Accessed 13 Mar 2019

UN Office of the High Commissioner for Human Rights (OHCHR) (1976) International covenant on economic, social and cultural rights. Available at: www.ohchr.org/en/professionalinterest/pages/cescr.aspx. Accessed on 13 Mar 2019

UN Office of the High Commissioner for Human Rights (OHCHR) (2000, February 15) The right to development, GA res 54/175, UN GAOR, 54th sess, agenda item 116(b), UN Doc A/Res/54/175

United Nations, Water Scarcity. Last updated 24 Nov 2014. Available at: www.un.org/waterforlifedecade/scarcity.shtml. Accessed 13 Mar 2019

The USGS Water Science School (2016) Water questions & answers: how much water does the average person use at home per day? https://water.usgs.gov/edu/qa-home-percapita.html. Accessed 13 Mar 2019

Vann A et al. (2014) Hydraulic fracturing: selected legal issues, Congressional Research Service. https://fas.org/sgp/crs/misc/R43152.pdf

Vengosh A et al (2014) A critical review of the risks to water resources from unconventional shale gas development and hydraulic fracturing in the United States. Environ Sci Technol 48(15):8334–8348

Wall Street Journal (2014) Online list IDs water wells harmed by drilling, 28 August 2014, http://online.wsj.com/article/AP16a162b66b5946d0837c7395cab7a5f4.html. Accessed 5 Sept 2014

Warner NR et al (2013) Impacts of shale gas wastewater disposal on water quality in Western Pennsylvania. Environ Sci Technol 47(20):11849–11857

Robert Palmer is a central academic and lecturer in law for The Open University. He has spent his academic career researching the nexus between environmental law and human rights law, and focuses on immediate impact research and case impact studies. He has been a consultant for numerous environmental law projects, including the Blue Marine Foundation's fisheries policy reports and the Earth Community Trust. His current field of research centers on international environmental law, the right to water, and extreme energy. Palmer has a special interest in constitutional and administrative law and is presently consulting on legal challenges to the Brexit, including *Webster v The Secretary of the State for Exiting the European Union*. He was also centrally involved in the initial design, construction, and population of the virtual research repository, the Global Network for the studies of Human Rights and the Environment (GNHRE). Palmer is Associate Editor of the International Journal of Human Rights. Following extensive research undertaken for Eradicating Ecocide and the Earth Community Trust, he currently supports Dr. Damien Short's project advising English local anti-fracking groups on the interplays of environmental rights, private rights, and human rights in relation to unconventional (extreme) energy extraction technologies.

Damien Short is Director of the Human Rights Consortium (HRC) and a Reader in Human Rights at the School of Advanced Study. He has spent his entire professional career working in the field of human rights, both as a scholar and human rights advocate. He has researched and published extensively in the areas of indigenous peoples' rights, genocide studies, reconciliation projects and environmental human rights. Short is currently researching the human rights impacts of, what he calls, 'the process of 'extreme energy.' Short is a regular academic contributor to the United Nation's 'Expert Mechanism on the Rights of Indigenous Peoples' and an academic consultant for the 'Ethical Trade Task Force' of the Soil Association. He is also Editor in Chief of the *International Journal of Human Rights* (Taylor and Francis) and convenor of the British Sociological Association's Sociology of Rights Study Group and an active member of the International Network of Genocide Scholars. Short has worked with a variety of NGOs including Amnesty International, War on Want, Survival International, Friends of the Earth, Greenpeace and the International Work Group for Indigenous Affairs; and with a range of campaign groups including Eradicating Ecocide, Biofuelwatch, Climate Justice Collective, and the UK Tar Sands Network. He currently advises local anti-fracking groups in the UK and county councils on the human rights implications of unconventional (extreme) energy extraction processes.

Ted Auch is the Great Lakes Program Coordinator for FracTracker Alliance, where his primary responsibilities include mapping and bringing to light data gaps associated with the waste, water, and land-use footprint of unconventional oil and gas operations across the Midwest/Great Lakes region of North America. Ted's primary interests include frac sand mining, watershed security/resilience, the food-energy nexus, and oil and gas waste production, transport, and disposal. Previously, Auch was a Cleveland Botanical Garden postdoc fellow quantifying the Great Lakes Basin's (GLB) vacant lot portfolio, constructing various Vacant Land Repurposing (VLR) scenario models, and working with institutions, urban planners, and community groups to understand the cost and benefits associated with VLR from an economic, social continuity, and environmental perspective. Prior to that, he completed a postdoc at Green Mountain College. Auch is also Adjunct Faculty at Cleveland State University. He holds a PhD from the University of Vermont and a M.S. from Virginia Tech University. His PhD dissertation was titled "Modeling the interaction between climate, chemistry, and ecosystem fluxes at the global scale."

Chapter 4
Global Conflicts Surrounding Hydraulic Fracturing and Water

James D. Bradbury and Courtney Cox Smith

Abstract In a little more than a decade, hydraulic fracturing has unlocked significant worldwide reserves of hydrocarbons, increased the stability of energy supplies, and generated billions in economic returns, but has also delivered one other aspect: widespread controversy over the process and its potential consequences. For industry, hydraulic fracturing represents an unprecedented technological evolution that has forever changed energy production. But for many, it underscores a growing concern over the impacts of oil and gas production on water, and public opposition is often closely linked to water risks. This opposition is now global in its influence, which cannot be ignored by policymakers or industry. This chapter analyzes the primary areas of conflict and public concern over hydraulic fracturing, as well as regulations and mechanisms for resolution. Among the topics discussed are trends towards the complete ban of hydraulic fracturing. Bans, while often overturned, illustrate the intensity of the conflict and the risks of failing to understand the driving elements of these efforts and potential resolutions. This chapter also considers issues related to use of water in arid environments, mandatory disclosure of water volumes and chemical components, baseline testing of local water supplies, induced seismicity, contamination risks related to hydraulic fracturing and disposal of waste materials, noise pollution and residential proximity to operations, and methods of resolving or minimizing conflict and opposition. Controversies are examined in Texas, Colorado, New York, the United Kingdom, and Spain. Addressing the sources of social resistance and resolving them through meaningful and transparent policy mechanisms are critical elements of continued worldwide production. The reputation of the process is critical given the scope of production, social media, and the consequences for cities, industry, and citizens living in and around production zones.

J. D. Bradbury (✉)
Texas A&M University School of Law, Fort Worth, TX, USA
e-mail: jim@bradburycounsel.com

C. C. Smith
James D. Bradbury PLLC, Fort Worth, TX, USA

© Springer International Publishing AG 2020   69
R. M. Buono et al. (eds.), *Regulating Water Security in Unconventional Oil and Gas*, Water Security in a New World,
https://doi.org/10.1007/978-3-030-18342-4_4

Keywords Public participation · Prohibitions · Disclosure · Social resistance · License to operate

4.1 Introduction

One does not have to look far to find conflict over fracking and water. It is omnipresent on live television, newspapers, radio and dinner discussions. This chapter takes a closer look at select geographic locations in the United States and Europe to explore the conflicts that exist in those communities over fracking, the regulatory framework in those locations, and proposed solutions that have succeeded and failed. Despite significant cultural differences, the conflicts essentially mirror one another. The various locations cover the spectrum of fracking conflict from outright bans on fracking as seen in New York and Scotland to broad regulations that go largely unenforced and ongoing battles between federal, state, and local governments over regulating fracking. After looking at these individual case studies, this chapter will explore options for addressing the conflict between fracking and water and finding the elusive balance that will foster growth while preserving water resources for future generations.

Hydraulic fracturing is simply a method of well completion. The use of water, sand and horizontal wellbores has become a holy trinity of unconventional resource production. But, water is essential to every person, city and business. For drought-ridden countries, the security of water is apparent as years of hard drought take hold. Concurrently, energy development is in its zenith. From oil and gas to electric power generation, water is the vital element. So too, the continued supply of fresh water to metropolitan areas globally depends on these energy sectors. The two are bound together. Growing concerns over the quantity and quality of water raises significant concerns for regulators, energy companies, and citizens. Population growth estimates predict exponential growth in coming decades, placing higher demands on both energy and clean water. It is the essential thread of conflict.

Against this backdrop, the discovery of shale plays worldwide have oil and gas drilling in an historic boom resulting in a proverbial face-off between concerns of big business and the economy against environmental protection, water conservation, and public perception. The emergence of hydraulic fracturing to access these newly discovered shale plays further complicates the matter. Sparking controversy across the globe, hydraulic fracturing has sparked a firestorm, creating conflicts between state and local governments, placing landowners at odds with industry, forcing communities to choose between cheaper fuel and energy independence or clean water. Faced with outdated, nonexistent, or unenforced regulations, some communities (and countries like France) have responded by passing moratoriums or outright bans on oil and gas development, exploration, or hydraulic fracturing. Yet, discord has increased with time. Citizens of the world are now faced with an important quandary: how to find the elusive balance between water and energy that fosters environmental stewardship and technology to allow business to grow while protecting water resources. Like all significant conflicts, the competing forces must identify and pursue methods of resolution and avoid consequences of a win at any cost approach.

4.1.1 Water Status Globally

In 2017, the world population numbered nearly 7.6 billion. Yielding an additional 83 million people annually, the population is expected to increase by more than 2 billion people by 2050 and further to 11.2 billion by 2100. More than half of this growth is expected to occur in Africa, where clean water resources are already scarce in many places (UN Revision 2017). In the United States, significant growth is also predicted. Texas, where hydraulic fracturing was perfected in the 1990s, is a state that has faced significant drought in the past decade, and the population is expected to grow by 70% between 2020 and 2070 (Revkin 2013; Texas Water Development Board 2017). As populations grow, demand for water will increase not only for drinking water but also for power generation. The predicted expansion of the global population, therefore, could immeasurably strain water resources in the future. Add to this strain, diminishing sources of water, scarcity due to drought, over-allocation, or contamination, and it is easy to understand the intensity of the conflict over hydraulic fracturing and water security.

4.1.2 Uses of Water in Hydraulic Fracturing

For hydraulic fracturing, more than conventional oil and gas production, water is indispensable. The process wholly depends on access to significant amounts of water and other fluids under pressure to fracture or crack the rock and release the oil and gas. Water use is, in many respects, central to shale drilling and exploration. It is a pure trade of water for oil and gas. And too often rich reserves are located in areas with little or no available water.

4.2 Sources of Conflict Surrounding Fracking and Water

4.2.1 Water Allocation in Arid Environments

The conflicts between water and fracking are particularly acute in arid climates where water availability is scarce. In the western United States, which average less than 20 inches of rain per year (or 508 mm/year), water is a valuable commodity that is spread between many uses, including public supply, irrigation, industrial, aquaculture, livestock, and mining (Maupin et al. 2014). In Colorado and Texas, two western states with large shale plays, many of the unconventional reserves are located in areas of high water stress. In fact, 47% of fracking wells in Texas and 92% of wells in Colorado were in regions of "extremely high" water stress (Freyman and Salmon 2013). Many of these western states are also substantial agriculture producers, with irrigation being the single largest use category. While irrigation constitutes an estimated 33% of total water usage in the U.S., 83% of these withdrawals and 74% of the acres irrigated were in western states (Maupin et al. 2014).

While water used for fracking pales in comparison to that used for agricultural irrigation, as energy exploration and production rises and clean water resources diminish, agriculture and energy find themselves increasingly at odds over water. In Texas, for example, during the drought of 2011, some agricultural irrigation water rights were suspended by regulatory authorities in favor of other uses such as power production. This has been the source of great controversy and spawned litigation. Conflicts like this will undoubtedly be more common as communities and regulatory agencies in arid climates are faced with choosing the most "beneficial use" for water resources.

4.2.2   Mandatory Disclosure of Water Volumes and Chemical Components

Another source of conflict concerns data transparency and the disclosure of water volumes and chemical components used in fracking. In many areas, fracking companies are not required to disclose what chemicals are used in drilling operations because the extracting process is considered "confidential business information" (Hunter 2017). The Groundwater Protection Council and the Interstate Oil and Gas Compact Commission manage a publicly-accessibly website called "FracFocus" that is a chemical disclosure registry. Oil and gas production operators can disclose information about water and chemicals used in fracking through the website. Although a step in the right direction, the database provides an incomplete picture of all fracking due to the voluntary reporting in some states, omission of information on chemicals and mixtures that is claimed to be confidential, and invalid or erroneous information included in original disclosures (EPA 2015). As of 2015, only 14 states required operators to disclose the chemicals used in fracking fluids to either FracFocus or the state (EPA 2015). What began as a voluntary reporting site with only 37 participating companies has expanded to more than 1000 companies reporting chemical data (FracFocus 2017). Because of the website's success, as of December 2017, 23 states either require or allow companies to disclose chemical data via FracFocus. In Europe, the issue is no clearer. As of 2015, only Poland had a clear policy on the disclosure of chemicals used for fracking, and several other countries, including Spain, had partial measures that requires the disclosure to regulatory authorities but not the general public.

(Chemical Watch 2015). Importantly, if something goes wrong the burden rests with the landowner to prove injury, and this can be a difficult burden to sustain.

But the conflict does not stop there. In November 2017, policy group Partnership for Policy Integrity together with more than 100 first responders, health professionals and scientists from 21 states and the District of Columbia sent a letter to U.S. EPA Director Scott Pruitt requesting that the EPA disclose 41 chemicals that EPA regulators reviewed between 2003 and 2014. The chemical manufacturers have declared all or some of the identifying information confidential (Hunter 2017). Without the requested information, first responders, regulators, and the public has no knowledge of the precise identity or nature of the chemicals being used in fracking operations in the event of a spill or emergency.

4.2.3 Contamination Risks Related to Fracking and Waste Disposal

As fracking has become more prevalent and oil and gas operations have moved into urban areas, the proximity of wells and fracking to sources of public drinking water are a serious concern. The U.S. EPA estimates that between 2000 and 2013, approximately 3900 public water systems, which served more than 8.6 million people by 2013, had at least one fracking well within one mile of the drinking water source (EPA 2016). When fracking operations are located so close to public drinking water sources, the potential for impacts to those sources increases dramatically. This has created the need for baseline testing of local water supplies to determine the quality of water and evaluate potential impacts from fracking now and in the future. In most places these baseline studies are done by the oil and gas operator, but communities can do this if they choose.

Perceived negative impacts to water quality provide significant fuel for conflict over fracking operations. These impacts include potential contamination of groundwater aquifers from drilling operations and disposal and injection wells. Shallow water well contamination can occur during drilling and completion operations if surface casing or cementing is inadequate. Surface leaks or spills can also cause contamination (House Comm. on Nat. Res., Tex. H.R. 2013). Disposal of produced water and flowback further creates its own issues since the water used in fracking is rendered unusable after chemical additives are mixed with the water. Many operators opt to inject the wastewater from fracking into wells deep below the surface. However, in addition to turning millions of gallons of water into waste, the process if not done properly, can contaminate groundwater resources.

In 2016, the U.S. EPA issued a report on the impacts from the hydraulic fracturing water cycle on American drinking water sources. The fracking water cycle encompasses the use of water in fracking operations from the water withdrawals to make fracking fluids to mixing and injection of fluids into wells to collection and disposal or reuse of wastewater from the operations. In its report, the EPA listed several fracking activities that were most likely to result in more frequent or severe impacts to water:

- Water withdrawals for hydraulic fracturing in times, or areas, of low water availability, particularly in areas with limited or declining groundwater resources;
- Spills during the management of hydraulic fracturing fluids and chemicals or produced water that result in large volumes or high concentrations of chemicals reaching groundwater resources;
- Injections of hydraulic fracturing fluids into wells with inadequate mechanical integrity, allowing gases or liquids to move to groundwater resources;
- Injection of hydraulic fracturing fluids directly into groundwater resources;
- Discharge of inadequately treated hydraulic fracturing wastewater to surface water resources; and
- Disposal or storage of hydraulic fracturing wastewater in unlined pits, resulting in contamination of groundwater resources (EPA 2016).

Notably, these activities cover nearly the entire water cycle and process of fracking (Schwackhamer 2014).

With well operations on the rise, particularly in or near heavily populated, urban areas, governments, industry, and citizens alike are motivated to find practical alternatives to water usage that facilitate growth while minimizing the environmental footprint left behind by oil and gas operations. Three options that are readily considered by industry and regulators are: (1) brackish water, (2) recycling and reusing water from fracking operations, and (3) developing new technologies that minimize (or eliminate) the need for water at all. In short, technological advancement to address social risks. Brackish water or seawater is in abundant supply. Although certain additives can reduce the salinity in brackish water sufficiently to make it useable for fracking, concerns over contamination and well corrosion remain. Water recycling can occur by a variety of methods including microfiltration, vapor recompression and other methods. Once recycled, the water is then free to be returned to the water cycle or reused in future fracking operations. Trucking and disposal rates are reduced by more than 60%, but recycling is considerably more expensive than current practices of deep well disposal. Some companies are exploring new technologies that minimize or eliminate the use of water in oil and gas operations. One company has developed a process utilizing gelled Liquified Petroleum Gas (LPG) in place of customary fracturing fluids. Use of this gel process eliminates the need for water entirely as the gel LPG maintains its fluid state through the fracturing process and nearly all of the fluid can be recovered (House Comm. on Nat. Res., Tex. H.R. 2013). These water alternatives are often overshadowed by the conflict surrounding fracking and water usage but are a key part of the long-term solution.

4.2.4   Induced Seismicity

One source of conflict that is on the rise in areas seeing increased fracking activity is the issue of induced seismicity. Induced seismicity is the occurrence of earthquakes caused by human activity instead of tectonic forces (U.S. Department of Energy 2016). The occurrence of induced seismicity is not new and has been recorded in the U.S. since the early 1930s (U.S. Department of Energy 2017). While the cause of induced seismicity varies widely by region, the source of induced seismicity near fracking wells is often not from the fracking process itself but instead results from the disposal of produced water and flowback through injection into deep subsurface rock formations (U.S. Department of Energy 2016). For other areas, the earthquakes are triggered by fluid extraction. Regardless, earthquakes of magnitude 3 or higher in Oklahoma and Texas have significantly increased in recent years in direct correlation with wastewater disposal operations related to fracking operations (U.S. Department of Energy 2016). In 2010, Oklahoma experienced 41 magnitude-3 or higher earthquakes. In 2015, the number jumped to 903 (earthquakes.ok.gov 2017). The USGS released a report in 2016 showing that the risk of experiencing a damaging earthquake has increased tenfold since 2008 in areas like Dallas and Fort Worth (Kuchment 2016). Impacts from these earthquakes go beyond unexpectedly feeling the ground shake but have resulted in significant property damage to homes and businesses. It therefore has the public's attention.

4.3   Conflict and Resolution

The sources of conflict surrounding fracking and water are varied, significant, and on the rise as unconventional oil and gas operations grow into densely populated urban areas. These conflicts emphasize the need for an open dialogue between citizens, governments, and industry about effective regulation. Resolving these conflicts is inherently difficult because it poses a public unfamiliar with production against an industry unfamiliar with dealing with public concerns. Finding the balance between water and energy demands creativity, innovation, and a commitment to finding solutions. Key aspects to consider are the role of social media and industry marketing and education. A thorough examination of conflicts in specific geographic locations and the impacts, successes, and failures of fracking bans are also essential to finding meaningful and effective solutions.

4.3.1 Social Resistance: The Role of Social Media and Industry Education and Marketing

The twenty-first century is the age of technology. The impacts of technology on education and marketing must not be underestimated. People are constantly confronted with information on all types of subjects. The "truth" is online daily. Turn on the television in the United States and people are constantly overwhelmed by pharmaceutical commercials, political advertisements, entertainment ads, and more recently, energy marketing and education. With the rise of fracking, the industry has heralded a new approach to the public using carefully orchestrated commercials, casting operators as friends of the environment, using beautiful scenery, and demonstrating how much they care about the local communities in which they operate. These commercials tout their positive impacts to the economy, quality of life, and even the environment. The oil and gas industry devotes significant amounts of money to fund aggressive and visually appealing marketing and education campaigns, relying mostly on television as its preferred method of disseminating information. They place commentators in position to sell the oil and gas message as good stewards of community, land, water, air, and health (API 2018).

Along with the steady barrage of information facing consumers today, the rise and power of social media is no less significant. Facebook, Twitter, Snapchat, Youtube, and other sites draw in millions of followers who spend sizeable amounts of time viewing information, materials, videos, reports, and articles concerning a wide range of topics, including fracking. The rise of smartphones and tablets enabling consumers to access infinite amounts of information at their fingertips has only fueled the rise of electronic and social media, expanding its scope of influence exponentially. People are now connected with vast numbers of others that share similar perspectives.

In 2014, 57% of U.S. consumers considered fracking one of the top three environmental issues (Mitchell 2014). Nearly 80% of respondents to a recent survey reported hearing about fracking from Internet news sites and social media (Mitchell 2014). Between January and July 2014, there were 1.3 million references to fracking on Twitter, and of those tweets, anti-fracking activities were responsible for 2000% more of those messages than groups supporting fracking (Mitchell 2014). In contrast, oil and gas companies primarily rely on television for their marketing and education campaigns. Interestingly, only 18% of respondents surveyed in 2014 stated that they heard about fracking from television (Mitchell 2014). Anti-fracking organizations understand the value and power of social media for spreading their message and fundraising, and this has fueled conflict over fracking operations. While the ease of social media is appealing, it has opened the door to content that can be less reliable, with no cost, no admission, no verification, and no accountability. These unfiltered exchanges can reach substantial numbers of people and be shared over and over with the touch of a button.

4.3.2 Hydraulic Fracturing Bans

As tensions rise over fracking and regulation seems to be more and more about politics than science, many communities, citizens, and environmental groups have turned to outright bans on fracking or fracking-related activities and operations. Each side wants to win outright. These bans have emerged internationally from Colorado, Texas, and New York to Scotland or France. The success of these bans varies widely with some bans surviving significant political and judicial challenges and others being overturned or invalidated through legislative or judicial efforts, supported and often funded by the oil and gas industry (Corriher 2017; Minor 2014).

Not all bans are created equal. In some locations, the bans only touch on a portion of the fracking process, banning only deep well injection disposal, of fracking fluids, for example, while allowing the actual process of fracking to continue. Other bans attempt to forestall any *new* wells after a certain date. Still other bans attempt to attack the issue by instituting a temporary ban or moratorium on fracking for a period of years, intending to research and promulgate sufficient regulations to address the concerns over fracking. At the heart of these bans is an effort to slow the process and progression of fracking. The absence of regulation, outdated regulations, and the increase in fracking operations in metropolitan areas under a typical "boom or bust" cycle have fueled the desire to slow the progression of fracking until impacts can be adequately researched, understood, and addressed through regulation. On the far end of the spectrum are the outright bans on fracking, which prohibit any fracking activity or operation of any kind for all time. Total bans are often seen in communities that have experienced negative impacts from fracking, have confirmed risks to public health and the environment, and have determined that the only solution to ensure public safety and protect the environment is to prohibit fracking entirely.

The rise of fracking bans highlights both the prevalence and the intensity of the conflict and controversy over fracking. Despite the passage of time, these conflicts seem only to intensify not wane. Before solutions can be found, however, consideration of specific community responses to fracking—through bans and other means—must be considered and analyzed. In looking at these specific case studies, certain trends are revealed that will necessarily inform any solutions.

4.3.3 Specific Conflicts—Case Studies in the United States and Europe

4.3.3.1 Colorado

Colorado is no stranger to mineral extraction. Its very existence, history, and culture is inextricably connected to mining gold and other minerals. In 2012, Colorado was the sixth leading producer of natural gas and ninth leading producer of oil in the

U.S. (Weiner 2014). Over 90% of new oil and gas production in Colorado is accomplished through fracking (Weiner 2014). What was once activity conducted on the open range along sparsely populated rural areas, however, has now expanded to the densely-populated urban areas, changing the entire debate.

Oil and gas operations are regulated by the Colorado Oil and Gas Conservation Commission (COGCC). Lenient regulations on the state and federal level have encouraged the fracking boom and allowed it to go on largely unchecked (Minor 2014). As one author described it "fracking currently operates within a sixty-year old regulation system that was designed to maximize production with little regard for environmental or health impacts, and without anticipation that this industry would commonly intrude on heavily populated areas." (Toan 2015). As a result, environmental contamination resulting from fracking operations is on the rise. In 2011, there were 513 spills reported in Colorado, 26% of which contaminated surface or groundwater (Minor 2014). Further, as a western state where water availability is a concern, water usage in fracking is also a source of conflict. Between 2010 and 2015, fracking operations used approximately 32 billion gallons of water (or over 121,000,000 m^3) (Minor 2014). Other impacts abound as well, including quality of life impacts from 24-h drilling, noise and light pollution, and increased prevalence of depression and substance abuse common in "boomtown" communities (Minor 2014).

Longmont, Colorado, is home to one of the more active oil and gas areas in the state. The town straddles two counties—Weld and Boulder—with drastically different policies concerning fracking (Toan 2015). Weld County is home to over 18,000 wells, which is the most wells of any county in the United States. Boulder County, on the other hand, has instituted a moratorium on all new oil and gas development through 2018 (Toan 2015). Longmont passed an ordinance in 2012 that prohibits the issuance of permits for oil and gas operations in residential areas, among other things (Toan 2015). The COGCC quickly sued Longmont to overturn the ordinance. In response, over 80 local government officials submitted a letter to the Governor of Colorado asking for the withdrawal of the lawsuit against the ordinance (Toan 2015). For many residents of Longmont, the city's ordinance did not go far enough. In 2013, Longmont residents voted to ban all fracking within the city, with the initiative carrying 60% of the vote (Toan 2015). COGCC filed another lawsuit seeking to overturn the voter's ban. In July 2014, the Boulder County district court struck down the voter ban on fracking finding that it was preempted by state law. In May 2016, the Supreme Court of Colorado affirmed the ruling, overturning the ban. The Court reasoned "Longmont's ban, if left in place, could ultimately lead to a patchwork of regulation that would inhibit the efficient development of oil and gas resources." *Longmont v. Colo. Oil & Gas Ass'n*, 369 P.3d 573, 581 (Colo. 2016). The Court also struck down a temporary moratorium against fracking that was implemented in Fort Collins. *Fort Collins v. Colo. Oil & Gas Ass'n*, 369 P.3d 586 (Colo. 2016).

During the controversy over the Longmont and other city bans and moratoriums on fracking, anti-fracking advocates, including U.S. Representative Jared Polis supported ballot initiatives restricting fracking statewide. To avoid the initiatives making the 2014 ballot, the state reached a compromise with Representative Polis and others to drop the initiatives, in exchange for certain concessions. Chief among these concessions was the formation of a statewide task force to analyze the problems with fracking in Colorado (Newton-Small 2014). In this report, the task force's recommendations focused on (1) enhancing local governmental involvement with oil and gas operations located within the municipality, (2) increasing the number of staff responsible for inspecting active wells, and (3) creating a clearinghouse to more effectively communicate "unbiased information" to the public related to Colorado's oil and gas industry (Colorado Task Force Report 2015).

Other cities in Colorado have attempted to regulate fracking in other ways including Boulder, Greeley, Fort Collins, Lafayette, among others. Many cities and counties have oil and gas regulations that existed before the fracking boom (Minor 2014). Commerce City has taken a case-by-case approach that bans fracking near certain wildlife areas. For other areas, operators must negotiate specific agreements with the City on a case-by-case basis (Minor 2014). Commerce City's regulation has not yet been challenged in court. Other cities that have passed less restrictive ordinances than outright bans have survived without judicial challenge. It is important to note, however, that no matter the regulations in place, the government must be willing to enforce the regulations (either politically prepared or fortified to enforce) for them to have any effect, and this continues to be an ongoing issue for many communities in Colorado, on both the state and local level.

4.3.3.2 Texas

Like Colorado, Texas history is deeply intertwined with oil and gas. The West Texas landscape has long been adorned with oil rigs and pumps. More recently, with the discovery of several large shale plays, natural gas drilling has taken off and expanded into heavily populated urban centers like the Dallas-Fort Worth metroplex. Denton, a city of 115,000 people north of Fort Worth, is part of this metroplex and by 2014 had more than 200 wells in its borders. As the gas drilling boom grew, citizens grew concerned about the environmental and health impacts. Many of these wells were springing up in residential areas. Rather than the far, distant rural rigs that were surrounded by little other than dry, flat land, these new wells were now coexisting with humans and wildlife like never before. Conflicts over the operational practices of operators and proximity to homes and schools grew into an anti-fracking grassroots movement. In 2014, 59% of Denton voters passed a full ban on fracking (Roth 2016). Notably, the terms of the ban did not preclude drilling and other types of exploration, only fracking. The day after the ban passed, the oil and gas industry sued to prevent the ban from going into effect, and the State of Texas also joined. Along with the lawsuit, industry representatives began lobbying the Texas Legislature to pass a law to overturn the ban and any others like it in the future.

Oil and gas operations are regulated by the Texas Railroad Commission, but like Colorado, Texas is an oil and gas friendly state, and the regulations tend to be lenient (Sunset Report 2017). Seeking to avoid that "patchwork of local regulations," in

2015, Texas legislators overwhelmingly passed House Bill 40 that now preempts a wide array of drilling activities, including a ban on fracking (Malewitz, May 2015). The bill provided certain areas where local governments could still regulate oil and gas drilling, including fire and emergency response, traffic, lights and noise, but only if such regulation was "commercially reasonable" (Malewitz, May 2015). This legislation impacted not only Denton, but countless other Texas cities that regulate production within their borders. In response to the House Bill 40, the City of Denton had no choice but to repeal its ban (Baker 2015). Fracking resumed in Denton in June 2015 (Malewitz, September 2015).

On August 4, 2015, Denton amended its regulations concerning oil and gas drilling (Ordinance 2015–233, 2015). The new ordinance seeks to place certain restrictions and regulations on operations in Denton that fall short of prohibited conduct in House Bill 40. The new ordinance provided new setback requirements for well sites, amended zoning restrictions, required operations to create and use a site plan, among other provisions. This ordinance has not been challenged judicially.

4.3.3.3 New York and Delaware River Basin

In contrast to Texas and Colorado, the Northeastern United States has experienced greater traction in banning fracking and drilling operations. In June 2015, New York banned fracking after a seven-year review of the practice (Department of Environmental Conservation 2015). The primary source of conflict and concern was water quality and supply. The findings of the Department of Environmental Conservation concluded that there were "no feasible or prudent alternatives that adequately avoid or minimize adverse environmental impacts and address risks to public health from this activity" (Department of Environmental Conservation 2015). Notwithstanding the public support for the ban, controversy remains. Since the ban has been in effect, New York has imported shale gas from neighboring Pennsylvania and is planning to process it in a new power plant under construction outside New York City (Lombardi 2017). By one estimate, "drillers outside the state would have to tap 130 wells each year, on average, to supply the plant with enough gas to operate. That translates into thousands of fracked wells over the 40-year lifetime typical for such a facility" (Lombardi 2017).

In the nearby Delaware River Basin, which provides drinking water for 15 million people in New York, New Jersey, Pennsylvania, and Delaware, the Delaware River Basin Authority (DRBA) implemented a *de facto* fracking moratorium beginning in 2010 (Phillips 2017). It has faced its own controversy because part of the river basin lies in Pennsylvania, a state that openly allows fracking. The DRBA ban, however, applies in the two Pennsylvania counties within the river basin area, demonstrating that it is the water that matters and the political boundaries mean less. In opposition to the ban, a Wayne County landowner group filed a lawsuit against the DRBA and the ban in May 2016.

While the ban has been in place for over 7 years, with lawsuits like the Wayne County lawsuit, concerns abound that it might be lifted since it was not made permanent. (Whitehead 2017) In 2017, the DRBA has approved the rulemaking process to formally ban fracking (Phillips 2017). Yet, environmentalist groups remain concerned that in view of the absence of a statewide drilling ban in Pennsylvania, the proposed regulations will allow water withdrawals and waste-water disposal for drilling in other areas (Maykuth 2017). Like New York, while the bans remain in place, the potential environmental impacts are not eliminated because of drilling in other nearby areas.

4.3.3.4 Europe: The United Kingdom and Spain

Although Europe has more recoverable shale gas than the U.S., fracking in Europe, where Russia provides one third of Europe's gas, has not enjoyed the success of fracking in the U.S. (Gilblom and Patel 2016). Scotland recently banned fracking (Nicolson 2017). Touted as a social movement, anti-fracking advocates took hold in Scotland as a political shift occurred with leaders that were intent on putting "as much distance between themselves and that of the Conservative-led, pro-fracking, UK Government" (Young and Lander 2015). The anti-fracking movement wrote letters, lobbied political leaders, organized events and conferences, and made use of the Scottish media (Young and Lander 2015). The result was a moratorium that became a full ban on fracking.

Scotland is not alone in its ban of fracking. Germany, France, the Netherlands, and Bulgaria all ban fracking as well (Gilblom and Patel 2016). Environmentalists contend that the failure of shale gas in Europe is largely based on the dense population and higher environmental standards (Neslen 2016). Much of the anti-fracking movement in Europe is supported by "not in my backyard" type grassroots opposition. Finding sufficient amounts of water for fracking operations is yet another problem stalling the fracking movement in Europe. But while not documented, it is most certain that the emergence of fracking opposition in the U.S. influenced these conversations in Europe.

In Spain, some regions are estimated to contain enough gas potential equal to 70 years of domestic consumption (Planelles 2017). In 2014, a Sustainability Assessment was completed in Cantabria, Spain to determine the environmental impacts from fracking (Ferreras 2014). Like research and studies in the U.S., the Spain study concluded that the possibility for water contamination exists at virtually all stages of the fracking water cycle (Ferreras 2014). Spanish law also provides for tax benefits for local communities from fracking (Buono et al. 2017).

The anti-fracking message has been widely accepted in Spanish society, and while Spanish law does not outright ban fracking, it places such high environmental standards on operators that the practice may as well be banned (Planelles 2017). "Reluctance, distrust and opposition to hydraulic fracturing—fostered by alarmist messages, the reputation of the oil and gas sector, and the fear of unknown change— have gained ground among the Spanish population. Strong and effective campaigns by ecological and resistance groups, aided by alarmist literature from the U.S, has resulted in hydraulic fracturing being widely associated with water threats in Spain. Meanwhile, water is an extremely sensitive issue for Spanish culture and society" (Buono et al. 2017).

Several local governments attempted to ban fracking, but these bans were struck down by the Constitutional Court on grounds that the authorities were overstepping their powers (Planelles 2017). Like Scotland, social pressure has been an important part of the anti-fracking movement. "Spain [] faces more cultural opposition to hydraulic fracturing and enjoys less institutional momentum and, potentially, fewer economic benefits" (Buono et al. 2017).

4.4   A Third Way—Finding a Solution

As the case studies demonstrate, the response to fracking can vary widely across the geographic and global landscape. Even in those areas where fracking is banned, the impacts may not be reduced to the extent desired as many of these areas import shale gas from other areas or supply water for other drilling operations. What can be done to find the balance?

Another way must be charted, and it will require the cooperation of lawmakers, regulators, industry, environmentalists, and citizens alike. There must be open and transparent policies. A process must be developed that is clear, consistent, and reputable. The laws and regulations addressing environmental impacts must be fueled and built upon science, not politics or interests.

Local governments should be empowered to enforce state regulations if state regulators are unable or unwilling to enforce the law. Local governments could consider implementing regulations where each operator is required to negotiate its own terms on a case-by-case basis. Companies must be accountable to disclose all chemicals and water volumes used in the process without exception. A focus of regulation should be on promoting the use of water alternatives in the fracking process. Local regulations should also address conflict resolution, and to that end could include the opportunity for a review panel or forum to consider disputed scenarios between operators and citizens. The City of Fort Worth, Texas included a provision creating a Gas Drilling Review Committee in their drilling ordinance and covers certain disputed scenarios, providing an opportunity for City staff, operators and citizens to come together and attempt to resolve conflicts.

Additional steps that can go a long way in resolving conflicts between the public and operators include increased scientific research, development of a global standard of best management practices for fracking, well disposal, and other drilling operations, and water protection measures. Finally, operators can obtain a "social license to operate," which attempts to garner public support for the project through "ongoing communication with communities, transparency and engagement in decision-making, and the establishment of effective conflict resolution mechanisms" (Smith and Richards 2015). The social license to operate reflects "an ongoing and negotiated process" where a community can feel free to object to portions of a project without withdrawing support for the entire project (Smith and Richards 2015). It is an empowering tool for operators to manage risk on a project by project basis, assessing and taking measures to reduce risk. (Smith and Richards 2015). Further, one way some European countries have found to make fracking more palatable at the local level is to ensure that some of the profits generated by the projects are returned to the local community.

There is no single answer to the conflicts that go together with fracking today. Finding a balance that is protective of water and the environment, while allowing growth in business requires that the conflicts take a backseat to open, constructive conversation focused on finding ways to make fracking safer and minimizing its impacts on water. Addressing the sources of social resistance to resolve them through meaningful and transparent policy mechanisms are critical elements of continued worldwide production. Water and energy are interdependent resources that are both essential to life on this planet. Finding a path through the conflict between fracking and water is no less essential. Balancing regulation with production could provide the way forward.

References

API. http://www.api.org/news-policy-and-issues/media/industry-advertisements (last visited February 2, 2018)

Buono RM, Mayor B, Lopez-Gunn E (2017) A comparative Study of water related issues in the context of hydraulic Fracturing in Texas and Spain. Environ Sci Policy. Available at https://doi.org/10.1016/j.envsci.2017.12.006

Chemical Watch (2015) Few EU countries clear on disclosure of fracking chemicals, March 11, 2015. Available at https://chemicalwatch.com/23097/few-eu-countries-clear-on-disclosure-of-fracking-chemicals

Colorado Oil and Gas Task Force Final Report, February 27, 2015. Available at http://cred.wpengine.com/wp-content/uploads/2015/04/OilGasTaskForceFinalReport.pdf

Corriher B (2017) Big money courts decide fate of local fracking rules, Center for American Progress, January 9, 2017. Available at https://www.americanprogress.org/issues/courts/reports/2017/01/09/296113/big-money-courts-decide-fate-of-local-fracking-rules/

Denton Ordinance 2015-233, August 4, 2015. Available at https://www.cityofdenton.com/CoD/media/City-of-Denton/Government/Ord_2015-233__Gas_Well_Drilling_and_Production_-_current.pdf

Earthquakes.ok.gov (2017) What we know

EPA (2016, December), Executive summary, Hydraulic Fracturing for Oil and Gas: Impacts from the Hydraulic Fracturing Water Cycle on Drinking Water Resources in the United States 4

EPA, Analysis of Hydraulic Fracturing Fluid Data from the FracFocus Chemical Disclosure Registry 1.0 8 (March 2015). Available at https://www.epa.gov/sites/production/files/2015-03/documents/fracfocus_analysis_report_and_appendices_final_032015_508_0.pdf)

Ferreras JAF (2014) Hydraulic fracturing sustainability assessment: case study of Luena (Cantabria, Spain) July, 2014

FracFocus (2017) FracFocus Celebrates Its 5th Anniversary, http://fracfocus.org/node/358. Last visited December 18, 2017

Freyman M, Salmon R (2013) Hydraulic fracturing & water stress: growing competitive pressures for water 6. Available at https://www.ceres.org/resources/reports/hydraulic-fracturing-water-stress-water-demand-numbers

Gilblom K, Patel T (2016) Fracking in Europe, Bloomberg, (November 22, 2016)

House Comm. on Natural Res., Tex. H.R., 83rd R.S. (2013), Interim report 2012 66 (January 2013)

Hunter J (2017) First responders ask for transparency for fracking chemicals, The Weirton Daily Times, December 2, 2017

Kuchment A (2016) UT Study: Fracking-related activities have caused majority of recent Texas earthquakes, Dallas Morning News (May 2016)

Lombardi K (2017) New York's heralded fracking ban isn't all it's cracked up to be, StateImpact (December 8, 2017)

Malewitz J (2015) Curbing local control, Abbott signs "Denton fracking bill", the Texas tribune (May 18, 2015)

Malewitz J (2015) Texas drops suit over dead Denton fracking ban, The Texas Tribune (September 18, 2015)

Maupin MA, et al (2014) Summary of estimated water use in the United States in 2010: U.S. Geological Survey circular 1405, 7 (2014). Available at https://pubs.usgs.gov/fs/2014/3109/pdf/fs2014-3109.pdf

Maykuth A (2017) Up in arms on the upper Delaware over a proposed fracking ban, The Inquirer, (September 28, 2017)

Minor J (2014) Local government fracking regulations: a Colorado case Study, 33 Stan Envtl LJ 59 (January 2014)

Mitchell ES (2014) STUDY: social media is winning PR war for anti-fracking groups, Adweek, (August 12, 2014)

Neslen A (2016) The rise and fall of fracking in Europe, The Guardian (September 29, 2016)

New York Department of Environmental Conservation (2015), New York state officially prohibits high-volume hydraulic Fracturing, Press Release (June 29, 2015)

Newton-Small J (2014) Democrat Jared Polis withdraws support for Colorado fracking initiatives, Time, (August 4, 2014)

Nicolson S (2017) Scotland and fracking: how did we get here? BBC News (October 3, 2017)

Phillips S (2017) DRBC takes a step toward banning fracking in Northeast Pa., StateImpact, (September 13, 2017)

Planelles M (2017) Spain's fracking bubble bursts, El País, (March 16, 2017)

Revkin A (2013) Dot Earth: Daniel Yergin on George Mitchell's Energy Innovations and Concerns, NY Times, July 29, 2013. Available at http://cgmf.org/blog-entry/69/Dot-Earth-Daniel-Yergin-on-George-Mitchell's-Energy-Innovations-and-Concerns.html

Roth Z (2016) What happened in Denton: the war on local democracy, NBC News, (August 2, 2016)

Schwackhamer DL, Ph. D., M.S. (2014) Potential impacts of hydraulic fracturing on water resources, Water Quality-Health Impact Assessment of Shale Gas Extraction

Smith DC, Richards JM (2015) Social license to operate: hydraulic Fracturing-related challenges facing the oil and gas industry. Oil and Gas, Natural Resources and Energy Journal 1(2)

Sunset Advisory Commission (2017) Staff report with final results- railroad commission of Texas 34–35, June 2017. Available at https://www.sunset.texas.gov/public/uploads/files/reports/Railroad%20Commissio%20of%20Texas%20Staff%20Report%20with%20Final%20Results_6-21-17.pdf

Texas Water Development Board (2017) Water for Texas 2017 State Water Plan 3

Toan K (2015) Not under my backyard: the battle between Colorado and local government over hydraulic fracturing, 26 Colorado Natural Resources, Energy, & Environmental Law Review 1 (Winter 2015)

U.S. Department of Energy, Energy Efficiency & Renewable Energy (2017). http://esd1.lbl.gov/research/projects/induced_seismicity/oil&gas/. Last visited December 18, 2017

U.S. Department of Energy, Office of Oil & Nat'l Gas, Induced Seismicity 1 (2016). Available at https://energy.gov/sites/prod/files/2016/08/f33/Induced%20Seismicity.pdf

United Nations (2017) World population prospects: the 2017 revision

Weiner C (2014) Oil and gas development in Colorado-10.639, Colorado State University Extension, (September 2014).

Whitehead S (2017) Activists seem worried delaware river basin fracking moratorium will be lifted—and rightfully so, Energy In Depth (August 2, 2017)

Young E, Lander R (2015) Going on the offensive—a picture of Scotland's anti-fracking movement. Green European Journal, 10 (September 1, 2015)

James D. Bradbury operates his own law practice with offices in Fort Worth and Austin, Texas, and focuses on environmental, energy, water, and regulatory matters, including the imminent water policy issues facing Texas. He serves as a policy advisor to statewide groups and elected officials on environmental and natural resource issues, particularly as they affect energy. Bradbury also has significant experience in the environmental aspects of hydraulic fracturing production. He served on the City of Fort Worth Urban Gas Drilling Task Force that developed the Fort Worth Shale Drilling and Pipeline Ordinance, a nationwide model for regulating production in urban environments. He has been a speaker and advisor on unconventional production regulations for the Departments of State, Commerce and various foreign countries, including China, Argentina, and South Africa. Bradbury is a consultant with Blue Wind Partners, an entity he founded to provide natural resources policy and communication services to private and governmental entities. He is also an adjunct professor at Texas A&M University School of Law teaching courses in natural resources law and related areas. Bradbury holds a bachelor's degree in agricultural economics from Texas A&M University and a J.D. from the University of Idaho School of Law.

Courtney Cox Smith is an attorney with the firm James D. Bradbury, PLLC, where she focuses her practice in the areas of environmental, water, energy, business and regulatory and policy matters. Previously, she was an associate with Jackson Walker L.L.P. in Fort Worth and Austin, Texas, focusing her practice in the areas of environmental, water, and administrative law. Smith assists clients with matters before state and federal courts and state agencies. Smith holds a bachelor's degree in American studies from Baylor University and a J.D. from Boston College Law School.

Part II
Acquiring Water for Fracturing: Conflicts and Regulatory Issues

Chapter 5
Frac Water Acquisition in the Major U.S. Unconventional Oil and Gas Plays

Gabriel Collins and Julie A. Rosen

Abstract This analysis fills a vital gap in the existing literature by examining in detail methods and pathways by which unconventional oil & gas producers in the U.S. commonly source the water used in hydraulic fracturing completions. The authors leverage large data sets, their practical professional and research experiences, and conversations with well-placed industry sources to describe the prevalent business models and contractual structures under which oil and gas producers obtain frac water. Geographically, the study focuses on five world-scale unconventional oil and gas basins: the Bakken play in North Dakota, the Niobrara in Colorado, the Eagle Ford in south Texas, the Marcellus in Appalachia, and the Permian Basin in West Texas and Eastern New Mexico.

Keywords Hydraulic fracturing · Water acquisition · Water midstream · Contract · Permian · Bakken · Niobrara · Marcellus · Eagle ford · Water rights · Groundwater

5.1 Introduction

Frac water acquisition is a fascinating and important subject that, to date, has not received public discussion nearly commensurate with its importance to U.S. energy security at the national level, and water resource security at the local level. Headlines often focus on the perceived harms of oilfield water use, despite the fact that data show in many cases the impacts of even intensive frac development are far less water intensive than common farming activities (Galbraith 2013; for a sample estimate of field-level frac water use, see: Collins 2017a, b). Yet the reality on the

G. Collins (✉)
Baker Institute for Public Policy, Rice University, Houston, TX, USA
e-mail: gabe.collins@rice.edu

J. A. Rosen (✉)
Ryley Carlock & Applewhite, Denver, CO, USA
e-mail: jrosen@rcalaw.com

© Springer International Publishing AG 2020
R. M. Buono et al. (eds.), *Regulating Water Security in Unconventional Oil and Gas*, Water Security in a New World,
https://doi.org/10.1007/978-3-030-18342-4_5

ground is much more nuanced and interesting, points that this study aims to illustrate in an ideologically-neutral, data-driven manner.

Water transactions in the most actively drilled states—especially Texas—typically occur between private parties and the details are rarely revealed to the public. This analysis fuses a range of disparate data sources that speak to specific aspects of the frac water supply chain, along with deep primary research and data analysis, including interviews of those who actually sell frac water to energy companies, and weaves the data mass together into an account aimed at providing a useful resource to experienced oilfield hands and the uninitiated alike.

One upfront caveat: many of the plays discussed in this analysis, including the Bakken, Marcellus, Niobrara, and Permian cross state lines. Accordingly, for legal and regulatory questions, we focus on the main jurisdiction each play occurs in. In other words, the Bakken discussion will focus on North Dakota despite some level of Bakken development activity in Montana, the Niobrara discussion will focus on water issues under Colorado law, the Marcellus will focus on Pennsylvania, and the Permian Basin analysis will focus on Texas even though a sizeable, albeit minority share of Permian production comes from wells in New Mexico.

5.2 How Is Water Used in a Frac and How Much Is Being Used Nationwide and in Key Plays?

Hydraulic fracturing of oil and gas-producing rock, otherwise known as "fracing" involves pumping a fluid—typically water—into the formation under high pressure to create fractures in the rock surrounding the wellbore and free oil and gas molecules so that they can flow into the wellbore and be recovered. The water mixture pumped downhole contains what is known as "proppant" (i.e. sand or ceramic beads) that remains behind in the cracks and "props" them open to allow production, hence the term "proppant." The rock layers being fractured are in many cases even *less* porous than concrete (King 2012).

At its near-term peak in the fall of 2014, daily frac water demand across the U.S. unconventional plays was 8.3 million barrels per day, as calculated from FracFocus disclosure data. To put that number in perspective, it is approximately 3.5 times the daily average water use of Washington D.C., or enough to fill NRG Stadium (home of the Houston Texans) in 2 days (Texas Football Stadium Database 2018). Texas is the primary U.S. frac water demand driver, accounting for approximately half of national frac water usage reported to FracFocus at any given point between January 2013 and February 2017. Frac water demand in Texas is increasingly driven by the Permian Basin, a world-class resource base that now produces roughly as much oil per day as Kuwait and more gas per day than Australia (U.S. Energy Information Administration 2018; Opec 2017; BP Statistical Review of World Energy 2017) (Fig. 5.1).

Along with other U.S. unconventional plays, the Permian Basin saw rig counts drop significantly beginning in late 2014 as oil prices fell. However, water and sand

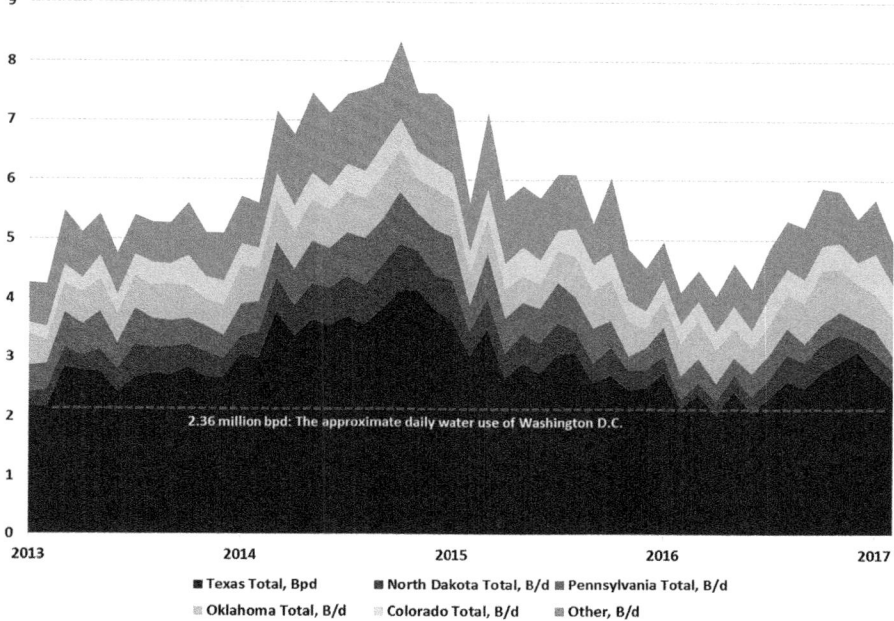

Fig. 5.1 U.S. National frac water usage by state, million barrels per day. (Source: FracFocus, DC Water Utility)

intensity per well rose dramatically as producers begin pumping much larger completions in order to maximize capital efficiency. Using data from FracFocus, we singled out the "monster fracs," defined as those which used 400 thousand barrels or more of water per completion. To put that volume in perspective, 400 thousand barrels of water could fill about 17 Olympic swimming pools. In the Permian Basin, such wells only accounted for 0.3% of fracs reported to FracFocus in 2013, but by 2016, the proportion climbed to nearly 43% and for the first 2 months of 2017, nearly 60% of reported completions were "monsters" (Fig. 5.2).

The prevalence of larger frac completions also shows up in their geographical proliferation across the Permian Basin. The heatmaps presented in Fig. 5.3 shade in areas in the Permian Basin (Texas and New Mexico) cumulatively based on the size of fracs conducted there. The color scheme is as follows: the lightest dots indicate fracs that used less than 500 thousand barrels of water, and the darkest dots represent wells that used more than 500 thousand barrels of water.

The heatmap shows where frac water sourcing activity will be the most intensive. Even if WTI crude oil prices remain around $50/bbl and Henry Hub natural gas prices remain in the $3/mmbtu range, drilling and completions activity is likely to remain robust, particularly in the "sweet spots" where the best geology is. Not coincidentally, these sweet spots tend to lie in the areas with a lot of red, as oil producers enjoy significant returns from each incremental unit of water used

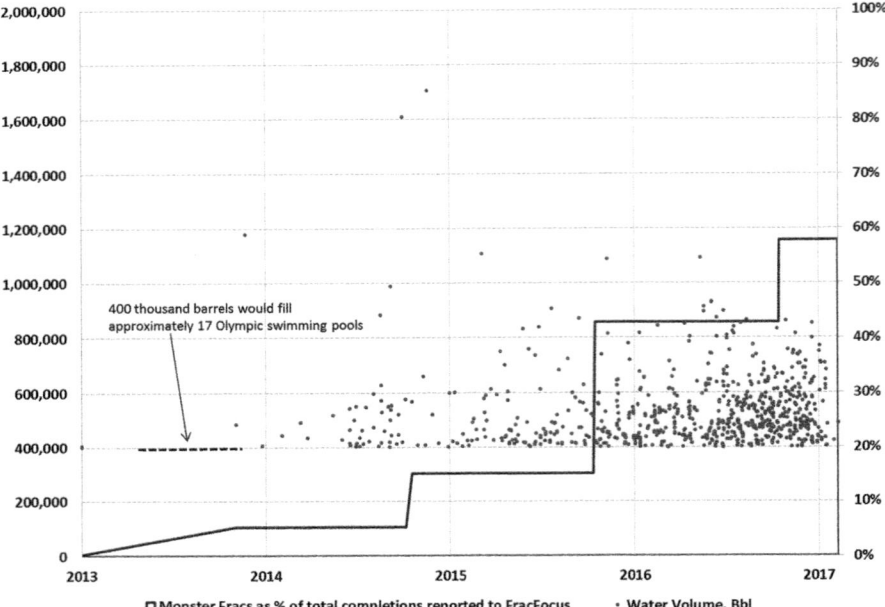

Fig. 5.2 Wells using 400 thousand barrels or more of water now account for the majority of Permian Basin frac completions. (Source: Frac Focus and Authors' Estimates)

as they move to frac designs that are more water-intensive (i.e. more water per foot) and do so on wells with longer laterals, which require more water than their shorter predecessors.

The likely practical impact is that once operators experience the increasing marginal returns of using more water per lateral foot, they are unlikely to scale back. Similar dynamics have unfolded on the proppant front (frac sand), where despite the rig count in the U.S. overall standing at less than half its 2014 level, sand use has nonetheless risen by nearly 50%. In addition, multiple operators, including Concho Resources, Laredo Petroleum, and numerous others are amassing contiguous acreage blocks so that they can drill longer laterals, extract more oil & gas, and maximize returns on capital invested (Laredo Petroleum 2017).

Operators are also moving towards greater use of "pad drilling," where multiple wells are drilled and completed from a common location in a short period of time (Collins and Medlock III 2017). Pad drilling creates intense local water supply demands, and puts a premium in dependable supplies, since a large pad drilling operation can require millions of barrels of water in a relatively short time span (International Association of Directional Drilling 2018).

The Permian Basin is actually a relative latecomer to many of the enhanced efficiency techniques such as pad drilling, centralized water sourcing, and large fracs. These were largely pioneered in the Marcellus during its boom and the subsequent

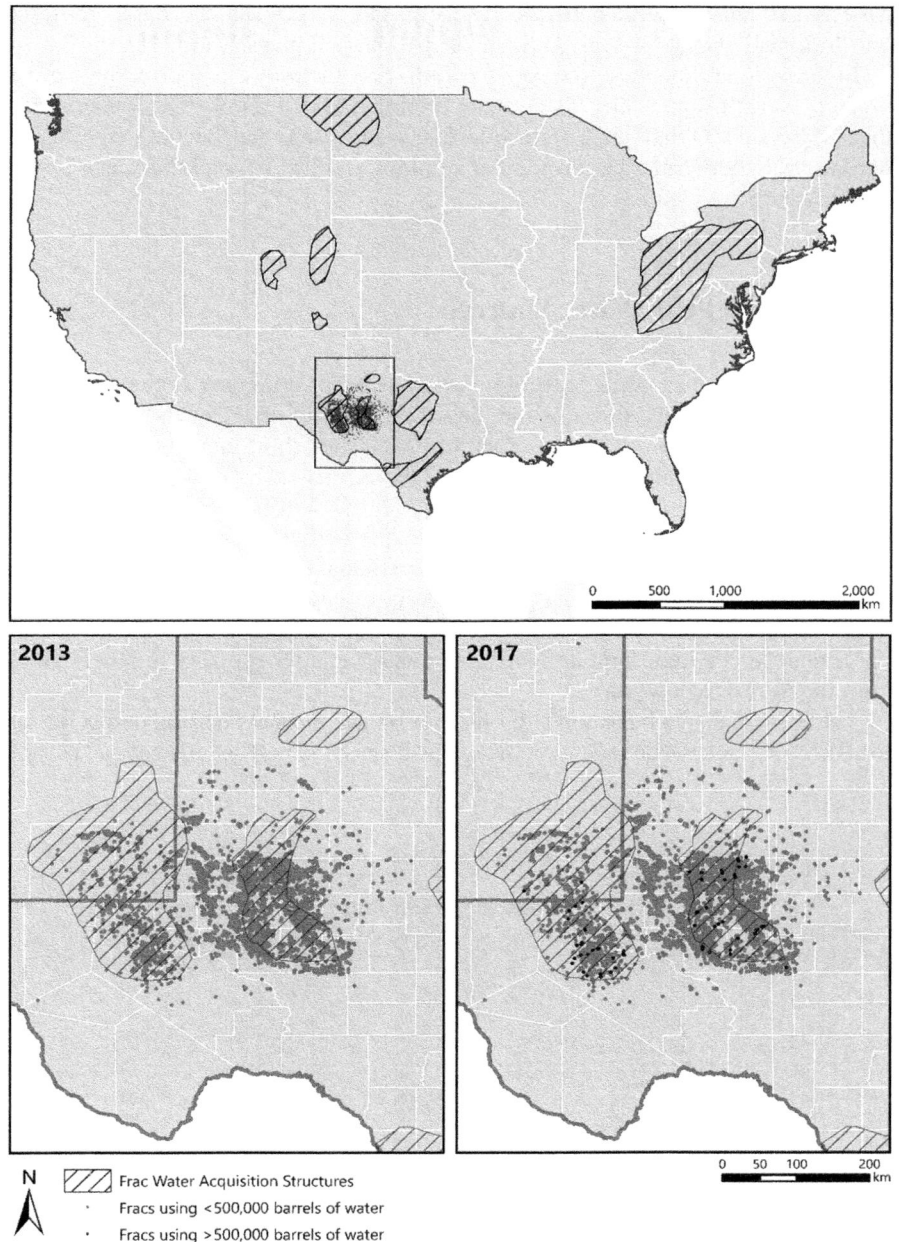

Fig 5.3 Permian Basin fracs mapped out. Major increase in water intensity between 2013 and 2017 (Callum Foster, UWE, Bristol Cartography)

gas price collapse in 2008 and 2009, as well as in the Bakken and Eagle Ford, the two "original" industrial-scale U.S. unconventional oil plays.

In terms of architecture, the paper will discuss the plays in approximate order of their water usage scale. This means Permian Basin, Niobrara, Bakken, and then Marcellus. Oklahoma's statewide frac water usage has become significant, but is spread between a larger number of plays and lies beyond the scope of our analysis here.

5.3   How Is Frac Water Sourced?

Several core criteria come into play when an E&P company looks to acquire water for fracing, regardless of the play it is operating in. Dependability is the most important factor, since water supplies that fall short can seriously disrupt a drilling and completions program. If this happens, the company and other investor parties involved face the risk that capital will be stranded if access to the oil and gas reserves they seek to produce is delayed or otherwise impeded. The stakes are also high for employees on a personal level, since a water manager who fails to procure secure supplies is likely to end up being a water manager with very short tenure in his/her position. Price is generally a secondary factor, but remains very important and has come under greater scrutiny in recent years with the oil price downturn.

The basic decision tree shown in Fig. 5.4 outlines the key decision points that an E&P needing frac water is likely to face. The discussion below will analyze, in turn,

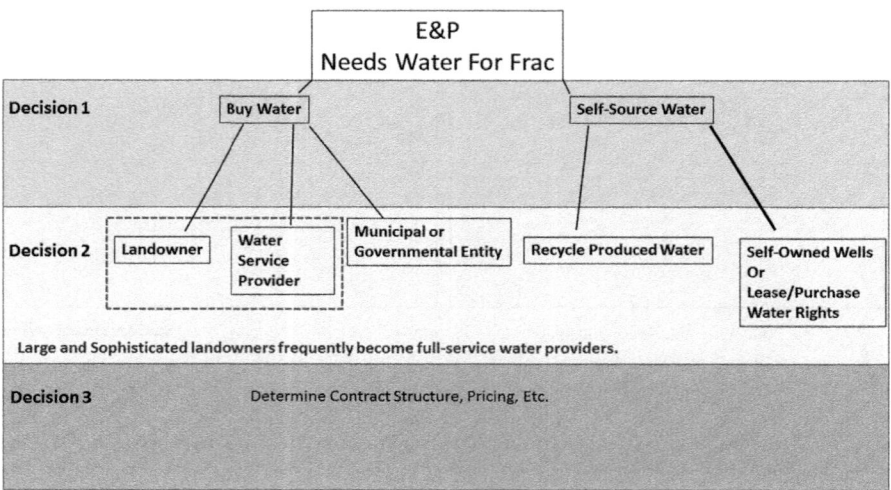

Fig. 5.4 Frac water acquisition overarching architecture

how this process plays out in various jurisdictions and what the capital investment and legal implications are of various strategies.

Decisions on water acquisitions are also driven by factors that vary among the plays in the U.S. This is due to the diverse hydrology across the U.S. and the federalist structure whereby individual states have devised their own laws and systems governing landownership and water resources. Relevant factors include geological and hydrological characteristics, landownership rights, state water use rights and water rights transfer laws, as well as other considerations, such as water transport options and the distance between the water source and E&P site.

As a result, the most common water source pathways vary by plays, and within each play, by state. Different water access models are shown in Fig. 5.5. In wetter areas like Pennsylvania and Ohio, a significant portion of frac source water is either (1) obtained from local surface water resources or (2) derived from recycled flowback and produced water, since a lack of local disposal well capacity makes it far more cost-effective to recycle water than to truck it dozens of miles or more to injection sites.

In North Dakota, the purchase of water from water depots and obtaining water resources from surface water or groundwater are the most common water sourcing methods. E&Ps in Colorado source the majority of their water via leased or purchased water from municipalities or farmers. Other sources of water for E&P in Colorado are recycled produced water, imported water from out-of-state, and in rare cases, surface water diversions by the E&P company.

In Texas, the majority of frac water—especially in the booming Permian Basin—is obtained from groundwater, with a smaller portion coming from recycled produced water and treated effluent purchased from cities located near hubs

Ground Water Formation	Primary Frac Water Source	Groundwater Ownership	Surface Water Ownership	% of Private Surface
Permian	GW	Private	State	97%
Marcellus	SW		Riparian rights	85%
Bakken	Mixed	State		94%
Eagle Ford	GW	Private		97%
Niobrara	Mixed	Public if Tributary/ Private if Non-tributary	Public, subject to prior appropriation	63%

Fig. 5.5 Frac water acquisition pathways, by state

of oilfield activity. The water is purchased from a combination of (1) landowners who host drilling activities on their surface, (2) water owners and lessees that sell water for fracing operations on tracts other than those from where the water originates, and (3) may also be self-sourced from wells that the E&P company owns, leases, or otherwise operates. In some parts of the Permian Basin, produced water recycling activity also accounts for a growing volume of total frac water supply.

The most common method of obtaining frac water at present in the Permian Basin is to purchase it from third-party suppliers, be they farmers, ranchers, or service companies specializing in providing oilfield water supplies. Under Texas law, groundwater belongs by default to the surface owner, so unless the surface owner has sold or leased his/her groundwater estate, they are the water owner and the party that an E&P or any other entity seeking water will contract with (*Edwards Aquifer Auth. v. Day* (2012) 369 S.W.3d 814, 832).

In some of these cases, landowners will lease or sell groundwater for a royalty paid "at the wellhead." This can be done for both on-tract and off-tract sales. Many landowners, particularly in the Permian Basin, now include clauses in their surface damage agreements that require an E&P drilling on their land to purchase water exclusively from that landowner, unless the E&P's water needs exceed what the landowner's tract can reasonably supply. Larger landowners in the Permian Basin also sometimes become "full-service" water providers with their own wells, storage facilities, and pipelines or layflat hose for delivering water directly into customers' frac pits (Personal communications with Delaware Basin frac water sellers in July 2016 and October 2017).

The amounts of money at stake in such situations can be significant: a landowner who receives a royalty of $0.50 per barrel of water produced and sold under a surface damage agreement like that described above can make $250,000 from supplying a single frac job [$0.50/bbl X 500,000 bbl per frac]. For ranchers in the Delaware Basin, a quarter-million dollar frac water sale—which is pure profit if the sale was conducted on a royalty basis—can be the financial equivalent of selling 1000 head of feeder cattle.

In the Marcellus shale and the nearby Utica, frac water is generally sourced from surface water, and to a lesser, but still meaningful extent, recycled produced water. The Permian Basin and the Eagle Ford present a very different set of hydrological circumstances than the Marcellus does. Both plays lie in arid zones where there is very little permanent surface water and those sources that do exist—namely the Pecos River in West Texas and the Rio Grande and certain other rivers in South Texas are problematic from a water sourcing perspective because they often suffer from low flows, lie far from the frac activity epicenters, and carry water that has already been appropriated by farmers and other historic users along the river. Therefore, groundwater supplies dominate the frac water pool in both major Texas unconventional basins.

In North Dakota, water for E&P operations can be purchased or leased from landowners or other water rights holders. Water depot operators purchase individual surface water rights and offer a variety of water contracting methods. E&P operators contract with the water depots for a supply of water, which is hauled from a depot location to well sites. These water contracts can provide for a subscription allowed continued acquisition of water over a certain time period, direct water transfers, or a pay-as-you-go self-service arrangement (Golder Associates 2014).

North Dakota implements a permitting system to allocate water use rights (North Dakota's Appropriation of Water Statutes, N.D. Cent. Code § 61-04, *et seq.*). If an existing water permit does not allow for industrial use of the water, a change of water use must be obtained from the state regulatory agency—the North Dakota Office of the State Engineer—in order to transfer such water rights for E&P use (North Dakota's Appropriation of Water Statutes, N.D. Cent. Code §§ 61-04-02 and 61-04-15.1).However, North Dakota law limits the extent which a water right may be transferred. When there a competing rights to water from the same surface or groundwater source in North Dakota, which is a common scenario, preference is given to non-industrial uses, such as domestic, municipal, livestock and agricultural (North Dakota's Appropriation of Water Statutes, N.D. Cent. Code §§ 61-01-01.2 and 61-04-06.1). This policy makes it generally infeasible to transfer water rights for E&P use (Beck 2011).

All water acquisitions in Colorado are subject to the state's complex laws on water allocation and use, administered by the Colorado Division of Water Resources and overseen and adjudicated by Colorado "water courts" (Colorado's Water Right Determination and Administration Act of 1969, C.R.S. § 37-92-102). Surface water and tributary groundwater rights in Colorado are acquired under a system of prior appropriation, in which water users with a more senior decreed appropriation date have first priority to available water. Most Colorado rivers and streams are "over-appropriated," as such, depletions by more junior water rights are often curtailed to satisfy senior water rights holders, and little or no water is available to support new water rights.

Therefore, the security of water for E&P depends on the source of the water right and its place in the appropriation system. Water rights in Colorado are generally not tied to surface landownership. An exception to this general rule, for instance, is where a surface landowner has obtained a right to access nontributary groundwater below their property. Nontributary groundwater rights are generally not allocated through the prior appropriation system mentioned above. In some instances nontributary groundwater rights attendant to land ownership, may be severed from the property and sold for frac water, or, more commonly, may be "leased" subject to certain pumping restrictions. In Colorado, E&P operators commonly obtain water from municipalities and farmers and sometimes ditch companies[1] (Colorado's Ditch and Reservoir Companies Statutes, C.R.S. § 7-42-101, *et seq.*; *Jacobucci v. Dist. Court of Jefferson* (1975), 541 P.2d 667).

---

[1]A ditch company is a corporate entity under Colorado law created for the purpose of providing water to shareholders, who are primarily irrigators and municipalities, and sometimes other types of water users, such as industrial companies. Individual shareholders own shares that represent a certain defined interest in the water rights held by the ditch company.

As an example, during average and wet years in Colorado, municipalities and irrigators, including farmers and shareholders of ditch companies, may have excess water for lease. To the extent that a municipality has water rights in amounts that exceed their water demands, one can expect the municipality will capitalize on the E&P needs and lease excess water at a premium. For treated water, oil and gas companies paid more than 65 times the amount paid by farmers in average years for non-treated water (Healy 2012). For instance, Colorado experienced an "epic water year" in 2011, during which a Northern Colorado municipality situated above the Niobrara leased approximately 326 million gallons of water to oil and gas operators, with such transactions ringing in $1.5 million for the city (Finley 2011). In 2012, a suburb of Denver, Colorado approved an agreement to lease 2.4 billion gallons, for $9.5 million, to Anadarko Petroleum for a duration of 5 years (Healy 2012). The agreement between Anadarko and the suburb near Denver provided for $1200 per acre foot—nearly four times the market rate of approximately $350 per acre foot (Castellanos 2012).

To secure a more reliable water source, E&P operators attempt to lease water from a water rights holder with seniority where possible. Transferred water rights, via lease or purchase, maintain their original priority. Water rights acquired through a lease are subject to the terms of the original water decree, which govern the amount of water that may be used, how it may be used (industrial, residential, irrigation, recreation), and whether and how water must be augmented. Alternatively, nontributary groundwater rights, when they can be found, can provide a secure source for frac water since they are generally conveyed outside of the prior appropriation system.

Importantly, in Colorado, the process for water rights transfers varies based on the source and nature of the water right, which is a unique factor that affects decision-making by E&P operators in Colorado as compared to other jurisdictions. Some water rights transfers to E&P operators necessitate approval through an adjudicatory process in a Colorado Water Court or regulatory approval by the State Engineer's Office. For example, surface water and tributary groundwater rights subject to the prior appropriation system in Colorado are decreed for specific beneficial uses, a category in which oil and gas activity has now been included (*Vance v. Wolfe* (2009), 205 P.3d 1165; a prevailing Colorado Supreme Court case relied on by E&P operators claiming frac water use is a beneficial use). To use a water right for E&P uses, the water right must be decreed to allow for such use. If the water right transferred to an E&P operator is not decreed for a use that would cover E&P operations (such as E&P, industrial, or augmentation or replacement uses), a "change of use" may need to be obtained by water court through an adjudicatory process. Some water right conveyances in Colorado, particularly non-tributary groundwater, may be achieved by only a transaction between the conveying parties. The latter are highly sought after where speed and discretion are desired.

5.4 Who Supplies Frac Water? Core Categories of Water Suppliers

While oil companies still purchase meaningful volumes of water from small "mom & pop" suppliers, the trend in multiple basins is to emphasize large-volume suppliers. This primarily stems from the fact that many operators are now pumping frac jobs as large as 700,000 barrels in size (enough to submerge an American football field under about 80 feet of water) and need commensurately large supply sources to manage and offset economic and operational risks.[2]

This paper focuses on the high-volume frac water supplies. In doing so, it maintains analytical quality because (1) the largest suppliers account for a disproportionate volume of total water supply capacity in the Permian Basin; (2) in prior appropriations states like Colorado and North Dakota, certain suppliers with large/more senior water rights are positioned to dominate supplies in their respective zones of influence; and (3) the basic operational dynamics and business structures under which large-volume providers supply water broadly characterize the smaller sellers as well.

5.4.1 Operators Seek Diversity in Frac Water Supply, But Favor Large, Dependable Suppliers

Having a roster of multiple large-volume sellers helps operators work with maximum speed on completions during period of high oil and gas price volatility since all else held equal, filling frac pits faster means that jobs can be pumped more quickly. This consideration is especially important for instances where an operator seeks to bring drilled-but-uncompleted wells ("DUCs") online in response to higher prices. Large-volume suppliers also reduce water costs because they mean (1) a pit can be filled through fewer channels, thus decreasing per barrel water transfer costs and (2) because filling pits more quickly reduced evaporation losses, a key consideration in dry areas such as the Permian Basin.

Evaporation losses can be costly—one vendor who sells frac pond covers in the Permian basin reports that many operators say that during the summer they can lose from 1 to 4 inches per day of water from pits due to evaporation (BigD Companies 2017). At least one operator actually hired a night watchman because they thought someone was stealing water from their pit (Wiseman 2012). The authors' model suggests that a million barrel pit with dimensions common in the Midland area of

---

[2]Insights derived from FracFocus data, as well as company reports. See, for instance, Antero Midstream Services Sept 2017 Investor Presentation (showing water use trends for wells in the Marcellus shale) and Continental Resources Investor Presentation from September 2017 Barclays CEO Energy—Power Conference (data from NUZUM 1-1-12XH and IKE 1-20-17XH wells with 10,000 ft. laterals and completions requiring approximately 422 thousand and 505 thousand barrels of water, respectively).

the Permian Basin—approximately 715 feet X 715 feet in size—can lose nearly 17,000 bpd of water during hot, dry, and windy conditions. At a $1.50/bbl cost for water delivered into the pit, this represents a loss of approximately $25,000 per day worth of water: more than the dayrate of a top of the line horizontal drilling rig. And finally, large volume suppliers reduce an operator's most feared risk of all: running short of water and having to shut down a frac mid-job.

5.4.2 Key Frac Water Supplier Types, by Basin

Several basic categories dominate the high volume water sellers in the Bakken, Eagle Ford, Marcellus, Niobrara, and Permian Basin. For readers' convenience, we highlight in bold before each section which frac plays it is most relevant to.

5.4.2.1 Temporary Sales by Farms

Permian Basin, Eagle Ford, to a Lesser Extent, the Niobrara First are the farmers who decide to temporarily cease farming and enter the water business. This has become increasingly common in West Texas, where farmers own the water under their tract as real property. In prior appropriations jurisdictions, water is not explicitly tied to the land and farmers who seek to temporarily cease farming and sell water to oilfield users often face an onerous and expensive permitting process, particularly if they seek to reclassify water rights.[3]

Some of these entities can likely supply more than 100 kbd of water on a sustained basis. Farmers typically sit in the catbird seat. A farmer located near Saragosa, TX (the Southern Delaware Basin) tells this author that as a rule of thumb, it takes at least 1000–1200 gallons per minute of sustainable water supply to operate a center-pivot sprinkler system. Translated into barrels per day, that means a single such well could potentially supply as much as 35 thousand barrels per day of water throughout the year if the farmer fallows his/her fields to devote water exclusively to oilfield sales, or during the winter otherwise. The author's past interviews with farmers in the vicinity of Pecos, TX suggests that their water sales can sometimes have a seasonal dimension. In short, some farmers have largely foregone crop cultivation in order to focus on higher value water sales to the oilfield, while others reduce frac water sales during the growing season in order to ensure adequate supplies for their crops.

---

[3] Re-classification also poses potential legal risks. One of the authors comes from a family with a 40 acre farm in the Pecos River Valley of New Mexico, a prior appropriations state. While there is robust frac activity downstream, the family ultimately decided that the time, expense, and potential loss of water rights due to non-use on the farm outweighed the potential economic upside from reclassifying the rights and leasing them to oilfield users in the region.

5.4.2.2 Converted Farms

Permian Basin The second group of high-volume sellers are the converted farms— and at least one ranch with access to prolific groundwater. One example is Layne Christensen's converted cotton farm just west of Pecos, TX, which currently has 100 kbd of water supply capacity through a 22-inch HDPE pipeline serving fracers between Pecos and Orla, approximately 25 miles to the north. This author strongly suspects that based on the potential for frac water demand growth in the area and the fact that Layne's property sits atop some of the most prolific portions of the Pecos Valley Aquifer, the company will likely expand its capacity in coming quarters. A second example is Pecos SS, whose land is an old farm northwest of Fort Stockton that has 6 existing wells that have been flow tested at a combined 357 kbd of brackish water (3800 ppm total dissolved solids) (Water in West Texas, date unknown).

Pecos SS is located further from existing frac water demand centers than Layne and faces steeper logistical challenges to consistently consummate bulk water sales. Nonetheless, it has secured a permit from the Middle Pecos Groundwater Conservation District enabling it to export 212 kbd of water, a volume that would be sufficient to simultaneously supply multiple fracers.

5.4.2.3 Large Land Holders

Permian Basin A prime example example comes from Agua Grande, LLC, an oilfield-focused water supply venture owned by oil magnate Dan Allen Hughes, Jr. located on the Apache Ranch north of Van Horn. The ranch's 140,000 acre tract is larger than the combined area of the cities of Lubbock and Midland. Agua Grande recently received a permit from the Culberson County Groundwater Conservation District to export 6000 acre-feet per year (~128 kbd) of Capitan Reef water (Marfa Public Radio 2017).

Ranch operations also are important frac water providers, especially in the Northern Delaware Basin, in both Texas and New Mexico. Whereas farmers are leveraging advantageous positions over the most prolific water resources, ranches tend to lie atop less productive water resources and instead leverage their large surface land positions to exclude potential water sales competitors through exclusive water purchase provisions in surface use agreements and by levying "trespass fees" upon any oilfield party that obtains water elsewhere and seeks to transport it across the ranch's surface.

5.4.2.4 Dedicated Greenfield Water Deposits

Permian Basin The fourth category of high-volume oilfield water suppliers are those who have, through quiet and diligent hydrological and land work, managed to secure a privileged position atop a large-scale groundwater reserve. Wolfcamp

Water Partners offers an example of this, with its 31,000 acre lease at the foot of the Davis Mountains near Balmorhea (Hunn 2017). Wolfcamp seeks to drill into the Capitan Reef Aquifer complex and supply approximately 200 kbd of water to oilfield customers.

5.4.2.5   Large-Scale Surface Water Procurement

Operators in the Marcellus Shale almost exclusively use either freshwater from surface sources or recycled flowback and produced water to make up their frac fluid. For example, Antero Midstream Services, one of the largest water suppliers in the Marcellus and Utica shale region, reports that as of September 2017 it supplied on average 173 kbd of freshwater (Antero Midstream Services 2017). Of that, approximately 145 kbd is sold in the Marcellus shale area, a volume equivalent to approximately 40% of total Pennsylvania frac water usage between February 2016 and February 2017. This water is sourced from the Ohio River and a number of local lakes and other rivers and is all sold under fixed fee long-term contracts, according to the company (Antero Midstream Services 2017). Likewise, Rice Midstream Partners, another large Marcellus and Utica midstream services provider, sources its freshwater from the Ohio River for its Ohio water supply operations and from the Mononghahela River for its Marcellus operations in Pennsylvania (Rice Midstream Partners 2017).

In the Bakken, operators can source water from the Missouri River system or the Yellowstone River, depending on their relative location within the play. For instance, Caliber Midstream sources water from the Yellowstone River and can supply up to 55,000 bbl of water per day through its pipeline system, according to the company's website (Caliber Midstream 2018).

5.4.2.6   Public/Private Partnerships with Governmental Entities

Permian Basin Sixth, some suppliers have struck agreements with municipal and governmental entities that effectively amount to public-private partnerships. Under such arrangements the water developer can obtain reliable, large-volume water supplies at a defined price in exchange for funding and/or building municipal water infrastructure for the city, which reverts to city ownership when the parties' agreement expires. For instance, WaterBridge Resources, LLC signed an agreement with the City of Fort Stockton in July 2017 under which WaterBridge (1) obtains the exclusive right to purchase up to 18,000 acre-feet per year (390 kbd) of water from the Blue Ridge, Stockton, and Riley Farms and all brackish water from Belding Farms at the City's commercial and governmental rates, which for the volume in question will average approximately $2.86 per thousand gallons or $0.12 per barrel and (2) agrees to build out water supply infrastructure at the farms named above, as well as invest in pipeline capacity linking them to the City and will turn this infrastructure over to City ownership once the parties' agreement expires.

Pioneer Natural Resources' 2014 deal to source treated municipal effluent from the City of Odessa's Bob Derrington Treatment Plant offers a second example of frac water supplies derived from a public-private partnership. The parties' agreement is a "take or pay" structure in which Pioneer agreed to pay for and Odessa agreed to provide a guaranteed average annual minimum volume of approximately 85 kbd of treated effluent. After the first year of the agreement, Pioneer must then pay for a guaranteed annual water supply volume of 119 kbd. Pioneer agreed to help pay for improvements to the City's water delivery infrastructure and agreed to pay $0.25 per barrel for water during the first 6 years of the agreement and approximately $0.30/bbl in years 6 through 10 of the contract. For years 11 through 15, the price will be $0.35/bbl and in years 16 through 20, approximately $0.40/bbl.

In North Dakota, the Western Area Water Supply Project (WAWSP) has an allotment of up to 7500 acre-feet per year of water that can be sold to industrial users. This water comes from the Missouri River and local aquifers (Western Area Water Supply Authority 2018). The WAWSP sells its treated water at prices ranging from $0.40 to $0.84/bbl, delivered to the customer and is on track to sell 3500-to-4000 acre-feet (27-to-31 million barrels) of water in 2017 at an average price of roughly $0.53 per barrel (Author's personal communication with WAWSP representative, 9 November 2017). The entity has close to 200 miles of trunk pipeline and a substantially greater mileage of distribution pipeline covering a significant portion of the Bakken play (Author's personal communication with WAWSP representative, 9 November 2017). WAWSP sells water to oilfield customers under a range of commercial arrangements, including short-term sales and longer term minimum volume or even take-or-pay contracts for customers who seek assured long-term water supplies at the lowest price (Author's personal communication with WAWSP representative, 9 November 2017). WAWSP fits into the "public/private partnership" category because the premium prices that industrial users pay for their water are helping the utility accelerate the paydown of its debt and reduce the water rates it needs to charge its other customers.

5.4.2.7 Self-Sourcing by E&Ps and/or Their Captive Midstream Subsidiaries

Seventh, a number of operators are choosing to self-source a material portion of their frac water. Part of the supply stream comes from wells the company drills and owns, with the remainder coming from produced water recycling, which will be discussed in Sect. 5.4.2.8 below). Three core factors generally motivate water self-sourcing programs.

The first is a push to lower costs. As an operator scales up its drilling and completion program, the amounts of water required—and those water volumes' contribution to final completed costs of wells—can become significant. A well that requires 650,000 bbl of water to frac in the Permian Basin can experience total water supply costs of nearly $500,000, assuming a delivered water price of $0.75/barrel ($0.50 to purchase the water and another $0.25/bbl to move it into the frac pit).

Self-sourcing can avoid much of this cost burden. For instance, Jagged Peak Energy, a medium-sized producer focused on the Delaware Basin, reported in September 2017 that using its proprietary water wells and water pipelines, it was able to achieve an average water acquisition cost of only $0.10/bbl (Jagged Peak Energy 2017). At such cost levels, a large completion could save $422,500 [650 kb of water X ($0.75/bbl of water—$0.10/bbl of water)]. What this means in practical terms is that if all other costs are held equal and one assumes a production value of $27/barrel of oil equivalent at average prices for the first half of 2017, the water sourcing cost reduction is effectively equivalent to increasing the well's lifetime hydrocarbon production by 15,648 barrels of oil equivalent.[4] In oil-rich plays, this economic booster effect would be more pronounced, since oil is premium priced relative to natural gas and natural gas liquids.

A second—and related—driver is the desire to gain price negotiation leverage. An operator fully dependent on external water suppliers has little ability to negotiate more favorable prices and contract terms. In contrast, an operator with access to well capacity capable of covering a meaningful portion of frac water needs has breathing room to negotiate better terms. These benefits flow in part from the credible threat that if suppliers ask too high a price for water, the company can delay the completion and make up the difference with its own wells. Likewise, water wells and pipelines are cheap relative to the costs incurred in drilling and completing horizontal oil and gas wells and can be installed rapidly. This generally means that so long as the local aquifer can support additional extraction, the operator can rapidly scale up its proprietary water production capacity if external suppliers are squeezing it on price.

The third driver is that self-sourced water not only can insulate an operator from cost pressures, but also can provide a hedge against water supply disruptions. The one thing that is significantly worse than overpriced water from an operator perspective is water that does not arrive in the frac pit when needed, and either forces a delay or shutdown of a frac completion. Self-sourced water can help mitigate this problem because, even if insufficient to cover the entire needs of a completion program, water produced in house can offset what might have been purchased from less reliable parties and help the operator prioritize its external sourcing relationships with the most reliable suppliers.

5.4.2.8 Produced Water Recycling

Finally, an increasing number of operators are deriving material portions of their completion fluid from recycled produced water. Recycling has played an important role in the Marcellus for years now, since a lack of disposal options and extremely high water transport costs to distant disposal facilities renders even expensive recycling practices economically competitive.

---

[4] Economic effects calculated using actual production data from sample Delaware Basin hz wells in Lea County, NM and price information from the EIA for January through June 2017.

This trend is gaining strength due to the operational cost savings recycling enables, with economic benefits turbocharged by advances in frac fluid chemistry that allow operators to lightly treat water and thus minimize costs (Collins 2017a, b). Greater use of recycled produced water also helps operators optimize their fresh water sourcing portfolios by (1) reducing reliance on freshwater and creating space for more selective engagement with freshwater suppliers and (2) empowering operators with greater latitude to negotiate favorable freshwater pricing (for example, Matador Resources Co. 2017).

Approach Resources, which operates in the Southern Permian Basin, stated in April 2017 that its recycled water supplies cost between $0.50 and $0.80 per barrel (Approach Resources Inc. 2017). Freshwater supplies in the vicinity of Approach's acreage cost approximately this much, but by recycling, the company minimizes other water-centric cost drivers such as trucking and disposal, and reports that it ultimately saves from $3.20 to $4.50 per barrel of water by using recycled water processed through the firm's own infrastructure (Approach Resources Inc. 2017).

Savings of this magnitude in programs drilling Delaware Basin horizontal wells using 500 thousand barrels of water per frac could reduce completed well cost by as much as $2.25 million per well as a project proceeds and the operator begins reaping the benefits of minimizing/avoiding produced water disposal costs. If those savings were re-invested back into drilling, an operator could potentially drill at least three additional Wolfcamp wells per $100 million in capital spent—a 23% increase in potential productive assets.[5] Depending on the productivity of the formations in a given area, three additional wells per $100 million in CAPEX spend could add 3 million barrels of oil equivalent—or more—in ultimately recoverable reserves.

Produced water recycling can push producers onto a lower position on the global production cost curve. This is an advantageous place to be with commodity price volatility and as OPEC producers struggle to balance budgets at current oil price levels. Whether the market is "forties forever," "lower for longer," or "fifties for a while," one thing is certain—the lowest cost producers globally will be the best off. In the forest, you don't need to be faster than the bear that is trying to eat you—you just need to be faster than the other people.

---

[5] Calculated based on EOG Resources 2Q2017 reported completed well cost of $7.6 million per well in the Delaware Basin Wolfcamp play. I adjusted the $2.25 million per well potential cost-savings from water recycling down to $1.5 million per well (yielding a per well cost of $6.1 million), as the top-tier operators like EOG are already applying cost-saving approaches across their water value chains and I don't want to over-estimate the potential economic impact of produced water recycling on their operations. The 23% figure is based on the fact that including the $1.5 million per well in savings, $100 million in a development budget can now go for 16 wells instead of 13, as would be the case with a cost per well of $7.6 million. Original data source: EOG Resources 2Q2017 Quarterly Presentation, available at http://investors.eogresources.com/Presentations-and-Events

5.5   Deal Structures Under Which Frac Water Is Purchased

Frac water sales is procured under three basic approaches (Fig. 5.6), with many variants on the core frameworks. Approach one entails a spot purchase—in other words, I buy water from you for this particular completion and maybe we do business again or maybe we don't. There is no long-term commitment, it is a cash and carry model. The second approach involves term deals, which define minimum volumes that a supplier will make available for an operator to purchase—loosely akin to a right of first refusal—but often without the binding strictures imposed by a true "take or pay" contract structure. In the authors' experience, such terms deals in the Permian Basin of Texas and New Mexico generally place more risk on the seller of the water, since a buyer often does not have to pay if for some reason they stop taking water. In some cases, these contracts can take the form of a "flexible take or pay" or "water reserve contract" that entitles the buyer to a defined volume of water, but allows them to stop taking water but not have to pay, provided that they give the seller sufficient advance notice to allow it to locate alternate buyers (Author's personal communication with a Delaware Basin frac water supplier, July 2017). Or, if alternate buyers cannot be timely located, the buyer can push its required offtake volume into the future, so long as it ultimately takes delivery of the water or a buyer for all of the requisite volume at that point in time can be located (Author's personal communication with a Delaware Basin frac water supplier, July 2017). With the time value of money, such a structure generally places significantly more financial risk on the seller of the water.

Water suppliers—particularly those with prolific groundwater supplies—but who may be located some distance from current completion activity hotspots will seek longer term "take or pay" deal structures to give them a committed cash flow

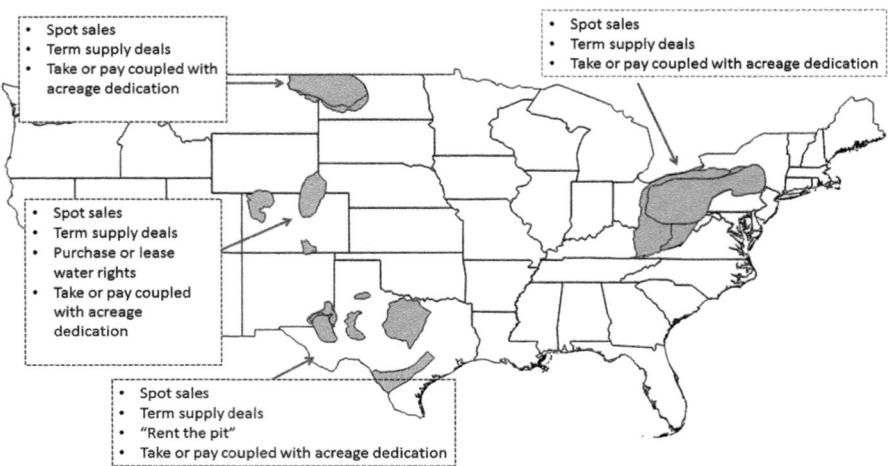

Fig. 5.6 Frac water acquisition transaction structures

they can use to attract project finance capital. Water users, in contrast, face high oil and gas price volatility and generally speaking seek to avoid the sustained balance sheet burden that take or pay agreements can pose if prices fall and their demand for water falls below the minimum volume required under the contract. In capital terms, financial commitments for take or pays also compete with self-owned water prod and water-focused midstream infrastructure that may have lower capital costs per bpd of water production capacity and which can be operated in a manner more tightly synchronized with the water user's actual needs at a given time.

The third structure is the true "take or pay" agreement, under which a buyer subscribes to a minimum offtake volume and pays regardless of whether it takes delivery of the water. Examples the authors locate suggest that for frac water supplies, take or pays are typically priced as long-term fixed fee structures that are coupled with acreage dedications and minimum volume guarantees, generally exist when an E&P is dealing with a midstream service provider that is, for practical purposes a quasi-captive subsidiary of the E&P. For instance, Antero Midstream has a 20-year exclusive agreement to supply water to operations on Antero's 616,000 net acre position in the Marcellus and Utica region.

Antero Midstream also has a "right of first offer" that effectively functions as a right of first refusal for becoming the midstream water services provider if Antero expands or otherwise alters its acreage position. The right of first refusal provisions governing Antero Midstream's relationship with Antero Resources reflect privileged positions and are broadly repeated in the relationships between other E&Ps and their preferentially-situated midstream providers. This holds true in the Bakken, DJ/Niobrara, and now the Permian Basin as well for multiple operators.

Numerous E&Ps are investing substantial sums in developing captive water handling loops, which in many instances reflect an integrated approach that brings together freshwater supply, produced water handling, water recycling, and in some cases, sourcing of non-potable water sources such as brackish groundwater or treated municipal effluent (as discussed above in the section on public/private partnerships).

It is difficult to ascertain the amount of capital being invested in self-sourced water infrastructure and the recycling systems that often accompany it. That said, bits and piece of investment data disclosed by major publicly-traded operators suggest the CAPEX is substantial. For instance, Pioneer Natural Resources has said it is on track to spend approximately $150 million on water infrastructure in 2017 and that it plans to spend about the same amount in 2018 (SeekingAlpha, 3Q2017).

Among these operators, Anadarko, Apache, and Pioneer Natural Resources (and likely multiple other operators) are creating sufficiently large systems that they will have the physical capability to offer third-parties commercial access to their water handling networks, which are currently proprietary. If the commercial logic of doing this is sufficiently strong for multiple large E&Ps to publicly disclose such intentions, other operators building these systems may consider a similar path to maximally monetize their midstream assets. For instance, Anadarko secured its first third-party produced water disposal customer in the second quarter of 2017 (SeekingAlpha, 2Q2017).

5.6   Water Services Go Midstream

Emerging independent water midstream services providers add a new dynamic to frac water sourcing. Many of these sophisticated, private equity-backed entities originally focused on water disposal assets, since these involved a midstream business model akin to that used by the gas, crude, and products-focused ventures many management teams cut their teeth in.[6] Some are now also considering greater involvement in providing frac fluids, since doing so effectively makes them "one stop shops" to which operators can outsource their full cycle water management.

From the perspective of a midstream services provider, integrated water systems provide an opportunity rarely found in the oilfield: the ability to sell the same barrel multiple times. A provider with integrated water supply, produced water gathering, water supply wells, and recycling infrastructure can (1) sell water to a frac'er; (2) collect a fee for gathering and taking away the flowback water after the completion is done (some of which was originally sold as frac water); and (3) either inject the water down an SWD for a fee or treat it and re-sell it to another E&P needing water for well completions.

Many of the midstream operators have a presence in all of the major basins we analyze and that best practices developed in one area are likely to rapidly proliferate into the other plays, providing that local conditions permit.

5.7   Conclusion

Frac water sourcing practices vary between basins, but follow a handful of core pathways. In the Permian Basin—the premier global unconventional liquids play—produced water recycling is increasingly taking hold. Simultaneously, multiple operators and water midstream providers are executing projects that have, and will continue to, expand the supply of non-potable water sources **and** the infrastructure needed to move them to frac pits. Ultimately, the frac water business is one driven by relationships and most of all, by pipeline and infrastructure connectivity. Once produced water, municipal effluent, and other non-freshwater supplies factor into the equation, it becomes clear that in the Permian and other key basins, the chief problem is not water's absolute availability, but rather its location relative to demand hotspots. The overview of frac water acquisitions structures set forth in this paper

---

[6] Some examples include Solaris Midstream, backed by Trilantic Capital Partners, http://www.solarismidstream.com/and WaterBridge Resources, backed by Five Point Capital Partners, http://h2obridge.com/. Both companies are based in Houston.

will hopefully help lay the foundation for further development of the physical pipe infrastructure and accompanying legal and economic frameworks needed to ensure that frac fluid can reach the places where it is needed. Delivering frac water timely, at a competitive price, and in quantity creates opportunities for innovative operators and investors to continue unlocking economic value and making U.S. unconventional oil and gas production increasingly competitive in the global marketplace.

References

Antero Midstream Services (2017, September) Investor presentation. Accessed 26 Nov 2018

Approach Resources Inc (2017) DUG water conference 2017 presentation. Available via http://ir.approachresources.com/phoenix.zhtml?c=214016&p=irol-presentations. Accessed 26 Nov 2018

Beck R (2011) Water resources and oil and gas development: a survey of North Dakota law. North Dakota Law Review 87:507

BigD Companies (2017) Floating frac pit evaporative covers. Available via http://www.bigdco.com/pit-lining/floating-frac-pit-covers. Accessed 26 Nov 2018

BP Statistical Review of World Energy (2017) Available via https://www.bp.com/en/global/corporate/energy-economics/statistical-review-of-world-energy.html. Accessed 26 Nov 2018

Caliber Midstream (2018) Services. Available via http://www.calibermidstream.com/services. Accessed 26 Nov 2018

Castellanos S (2012) Aurora OKs $9.5 million fracking water deal with Anadarko. Aurora Centennial. Available via http://www.aurorasentinel.com/news/aurora-oks-9-5-million-fracking-water-deal-with-anadarko/. Accessed 26 Nov 2018

Collins G (2017a) Frac ranching vs. cattle ranching: exploring the economic motivations behind operator-surface owner conflicts over produced water recycling projects. Rice University's Baker Institute for Public Policy. Available via https://www.bakerinstitute.org/media/files/files/448a93b3/BI-Brief-101717-CES_Ranching.pdf. Accessed 26 Nov 2018

Collins G (2017b) How much water does Apache potentially need to develop alpine high? Texas Water Intelligence. Available via https://texaswaterintelligence.com/2017/06/19/how-much-water-does-apache-potentially-need-to-develop-alpine-high/. Accessed 26 Nov 2018

Collins G and Medlock III K (2017) Assessing shale producers' ability to scale-up activity. Rice Univ's Baker Inst for Public Policy. Available via https://www.bakerinstitute.org/files/11282/. Accessed 26 Nov 2018

Colorado's Ditch and Reservoir Companies Statutes, C.R.S. § 7-42-101, *et seq*

Colorado's Water Right Determination and Administration Act of 1969, C.R.S. § 37-92-102

Edwards Aquifer Auth. v. Day (2012) 369 S.W.3d 814, 832

Finley B (2011). Fracking of wells puts big demand on Colorado water. The Denver Post. Available via http://www.denverpost.com/2011/11/22/fracking-of-wells-puts-big-demand-on-colorado-water/. Accessed 26 Nov 2018

Galbraith K (2013). West Texas oilfield town runs out of water. The Texas Tribune. Available via https://www.texastribune.org/2013/06/06/west-texas-oilfield-town-runs-out-water/. Accessed 26 Nov 2018

Golder Associates (2014) Oil and natural gas exploration and production water sources and demand study: 27. Available via www.westernenergyalliance.org/sites/default/files/WesternWaterUseStudy.pdf. Accessed 26 Nov 2018

Healy J (2012) For farms in the West, oil wells are thirsty rivals. NY Times. Available via http://www.nytimes.com/2012/09/06/us/struggle-for-water-in-colorado-with-rise-in-fracking.html. Accessed 26 Nov 2018

Hunn D (2017) As the oil patch demands more water, West Texas fights over a scarce resource. Available via http://www.houstonchronicle.com/business/article/As-the-oil-patch-demands-more-water-West-Texas-11724100.php. Accessed 26 Nov 2018

International Association of Directional Drilling (2018) A Permian factory drilling showcase by Encana. Intl Assoc of Directional Drilling. Available via https://www.iadd-intl.org/articles/a-permian-drilling-factory-showcase-by-encana/. Accessed 26 Nov 2018

Jacobucci v. Dist. Court of Jefferson (1975), 541 P.2d 667

Jagged Peak Energy (2017, September) Investor Presentation. Accessed 26 Nov 2018

King G (2012) Hydraulic fracturing 101: what every representative, environmentalist, regulator, reporter, investor, university researcher, neighbor and engineer should know about estimating frac risk and improving frac performance in unconventional gas and oil wells. SPE International. Available at https://fracfocus.org/sites/default/files/publications/hydraulic_fracturing_101.pdf. Accessed 26 Nov 2018

Laredo Petroleum. Corporate presentation (2017, November). Available via http://www.laredo-petro.com/media/205599/november-2017-corporate-presentation.pdf. Accessed 26 Nov 2018

Marfa Public Radio (2017) West Texas water board approves contentious export permit. Available via http://marfapublicradio.org/blog/west-texas-water-board-approves-contentious-export-permit. Accessed 26 Nov 2018

Matador Resources Co. (2017) Investor presentation: slide 23. Available via http://investors.matadorresources.com/phoenix.zhtml?c=248247&p=irol-presentations. Accessed 26 Nov 2018

North Dakota's Appropriation of Water Statutes, N.D. Cent. Code § 61-04, *et seq*

North Dakota's Appropriation of Water Statutes,. N.D. Cent. Code §§ 61-04-02 and 61-04-15.1

North Dakota's Appropriation of Water Statutes, N.D. Cent. Code §§ 61-01-01.2 and 61-04-06.1

OPEC monthly oil market report (2017, September). Available via http://www.opec.org/opec_web/static_files_project/media/downloads/publications/MOMR%20September%202017.pdf. Accessed 26 Nov 2018

Rice Midstream Partners (2017) Investor presentation:19-20. Available via http://s2.q4cdn.com/444921097/files/doc_presentations/2017/RMP-May-Presentation_vF.pdf. Accessed 26 Nov 2018

SeekingAlpha, 2Q2017. Western Gas Partners—earnings call transcript, ben fink (CEO). Available at: https://seekingalpha.com/article/4090888-western-gas-partners-wes-ceo-ben-fink-q2-2017-results-earnings-call-transcript?part=single

SeekingAlpha, 3Q2017.. Pioneer natural resources—earnings call transcript. Available at: https://seekingalpha.com/article/4119804-pioneer-natural-resources-pxd-q3-2017-results-earnings-call-transcript?part=single

Texas Football Stadium Database (2018) AT&T stadium. Available via http://www.texasbob.com/stadium/stadium.php?id=1285. Accessed 26 Nov 2018

U.S. Energy Information Administration (2018) Permian region drilling productivity report (Nov 2018). U.S. Energy Info Admin. Available via https://www.eia.gov/petroleum/drilling/pdf/permian.pdf. Accessed 26 Nov 2018

Vance v. Wolfe (2009), 205 P.3d 1165

Water in West Texas. Summary of wells to date. Available via http://waterinwesttexas.com/summary-of-wells-to-date. Accessed 26 Nov 2018

Western Area Water Supply Authority (2018) Overview. Available via http://wawsp.com/overview. Accessed 26 Nov 2018

Wiseman P (2012) Local company offers solution to frac water pit evaporation. Midland Reporter Telegram. Available via http://www.mrt.com/business/energy/article/Local-company-offers-solution-to-frac-water-pit-7427890.php. Accessed 26 Nov 2018

Gabriel Collins is the Baker Botts Fellow in Energy and Environmental Regulatory Affairs at the Baker Institute Center for Energy Studies at Rice University. He was previously an associate attorney at Baker Hostetler, LLP, and is the co-founder of the China SignPost™ analysis portal. Collins worked in the Department of Defense as a China analyst and as a private sector global commodity researcher, authoring more than 100 commodity analysis reports, both for private clients and for publication. His research focuses on legal, environmental and economic issues relating to water, as well as unconventional oil and gas development and the intersection between global commodity markets and a range of environmental, legal and national security issues. His analysis draws from a broad swath of geospatial and other data streams, and often incorporates insights from sources in Chinese, Russian and Spanish. Collins holds a B.A. from Princeton University and a J.D. from the University of Michigan Law School.

Julie A. Rosen is an attorney and shareholder at Ryley Carlock & Applewhite in Denver, Colorado, where she focuses her practice in the areas of environmental, water and energy resources. She represents a wide variety of interests, including oil and gas, electric utilities, mining, manufacturing, wastewater utilities, and agricultural, as well as municipalities, landowners, developers and financial institutions. After graduating from the top environmental law school in the U.S., Vermont Law School, Rosen developed expertise in issues concerning water quality, air quality, solid and hazardous waste, brownfields, National Environmental Policy Act, Wild and Scenic Rivers Act and Endangered Species Act matters. During law school, she clerked for the U.S. White House Council on Environmental Quality in Washington DC. Before turning to the law, Julie graduated from Arizona State University with a B.A. in journalism/public relations and worked as a lobbyist for the Arizona Fish and Game Department in Phoenix, Arizona. Rosen regularly writes articles and gives presentations on environmental, water, and energy issues.

Chapter 6
Access to Water for Hydraulic Fracturing in China

Libin Zhang, Sheng Shao, Fang Dong, and Jiameng Zheng

Abstract China is very ambitious in developing shale gas. To develop this unconventional natural gas resource requires hydraulic fracturing, which is a process to stimulate well production. Hydraulic fracturing operations use massive volumes of water, while most shale gas reserves in China are located where there are water shortage issues due to either seasonal drought or high demand for local domestic water. This Chapter discusses how shale gas developers may acquire water for hydraulic fracturing and some regulatory matters relevant to this water-energy nexus. Under the state ownership rule, shale gas developers may only have water use rights. Developers may acquire water by abstracting with a license or buying from water suppliers. The national water exchange can be another option to obtain water rights from other water abstraction license holders, although no such transaction has been made yet. Government may curtail water access for fracking and the public may challenge through different ways, but no such case is available. It is expected that a more robust regulatory practice is ahead along with the development of industry as well as the growth of water-energy tension.

Keywords China · Hydraulic fracturing · Water acquisition · Scarce water

L. Zhang
Broad & Bright, Beijing, China
e-mail: libin_zhang@broadbright.com

S. Shao (✉)
Sunshine Law Firm, Hangzhou, China
e-mail: shaos@sunshinelaw.com.cn

F. Dong
PowerChina Resources, Ltd., Beijing, China
e-mail: dongfang@powerchina.cn

J. Zheng
The University of Texas at Austin, Austin, TX, USA
e-mail: jiameng.zheng@utexas.edu

© Springer International Publishing AG 2020
R. M. Buono et al. (eds.), *Regulating Water Security in Unconventional Oil and Gas*, Water Security in a New World,
https://doi.org/10.1007/978-3-030-18342-4_6

6.1 Introduction

Hydraulic fracturing, also referred to as "fracturing" or "fracking", is a process to stimulate well production by pumping fluids into shale to crack rock formations and release trapped hydrocarbons. This has made many shale gas reserves economically recoverable, which has directly helped the U.S. become the world's top oil and gas producer (EIA 2016). China has set aggressive goals for shale gas development, driven by the growing energy demand and the challenging environmental conditions. Today, China is still in its early stage of shale gas development, mainly including resource assessment and exploration (Dong et al. 2016).

Water availability is a fundamental factor in successful shale gas development, as a massive amount of water is required for production. Shale gas production uses water mainly for drilling and fracking, and the fracking requires much more water (Guo et al. 2016). To fracture one well in a U.S. shale formation may require 3–5 MM gallons of water (Krupnick et al. 2014). Sinopec, one of China's major national oil companies (the "NOCs"), kicked off the State's first large-scale commercial shale gas production at Fuling field in 2014 (Sinopec 2016). Studies suggest that a Fuling well on average uses almost 300% of fracking water compared with that required for fracking a Barnett well in the U.S. (Guo et al. 2016).

China owns the largest technically recoverable shale gas resources, possibly as much as 2/3 again greater than North America's (EIA 2013). However, most shale gas areas have existing water shortage issues (Han and Xiao 2015). Some have concluded that this may be the key barrier to China's shale gas development (Luo 2014). Others suggest that water supply should be of less concern in the meantime (Sandalow et al. 2014). Either way, the increase in the number of wells will cause a significant increase not only in China's fracking water demand, but also in the pressure on local water supply (Marsters 2013). A study shows that, if China is to achieve the shale gas production goals by 2020, the nation-wide water consumption for shale gas development will peak in 2019 (Guo et al. 2016).

This Chapter focuses on the water acquisition for fracking and associated regulatory matters in China. In Sect. 6.2, we briefly introduce China's shale gas resources and challenges associated with the water supply. In Sect. 6.3, we review how shale gas developers may and have acquired water for fracking. In Sect. 6.4, we discuss the potential restrictions and public participation matters involved in water acquisition for fracking.

6.2   Water Availability Challenges to China's Fracking

6.2.1   Rich Reserves and Ambitious Goals

Resources Overview China owns the largest technically recoverable shale gas resources in the world (WRI 2014). According to the U.S. Energy Information Administration (the "EIA"), China has an estimated 1115.2 trillion cubic feet (the "Tcf") of risked, technically recoverable shale gas (EIA 2013). China itself has a smaller estimate of 769 Tcf (Shale Gas Plan). Seven basins containing prospective shale gas formations (see Fig. 6.1) are Sichuan, Tarim, Junggar, Songliao, Yangtze Platform (South China), Jianghan, and Subei (EIA 2015).

The Sichuan Basin in the Southwest is China's largest shale gas region, and it is estimated to contain over 55% of the State's risked, technically recoverable shale gas reserves (EIA 2015). The Fuling field, which is the State's first large-scale shale gas commercial production project, is located in the Sichuan Basin (Sinopec 2016). As for the other six basins, they currently remain at the exploration and evaluation stage (Guo et al. 2016). In regards to these six basins, it is unlikely for them to commence large-scale development in the near future partly due to geology complications (Krupnick et al. 2014).

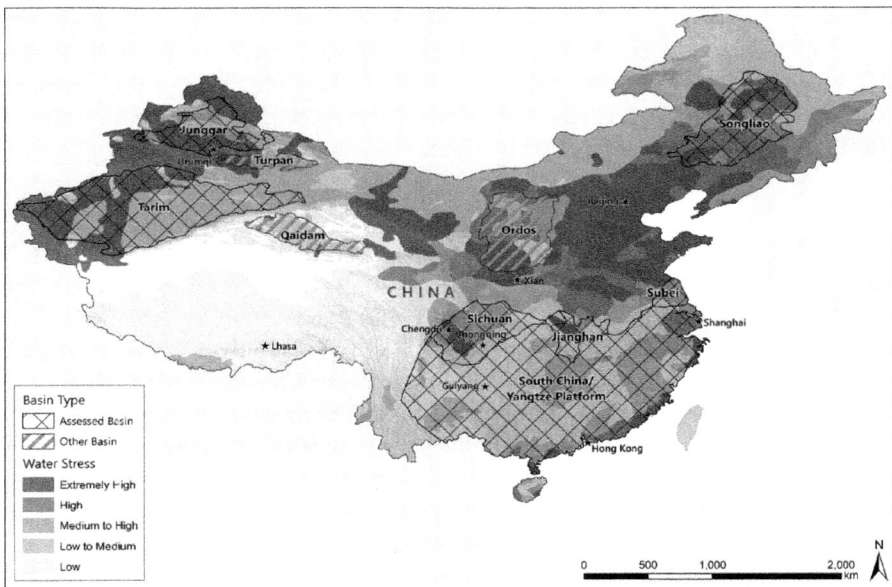

Fig. 6.1 Major basins with prospective shale gas formations in China (Callum Foster, UWE Bristol Cartography, based on ARI 2013 and other sources)

In comparison with the U.S., China's shale gas accumulation and geological formation are much more complicated (Zhao et al. 2013). A large portion of shale in China is continental shale, which has never been proven to be commercially viable anywhere (Marsters 2013). The high clay content also makes many of China's shale formations more pliable and less apt to fracking (Tollefson 2013). Furthermore, many of China's shale gas reserves are buried deeper than that of the U.S., which leads to more difficulties in development (Wang and Wang 2016). For instance, over 50% of shale gas reserves in the southern part of Sichuan Basin are buried over 3500 m (2.17 mi.) underground, for which developers still lack proper technologies and techniques (Shale Gas Plan).

Another unsolved challenge for many basins in China is water shortage. Hydraulic fracturing consumes large amounts of water, so it is hard to exploit shale gas where there is a water supply problem, such as the Tarim Basin in Northwest China (see Fig. 6.1). The Tarim Basin is the second largest shale gas region in China (Wang 2015). However, it is subject to extremely arid conditions, which will pose big challenges for fracking operations (WRI 2014). The water shortage issue is further discussed in Sect. 6.2.2.

Development Goals Driven by the growing energy demand and worsening environmental conditions, Chinese government has set ambitious goals for shale gas development. The production was set to reach 30 billion cubic meters (the "Bcm") (1059.44 Bcf) in 2020, but the figure for 2015 was only 4.6 Bcm (162.45 Bcf) (Shale Gas Plan). Some other goals can be found in several laws and regulations, including the Energy Development Strategy and Action Plan for 2014–2020 (2014), the 13th Five-Year Plan of Natural Gas Development (2017), and the Shale Gas Development Plan for 2016–2020 (2016) (the "Shale Gas Plan"). For example, according to the Shale Gas Plan, the State expects an 80–100 Bcm (2825.1-3531.4 Bcf) shale gas production in 2030.

China plans to accomplish technology breakthroughs by 2020 in the exploration and production of shale gas reserves buried over 3500 m (2.17 mi.) underground (Shale Gas Plan). This may significantly affect the overall scale of shale gas production for the 2016–2020 period, since over 50% of the shale gas reserves in the south part of Sichuan Basin are buried there (Shale Gas Plan).

After comprehensive geological evaluation, exploration evaluation, and pilot development, the National Energy Administration of China (the "NEA") has targeted five key development plots as main drivers of shale gas production (Shale Gas Plan). All of these are believed to have the greatest potential to produce commercial volumes of shale gas (see Table 6.1).

Table 6.1 Geological resource potential of China's five key shale gas plots with production potential (Shale Gas Plan)

Development plot	Basin	Title	Burial Depth (m/mi.)	Geological Resource Potential (Bcm/Bcf)
Fuling	Sichuan	National demonstration plot	< 4000/2.48	476.7/16.834.5
Changning	Sichuan	National demonstration plot	< 4000/2.48	1900/67,097.87
Weiyuan	Sichuan	National demonstration plot	< 4000/2.48	3900/137,727.2
Zhaotong	Sichuan	National demonstration plot	N/A	496.5/17,533.73
Fushun-Yongchuan	Sichuan	Sino-foreign cooperation plot	N/A	500/17,657.3

6.2.2 Water Use Challenges Affecting Hydraulic Fracturing

Water shortage is a big challenge for shale gas development because hydraulic fracturing operations consume large volumes of water. In order to complete each well, including drilling and fracking operations, approximately 0.8–5 MM gallons of water on average could be used (EPA 2015). China has existing water shortage issues (Liu and Yang 2012). Home to 21% of the world's population, China only has 7% of global freshwater supply (Lamb 2016). Water acquisition for hydraulic fracturing would potentially increase pressure on the local water supply, especially during drought seasons or in arid areas (Zhang and Yang 2015). China has to find a balance between aggressive shale gas development and sustainable water resources utilization, which can be a major challenge for this water-deprived country (Wang et al. 2014).

Geographic Distribution Mismatched China has enormous water resources, but they are unevenly distributed in space (Shen 2015). The southern region of China has larger water resources than the north (WRI 2014). For example, water capacity per capita in the south is 500% greater than that of the north (Wang et al. 2000). In addition, about 85% of water resources are located in the Yangtze River basin and its southern area, covering 15 south provinces (Wang et al. 2000).

However, as Table 6.2 indicates, over 40% of technically recoverable shale gas resources are located in the North, Northeast, and Northwest China (Pi et al. 2015). Thus, water shortage can be a barrier to the commercial production of China's technically recoverable shale gas. The mismatched geographical distribution of energy and water resources adds to the pressure on water supply in exploiting natural resources in China (Hu and Xu 2015). Uneven geographical distribution of water resources between the north and the south has also been a problem for shale gas development.

Table 6.2 Distribution of technically recoverable shale gas resource potential in China (Source: Pi et al. 2015)

Area	Technically recoverable resources (TCF)	Percentage (%)
Upper Yangtze and Dian-Qian-Gui	351.02	39.62
North and Northeast China	236.60	26.70
Mid-Lower Yangtze and Southeast	163.86	18.49
Northwest China	134.54	15.18
Total	886.02	100

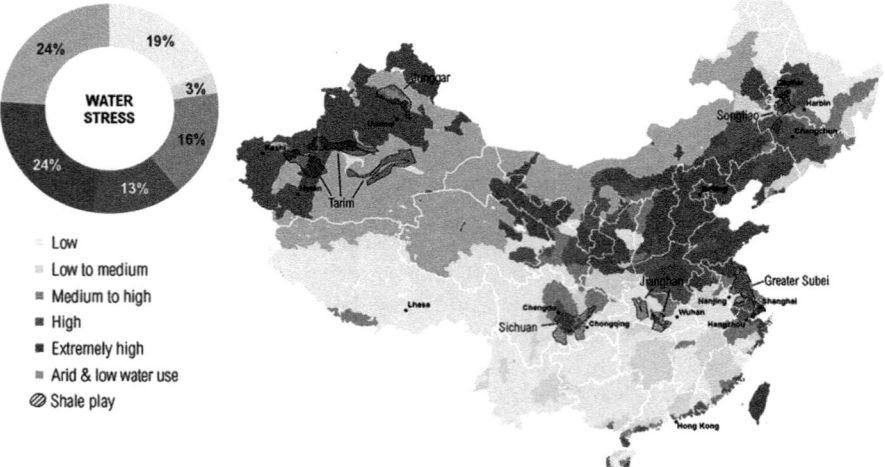

Fig. 6.2 China's shale plays and baseline water stress (Reproduced from the WRI www.wri.org 2014)

Local Water Stress According to the World Resources Institute (the "WRI"), over 60% of the shale gas resources are in areas of high or extremely high baseline water stress or arid conditions (Maddocks and Reig 2014). The WRI defines "baseline water stress" as the ratio of total water withdrawal by residential, industrial and agricultural users to the available supply (Gassert et al. 2013). This helps indicate the extent of competitions among water users.

WRI's (2014) map describes the baseline water stress in different regions of China, and identifies major shale plays on the map (see Fig. 6.2). The darker colored regions represent a higher level of water stress, which is caused by a higher number of water acquisition competitions and a greater depletion of water resources (Maddocks and Reig 2014). Figure 6.2 clearly shows that most shale gas plays are located in the areas with medium-extremely high water stress. Some estimates suggest

Table 6.3 Shale gas development's water use in Chongqing, China (You et al. 2015)

Local area	Ratio of shale gas water consumption to available water resources (%)	Ratio of shale gas water consumption to total water use (%)
Qianjiang	2.09	19.69
Chengkou	1.74	19.79
Xiushan	1.54	37.11
Youyang	2.01	36.33

that, in order to achieve the 2020 production goal set by law, the national water consumption for shale gas development will peak in 2019 and that both Tarim and Jungger Basins' water supply will fail to meet the demand (Guo et al. 2016).

Some have studied local water stress in Chongqing to see how shale gas development would affect local water stress (You et al. 2015). Chongqing is a major city in the east Sichuan Basin, and has Fuling field, which is the State's first large-scale commercial production field. As Fig. 6.1 shows, Chongqing has a medium level of water stress. Moreover, the city's water resources quantity varies seasonally (Long 2016).

You et al. (2015) collected data from four local areas in the city where there are ongoing shale gas development activities, and compared the assumed water consumption for shale gas development with the other two figures, which are the quantity of available water resources and total water used respectively. The result (see Table 6.3) indicates that the water use for shale gas would take a small portion of the available water resources (1.54%–2.09%), but it would take a much bigger portion (19.69%–37.11%) of the total water use quantity. This suggests that shale gas development would probably be a challenge to local water supply, especially during the drought seasons (You et al. 2015).

In the Fuling field, Sinopec transported water through pipelines from the 20 km-away Wujiang River Industrial Park in order to avoid interrupting local people's water use (Sinopec 2014). It is unknown, however, whether the Fuling field's fracking water demand would in any way place any pressure on the Wujiang River Industrial Park's water availability.

6.3 The Rights to Acquire Water for Hydraulic Fracturing

6.3.1 Water Rights Overview

State Ownership The Water Law of 2002, as amended in 2016, (the "Water Law") is the principal law governing the development, utilization, conservation, protection, and administration of both surface and underground water resources within China (Water Law, Art. 2). Under Art. 3, the State owns water resources, and the State Council exercises such ownership on behalf of the State except in certain circumstances.[1] The state ownership of water resources is also provided by Art. 9 of the Constitution of 1982, as amended in 2004 (the "Constitution"), and Art. 46 of the Property Law of 2007 (the "Property Law").

Given the general state ownership rule, water right in China is in fact limited to the right to use water (Jia 2014a, b). One other than the State may only hold water rights in the *usufruct* nature, which are also referred to as "water use rights" (Wang 2011). The Property Law also provides that the right to abstract water is of the *usufruct* nature (Property Law, Art. 123).

The State has also adopted a water use planning and control regime including, among others, the water resources allocation plans (the "WRAPs"), the industry-specific quotas, and the region-based total quantity control (Water Law, Art. 45 & 47). Therefore, a shale gas developer, like other industrial water users in China, may only have the rights to acquire and use water subject to certain allocation plans, but cannot own the resource.

Water Rights Allocation There are three types of water administrations involved in the water rights allocation, which are the Ministry of Water Resources of China (the "MWR") in charge of nation-wide water resources administration, the river basin agencies (the "RBAs") authorized by the MWR to manage water resources in certain important rivers and lakes, and the water administrations of local governments (county-level or above) (the "local water administrations"). Figure 6.3 shows such institutional framework.

Water rights are basically allocated in a two-level system (see Fig. 6.4). At the first level, water rights or resources are allocated from the national river basins to multiple levels of administrative regions, such as the provinces, cities, and counties (Shen 2015). This is a process for the State to allocate water resources to local governments through certain plans, including, among others, the water resources allocation plans (the "WRAPs"), drought contingency plans (the "DCPs"), industry quotas, and water use plans (the "WUPs").

The trans-regional WRAPs, as well as the DCPs, are prepared by certain RBAs or local water administrations in negotiation with the local governments (Water Law, Art. 45). Once approved, RBAs or the local water administrations should also prepare the annual WRAPs accordingly. Provincial-level governments prepare the industry-specific water use quotas applicable in their respective regions (Water Law, Art. 47). In accordance with WRAPs and industry quotas, local governments (county-level or above) should further prepare the annual water use plans (the "WUPs") that cap the total amount of water consumed within their regions (Water Law, Art. 47).

---

[1] There are exceptions where the rural collective economic organizations (the "rural collectives") are involved. Under Art. 3 of the Water Law, the water in the ponds of rural collectives and that in the reservoirs, constructed and managed by rural collectives, shall be used by those organizations. This means that the rural collectives may utilize such water without obtaining an abstraction license for daily or agricultural purposes.

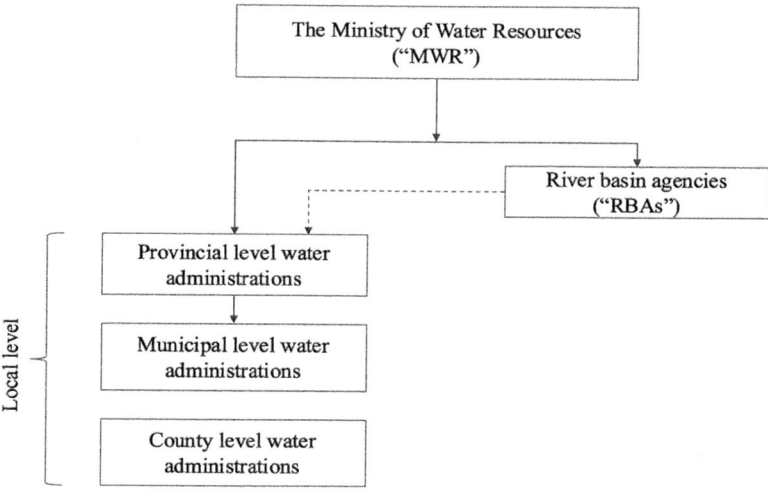

Fig. 6.3 Instiutional framework of China water resources administration (Reproduced from JICA 2006)

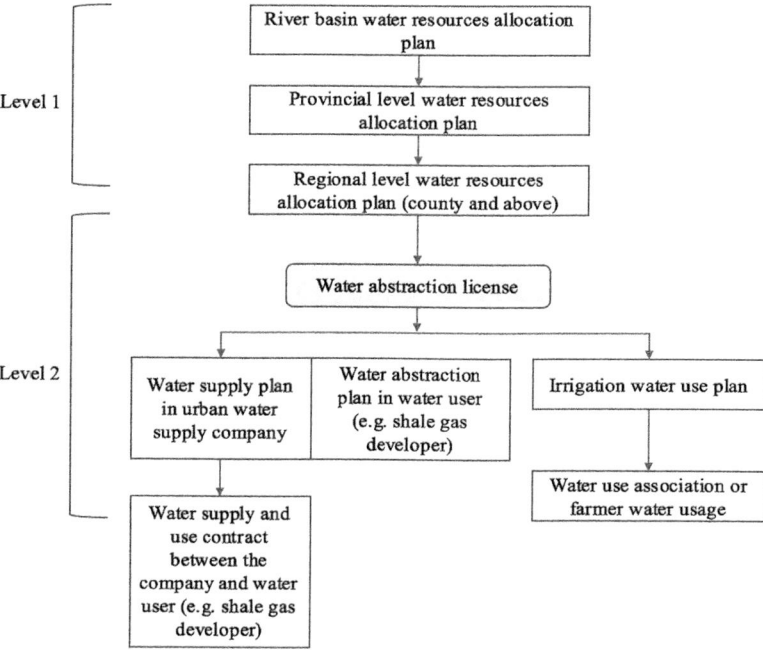

Fig. 6.4 Water resources allocation framework in China (Reproduced from Shen 2015)

At the second level, water rights are allocated from administrative regions to the actual water users through the water abstraction license system (JICA 2006). This allocation reflects the grant of water rights from the government to the consumers for industrial, agricultural, or residential use purposes. This is also where shale gas developers may acquire water for fracking (see Sect. 6.4).

6.4   Ways to Acquire Water for Fracking

Abstract with a License   In order to acquire enough water for fracking, a shale gas developer must apply for a water abstraction license (the "WAL") and pay the government water resources fee for the water abstracted (Water Law, Art. 7). A WAL, issued by the water administration, is an administrative license that allows the license holder to utilize the water resources under the general state ownership rule.

With a WAL, a shale gas developer may directly abstract water from rivers, lakes and underground water resources by using water abstraction engineering projects or facilities (Water Law, Art. 48). These water abstraction projects or facilities may include sluices, dams, canals, water pumps, water wells, and hydropower stations (SC Decree 460, Art. 2). For instance, at the Wei-202/204 blocks in the Weiyuan field, the developer constructed pipelines, impounding reservoirs and other auxiliary facilities to pump fracking water from a sluice, a reservoir, and two rivers nearby (CNPC 2015).

Under the law, if a shale gas developer wants a WAL, it must first apply for the water abstraction approval documents, then have the water projects or facilities -- either new or pre-existing – be reviewed by the approving authority. Once the projects or facilities have been approved, the authority will issue the license.

The application for a WAL must specify, among others, why, when, where, and how the developer would abstract water, as well as the abstraction quantity and the monthly consumption within the year (SC Decree 460, Art. 12). In addition, because construction activities are involved, the shale gas developer must also submit a Water Resources Assessment for Construction Project (the "WRACP") (SC Decree 460, Art. 11). This construction-related water resources assessment should, among others, study the water source for abstraction, justify the water use's rationality, and evaluate ecological and environmental impacts of water disposal (MWR Decree 15, Art. 6). The environmental impact assessment report (the "EIA Report") may rely on the WRACP's analysis and conclusions.

The content of a WAL specifies, among others, the term of license and the maximum quantity of water to be abstracted (SC Decree 460, Art. 24). The term is usually from 5 to 10 years, and is renewable upon government approval (SC Decree 460, Art. 25). Before the term expires, alterations made to certain contents of the WAL, such as the water source and purposes of water use, require a new application (MWR Decree 34, Art. 29). The maximum quantity of water abstraction specified on the WAL represents an average cap of water abstraction within each year (SC Decree 460, Art. 24).

It should be noted, however, that the maximum abstraction quantity printed on the WAL might be different from the volumes that one may actually abstract in a given year. It is the annual water abstraction plan (the "WAP") that governs how much water may be abstracted within a given year. The annual WAP applicable to a shale gas developer is prepared and issued by the WAL approving authority on a yearly basis, in which both the region's annual WAP and the developer's annual WAP proposal would be considered (SC Decree 460, Art. 40).

The shale gas developer should pay a water resources fee for the volumes that it actually abstracts (SC Decree 460, Art. 28). If the actual volumes exceed the limits set by the annual WAP or industry quota, the developer should pay for the exceeding volumes on a progressive basis (SC Decree 460, Art. 28).

Acquire from Suppliers   Besides abstracting water with a WAL, a shale gas developer may also acquire water from water suppliers by paying water bills under Art. 55 of the Water Law. According to a shale manager, some shale gas companies in the Sichuan Basin bought water from localities through negotiations (Marsters 2013). For instance, according to an EIA Report prepared for a shale gas well in the Sichuan Basin, Sinopec acquired water from nearby towns and transported it in tankers (CHALIECO 2016). Since the EIA Report does not refer to any WRACP which is required for obtaining a WAL, it can be inferred that Sinopec in the project doesn't need WALs for water access (SC Decree 460, Art. 11; MWR Decree 15, Art. 6). This means that Sinopec simply acquired water from local water suppliers by paying water bills.

This approach may lead to an issue when the water supplier provides water from a reservoir. According to The Supreme People's Court of China (the "SPC") (2004), if a reservoir is designed for purposes other than water supply, such as flood control or power generation, the user must obtain a WAL and pay the water resources fee. However, if the reservoir's management entity (water supplier) has already received a WAL and paid a water resources fee, and if the reservoir's designed functions have included water supply, then the user only needs to pay bills to the supplier (SPC 2004). In other words, in order to use the water stored in a reservoir, a WAL must be in place and held by either the water supplier or the user. Therefore, a reservoir's designed function decides how a shale gas developer may utilize the water stored therein, i.e. whether the developer should apply for a WAL.

Water Rights Exchange   A shale gas developer may acquire water resources from other water users through water rights transactions. Under the law, a water rights holder may transfer the water resources that it has saved to other entities upon government approval, so as to completely or partially transfer its right to abstract water to the other water users (SC Decree 460, Art. 27).

A shale gas developer may also purchase water rights on an exchange platform. The Interim Measures on Water Rights Exchange Administration (the "Interim Measures"), promulgated by the MWR in 2016, governs the transaction of water rights in China, and specifies how to transfer water rights between water users. Under the Interim Measures, a water user may transfer its water rights to other users

through either exchange platforms or direct transactions, but those cross-region or
large-scale deals must be done via exchange platforms (Interim Measures, Art. 7).
Because the fracking operations usually consume large volumes of water, relevant
water rights transfers should probably be made on an exchange platform. An
unsolved issue, however, is that the law has not set a specific threshold for a water
rights transaction to be regarded as large-scale.

The China Water Exchange (the "CWEX") is the exchange platform where shale
gas developers may seek to purchase the right to abstract water from transferors.
The CWEX, jointly set up by the MWR and Beijing Municipal Government, is a
national market where water rights can be transferred between different regions and
water users (CWEX 2016). Three forms of transfer are available in this exchange,
including the cross-region transaction, water abstraction right transaction, and irrigation right transaction (CWEX Trial Rules, Art. 5). The right to abstract water can
be transferred through either public listings or private negotiations (CWEX Trial
Rules, Art. 8).

Nevertheless, it is too early to tell how an exchange of water rights may help
shale gas developers acquire enough water. As of today, only ten deals have been
closed on the CWEX since its opening in mid-2016, and none of them is fracking-
related. Academics have proposed to establish such a transfer mechanism to better
allocate scarce water resources thus helping shale gas projects acquire enough water
(Hu and Xu 2013). Today, we have the CWEX, but there has yet to be a transfer for
fracking purposes to be closed.

6.5 Restrictions and Claims

Restriction and Withdrawal When the shale gas developer abstracts water for
hydraulic fracturing operations with a WAL, an annual WAP caps the maximum
quantity of water that may be abstracted in a given year (SC Decree 460, Art. 28).
In addition, the government may restrict or curtail a shale gas developer's WAL-
based water access under certain circumstances.

The government may only restrict or withdraw one's right to abstract water when
the underlying laws or conditions have materially changed, and must do so for public interest (ALL, Art. 8). In other words, the government may not restrict or withdraw a WAL unless the basis for the license's issuance has disappeared or materially
changed, such as when the underlying laws and regulations are modified or abol-
ished, or when the underlying conditions on which the WAL is issued have materially changed (ALL, Art. 8).

These material changes made to conditions are further specified in the Art. 41
and 44 of the State Council Decree No. 460. Under Art. 41, the approving authority
may restrict, with a prior written notice in timely manner, the shale gas developer's
annual quantity of water abstraction under certain circumstances. These circumstances include, among others, that the region's normal water supply is at stake due

to natural conditions, that the abstraction or return of water seriously harms the water's capabilities, ecology, and environment, and that the underground water abstraction causes geological hazards such as land subsidence (SC Decree 460, Art. 41). Moreover, if there is heavy drought, an urgent measure may be taken to curtail the water abstraction quantity without giving any prior written notice (SC Decree 460, Art. 41).

Under Art. 44, the approving authority may withdraw the WAL if the abstraction activities have consecutively ceased for 2 years. However, if the two-year cessation is due to *Force Majeure* or material technological transformation, the shale gas developer may retain the WAL upon authority's consent (SC Decree 460, Art. 44).

So far, we have not found any public information indicating that the shale gas developers' access to water with a WAL has been restricted or withdrawn for any reason. This is probably because the shale gas industry enjoys strong policy supports and the government monitoring is arguably less efficient (Zhang 2014).

Petition and Claim An issue is the approving authority's discretion in restricting and curtailing the annual quantity of water abstraction. Under the law, the restriction on volumes abstracted may be applied where any special circumstance so requires (SC Decree 460, Art. 41). This gives the approving authority a fairly big discretion in deciding whether there is an appropriate ground for taking restrictive measures.

In order to challenge this, a shale gas developer may file a petition for administrative reconsideration, requesting the approving authority to review its specific administrative decisions on restrictive measures and compensations (ARL, Art. 30). If the developer refuses to accept the decision of reconsideration, it may bring administrative proceedings to court (ARL, Art. 30). However, the developer may not further challenge a decision of reconsideration in front of the courthouse, if such decision of reconsideration on confirming water rights is made by provincial-level governments according to the state- or provincial-level governments' decisions on administrative division or land expropriation (ARL, Art. 30).

Another issue is about the administrative compensation. If the restriction or withdrawal of a WAL causes property damages to the developer, how can they receive administrative compensation? The answer depends on whether the restriction or withdrawal of the WAL is legal.

If the WAL is illegally restricted or withdrawn, the developer may request administrative compensation under the State Compensation Law of 1995, as amended in 2012 (the "SCL"). Under the law, the developer should first request compensation from the relevant authority, and then may decide to file a lawsuit against the authority if it disagrees with the authority's decision (SCL, Art. 14).

Even if the administrative decision is legal, the developer should be compensated for its property losses under Art. 8 of the Administrative License Law of 2003 (the "ALL"). This is because the license holder has paid water resources fee as a consideration for the grant of the water use rights. Nevertheless, a problem is that no law or regulation has further provided any specific approaches to calculating or requesting this administrative compensation in the scenario where a WAL is legally restricted or withdrawn (Tang and Liu 2008).

6.6 Public Participation

6.6.1 Hearing on the Application for a Water Abstraction License

The development of shale gas would possibly affect the local community's access to water resources, because fracking operations require large volumes of water while many shale gas areas in China face water scarcity issues (see Sect. 6.2.2). In order to better protect their own water interest, localities need to learn about the project's potential impacts on water availability and express their own needs to the government. Some have concluded that the State doesn't provide appropriate channel for the public to participate in the shale gas development's decision making process (Liu 2015). However, this is incorrect.

Under the law, one may request a hearing on the shale gas developer's application for a WAL, as long as it has material interest in the outcome (SC Decree 460, Art. 18). Such a hearing upon request gives the interested parties a chance to look into the project and to defend their own interest. However, the law has neither provided nor defined what material interest is. Instead, the law simply leaves relevant authorities the discretion in deciding whether the interest is material (ALL, Art. 47; Zhang et al. 2003). This brings uncertainties to the interested parties' actual chances of successfully requesting a hearing.

Even if one does not have any material interest recognized in the shale gas developer's WAL application, it may sit in the hearing when the developer's water acquisition is related to the public interest (SC Decree 460, Art. 18). An issue, however, lies where trade secrets are involved. Under the law, a hearing may not be available to the public where state secret, trade secret, or personal privacy is involved (MWR Decree 27, Art. 9). Many technologies applied in fracking operations, especially the components of fracking liquid, have been protected by oil companies as trade secrets (Luo 2015). Therefore, a hearing involving key information about fracking may not be open to the public who lack material interest therein.

6.6.2 Public Participation in the Environmental Impact Assessment

The Environmental Protection Law of 2014 (the "EPL") is a principal law governing the environmental protection in China. The EPL has recognized the public's rights to participate in and supervise the environmental protection (EPL, Art. 53). As to water acquisition for fracking, one mechanism for public participation is via the environmental impact assessment (the "EIA") process.

Under the EPL, in a construction project where an EIA Report is required, the developer must consult with the public under potential influence during the preparation of the EIA Report (EPL, Art. 56). A regulation requires all shale gas development projects to prepare an EIA Report (MEP Decree 33, Annex). Therefore, a shale gas developer must consult with the local community when it prepares the EIA Report. This enables the public to learn about the developer's water acquisition activities and to express their own opinions during the EIA process.

For example, in the Weiyuan 9# Drilling and Production Platform Project (the "Weiyuan 9# DPP"), the developer engaged a third party to prepare the EIA Report and to consult with local communities (CHFC 2016). According to the EIA Report, public participation was carried out through two announcements of project information and a questionnaire survey (CHFC 2016).

A potential issue, however, is the credibility of such an EIA Report. In the EIA Report for Weiyuan 9# DPP, some contradictions can be found which threaten the credibility of the report. According to the EIA Report, two project information announcements were published on the local government website for public comment, but there was no feedback (CHFC 2016). However, the local community which included 50 individual residents and three groups did express some concerns in the questionnaire survey (CHFC 2016). According to the questionnaire survey, 82% of the individuals believed that the water pollution might be a relatively obvious environmental issue brought by the project, and 84% held the opinion that the project would have comparatively great or serious impacts on local water quality (CHFC 2016). Although 96% of those that participated in the survey were farmers and even though over 82% believed that the project would have a big or serious impact on water quality, 76% believed that their daily life would be under little influence and 54% believed that farming would be under little influence (CHFC 2016).

6.6.3 Environmental Public Interest Litigation

One of the most controversial issues in revising the EPL is about who may file an environmental public interest litigation (Cai 2013). Since January 2015 when the revised EPL came into effect, certain types of non-governmental organizations (the "NGOs") have been able to file an environmental public interest litigation (the "EPIL") to challenge the activities that negatively affect the ecology or environment in China (CPL, Art. 55; EPL, Art. 58). This is another way by which the public may protect their own water access interest in the big wave of shale gas development.

There are two types of EPIL: civil and administrative. A civil EPIL is against the non-government parties, either individuals or organizations, who have harmed the ecology and environment (EPL, Art. 58). By contrast, an administrative EPIL is against the government bodies or officials whose illegal administrative activities or nonfeasance have caused harms to the State's and the society's interest in environment and resources conservation (SPP Judicial Interpretation No. 6, Art. 28).

Civil Environmental Public Interest Litigation Currently, the NGOs in China may only file a civil EPIL (Wang and Zhang 2016). In order to file a civil EPIL, an NGO must have been registered with the civil affairs department on a certain municipal level, and must have a record of environmental protection operations without breaking any law over a span of five consecutive years (EPL, Art. 58). These legal requirements are believed to have made 90% of the Chinese environmental NGOs unqualified to file an EPIL (Yang and Xie 2015).

Not many environmental NGOs seem to be interested in filing an EPIL. A survey shows that only 30% of the environmental NGOs are willing to take civil EPIL as the primary approach to environment defense (Jiang 2015). In fact, among the 700 environmental NGOs qualified for civil EPIL, only nine brought such cases to court in 2015 (Ye 2017). This is possibly because many environmental NGOs are industry associations that are associated with or backed by the government (Yang and Xie 2015).

Financial viability can also be a major hurdle for environmental NGOs in filing a civil EPIL (Boer and Whitehead 2016). NGOs are prohibited by law from obtaining any financial benefits in the civil EPIL proceedings (EPL, Art. 58). Without the right to sharing any financial outcome of the proceedings, the litigation costs and expenses would put a huge burden on the NGOs. In an EPIL case, the plaintiff-NGO requested the defendant to cover 400,000 RMB (62,000 USD) in attorney fees, but the court rejected this claim on the ground that the NGO had not actually paid such fees to attorneys (ACEF v. Zhenhua).

Administrative Environmental Public Interest Litigation Certain procuratorates may file an administrative EPIL against the government bodies or officials when their illegal administrative actions or nonfeasance have caused damages to the environment or resources conservation (SPP Judicial Interpretation No. 6, Art. 28). So far, the administrative EPIL has been available in 13 provincial-level administrative regions of China, under a pilot program adopted by the national legislature. (SPP Judicial Interpretation No. 6, Art. 29). In a recent case, a coal company kept its mining operations with an expired mining license, but the land resources administration failed to take active action to stop the illegal mining (SPP 2017). In effect, the procuratorate had suggested that the land administration should investigate the case, but the administration showed minimal concern (SPP 2017). Therefore, the procuratorate filed an administrative EPIL to challenge the land administration's nonfeasance (SPP 2017).

However, the public may not take this option, as neither governing law nor the pilot program provides that the public or any NGOs may file an administrative EPIL (Wang and Zhang 2016). In other words, the public whose interest is not directly affected is ineligible to either challenge the executive authority's illegal actions or urge its implementation of duties. This weakens the public participation in fracking, as the public may only challenge the developers, but not the authority who actually issues the mining or water abstraction licenses.

6.7 Concluding Comments

China is estimated to own the richest shale gas reserves in the world, but the development is difficult due to, among other issues, significant water shortages in most shale gas areas. This is because hydraulic fracturing as a necessary process in developing shale gas consumes huge volumes of water. For shale gas developers in China, figuring out how to acquire enough water for fracking can be quite a challenge.

Because of the state ownership rule, shale gas developers may only obtain the right to use water. A water abstraction license (the "WAL") is essential in using water resources. A developer may apply for a WAL in order to abstract water or it may acquire water from water engineering suppliers who have already been issued a WAL. Another way is to buy the water abstraction rights, through a water exchange, from a transferor who holds a valid WAL and has saved a certain amount of water thereunder. In practice, only the former two approaches have been applied, but it is still too early to evaluate how a water rights exchange may help shale gas developers acquire water.

The authority may curtail or withdraw a shale gas developer's water use rights under certain circumstances in the public's interest. Most of the provided circumstances are related to local water supply hardship or environmental harms. Under the law, the authority may also place curtailments when it decides that the case at present so requires, which in fact leaves the authority a huge amount of discretion. In response to curtailment or withdrawal decision, a developer may file a petition for administrative reconsideration first and then bring administrative litigation. Administrative compensation may be available when the illegal curtailment or the withdrawal has caused property damages to the developer. So far, we have not found any applicable case, which not only indicates the policy preference in favour of industry, but also the arguably less efficient government monitoring.

The public may seek to affect the shale gas developer's water acquisition activities in several ways. They may attend the hearings on the application for a WAL, but the material interest requirement and trade secrets may possibly keep many of the public out of the hearing room. They may also present opinions during the project's environmental impact assessment. Furthermore, amongst the hundreds of NGOs that are eligible to bring a civil environmental public interest lawsuit, fewer than ten have done so, although none is related to fracking water. Among the many reasons as to why this is the case include some NGOs' ties to the government and financial viability issues.

With the ambitious goal of shale gas development, the water-energy stress may increase and create tension or disputes among government bodies, developers and the public. A more robust regulatory practice is expected to be in place to tackle future challenges.

References

Boer D, Whitehead D (2016) Opinion: the future of public interest litigation in China. ChinaDialogue, 11 August. https://www.chinadialogue.net/article/show/single/en/9356-Opinion-The-future-of-public-interest-litigation-in-China. Accessed 7 Jan 2017

Cai S (2013) 从环境权到国家环境保护义务和环境公益诉讼 (From environmental rights to environmental protection obligations of the state and environmental public interest litigation). Modern Law Science 35(6):3–21

CHALIECO/China Aluminum International Engineering Corporation Limited (2016) 元坝 27–4 井钻井工程 环境影响报告书 (Environmental impact assessment report on the Yuanba 27–4 well drilling project). The People's Government of Cangxi. http://www.cncx.gov.cn/pub/PublicPage/GetArticAnnex.aspx?AnnexID=20160823171730629. Accessed 13 Oct 2016

CHFC/Sichuan Radiation Detection and Protection Institute of Nuclear Industry (2016). 威远 9#平台钻采工程环境影响报告书 (Weiyuan 9# drilling and production platform project environmental impact assessment report). http://www.schj.gov.cn/gsq/jshp/slgg/201609/W020160927366746757668.pdf. Accessed 13 Oct 2016

China National Petroleum Corporation (2015) 长宁、威远、昭通三个区块页岩气开发产能建设环境影响报告书 (Environmental impact assessment report on the development productivity construction projects in Changning, Weiyuan, and Zhaotong shale gas block). Sichuan Environmental Protection. http://www.schj.gov.cn/gsq/jshp/slgg/201512/W020151208609726725527.pdf. Accessed 14 Oct 2016

China Supreme People's Procuratorate (2017) 贵州省检察机关提起四起行政公益诉讼 (Guizhou province's procuratorates filed four administrative public interest litigations). Supreme People's Procuratorate, 20 January. http://www.spp.gov.cn/xwfbh/wsfbh/201701/t20170120_179228.shtml. Accessed 27 Jan 2017

China Water Exchange (2016) 中国水权交易所正式开业运营 (China water exchange officially goes into operation). http://cwex.org.cn/2016/news_0630/74.html. Accessed 6 Jan 2017

Dong D, Wang Y, Li X et al (2016) Breakthrough and prospect of shale gas exploration and development in China. Natural Gas Industry B 3(1):12–26

Gassert F, Reig F, Luo T et al (2013) A weighted aggregation of spatially distinct hydrological indicators. World Resources Institute. http://www.wri.org/sites/default/files/aqueduct_coutnry_rankings_010914.pdf. Accessed 12 Sept 2016

Guo M, Lu X, Nielsen CP et al (2016) Prospects for shale gas production in China: implications for water demand. Renewable and Sustainable Energy Review 66:742–750

Han X, Xiao G (2015) 页岩气开发中的水资源法律与政策研究 (Research on water resources law and policy in shale gas development). Journal of Xi'an Jiaotong University (Social Sciences) 35(4):120–124

Hu D, Xu S (2013) Opportunity, challenges and policy choices for China on the development of shale gas. Energy Policy 60:21–26

Hu D, Xu S (2015) 能-水关联及我国能源和水资源政策法律的完善 (On energy-water nexus and perfection of China's energy and water policy and law). Journal of Xi'an Jiaotong University (Social Sciences) 35(4):115–119

Japan International Cooperation Agency (2006) 中华人民共和国水权制度建设研究项目最终报告书第4卷附属报告书 (The People's Republic of China water rights system building research project final report vol 4, supplementary report). http://open_jicareport.jica.go.jp/pdf/11836843_01.pdf. Accessed 4 Oct 2016

Jia S (2014a) 中国水权制度改革 (The reform of China's water rights system). Lecture presented at the water governance research program, The University of Hong Kong, 24 July 2014

Jia S (2014b) Water rights in China. China Water Risk, 17 November. http://chinawaterrisk.org/interviews/water-rights-in-china/. Accessed 17 Jan 2017

Jiang Y (2015) 环境公益诉讼前路坎坷 (Environmental public interest litigation meets challenges ahead). Economic Daily, 31 March, Green Weekly section, 13

Krupnick A, Wang Z, Wang Y (2014) Environmental risks of shale gas development in China. Energy Policy 7

Lamb C (2016) How should business react to China's water crisis? World Economic Forum, 21 July https://www.weforum.org/agenda/2016/07/what-china-s-new-approach-to-water-means-for-business/. Accessed 13 Oct 2016

Liu C (2015) 页岩气勘探开发中信息公开制度之疏失与完善 (The defects and perfections of the information disclosure system in the exploration and exploitation of shale gas). Journal of Xiangtan University (Philosophy and Social Science) 39(3):48–51

Liu J, Yang W (2012) Water sustainability for China and beyond. Science 337:649–650

Long D (2016) 我市实行最严格水资源管理制度成绩突出 (The city accomplishes a strong record in the implementation of the strictest water resources administration). Chongqing Daily, 22 November, Front page, 1

Luo J (2014) 中国页岩气开发的水资源挑战有多大 (How big challenges does China shale gas development face). China Energy News, 27 January, Observation section, 5

Luo M (2015) 环境保护法中关于页岩气开发中水资源保护的立法现状与不足 (The status and defects of environmental protection law in the water resources protection related to shale gas development). Legality Vision no 4:259

Maddocks A, Reig P (2014) A table of 3 countries: water risks to global shale development. World Resources Institute, 5 September. http://www.wri.org/blog/2014/09/tale-3-countries-water-risks-global-shale-development. Accessed 17 Jan 2017

Marsters P (2013) A revolution on the horizon: the potential of shale gas development in China and its impacts on water resources. In: Turner J (ed) China environment series, vol 6. Wilson Center, Washington, DC, pp 35–47

Pi G, Dong X, Guo J (2015) The development situation analysis and outlook of the Chinese shale gas industry. Energy Procedia 75:2671–2676

Sandalow D, Wu J, Yang Q et al (2014) Meeting China's shale gas goals: working draft for public release. Columbia University, School of International and Public Affairs, Center on Global Energy Policy. http://energypolicy.columbia.edu/sites/default/files/energy/CGEP_American%20Gas%20to%20the%20Rescue%3F.pdf. Accessed 12 Sept 2016

Shen D (2015) Water rights development in China. The Hikone Ronso 403:62–79

Sinopec/China Petroleum and Chemical Corporation (2014) 中国石化页岩气开发环境社会治理报告 (Sinopec shale gas development environmental, social and governance report). http://wenku.baidu.com/link?url=H1LO74d4My_bvx7na1Z2AxbDzFvxLFauwSihpkVY-6eTROI3htThkWG5_FhIzOal12mkERjqA8TA8_mD7xUTWACHVjRYvtlF5iDGiJKE3y6m. Accessed 12 Sept 2016

Sinopec/China Petroleum and Chemical Corporation (2016) Sinopec corp 2015 communication on progress for sustainable development. http://english.sinopec.com/download_center/reports/2015/20160329/download/20160329001en.pdf. Accessed 13 Oct 2016

Tang X, Liu X (2008) 谈行政许可中的信赖利益保护 (On protection of reliance interests in administrative license). People's Judicature (11):100–102

Tollefson J (2013) China slow to tap shale-gas bonanza. Nature 494:294

US Energy Information Administration (2013) Technically recoverable shale oil and shale gas resources: An assessment of 137 shale formations in 41 countries outside the United States. https://www.eia.gov/analysis/studies/worldshalegas/archive/2013/pdf/fullreport_2013.pdf. Accessed 12 Sept 2016

US Energy Information Administration (2015) Technically recoverable shale oil and shale gas resources: China. https://www.eia.gov/analysis/studies/worldshalegas/pdf/China_2013.pdf. Accessed 12 Sept 2016

US Energy Information Administration (2016) United States remains largest producer of petroleum and natural gas hydrocarbons. http://www.eia.gov/todayinenergy/detail.php?id=26352. Accessed 12 Sept 2016

US Environmental Protection Agency (2015) Case study analysis of the impacts of water acquisition for hydraulic fracturing on local water availability. https://www.epa.gov/sites/production/

files/2015-07/documents/hf_water_acquisition_report_final_6-3-15_508_km.pdf. Accessed 4 Oct 2016

Wang H (2011) 论水权许可的私法效力 (On private law effect of water rights license). Journal of Comparative Law 113:43–54

Wang Z (2015) China's elusive shale gas boom. Paulson Institute. http://www.paulsoninstitute.org/wp-content/uploads/2017/01/PPEE_China-Shale-Gas_English_R.pdf. Accessed 20 Sept 2016

Wang Q, Wang L (2016) Comparative study and analysis of the development of shale gas between China and the USA. Int J Geosci 7:200–209

Wang X, Zhang C (2016) 完善我国环境公益诉讼制度的思考 (Thoughts of the improvement in China's environmental public interests litigation system). Academic Journal of Zhongzhou 231:49–54

Wang R, Ren H, Ouyang Z et al (2000) China water vision (executive summary): the eco-sphere of water, environment, life, economy and society. Paper presented at the 2nd world water forum's regional session on China water vision, IRC International Water and Sanitation Center, 17–22 March 2000

Wang C, Wang F, Du H et al (2014) Is China really ready for shale gas revolution—re-evaluating shale gas challenges. Environ Sci Policy 39:49–55

World Resources Institute (2014) Global shale gas development: water availability and business risks. https://www.wri.org/sites/default/files/wri14_report_shalegas.pdf. Accessed 12 Sept 2016

Yang W, Xie J (2015) 新环保法视角下环保NGO公益诉讼分析 (An analysis of pro-environment NGOs in public interest litigation from the perspective of the New Environmental Protection Law) Urban Insight no 2:69–75

Ye L (2017) 环境公益诉讼为何"遇冷"(Why does environmental public interest litigation "get cool"). Guangming Ribao, 24 January, science and education section, p 6

You S, Guo Q, Wu Y et al (2015) 页岩气开发中的水资源利用对策:以重庆地区为例 (Water utilization of shale gas development: a case of Chongqing). China Mining Magazine 24(S1):195–198

Zhang L (2014) 全国政协委员孙安民建议加强页岩气勘探开发的环境监管 (CPPCC member Sun Anmin suggests reinforce the environmental supervision over shale gas exploration and production). Ministry of Land and Resources of the People's Republic of China, 17 March. http://www.mlr.gov.cn/xwdt/jrxw/201403/t20140317_1307683.htm. Accessed 14 Jan 2017

Zhang D, Yang T (2015) 美国页岩气水力压裂开发对环境的影响 (Environmental impacts of hydraulic fracturing in shale gas development in the United States). Petroleum Exploration and Development 42(6):801–807

Zhang C, Li F, Li Y et al (eds) (2003) 中华人民共和国行政许可法释义 (interpretations of the administrative license law of the People's Republic of China). China Law Press, Beijing

Zhao X, Kang J, Lan B (2013) Focus on the development of shale gas in China–based on SWOT analysis. Renewable and Sustainable Energy Review 21:603–613

《中华人民共和国行政许可法》 [Administrative License Law of the People's Republic of China] (People's Republic of China) National People's Congress Standing Committee, Order No 7, 27 August 2003

《建设项目水资源论证管理办法》 [Administrative Measures on Water Resources Assessment for Construction Projects] (People's Republic of China) Ministry of Water Resources, Decree No 15, 24 March 2002

《中华环保联合会与德州晶华集团振华有限公司环境污染责任纠纷一审民事判决书》 [All-China Environmental Federation v Dezhou Jinghua Group Zhenhua Co Ltd], 山东省德州市中级人民法院 [Intermediate People's Court of Dezhou Municipality, Shandong Province, People's Republic of China], (2015) 德中环公民初字第1号 [(2015) Environmental Public Interest Civil Trial No 1], 18 July 2016

《建设项目环境影响评价分类管理目录》 [Classified Administration Catalogue of Environmental Impact Assessment for Construction Projects] (People's Republic of China) Ministry of Environmental Protection, Decree No 33, 9 April 2015

《中华人民共和国宪法》 [Constitution of the People's Republic of China]

《能源发展战略行动计划2014-2020》 [Energy Development Strategy and Action Plan for 2014-2020] (People's Republic of China) State Council General Office, Decree No 31, 7 June 2014

《中华人民共和国环境影响评价法》 [Environmental Impact assessment law of the People's Republic of China] (People's Republic of China) National People's Congress Standing Committee, Order No 77, 28 October 2002

《中华人民共和国环境保护法》 [Environmental Protection Law of the People's Republic of China] (People's Republic of China) National People's Congress Standing Committee, Order No 9, 24 April 2014

《水权交易管理暂行办法》 [Interim Measures on Water Rights Exchange Administration] (People's Republic of China) Ministry of Water Resources, Decree No 156, 19 April 2016

《人民检察院提起公益诉讼试点工作实施办法》 [Measures for the Implementation of the Pilot Program of Public Interest Litigations Filed by People's Procuratorate] (People's Republic of China) Supreme People's Procuratorate, Judicial Interpretation No 6, 24 December 2015

《水行政许可听证规定》 [Provisions on the Hearings for Water Abstraction Licensing] (People's Republic of China) Ministry of Water Resources, Decree No 27, 24 May 2006

《取水许可和水资源费征收管理条例》 [Regulations on the Administration of the Water Abstraction License and the Levy of Water Resources Fees] (People's Republic of China) State Council, Decree No 460, 21 February 2006

《页岩气发展规划2016-2020》 [Shale Gas Development Plan for 2016-2020] (People's Republic of China) National Energy Administration, Decree No 255, 14 September 2016

《页岩气产业政策》 [Shale Gas Industry Policy] (People's Republic of China) National Energy Administration, Notice No 5, 22 October 2013

《最高人民法院行政审判庭关于用水单位从水库取水应否缴纳水资源费问题的答复》 [Supreme People's Court Administrative Division's Response to the Question that Whether A Water Use Unit Shall Pay Water Resources Fees for Abstracting Water from A Reservoir] (People's Republic of China) Supreme People's Court, Administrative Case No 24, 25 April 2004

《天然气发展"十三五"规划》 [The 13th Five-Year Plan of Natural Gas Development] (People's Republic of China) National Development and Reform Commission, Decree No 2743, 24 December 2016

《中国水权交易所水权交易规则(试行)》 [Trial Rules on Water Rights Transactions at the China Water Exchanges] (People's Republic of China) China Water Exchange, 21 June 2016

《水法》 [Water Law of the People's Republic of China] (People's Republic of China) National People's Congress Standing Committee, Order No 74, 29 August 2002

Libin Zhang leads the Energy practice at Broad & Bright in Beijing, China. He has more than 20 years of experience in both private legal practice and in-house corporate experience. He started his legal career as associate with Paul, Weiss (New York and Beijing) and subsequently with White & Case (Shanghai). Prior to joining Broad & Bright, Zhang served as Head of Legal M&A, Asia and Australia at Siemens and supervised more than 20 M&A and JV projects in less than three years. Zhang is a senior member of the Energy Law Academy, China Law Society, and a member and distinguished lecturer for the Kay Bailey Hutchison Center for Energy, Law & Business at The University of Texas School of Law. He holds a J.D. from The University of Texas School of Law and a B.A. in economics from the University of International Business and Economics.

Sheng Shao is a consultant at Sunshine Law Firm's Beijing office, primarily working on power (renewables and nuclear) and oil and gas transactions as well as in commercial dispute resolution. He previously worked as an associate at Orchid Energy Group, an energy advisory firm in Texas, and interned at Environmental Defense Fund's Austin office. Sheng holds an LL.B. from Tianjin Normal University and an LL.M. from The University of Texas at Austin.

Fang Dong is an executive in the Legal and Risk Management Department of the Power Construction Corporation of China (POWERCHINA), a Fortune Global 500 company. Prior, she worked in an institute of water resources engineering exploration and design for 11 years as a major designer and project management planner. Dong also worked as a researcher of projects financed by Asian International Rivers Center (AIRC), where she took charge of writing the section of laws and regulations. She has written several important consultation reports for the China Europe Water Platform (CEWP), which was launched by Ministry of Water Resources of the People's Republic of China and the Presidency of the Council of the European Union. Dong's PhD focused in law and policy of the environment and natural resources, especially concerning water issues. She also holds a master's degree in water conservancy and hydropower engineering and a bachelor's degree in information and hydroelectric engineering.

Jiameng Zheng is a doctoral student at the LBJ School of Public Affairs at The University of Texas at Austin. Her research interests center on the economics of environment. The questions she seeks to study address the valuation of environment and natural resources, the evaluation of existing environment policies, and the health consequences of substances in water. Representative research projects include: (1) using hedonic method to evaluate homeowners' willingness-to-pay of improvement in water quality in Tampa Bay, Florida; (2) exploring the relationship between lithium in water and suicide rate in the United States; and (3) looking at the long-term consequences of early childhood exposure to lead from drinking water. Zheng holds a master's degree in public affairs from the Maryland School of Public Policy, University of Maryland, and a bachelor's degree in economics from Wuhan University, China.

Chapter 7
The Assessment and Acquisition of Water Resources for Shale Gas Development in the UK

Jenna Brown

Abstract Shale gas is conjectured to potentially improve the UK's security of natural gas supply's status by substituting up to half of natural imports by 2035. This paper explores the subsequent demands upon freshwater resources, the process of resource acquisition by operators and the prerequisite procedural of assessment. This is followed by a water management case study of Cuadrilla Resources, the leading shale gas operator in the UK before concluding comments.

Keywords United Kingdom · Shale gas · Water acquisition · Planning application

7.1 Introduction

Shale gas extraction has become an energy policy priority in the United Kingdom (UK) since 2012 with the government creating key economic drivers to encourage exploration (Cotton et al. 2014). The reasons to do so are primarily three-fold (Staddon et al. 2016). Geopolitically, to improve the natural gas security of supply status. Natural gas presently provides around 30% of the UK's electricity and 80% of heating (UKOOG 2017a, b; DECC 2016a, b); with native conventional gas supplies from the North Sea depleting at an average of 8% a year (Hardy 2015), the UK has become a net importer of natural gas, importing at least 50% of the 70–100 billion cubic meters (bcm) consumed annually since 2004 (DECC 2016a, b). Environmentally, natural gas is purported as a transition fuel towards a low-carbon economy in a bid to meet carbon reduction targets set out in the Climate Change Act (2008), displacing coal and supplementing renewable energy sources in the mitigation of climate change (Mackay and Stone 2013). Economically, the development of the apparent wealth of resources technically available could support 74,000 jobs and peak at £3.7 billion a year (IoD 2013).

J. Brown (✉)
University of the West of England, Bristol, UK
e-mail: jenna.brown@uwe.ac.uk

© Springer International Publishing AG 2020
R. M. Buono et al. (eds.), *Regulating Water Security in Unconventional Oil and Gas*, Water Security in a New World,
https://doi.org/10.1007/978-3-030-18342-4_7

However, support for shale gas is mixed: politically, only the Conservative Party (presently in power) are in favour, while a moratorium continues to be in place in Scotland and a precautionary approach has been adopted in Wales. This echoes public opinion – a 2017 poll revealed support was at its lowest in 5 years, with opposition to shale gas stemming from a combination of disapproval of continued dependence upon fossil fuels for energy and local environmental impact, including concerns for water resources (BEIS 2017).

As an island nation, one could be forgiven in assuming water resources were plentiful, however a high population density of 413 people/km^3 (ONS 2013) results in an internal freshwater availability of 2244 m^3 per capita, compared to 8846 m^3 per capita in the US (World Bank 2014). Furthermore, its global position straddling the mid-latitudes on the western seaboard of Europe places it in the path of the jet stream, generating a temperate climate subject to frequent pressure changes and consequently liable to extreme weather conditions from environmental drought to localised flooding. The protection of freshwater resources, both quality and quantity, is therefore paramount to ensure adequacy of supply for municipal, industrial and environmental needs.

Conscious of environmental concern and endeavouring to maintain a position as a *"world leader in well-regulated, safe and environmentally sound oil and gas extraction"* from over 50 years onshore experience (DECC 2015), the government continues to streamline policy and planning, developing strict requirements and safeguards through permitting and licencing, enforced by independent regulators. Campaigners warn that energy-sector leaders advising on policy can potentially result in conflicts of interest (Independent 2013).

This chapter describes the present status of shale gas development in the UK (dissolved governmental powers to Scotland and Wales and separate water availability data sets result in a focus upon England) and its projected contribution to energy security of supply status. It continues by describing present and projected water resource availability before systematically exploring the consequential freshwater demands, where resource availability is considered in the planning process and the process in which resources are assessed and acquired for shale gas development in England. This is followed by a water management case study of Cuadrilla Resources, the leading shale gas operator in the UK before concluding comments.

7.2   Shale Gas Development

7.2.1   Resource Estimates

The UK has potentially substantial volumes of prospective shale gas and shale oil resources within Carboniferous- and Jurassic-age shale formations distributed broadly in the northern, central and southern portions of the country. Three shale basins have been investigated by the British Geological Survey (BGS) in

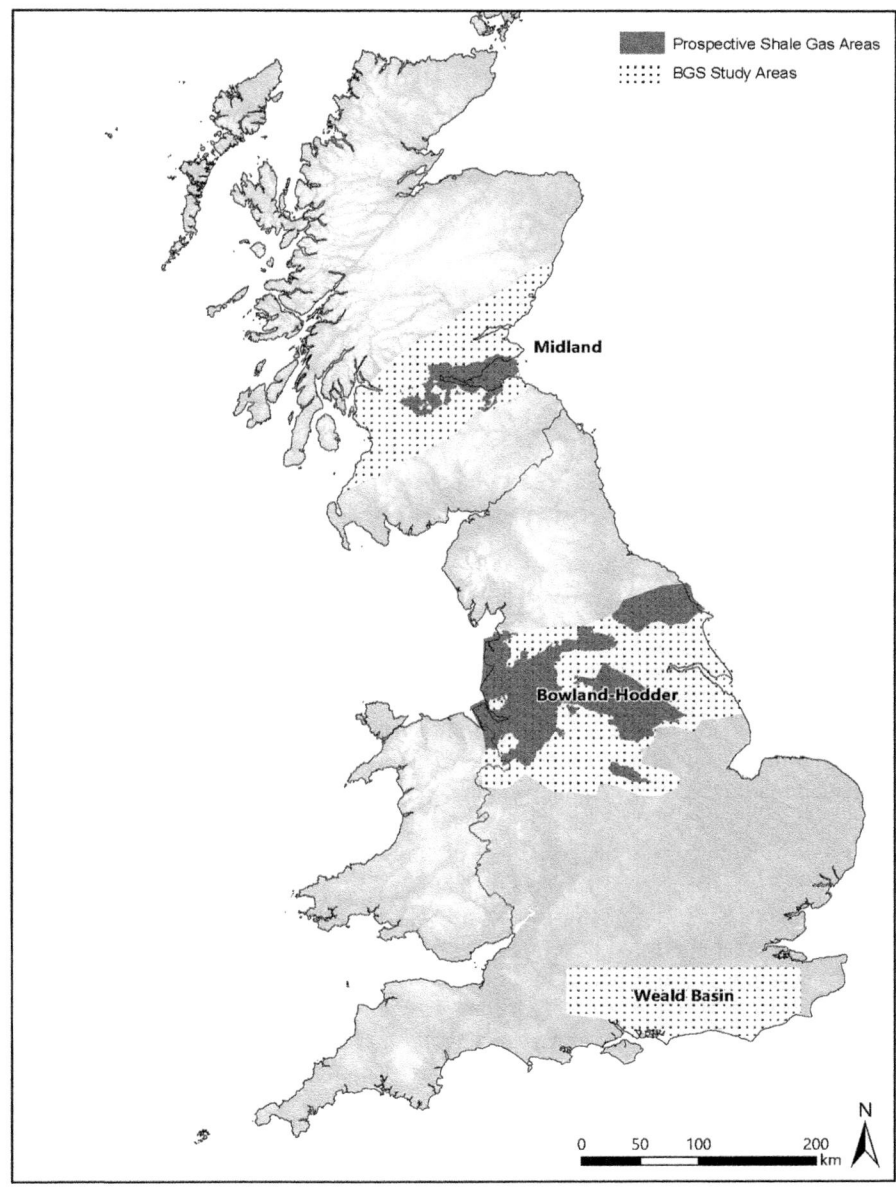

Fig. 7.1 Prospective shale gas basins in the UK. (British Geological Society 2015)

association with the Department of Energy & Climate Change (DECC), resulting in prospective areas being identified, shown in Fig. 7.1. The Bowland-Hodder in the north of England, estimated to contain a central estimate of 37,633 bcm (Andrews 2013); the Midland Valley in Scotland, contains a comparatively modest

2265 bcm (Monaghan 2014) and the Weald, located in the south-east containing shale oil (Andrews 2014). Estimates represent natural gas resources, i.e. 'gas in place' (GIP), an estimate of the amount believed to be physically contained in the source rock to 80% certainty (DECC 2013a, b). The volumes that could be recovered against predicted reserves are difficult to determine in the absence of exploratory drilling – figures of 10–20% of reserves could be recoverable although some instances show this figure could be lower (Hardy 2015) owing to the structural complexity of the UK's geology in comparison to the US (Harvey 2017). Industry estimates are more optimistic, with oil and gas company Cuadrilla suggesting the Bowland-Hodder could contain around 5.7 tcm GIP, with 15–20% of resources deemed technically recoverable (Whitton et al. 2017).

7.2.2   Licence to Develop

Although comparisons are often made between the U.S. and the UK in discussions of shale gas development, a key difference between the two countries makes exploitation far more attractive to the UK government. Unlike the US, the rights and ownership of hydrocarbon resources in Great Britain are not held by individual property owners but vested in the Crown by the Petroleum Act (1998) meaning the Treasury has a financial interest in the sector. A Petroleum Exploration and Development Licence (PEDL) typically measuring 10 km², allows an operating company to exclusively pursue oil and gas exploration activities, subject to necessary consents and planning permission. The opportunity to apply for licences is offered in rounds, managed by the Executive Agency at the Oil and Gas Authority (OGA) (Oil and Gas Authority 2015), a newly established department designed to simplify the planning process for deep drilling of shale gas, oil and geothermal energy sources with the 2014–2015 Infrastructure Bill. This appears to reflect the "significant development support" from the UK government (Whitton et al. 2017).

The 13th onshore round in 2006 saw several companies interested in applying for shale gas acreage in areas recognised for extraction potential, shown in light grey in Fig. 7.2. The 14th Onshore Oil and Gas licencing round was launched 28th July 2014 inviting companies to apply for PEDLs that included areas contained within the Strategic Environmental Assessment (area denoted by white squares), a document commissioned by DECC to identify and quantify potential environmental impacts, and identify measures for mitigation, with 57% of England and Wales included in the assessment (2013a, b). By 28th October 2014, there were 95 applications for 295 blocks. Following a process described by DECC as reviewing: *"applicants' competency, financial ability, environmental awareness and geotechnical analysis, and following the decision not to award licences to Scotland nor Wales"* (OGA 2015), the Oil and Gas Authority (OGA) formally offered 159 blocks to successful applicants in December 2015 (OGA 2015), shown in dark grey. The award of rights to a PEDL does not guarantee the presence of extractable oil or gas, that permission to extract shall be granted, nor that the PEDL has been purchased

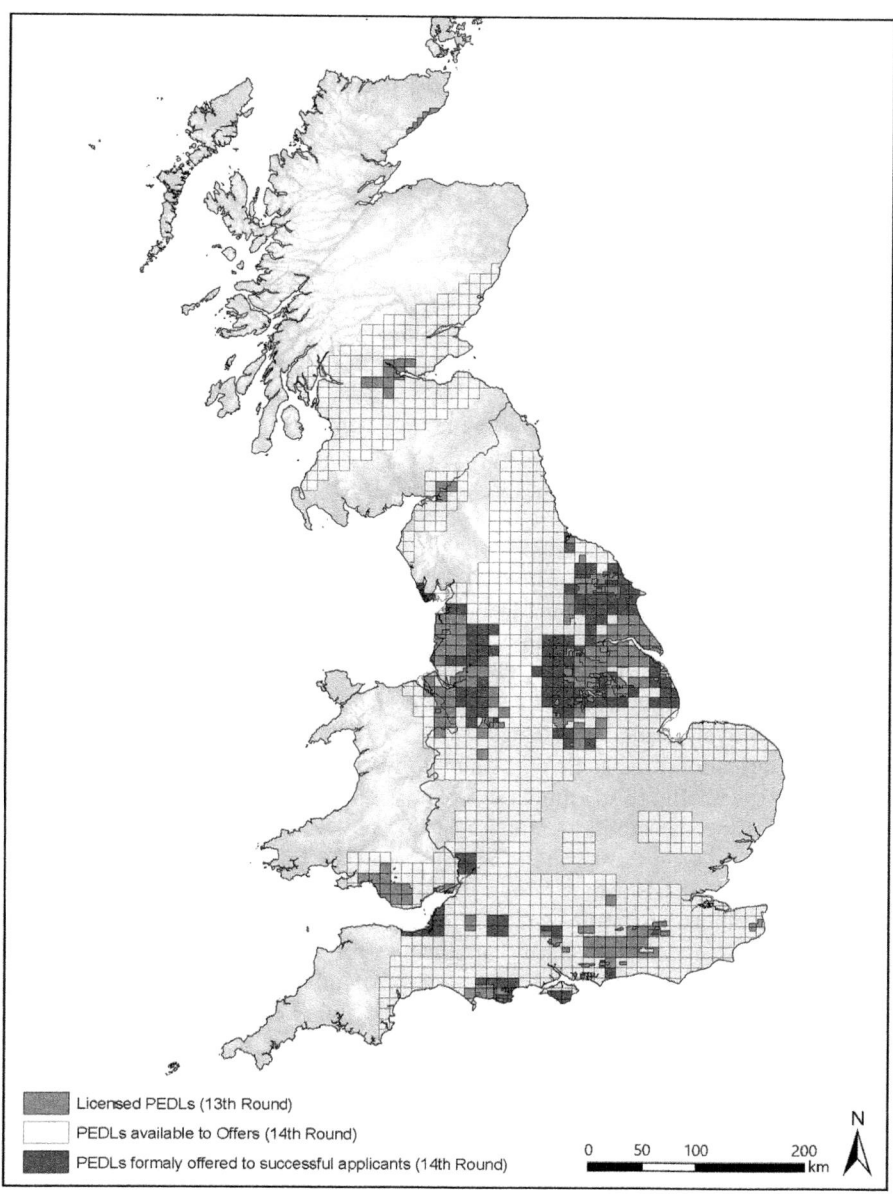

Fig. 7.2 Location of Petroleum Exploration and Development Licences (PEDLs) that are included in the Strategic Environmental Assessment (SEA), Licenced (13th round), formally offered to successful applicants (14th round) and available to offers (14th round). (Oil and Gas Authority 2015)

for shale gas development. Further, landowner and local planning and regulator permits are a prerequisite to exploratory drilling (Sect. 7.3). Should exploratory drilling prove encouraging, a re-application process to the former is required for commercial development of the site.

7.2.3 Development Status

Shale gas drilling in the UK remains at an exploratory phase. The slower progress than initially predicted is in part a result of the UK's geology. Unlike many North American shale regions, shale basins in the UK are generally not continuous structures but rather typically comprise a series of small fault-bounded sub-basins. Consequently, the structural complexity, coupled with the relatively small data base of onshore petroleum wells, makes resource assessments more difficult thus slowing the pace of shale exploration (EIA 2015a, b). In 2011 a moratorium followed investigations into two seismic tremors associated with drilling[1]. The events intensified industry scrutiny, by researchers and the general public. However, the release of updated resource estimates by the British Geological Society (Andrews 2013); industry-prepared investigations (de Pater and Baisch 2011) and a DECC commissioned independent report (Green et al. 2012) for seismic risks; and an influential *Royal Society and Royal Society for Engineering* report analysing the technical aspects of the environmental, health and safety risks associated with shale gas extraction (Royal Academy of Engineering 2012) led to new regulatory requirements in mitigation of seismic and water contamination risks. Additionally, financial support was offered to stimulate investment into the developing sector, with a political rhetoric of 'going all out for shale gas' declared by then Conservative Prime Minister David Cameron in January 2014 (UK Government 2014).

Drilling for the UK's first pilot well for horizontal hydraulic fracturing commenced on 17th August 2017 by Cuadrilla. The site has drawn national interest – in October 2016 Cuadrilla's planning application for exploration at two Lancashire sites was awarded by the Secretary of State for Communities and Local Government, Sajid Javid. This was a landmark decision, as it was previously rejected by Lancashire County Council, and led to anti-fracking campaigners seeking a judicial review. This too was dismissed by a High Court judge (Delebarre et al. 2017). It follows the first planning approval for fracking in the UK in five years (May 2015) granted to Third Energy, North Yorkshire. Despite continued local and national opposition, the government remains broadly supportive of shale gas development.

---

[1] May 2011. The first magnitude 2.3 ML shortly after Cuadrilla's Preese Hall well in the Bowland Shale was hydraulically fractured. The second, a magnitude of 1.5 ML occurred on 27th May 2011 following renewed hydraulic fracturing of the same well.

7.2.4 Development Scenarios

The UK's geology results in deeper shale gas reserve deposits than the US, an average 2387 m compared to 1317 m (Reig et al. 2014). This facilitates the development of 'pads' of wells, whereby up to 24 wells can be developed at a single site at interval depths (Department of Energy and Climate Change 2013a, b). The development of well pads maximises the efficiency of a site, ensuring maximum contact can be achieved between the fracturing fluid and the shale whilst minimising the surface footprint. The production of well pads is therefore considered as an economic advantage. Expected well pad designs vary significantly: DECC (2013a, b) consider 6–24 wells per pad with a surface footprint of 3 ha; UK Onshore Oil and Gas (UKOOG) (2017a, b) model a 2 ha surface footprint scenario of 10 well heads leading to 10 horizontal wells per pad with between seven and eleven pads per PEDL; whilst The Onshore Energy Services Group (2015) boast an expected 40 wells per pad.

How does this translate to gas production? United Kingdom Onshore Oil and Gas (UKOOG) forecast that the difference between natural gas supply and demand between 2017 and 2035 will be circa 850 billion cubic meters in that period, with import dependency increasing to 80% (UKOOG 2017a, b). Scenarios for the contribution shale gas could make to our energy security status also vary considerably. In their assessment of future energy scenarios, National Grid (2016) foresee four potential scenarios of which shale gas development only features in two. This is because they cite that the public perception of shale gas development does not align with the green ambition of society; the counter scenario 'Consumer Power' envisages that the support for conventional native gas is equalled for unconventional, resulting in native gas meeting 70% of supply. UKOOG (2017a, b) are confident in a growing sector – they predict a single well could produce over 1 million cubic meters (4 trillion cubic feet) of gas over a 20 year period (based on US production rates) and thus estimate that the production of 140 well pads by 2025 would reduce import dependency by circa 40%. Developing a further 260 pads (400 pads total) by 2035 would further reduce imports by 50% and a cumulative development of an additional 100 pads will maintain a 40% reduction in imports in 2050. Development scenarios therefore vary from shale gas making no contribution to our future energy security if public pressure to halt development continues, to halving of our import dependency.

7.3 Water Utilisation for Shale Gas Development

The relative abundance of water in combination with its thermal and solvent properties result in its utilisation throughout the drilling and slickwater hydraulic fracturing process. Water is utilised as a lubricant, protector and cooler to drilling apparatus during the drilling of the well. Gas-containing shale strata can be thin, therefore a combination of vertical drilling into the strata (in excess of 1 km) followed by horizontal or directional drilling through the target shale (up to 1500 m) achieves maximum contact. Multi-layered steel casing pipelines surrounded by cement are installed in the borehole to protect the adjoining geology, ground water reserves and improve the well's integrity, and reduce the risk of leakage of fluid or gas, as required by the Health and Safety Executive and Environment Agency to protect water reserves.

Fracturing fluid (slickwater) consists of a water base (>98%) with a proppant, typically sand (<1%) to hold fractures open in addition to a combination of chemical additives, a function of the local geology. Chemical additives can include: gels, to increase the fluid viscosity; acids, to help remove drilling mud near the wellbore; biocides, to prevent microbial growth; scale inhibitors to control precipitates, and surfactants to increase the injected fluid recovery. The combination and volumes of chemical additives are classed as commercially sensitive in the US with a voluntary code for their publication (Whitton et al. 2017). In the UK, all additives must be listed on the Registration, Evaluation, Authorisation and restriction of Chemicals (REACH) list as per the requirements of the Health and Safety Executive, and further declared to the Environment Agency for permission to drill to be considered as per The Infrastructure Act 2015 (44/1/d).

Hydraulic fracturing is performed in stages along the horizontal well. Perforations are made to the steel pipeline into which the fluid is injected at a pressure greater than the surrounding geology (Kargbo et al. 2010). It is an effective combination, maximizing the contact surface by generating a complex fracture network in the shale unit, enabling the gas to be mobilised (Johnson and Jonson, 2012). The quantities of fluids required for shale gas development are a function of geology, explained as: *"the lithology, petrophysical and geomechanical properties of the rock and hence, the pressure necessary to fracture the shale, the shale depth, length of laterals, the stimulation technique used, the number of fracturing stages per well and anticipated water returns"* (Rivard et al. 2014; Johnson and Jonson 2012). As a result, the water demand for shale gas varies geographically between countries, shale plays and well pads. A literature review of UK-applicable estimates continues to indicate a wide range of values from 10 to 34 Ml per well and subsequent flowback rates of between 27 and 75% (Table 7.1).

As a result of the significant uncertainty of how the shale gas industry may develop, how it may differ from existing international examples and in the absence of exploratory drilling in the UK, it is difficult to estimate the total quantity of water required for UK-based development. The only practical way of considering implications is to examine scenarios.

DECC's Strategic Environmental Assessment (2013a, b) considered two scenarios: high and low development, each with a range of assumptions and estimates of production wells between 180–2880, each requiring one re-fracture in their lifetime. The resulting water requirement was between 3.6 and 144 million m^3. The Chartered Institute of Water and Environmental Management's (CIWEM) (2016) estimated range is narrower, calculating that in order to meet 10% of the UK's natural gas demand (9 billion cubic meters) over a period of 20 years requires 25–33 million m^3

Table 7.1 Water demand estimates per shale gas well

Water demand per shale gas well			
Total water (Ml)	Incorporating flowback recycling (Ml)	Flowback (%)	Source
10–25		30–75	AEA (2012)
7.25–27			CIWEM (2016) and BGS (2013)
11–22			CIWEM (2016) and UKWIR (2013)
22–34	18–27	40%	Cuadrilla Preston New Rd (2014)
10–30[a]		27–75%[b]	UKOOG (2016a, b)

of water, or 1.2–1.6 million m³ per annum. To contextualise, CIWEM equates this to less than 0.1% of total abstraction for industry and agriculture when compared to annual licensed water abstraction in England and Wales. Therefore, overall the water demand for shale gas could be relatively modest, the challenges consequently come from managing the cumulative impact of development and sourcing the water that causes minimal disruption to existing environmental, municipal and industrial requirements.

7.4   Water Availability

7.4.1   Baseline Water Availability

In Europe, water availability is declining as a result of the reduced quantity permitted to be removed from the environment in response to sustainability initiatives under the Water Framework Directive (2006/60/EC). Adopted in 2000, it provides a common framework for water management and protection in Europe. Its primary aim is to enhance the status of aquatic systems, ensuring 'Good' qualitative and quantitative status of all water bodies (surface water and groundwater) by 2027. It is the responsibility of the Environmental Regulator to implement the WFD through managing water resources and assessing their resource availability.

Work completed by the EA through its 'Catchment Abstraction Management Strategy' (CAMS) considers water availability at catchment scale. This is important given the precipitation, topographical and geological variance in England and Wales are reflected in the balance between water availability and surface/groundwater abstraction. For instance, in the South and East of England, chalk overlying superficial deposits result in groundwater providing 70% of drinking water. Conversely, in the North and West of England the surface geology is relatively impermeable, resulting in water reserves being predominantly surface water based, quickly replenished by rainfall.

Even with tidal-influenced catchments considered in the CAMS assessments, most catchments in the country were found to already be at or near maximum sustainable abstraction. Therefore baseline water availability could pose significant challenges in sourcing both sustainable and adequate quantities of water in some parts of the country (Brown et al. 2014).

For instance, PEDLs offered in England and Wales to operators in the 13th round totalled an area of over 13,000 km² (located predominantly in the North with some in South Wales and the South East of England): 45% had a water resource availability less than 50% of the time to new abstraction permits, and 31% of land awarded is situated in areas with a water resource availability greater than 95% of the time (Table 7.2).

Following the 13th round, a strategic environmental assessment (SEA) (DECC 2013a, b) assessed the potential impact of further onshore oil and gas development in England and Wales. The area included in the SEA totalled over 85,000 km² of which 43% of the total area had water availability less than 50% of the time (Table 7.2).

Over 11,000 km² of land has been formally offered to operator applicants in the 14th round of PEDLs. The PEDLs are predominantly situated in tidally influenced locations or areas that receive high seasonal rainfall, reflected in the assessment of water resource availability with 41% of PEDLs in areas maintaining a water resource availability in excess of 95% of the time. However, the sensitivity of water quality requirements for hydraulic fracturing results in demands for freshwater sources from mains or groundwater/surface water. Therefore tidal sources may not be suitable and thus catchment water availability may not correlate with the above.

Water resource availability is geographically variant and may act as a limiting factor in shale gas development. Where water stressed catchments coincide with shale gas licence areas, operators need to be aware of the inherent risk that water may not be available in the future. The Tyndal Centre first warned of water scarcity for shale gas development in 2011 (Broderick et al. 2011), expressing that water resources in many parts of the UK were already under pressure. Although the North West of England is less water stressed than the South East in terms of the overall supply-demand balance, early engagement with the environmental regulator and/or

Table 7.2 Water resource availability (% available of time to new applicants) in purchased and offered PEDLs (Environment Agency 2013a)

		Offered 13th round	Strategic environmental assessment	Offered 14th round
Total area (km²)		13,486	85,200	11,224
% of time water available to new permits	<30%	29%	29%	20%
	>30%	13%	14%	17%
	>50%	15%	20%	14%
	>70%	7%	8%	8%
	>95%	35%	29%	41%

local water company will assist early into planning, and will ascertain available resource volumes and thus project viability.

7.4.2   Future Water Availability

The 'Future Flows project' (an 11 member ensemble including Defra, British Geological Society, and the Centre for Ecology and Hydrology) produced the first consistent assessment of the impact of climate change on river flows and groundwater levels across England, Wales and Scotland. They modelled hydrological and hydrogeological scenarios at 1 km scale for the period 1950–2098. Whilst it produced 11 equally probable scenarios, most indicated decreases in flows, especially in the south and east (up to −80%); in the west and north changes could be slighter. However, seasonal variation was far more diverse, ranged from +40% to −20% for winter water availability and + 20 to −80% in the summer months (Centre for Ecology and Hydrology 2012).

The Environment Agency (2012) warn that continuing with the current approach to water resource management will compromise the environment, the economy, or society – either singly or in combination. With commercial shale gas development unlikely until the 2020s, the consideration of future water availability is paramount as both a limiting factor to development and as a potential threat to local water security.

Per capita water availability is declining as a result of the demands of a growing population – by the 2030s, the population of England and Wales is expected to grow by an extra 9.6 million people with scenarios predicting demand increasing by as much as 35% – and a reduced quantity permitted to be removed from the environment as a result of climate change related impacts and sustainability initiatives required by the Water Framework Directive (Chartered Institute of Water and Environmental Management 2016). The European Water Framework Directive (WFD) came into force in 2000 and was transposed into UK law in 2003. Its purpose is to enhance the status, and prevent further deterioration, of the ecology of aquatic ecosystems and their associated wetlands and groundwater. Reducing unsustainable abstraction is key to meeting WFD targets.

At present, over 21,000 water abstraction licences exist; the EA suggest that should all licences issued be used to full capacity, ecological and environmental drought would occur. The Department for Environment, Farming and Rural Affairs (Defra) and the Welsh Government consulted on the regime during 2013–2016. Reforms to the system are expected by 2020 with a focus on linking abstraction volumes more closely to water availability. Further WFD 'sustainability reductions' includes a requirement for water companies to take less water out of natural resources. This could be up to 8% per five year Asset Planning Cycle. In combination, there will be a likely reduction in water availability. The Adaptation Sub-Committee of the Committee on Climate Change (ASC) warned in (2012) that in the interim there is: "*a risk that policy decisions that are sensitive to water avail-*

*ability (such as in energy [...]) do not take full account of future water availability
or the underlying requirement to support the natural environment [...] that lead to
unsustainable levels of abstraction in the future."*

7.5   Consideration of Water Resources in Planning Process

Water resources are considered at all stages in the planning process. Figure 7.3 cat-
egorises this by stages: Environmental Assessment (Sect. 7.5.1), Water Resource
Assessment and Acquisition (Sect. 7.5.2), Planning and Permits (Sect. 7.5.3) and
Notification of Intent to Proceed (Sect. 7.5.4) with water-specific points shown in
black.

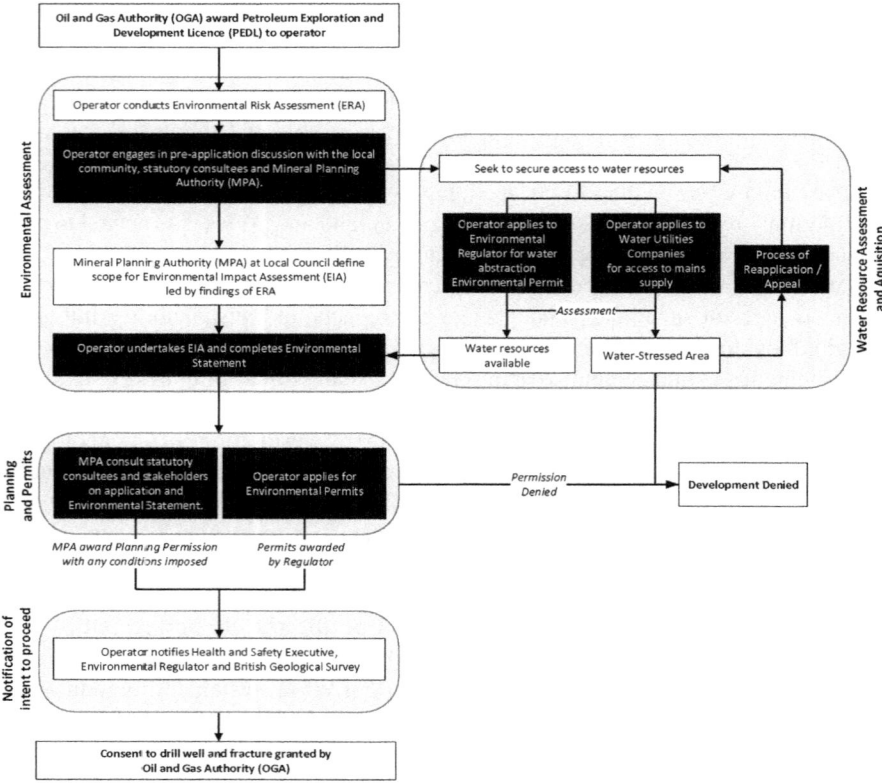

Fig. 7.3 The consideration of water resources during the planning process of shale gas
development

7.5.1   Environmental Assessment

Once issued a PEDL, an Environmental Risk Assessment (ERA) is required by the OGA as matter of good practice, with the acquisition of water resources listed as a risk factor (Environment Agency 2013b). Prior to a planning application to the Mineral Planning Authority (MPA) at the County Council, operators are encouraged to undertake a pre-application consultation with the MPA, the local community and key statutory consultees, which includes water and sewerage companies (Department for Communities and Local Government 2014). Pre-application consultation aims to address site issues and define arrangements for permits which include water withdrawal and wastewater management through the environmental regulator. The operator may also take the opportunity to enter discussions with the environmental regulator and/or water utilities companies to secure water supplies to the site. The MPA assess if an Environmental Impact Assessment (EIA) is required[2] and if so, define the scope and level of detail of an Environmental Impact Assessment, led by the findings of the ERA. The EIA draws together, in a systematic way, an assessment of the likely significant environmental effects of the proposed development. It includes baseline data in order to determine any environmental effects of development (baseline water availability and flood risk) and mitigation measures, estimated water demand per well and per site, the source/supplier of water utilised on site, and wastewater management plans. The content of the EIA is presented as an Environmental Statement (ES), the systematic content outlined in the Town and Country Planning Act (2017) which is submitted alongside the planning application to the MPA. The ES is shared with stakeholders, their comments – in support and in objection – are recorded by the MPA and considered in the planning decision.

7.5.2   Water Resource Assessment and Acquisition

In the UK, akin to hydrocarbon resources, rights to water do not follow land ownership. Shale gas operators have two options for sourcing water resources: (a) surface/ground water abstraction or (b) supplied by a water utilities company, delivered through main water pipelines or transported by tankers.

---

[2] Shale Gas developments feature in Schedule 2 of the Town and Country Planning (Environmental Impact Assessment) Regulations 2017 (England)) should the site exceed 0.5 ha and therefore it is responsibility of the MPA to screen for significant environmental impact, unless they expect to extract in excess of 500 tonnes/day in which case they feature in Schedule 1 and an EIA is mandatory.

7.5.2.1 Surface or Ground Water Abstraction

To abstract in excess of 20 m³ water per day from the environment (surface or groundwater), a water abstraction permit is required from the Environmental Regulator. The Environment Agency (EA) in England, Natural Resources Wales (NRA) and Scottish Environment Protection Agency (SEPA) regulate water abstraction, possessing the authority to issue a permit for a fixed time period (DECC 2013a, b) subject to resource assessment (see Sect. 7.5.2). Present water abstraction permits are known as water abstraction licenses. However, from 2020 all existing licenses will shift to a new system of water abstraction permits. Under the new permit regime, the volume of water that can be abstracted (surface water or groundwater) will depend on the source availability. A new charging system will mean that water taken from high risk/low resilience sources will cost more, water abstracted from abundant sources will be less expensive.

Water resources are assessed by the Catchment Abstraction Management Strategy (CAMS). The process informs future water availability, calculated at catchment scale in percentage units of time water is available to new abstraction permits. Reforms of abstraction management are presently underway to develop adaptive, sustainable abstraction. Should water resources be available for abstraction, the Environmental Regulator will award an Environmental Permit for abstraction.

7.5.2.2 Mains Supply

Operators retain the option to request access to mains water resources from a water utilities company, ensuring a consistent quality of water supply. Infrastructure to deliver water to site is the responsibility and at the expense of the operator (Chartered Institute of Water and Environmental Management 2016).

Under section 55(3) of the Water Industries Act 1991, a water company has a duty to provide non-domestic water subject to reasonable expenditure and the ability to meet existing/future supply obligations. It is the responsibility of the water company to ensure that water required for the activity fits within the conditions of their water availability and headroom, water resource plans and abstraction licences. However, should an operator's application be denied, they have the right to appeal to the ombudsman OFWAT for reconsideration.

Under April 2017 reforms, the market for water services has been opened up beyond the network of existing water companies (a programe called Open Water). It is hoped that new market entrants will allow businesses in England to strike better deals for their water services (OFWAT 2016). At time of writing, there are no working examples of such a transaction for the shale gas sector.

7.5.2.3 Recycled Water

It is widely recommended that the recycling and re-use of produced and flowback water should be utilised to offset freshwater usage in hydraulic fracturing (Chartered Institute of Water and Environmental Management 2016; Royal Academy of Engineering 2012). High concentrations of total dissolved solids (TDS), constituents and anions can interfere with the performance of the hydraulic fracturing fluid by producing scale or interfering with the chemical additives in the fluids (Wang et al. 2014). Therefore, following treatment, a new fracturing solution may be composed of treated produced water and fresh water with additional chemicals. However, one operator removed the option of treating and reusing flowback water as a result of uncertainty of flowback duration and the anticipated percentage of total flowback that intended to be recycled (EA 2017). As operations are presently in their infancy in the UK, it is hoped that experience and a subsequent growing data set shall inform future operations to enable the recycling of flowback and thus minimise water investment.

7.5.3 Planning and Permits

Companies seeking to undertake exploratory investigations and to subsequently test for and possibly extract shale gas must apply for planning permission from the MPA that has strategic planning authority (Town and Country Planning Act 1990). The application will include definitions of the minimum and maximum expected extent of operations (e.g. number of wells and duration), supported by the ES. When an application is received by the MPA, it will be assessed on its merits against the policies of the development plan and in light of advice from statutory consultees (which includes the Environmental Regulator). Water and sewerage companies were added as statutory consultees in the Infrastructure Act 2015 and implemented in The Town and Country Planning (Development Management Procedure (England) Order 2015. This enables water managers to be included in the consultation from an early stage.

It is considered best practice that the operator makes information of their plans and proposals available to the public through their website and at the local library, and the MPA shall also do so on their website.

If planning permission is approved, operators must serve a notice to the Environment Agency under Section 199 of the Water Resources Act 1991 to *"construct... a boring for the purposes of searching for or extracting minerals"*. Shale gas operators may also need to apply for environmental permits, with most falling under the Environmental Permitting Regulations 2010 (EPR), to allow drilling to take place. The Environmental Regulator outlines what permits apply to the operator to enable them to comply with regulations (CIWEM 2016a, b). Environmental permits may include:

- A groundwater activity – unless the regulator is satisfied there is no risk of inputs to groundwater
- A mining waste activity – likely to apply in all circumstances as a result of waste-water contents
- A radioactive substances activity – likely to apply in all circumstances as a result of wastewater contents
- A water discharge activity – if surface water run-off becomes polluted, for example due to flowback fluid breach
- A groundwater investigation consent – drilling and test pumping where there is the potential to abstract more than 20 m^3/day
- A water abstraction licence – if it is planned to abstract more than 20 m^3/day rather than purchasing water from a public water supply utility company
- A flood defence consent – if the proposed site is near a main river or a flood defence

7.5.4   Notification of Intent to Proceed

Following the award of planning permission and consent for environmental permits, the operator must inform the Health and Safety Executive (HSE) at least 21 days prior to drilling is planned. The HSE monitors shale gas operations from a well integrity and site safety perspective, under the Borehole Site and Operations Regulations 1995 and the Offshore Installations and Wells (Design and Construction, etc.) Regulations 1996. Together with the Environmental Regulator, the HSE must be satisfied that wells are designed, constructed and operated to standards that protect people and the environment. Further, the Coal Authority and British Geological Society are notified before the OGA finalises plans and awards permission to drill.

7.6   Water Management Case Study: Cuadrilla Resources

Cuadrilla Resources Ltd are the leading shale gas company in the UK. They own the rights to PEDLs in the North of England (Bowland-Hodder) containing shale gas and the South of England (Weald) for shale oil. They were the first company to begin exploring for shale gas which resulted in the seismic event associated with their operations leading to a national moratorium in 2011. They were awarded planning permission by the Secretary of State in October 2016 for their Preston New Road site and commenced drilling in August 2017.

The Preston Road site is located in an area not classed as water stressed by the EA with water availability 95% of the time. Water will be supplied to the site via a water supply pipe from the service mains of United Utilities Water Company. United Utilities are the water supplied for the North West, with their supply area within the Bowland-Hodder an area of prospective shale gas. United Utilities acknowledge that have a legal duty to offer a supply to any legally operating company that requests their services and are confident that they have available resources to supple the sector. In a statement they announced: *"Even under the most optimistic assumptions for shale gas production in the North West the water required for hydraulic fracturing would amount to less than 1% of our current water production. We are confident we can supply these volumes without compromising our ability to supply water to our existing customers."* (United Utilities 2015).

In the Environmental Statement for Preston New Road (Cuadrilla 2014), they expect to drill up to four exploration wells to a depth of c. 3500 m. The combined expected water demand is 24–35.5 Ml per well (19.8–28.8 Ml including re-cycling of flowback) (p. 602). Whilst they envisage re-using flowback where possible, it is more practical to re-cycle water during production.

7.7   Concluding Comments

Shale gas development in the UK is yet to get underway in earnest, nonetheless the production of shale gas in the UK is conjectured to potentially substitute for up to half of natural gas imports by 2035.

The subsequent water demand for shale gas both nationally and regionally does not presently cause concern to municipal supply providers and with water abstraction monitored by the environmental regulator, the government too are confident that both the natural and built environment are protected. However, densification scenarios of up to 40 wells per 2 ha licenced site in addition to potential significant increases in per well water consumption rates could stress local resources. This situation is further complicated by the recent Open Water changes in the retail water market which are already bringing in new 'middle man' operators. Intended to create opportunities for greater allocative efficiencies through competition, it is unclear how water-intensive sectors like shale gas may be affected. Although there are, at time of writing, no working examples in the industry, the concept of purchasing large, timely volumes of water from 'outside' of the watershed raises concerns for its sustainable management.

References

Andrews IJ (2013) The Carboniferous Bowland shale gas study: geology and resource estimation. British Geological Society for Department for Energy and Climate Change

Andrews IJ (2014) The Jurassic shales of the Weald Basin: geology and shale oil and shale gas resource estimation. British Geological Society for Department for Energy and Climate Change, London

Broderick J, Wood R, Gilbers P, Sharmina M, Anderson K, Footitt A, Glynn S, Nicholls F (2011) Shale gas: an updated assessment of environmental and climate change impacts, Tyndall Centre, University of Manchester, UK. Available online at: http://www.mace.manchester.

ac.uk/media/eps/schoolofmechanicalaerospaceandcivilengineering/newsandevents/news/research/pdfs2011/shale-gas-threat-report.pdf. Last accessed 02/11/17

Brown J, Staddon C, Hayes ET (2014) Water security and shale gas exploration in the UK. Environ Sci Water Secur 23(3):49–55. Available from: http://eprints.uwe.ac.uk/25125

Centre for Ecology and Hydrology (2012) Future flows and groundwater levels [online]. Available at: https://www.ceh.ac.uk/our-science/projects/future-flows-and-groundwater-levels. Accessed 02/11/17

Chartered Institute of Water and Environmental Management (2016) Shale gas and water: an independent review of shale gas extraction in the UK and the implications for the water environment. CIWEM, London. Available at: http://www.ciwem.org/wp-content/uploads/2016/02/Shale-Gas-and-Water-2016.pdf. Last accessed 02/11/17

Chartered Institute of Water and Environmental Management, (CIWEM) (2013) Shale gas and water. CIWEM, London

Cotton M, Rattle I, Van Alstine J (2014) Shale gas policy in the United Kingdom: an argumentative discourse analysis. Energy Policy [e-journal] 73(0):427–438

Cuadrilla Resources (2014) Environmental statement: Preston New Road. Available online at: https://cuadrillaresources.com/wp-content/uploads/2015/02/PNR-ES.pdf.   Last   accessed 02/11/17

de Pater CJ, Baisch S (2011) Geomechanical study of Bowland Shale Seismicity Cuadrilla

DECC (2015) Shale gas and oil policy statement by DECC and DCLG. Available online at: https://www.gov.uk/government/publications/shale-gas-and-oil-policy-statement-by-decc-and-dclg/shale-gas-and-oil-policy-statement-by-decc-and-dclg. Last accessed 02/11/17

Delebarre J, Ares E, Smith L (2017) Research briefings: shale gas and fracking. Available at: http://researchbriefings.files.parliament.uk/documents/SN06073/SN06073.pdf. Last accessed 02/11/07

Department for Business, Energy & Industrial Strategy (BEIS) (2017) Energy and climate change public attitudes tracker [online]. https://www.gov.uk/government/statistics/energy-and-climate-change-public-attitudes-tracker-wave-22. Last accessed 02/11/17

Department for Communities and Local Government (DCLG) (2014) Further changes to statutory consultee arrangements for the planning application process. DCLG, London

Department of Energy and Climate Change (DECC) (2013a) Resources vs reserves: what do estimates of shale gas mean? DECC, London

Department of Energy and Climate Change (DECC) (2013b) Strategic environmental assessment for further onshore oil and gas licensing – environmental report. DECC, London. Available at: https://www.gov.uk/government/uploads/system/uploads/attachment_data/file/273997/DECC_SEA_Environmental_Report.pdf. Last accessed 02/11/17

Department of Energy and Climate Change (DECC) (2016a) Digest of UK Energy Statistics (DUKES) Natural gas: Chapter 4 [on-line]. Available at: https://www.gov.uk/government/statistics/natural-gas-chapter-4-digest-of-united-kingdom-energy-statistics-dukes. Last accessed 02/11/17

Department of Energy and Climate Change (DECC) (2016b) Gas used by industry – supply and consumption of natural gas and colliery methane (DUKES 4.2) [online]. https://www.gov.uk/government/statistics/natural-gas-chapter-4-digest-of-united-kingdom-energy-statistics-dukes. Last accessed 02/11/17

Energy Information Administration (EIA) (2015a) Technically recoverable shale oil and shale gas resources: United Kingdom. U.S. Energy Information Administration

Energy Information Administration (EIA) (2015b) Annual energy outlook. EIA, Washington, DC

Environment Agency (2012) The case for change – current and future water availability Report number GEH01111BVEP-E-E

Environment Agency (2013a) Water resources reliability map data

Environment Agency (2013b) An environmental risk assessment for shale gas exploratory operations in England

Environment Agency (2017) Letter from third energy: waste management plan amendment. Available at: https://consult.environment-agency.gov.uk/onshore-oil-and-gas/third-energy-kirby-misperton-information-page/user_uploads/waste-management-plan-letter-9th-oct-2017.pdf. Accessed 02/11/07

Green CA, Styles P, Baptie BJ (2012) Preese Hall shale gas fracturing review & recommendations for induced seismic mitigation. Available online at https://www.gov.uk/government/uploads/system/uploads/attachment_data/file/48330/5055-preese-hall-shale-gas-fracturing-review-and-recomm.pdf. Last accessed 02/11/17

Hardy P (2015) Introduction and overview: the role of shale gas in securing our energy future

Independent (2013) Revealed: fracking industry bosses at heart of coalition. *Independent*, 13th July 2017 [online]. Available at: http://www.independent.co.uk/news/uk/politics/revealed-fracking-industry-bosses-at-heart-of-coalition-8707589.html. Last accessed 02/11/17

Institute of Directors (IoD) (2013) Getting shale gas working [online]. Available at: https://www.iod.com/Portals/0/Badges/PDF's/News%20and%20Campaigns/Infrastructure/Infrastructure%20for%20business%20getting%20shale%20gas%20working%20report.pdf?ver=2016-04-14-101231-553. Last accessed 02/11/17

Johnson EG, Jonson LA (2012) Hydraulic fracture water usage in northeast British Columbia: locations, volumes and trends. British Columbia Ministry of Energy and Mines

Kargbo DM, Wilhelm RG, Campbell DJ (2010) Natural gas plays in the Marcellus Shale: challenges and potential opportunities. Environ Sci Technol:5679–5684

Mackay DJC, Stone TJ (2013) Potential greenhouse gas emissions associated with shale gas extraction and use. Department for Energy and Climate Change, London

Monaghan AA (2014) The Carboniferous shales of the Midland Valley of Scotland: geology and resource estimation. British Geological Society for Department for Energy and Climate Change, London

National Grid (2016) Future energy scenarios 2016 [on-line]. https://www.google.co.uk/url?sa=t&rct=j&q=&esrc=s&source=web&cd=2&cad=rja&uact=8&ved=0ahUKEwiho92axazXAhVKiRoKHVhqBHQQFgguMAE&url=http%3A%2F%2Ffes.nationalgrid.com%2F&usg=AOvVaw0K4k6y64nW0Yfj5PomxvNM. Last accessed 13/2/18

Office for National Statistics (ONS) (2013) Population and migration [online]. Available at: http://webarchive.nationalarchives.gov.uk/20160106185934/http://www.ons.gov.uk/ons/guide-method/compendiums/compendium-of-uk-statistics/population-and-migration/index.html. Last accessed 02/11/17

OFWAT (2016) Supplementary guidance on whether non-household customers in England and Wales are eligible to switch their retailer [online]. Available at: https://www.ofwat.gov.uk/publication/supplementary-guidance-whether-non-household-customers-england-wales-eligible-switch-retailer/. Last accessed 13/2/18

Oil and Gas Authority (2015) New onshore oil and gas licences offered [on-line]. Available at: https://www.gov.uk/government/news/new-onshore-oil-and-gas-licences-offered. Accessed 12 Jan 2016

Reig P, Lou T, Proctor J (2014) Global shale gas development: water availability and business risks. World Resources Institute

Rivard C, Lavoie D, Lefebvre R, Séjourné S, Lamontagne C, Duchesne M (2014) An overview of Canadian shale gas production and environmental concerns. Int J Coal Geol [e-journal] 126:64–76. https://doi.org/10.1016/j.coal.207.12.004. Available through: ScienceDirect http://www.sciencedirect.com/science/article/pii/S0166516213002711

Royal Academy of Engineering (2012) Shale gas extraction in the UK: a review of hydraulic fracturing. RAE, London

Scottish Environment Protection Agency (2015) Moratorium on unconventional gas developments in Scotland [on-line]. Available at: https://www.sepa.org.uk/environment/energy/non-renewable/shale-gas-and-coal-bed-methane/. Accessed 2 Feb 2016

Staddon C, Brown J, Hayes E (2016) Potential environmental impacts of fracking in the UK. Geography 101(Part 2):60–69

The Onshore Energy Services Group (2015) Multi-well pad drilling [online]. Available at: http://oesg.org.uk/wp-content/uploads/2015/03/FAB-Facts-06-Multi-well-pad-drilling.pdf

UK Government (2014) Millions of pounds in business rates will be handed to councils in England which give the green-light to shale gas developments [on-line]. Available at: https://www.gov.uk/government/news/local-councils-to-receive-millions-in-business-rates-from-shale-gas-developments. Accessed 15/01/2014

United Kingdom Onshore Oil and Gas (UKOOG) (2016a) Water and soil. Available at: http://www.ukoog.org.uk/regulation/water-and-soil. Last accessed 02/11/17

United Kingdom Onshore Oil and Gas (UKOOG) (2016b) Drilling and the hydraulic fracturing (Fracking) process: 4. Production. Available at: http://www.ukoog.org.uk/onshore-extraction/drilling-process. Last accessed 02/11/17

United Kingdom Onshore Oil and Gas (UKOOG) (2017a) Developing shale gas and maintaining the beauty of the British countryside. UKOOG, London

United Kingdom Onshore Oil and Gas (UKOOG) (2017b) Natural gas uses [online]. http://www.ukoog.org.uk/onshore-extraction/uses. Last accessed 02/11/17

United Utilities (2015) Shale gas statement – October 2015 [on-line]. Available at: http://corporate.unitedutilities.com/documents/uu-shale-gas-statement.pdf. Accessed 14 June 2017

Wang Q, Chen X, Jha AN, Rogers H (2014) Natural gas from shale formation – the evolution, evidences and challenges of shale gas revolution in United States. Renew Sustain Energy Rev [e-journal] 30:1–28. https://doi.org/10.1016/j.rser.207.08.065. Available through: ScienceDirect http://www.sciencedirect.com/science/article/pii/S1364032113006059

WaterUK (2013) Shale gas: water UK to work with UKOOG [on-line]. Available at: http://www.water.org.uk/shale-gas-water-uk-work-ukoog

WaterUK (2014) Water UK response to the DCLG Consultation on statutory consultee arrangements in the planning process.

Whitton J, Brasier K, Charnley-Parry I, Cotton M (2017) Shale gas governance in the United Kingdom and the United States: opportunities for public participation and the implications for social justice. Energy Res Soc Sci 26:11–22

World Bank (2014) Renewable internal freshwater resources per capita [online]. Available at: https://data.worldbank.org/indicator/ER.H2O.INTR.PC. Last accessed 02/11/17

Jenna Brown is a PhD Researcher within the Faculty of Environment and Technology at the University of the West of England, UK. Here, she is a member of the International Water Security Network, a research group funded by Lloyd's Register Foundation. Her research explores the water-energy nexus of shale gas, specifically water stress mitigation policy and practice. She has been following the development of shale gas environmental policy in the UK for the past seven years, the subject of her MSc dissertation in 2013.

Chapter 8
"94% of the Water Flows into the Sea": Environmental Discourse and the Access to Water for Unconventional Oil and Gas Activities in Neuquén, Argentina

Joaquín Bernáldez and Rocío Juliana Herrera

Abstract The province of Neuquén in Argentina has public domain over one of the largest unconventional oil and gas reservoirs in the world. In 2012, with the promulgation of the Decree 1483/12, the provincial government made the first law for the exploration and exploitation of unconventional oil and gas in the country. In order to "prevent, mitigate and minimize environmental impacts" of hydraulic fracturing the Decree prohibits the use of underground waters for the exploration and exploitation of unconventional oil and gas. However, it provides for the use of surface water for these activities and argues that 94% of the volume of the main rivers in the region flows into the sea without being used. Through a political ecology perspective and using critical discourse analysis, the chapter intends to make sense of how the state shapes socio-natural relations and seeks to manufacture consent for hydraulic fracturing. The analysis shows that government and industry actors develop an environmental discourse centered on two main arguments: the need to exploit natural resources in order to reinforce development and the possibility of exploiting natural resources with environmental protection. This discourse thus aims to legitimize the exploitation of unconventional hydrocarbons. In particular, government and industry attempt to secure access to water for unconventional oil and gas activities by unfolding a sense of excess availability of surface water, which tend to ignore alternative ideas and usages of water.

Joaquín Bernáldez and Rocío Juliana Herrera equally contributed to this chapter.

J. Bernáldez (✉)
International Center for Development and Decent Work (ICDD), University of Kassel, Kassel, Germany
e-mail: joaquinbernaldez@icdd.uni-kassel.de

R. J. Herrera
Institute of Regional Science (IfR), Karlsruhe Institute of Technology (KIT), Karlsruhe, Germany

© Springer International Publishing AG 2020
R. M. Buono et al. (eds.), *Regulating Water Security in Unconventional Oil and Gas*, Water Security in a New World,
https://doi.org/10.1007/978-3-030-18342-4_8

Keywords Political ecology · Extractivism · Hydraulic fracturing · Water ·
Neuquén · Argentina

8.1 Introduction

In July 2011 Jorge Sapag, the governor of the province of Neuquén in Argentina,
inaugurated "the first multi-fractured horizontal well aiming at shale gas in Latin
America" (Instituto Argentino del Petróleo y del Gas 2011). By then, the potential
of Neuquén with regards to shale and tight gas was already a matter of public knowledge. A few months later it was announced that Neuquén also had important shale
oil resources (La Nación 2011). The government and companies claimed that the
unconventional hydrocarbons reservoirs localized in the Neuquén Basin (Fig. 8.1)
is one of the largest in the world (Bertinat et al. 2014).

Tight gas, shale gas and shale oil, among others, are generally labeled as *unconventional hydrocarbons* because they are deposited in sands and rocks with low
porosity and permeability. Unlike conventional hydrocarbons the extraction of
unconventional oil and gas requires a combination of vertical and horizontal techniques generally known as hydraulic fracturing. Hydraulic fracturing, also called
fracking, involves the injection of large amounts of water mixed with chemicals and
corrosive elements into the subsoil. The high pressure under which this fluid is
injected causes fractures increasing the permeability of the rock and enabling the
release of hydrocarbons to the surface (see Staddon et al. 2016 for more details on
the process).

In different parts of the world fracking has been controversial due to its actual
and potential environmental and health consequences. In some countries the method
has been prohibited or temporally suspended (Bertinat et al. 2014; Svampa and
Viale 2014). In Argentina the opposition against fracking has led to its banning in
more than 50 cities and towns. Recently, Entre Ríos became the first province to
prohibit unconventional methods of hydrocarbons extraction, including fracking
(Página 12 2017). Indigenous communities in Neuquén, who for many years have
claimed territorial rights and denounced environmental contamination from oil and
gas extraction, now also struggle against problems deriving from unconventional
exploitation. Further, organizations such as citizens' assemblies, trade unions,
NGOs, among others, have declared their opposition to fracking and carried out different kinds of actions and protests. The provincial government has argued that the
exploitation of unconventional hydrocarbons will create jobs and promote regional
development while minimizing the risk of environmental contamination.
Furthermore, the possibility of alternatives to the exploitation of unconventional oil
and gas has been denied while the environmentalist discourse has been portrayed as
fundamentalist and extreme and thus delegitimized (Svampa and Viale 2014).
Within this context, unconventional hydrocarbons have become a main driver of
social conflict in the region.

Fig. 8.1 Neuquén basin and other prospective unconventional basins in Argentina (Callum Foster, UWE, Bristol Cartography, with information from U.S. Energy Information Administration 2013)

The chapter intends to contribute to the knowledge on how the state[1] and capital seek to manufacture consent for unconventional hydrocarbons and, in doing so, shape socio-natural relations. It is argued that, through a discourse based on the necessity of natural resource exploitation with strict environmental control, government actors intend to sustain the development of unconventional hydrocarbons. This is reflected and reinforced by the discourses of oil industry actors. In doing so, they both unfold specific notions of the environment and water that exclude alternative ideas of the environment and uses of water present in civil society (for example, indigenous communities, citizens' assemblies and NGOs). Thus, government and oil industry actors do not problematize understandings of water but present their perception as the only legitimate one.

The outline proceeds as follows. First, an overview of the analytical and methodological framework guiding the analysis is provided. Second, the chapter offers a brief outline of the background of the emergence of unconventional hydrocarbons in Neuquén, Argentina. In the third part, the government and industry's discourses and meanings of unconventional hydrocarbons, the environment and water are presented. Finally, the chapter discusses and summarizes the findings and opens possible lines for further investigation.

8.2   Political Ecology and Critical Discourse Analysis

Two decades ago, Fernando Coronil (1997) argued that mainstream social theory was unable to explain the role of the periphery in the configuration of the world system because it neglected nature. Somehow, this observation remains valid today considering that the increasing attention paid to environmental issues in social sciences did not always involve a thorough theoretical engagement with the concept of nature. This has led much social theory to reinforce underlying assumptions about nature as external and opposed to society.

In an attempt to overcome this problem in our analysis, we draw on insights from a political ecology perspective. Political ecology is understood as a broad field of knowledge and action transcending scientific research (Leff 2006). In its academic perspective political ecology has been nurtured by different traditions, i.e. the Anglophone, the French and the Latin American, each of which has its own trajectories, styles and identities (Martín and Larsimont 2016). Moreover, a long-standing interaction between different disciplines has contributed to shape the field (Alimonda 2011). In this way, themes in political ecology are very diverse and include different theoretical and methodological approaches. All in all, political ecology attempts to go beyond nature-society dualisms and engage with conceptualizing their mutual dependence, i.e. the material production of society and the social production of nature.

---

[1] We understand the State as a social relation. For a more detailed definition on the State see Oszlak (2011) and Brand (2011).

Questions around the extraction of minerals and hydrocarbons are progressively more important in political ecology, although many questions and issues on this topic are still pending (Bebbington 2015). One of the most interesting contributions on the topic has to do with the synergy between resource-making and state power projects, or the so-called "resource-state nexus" (Bridge 2014). In reference to this issue, political ecologies of extraction highlights how state power can be rooted in underground resources, particularly oil. In oil countries state power rests upon the state's function as the administrator of underground resources (Coronil 1997). Yet, disputes over access to rents may generate political fragmentation and spaces of governance beyond the state, as was demonstrated by Michael Watts (2004) in the Nigerian context. Moreover, as owners and/or trustee of the subsoil, states provide the legal conditions for the extraction and commodification of minerals, oil and gas. Especially, when these resources represent an important source of revenues governments tend to legitimize extractive activities by asserting the necessity of exploiting natural riches for the sake of growth and the nation (Bebbington and Bury 2013).

A group of authors, although not necessarily speaking from a declared political ecology standpoint, have recently entered into discussions about extraction and the so-called "extractivism" (Gudynas 2010; Svampa 2011; Machado Aráoz 2012; Seoane et al. 2013). The concept not only includes traditional extractive activities but also other "activities which remove great quantities of natural resources that are not then processed (or are done so in a limited fashion) and that leave a country as exports" (Gudynas 2010: 1), such as agribusiness and minerals. Extractivism, then, can be said to be as old as colonialism. However, in the present it is an expression of the expansion of neoliberal capitalist economy and is driven by both conservative and progressive governments (Machado Aráoz 2012). This has led to questions about the role of the state in promoting capital accumulation and redistribution in different contexts. Although arguments differ among scholars, they share the notion that the state has an important role in assuring and legitimizing the expansion and intensification of natural resource extraction.

Water-related issues have been addressed extensively in political ecology. Especially since, in the late 1990s, Raymond Bryant advocated for the abandonment of "land-centrism" (Bryant 1998: 89) and the consideration of other issues related to water, air and cities (Köhler 2005: 28–29; Bryant und Bailey 2007: 192–193). An essential contribution and a significant feature of political ecologies of water is its specific understanding of the nature of water. Whether as the chemical formula H_2O, lakes, rivers or as drinking water, water is perceived, interpreted, defined and designed within the framework of socio-cultural, political and economic processes. Through these processes water is attributed special uses and characteristics. This means that, despite its indisputable materiality, water is socially constructed in a non-final process of constitution and transformation (Swyngedow 2004, 2009; Linton 2010; Bakker 2012). However, water is not just a passive object of social processes. Water has an impact on society, by either changing or consolidating ongoing social processes or generating new ones (Budds und Linton 2014). In this way, water reflects the intertwined relation between society and nature, the struggles over its meaning and the power constellations of society.

Out of this form of understanding water, a key concept emerged within the political ecology of water: the *hydrosocial cycle*. In contrast to the concept of hydrological cycle, the hydrosocial cycle focuses on the role of society (institutions, power relations, patterns of perception and interpretation) in water-related issues. It goes beyond the consideration of evapotranspiration, precipitation and runoff processes as well as physical, chemical and biological processes in the water (Linton 2010; Budds und Linton 2014; Wilhelm 1997), and includes social processes (for example the development of institutions and legislation regulating water uses or the construction of water-related infrastructure such as dams to cover energy needs). The underlying assumption is that water involves not only hydrological or hydraulic aspects but also social, political, economic and cultural aspects. The materiality of water as well as its construct character are internally connected, produced, reproduced and transformed in an iterative, unfinished process (Swyngedow 2004; Linton 2010; Budds und Linton 2014).

The latter understandings of water are based on a critique of the nature/society dualism characteristic of modernity. In this context, the idea of "hybrids" becomes essential. Hybridity means that water neither refers only to nature nor society, but both. Nature and society are not seen as two separate units. Rather, they are a kind of hybrid entity that arises, reproduces and changes in the same process of interaction. In this sense, water is understood to be a hybrid: water is both material and socially constructed, nature and society, thing and process (Swyngedow 2004; Linton 2010; Bakker 2012; Latour 2012; Budds and Linton 2014).

By influencing social processes, water also has a certain agency and is therefore both a subject and an object in social processes. Water is designed, assigned meanings and used as a power instrument, and in turn influences social processes, generates new power constellations and technologies (Swyngedow 2004; Staddon 2009).

In order to grasp the meanings of water and its social effects in the context of unconventional hydrocarbons development we employ different tools inspired by a critical discourse analysis approach. Critical discourse analysis studies how written and spoken language works within sociocultural practices (Fairclough 1995). In doing so, it draws attention to issues of power and ideology. Thus, the study of discourse may throw light on the "mechanisms at work in constructing and maintaining subjectivities within particular social contexts" (Joseph and Roberts 2004: 4). Indeed, discourses do structure socio-natural relations, but this does not rule out the fact that "the production of meaning is itself constrained by emergent, non-semiotic features of social structure" (Fairclough et al. 2004: 27). In this sense, socio-natural reality cannot be reduced to discourse, although discourses are a very important part of it. Moreover, "[a] focus on environmental discourse [...] does not imply the belief that environmental knowledge is unreal or imagined, but instead indicates an interest in how statements about the real world have been made, and with which political impacts" (Forsyth 2003: 14–15).

For the analysis we have selected a series of discursive events, i.e. the annual openings of parliamentarian sessions in the province of Neuquén between 2008–2015; the provincial decree 1483/12, which is the first environmental regulation for unconventional oil and gas operations in the country and two interviews conducted with oil industry actors in 2014 and 2016 in the context of broader PhD field research activities. The case for selecting different types of discourses has to do with the importance of intertextuality, which is neglected in other approaches to discourse analysis (Fairclough 1995). The annual openings of parliamentarian sessions are especially important in respect to political discourses. Political discourses are textual manifestations of the political system (van Dijk 1980). The first interview was conducted with two representatives of the Argentine Gas and Oil Institute (IAPG, according to its initials in Spanish), who also are engineers and professors at the University of Comahue in the region. The second interview was conducted with two members of the department of institutional relations of an important oil and gas company operating in the region (the name of the company cannot be given, for reasons of anonymity).

The different discursive events were analyzed with qualitative social research methods. The documents were analyzed with basic interpretative tools such as coding, searching relevant categories, classifications and relations. The context of production of the documents was taken into consideration for the analysis. For the interviews, a combination of thematic coding (*thematisches Kodieren*) (Flick 2014; Mattissek et al. 2013: 203–206) and the documentary method (*doukmentarische Methode*) (Bohnsack 2014; Przyborski und Wohlrab-Sahr 2014) was used.

8.3   Brief Overview of the Context of Unconventional Hydrocarbons in Neuquén, Argentina

There is a popular saying that considers Argentina not as an oil country but as a country with oil. In fact, there are ten provinces that produce hydrocarbons in Argentina. Together, they founded in 1986 the Federation of Hydrocarbon-producing States (OFEPHI, according to its initials in Spanish) in order to defend the interests of the provinces and to strengthen their participation in the design and planning of the public policies connected with the sector (Organización Federal de Estados Productores de Hidrocarburos 2016). Patagonia is the region that contributes the most oil and natural gas to the country. Neuquén, Chubut, Santa Cruz (all provinces in the Patagonia region) and Mendoza accounted for 85% of the country's total oil production in early 2000 and 70% of total natural gas production (Amar and Martínez 2015).

The history of hydrocarbons in Argentina commemorates the discovery of oil in Chubut in 1907 as its foundational milestone. In those days domestic oil comprised a very small part of the country's energy consumption, which was mainly satisfied through coal imports.[2] In the early 1920s, with the creation of the national company *Yacimiento Petrolíferos Fiscales* (YPF) the state took a leading role in the incipient oil industry. As the first state-originated oil company in the world YPF was an

---

[2] Nowadays, oil and gas are the main source of energy in Argentina. In 2011 they comprised up to 87% of the energy matrix (Giuliani 2013).

example for many national oil companies in Latin America. Since then, and increasingly with the development of the so-called import-substitution industrialization (ISI), oil had begun to gain importance as consumable for the emergent industry. In this way, oil and later natural gas became strategic resources for developing the domestic market. Hydrocarbons were extracted and processed in the country with the aim of fulfilling the domestic energy requirements. The state ensured the investment needed for obtaining resources, regulated prices and YPF was the leading company in the sector. Nevertheless, oil imports still had a relative weight in supplying the market. With some minor changes, the oil and gas sector remained strongly regulated by the state throughout the ISI period. The neoliberal reforms and regulations starting in mid-1970s and especially since 1989 radically changed that situation (Mansilla 2007).

In the course of the state reform process carried out in Argentina in the 1990s, hydrocarbons were federalized and YPF privatized (Law 24.145 of 1992). The constitutional reform of 1994 extended the principle of federalization to all natural resources (Article 124 of the Argentine Constitution recognizes to the provinces the original domain of natural resources lying in their own territories). The administration of resources was therefore transferred to the provinces by means of a process of decentralization, while YPF and other companies of the sector were transferred to foreign hands through a process of privatization (Campodónico 2004). By 1999 the Spanish company Repsol had bought 98.23% of YPF. Other Spanish companies and companies from France, England, the United States and Brazil had acquired almost all Argentinian companies in the sector by 2002, and thus significant control over the promotion and the sources of oil and natural gas passed out of Argentinian hands (Campodónico 2004). Over this period the role of oil and gas within the economic structure was redefined. Exports were encouraged without any strategy to compensate for the depletion of reserves, with companies not compelled to invest in risky exploration activities. In this way, from being consumable goods for the industry they turn to be exported goods without any aggregated value (Mansilla 2007). These changes were taking place in the context of economic restructuring where industrial branches producing goods with high aggregated value gave way to exporting industries (Pérez Roig 2016).

Later, in the so-called "post-convertibility" period beginning in 2002 the state implemented different instruments of intervention in the sector, although there was also much continuity with the policies of the 1990s. In particular, the exploitation of existing reserves and the lack of investment in exploration continued and resulted into decreasing productivity and drastic reduction of reserves. Even though crude oil exports also decreased over the period, they still were significant in absolute and historic terms. In order to face the problem of the increasing gap between energy supply and demand Argentina relied more and more on fuel imports. In this way, fuel imports increased from 482 million in 2002 to 9413 million USD in 2011 and created a major economic imbalance especially since 2007 (Pérez Roig 2016).

In 2012 the national congress passed the so-called "hydrocarbon sovereignty" law. The law 26.741/12 declares the self-sufficiency of hydrocarbons of national interest. For this purpose, it commands the state to recover control of YPF through the expropriation of 51% of Repsol's shares. In this way, a large part of oil and natural gas exploitation was again in hands of the national company (Amar and Martínez 2015). Importantly, the law promotes the exploitation of conventional and *unconventional hydrocarbons*. By promoting the exploitation of unconventional fields the government intended to reconcile the dilemma between strategic resources for the domestic energy demand and exportable commodities (Pérez Roig 2016).

By then the potential of Argentina with regards to unconventional hydrocarbons was already a matter of public knowledge. Repsol YPF had announced the discovery of important unconventional gas resources and unconventional oil resources in 2010 and 2011 respectively. Also in 2011 a report of the United States Energy Information Administration (EIA) placed Argentina as the third country with technically recoverable shale gas resources in the world (EIA 2011). The report was completed and extended in 2013. There, Argentina was also placed in the top 5 countries with technically recoverable shale oil resources among 41 countries in the world (EIA 2013).

The Neuquén Basin, located in the north of Patagonia (Fig. 8.1), is considered the main emerging area of shale gas and shale oil development in South America (EIA 2013). The Neuquén Basin was already one of the most important sedimentary basins in the country. It contributed with 54% of the gas and 43% of the oil to the total amount of hydrocarbons extraction in 2011 (Giuliani 2013). The province of Neuquén extends widely over the Neuquén Basin although three other provinces also cover it: Mendoza, La Pampa, and Río Negro. Oil was discovered in Neuquén in 1918, although it was not until the late 1970s with the discovery of an important gas field in *Loma la Lata* that the economy, politics and society of the province became highly dependent on the extraction of crude oil and natural gas (Favaro 2001).

In 2013 the provincial government of Neuquén paved the way for the implementation of an agreement between YPF and the multinational Chevron, aimed at the exploitation of unconventional hydrocarbons in one area of the province. With the ratification of the agreement the exploration of unconventional hydrocarbons in the Vaca Muerta Formation gained momentum (Fig. 8.2). However, the YPF-Chevron agreement has encountered significant opposition from within Neuquén. A broad coalition including feminist groups, indigenous communities, citizens' assemblies, autonomous trade unions, NGOs, and left political parties protested against it. Different views on the agreement, but also on the extraction of oil and gas in general, were confronted. On one hand, exploiting the unconventional riches of the subsoil meant an opportunity to countervail the energy deficit and promote the development of the country and the region. On the other hand, the agreement meant the continuation of the plundering of nature and environmental devastation. These sectors argued that while most economic benefits move abroad the environmental damage stays behind for the local population.

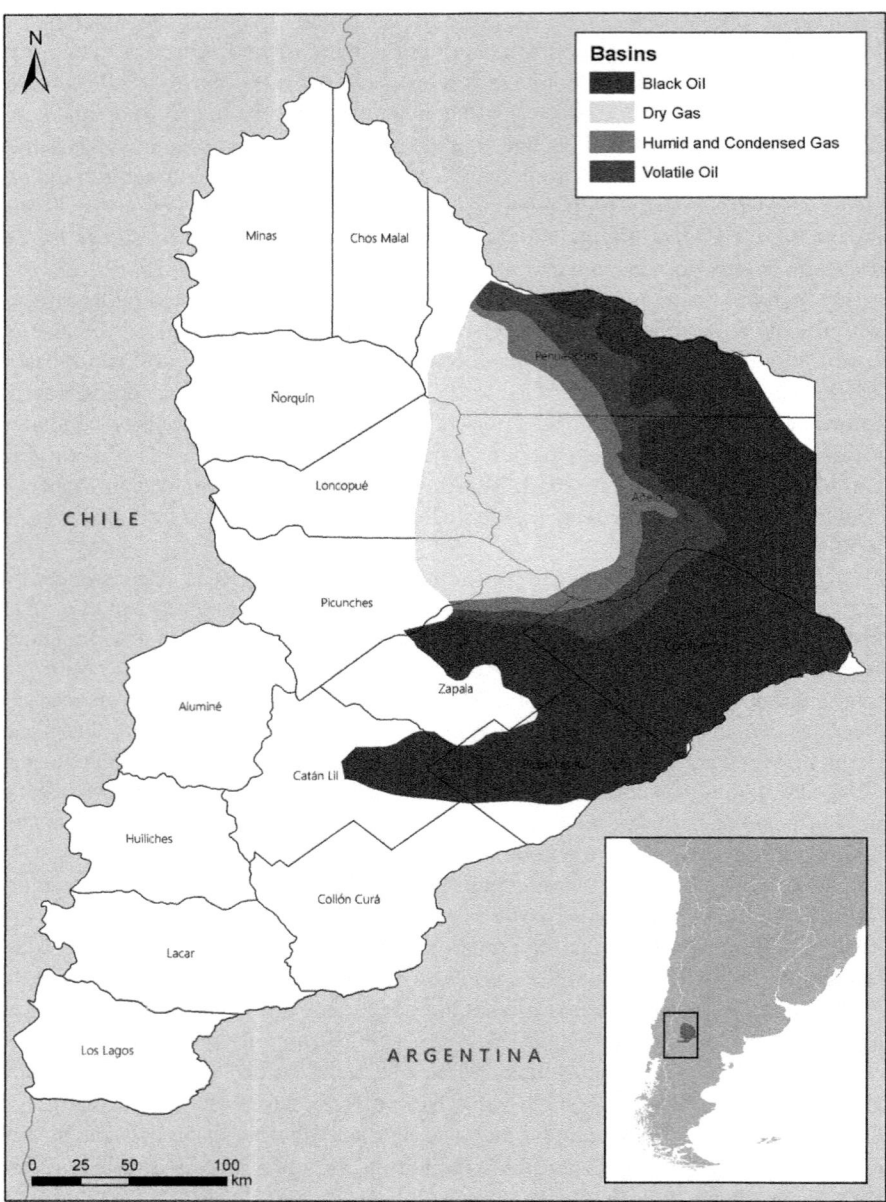

Fig. 8.2 Vaca Muerta Formation in Neuquén Province (Callum Foster, UWE, Bristol Cartography, with data from DGIyE, Province of Neuquén)

8.4   Unconventional Hydrocarbons, "Sustainable Development" and the Access to Water in Neuquén

8.4.1   The Necessity of Fossil Fuels and Unconventional Hydrocarbons

Based on the analysis of the interviews with oil industry actors and the governor's speeches it can be observed that unconventional hydrocarbons are considered as a *necessity* for economic and social progress. However, both industry and government argue about it in different ways, as we suggest below.

Among oil and gas industry actors, the arguments presented to justify this necessity refer to the domestic *use value* of fossil fuels (i.e. to the presence of hydrocarbons in multiple objects and processes of everyday life). This need for fossil fuels is assumed and not problematized. It is presented as an indisputable fact: *"It will be difficult to replace hydrocarbons. […] hydrocarbons have so many uses, the pharmaceutical industry would be paralyzed […] the packs of a coca cola are based on hydrocarbons, I mean, it would be a disaster […]"* (IAPG member, own translation). Another interviewee refers to how surprising it is that fossil fuel exploitation generates such a large rejection, as they are present in many every-day activities: *"[…] oil and gas are seen as a something different, external, something that should not be near us. Natural gas is indispensable, not only for heating but also for the development, for the growth of Argentina […]"* (Employee of a fossil fuel company, own translation).

This representation of the self-evident nature and the necessity of fossil fuels in our society serve as a kind of legitimation of the seemingly inevitable need to promote fossil fuels. The dependency of contemporary lifestyles and consumption patterns on fossil fuels is neither questioned nor critically considered. They are regarded as pre-given and inevitable. Within this mind-set the circle closes insofar as the solution to a growing fossil fuel demand is found in technological innovation, research and the expansion of the possibilities of promotion and exploitation of fossil fuels (for example through the introduction of unconventional methods) rather than in an examination of the current energy consumption pattern (Wissen 2016). In particular, the role of research and technological innovation is presented as an essential task of the fossil fuels industry. Not only because there are still plenty of resources to be promoted but also because companies need to increase their reserves in order to maintain the value of their shares in the stock market: *"From the 100% of the oil underneath, that is, in the reservoir, you can gain about 15% with primary oil recovery, right? […] with secondary oil recovery, you can still pull out 18% […] and the rest remains down, therefore, research and development is constantly being pursued, through universities, through own laboratories, etc. to see how you can go further […] because there is still a lot of it"* (IAPG member, own translation).

Unconventional hydrocarbons emerged strongly in the governmental discourse of Neuquén in the first period of Jorge Sapag's government (2007–2011). Within the governor's speeches, unconventional hydrocarbons are mentioned for the first time in the speech of 2010, when the governor calls attention to the opportunity that these resources represent for the province of Neuquén. Henceforth, this idea comes out in following speeches, including those of the second period of Sapag's government (2011–2015).

The argument emphasizes that the provincial subsoil contains great quantities of unconventional hydrocarbons. These hydrocarbons are increasingly gaining importance as sources of energy in developed countries and are expected to largely supply world markets in the near future. It follows that it is necessary to exploit these resources and therefore generate the conditions for that to happen: *"[…] So, I hope we are capable of creating the framework of opportunities and investment, so that this investment [in unconventional gas exploitation] […] is possible […]. I want to say to the representatives, and to the rest of the community through this session, that the province of Neuquén has in its subsoil […] gas reservoirs that have to be explored, that have to be extracted […]"* (Honorable Legislatura Provincia del Neuquén 2010, own translation).

The importance and thus the necessity of the exploitation of unconventional hydrocarbons is further justified and legitimized whilst they are presented as an opportunity to *"generate wealth, jobs, quality of life, progress and development over the soil [of Neuquén]"* (Honorable Legislatura Provincia del Neuquén 2011, own translation). Even more, unconventional hydrocarbons are presented as a milestone for the historical re-foundation of Neuquén as *"[…] subsoil riches also allow us [inhabitants of Neuquén] to experience foundational moments, because with the riches of the subsoil we can generate wealth from the soil and transform it into social rights, equal opportunities for everyone and a diversification of the economy"* (Honorable Legislatura Provincia del Neuquén 2014, own translation). In this way, the necessity and opportunity to exploit unconventional hydrocarbons is related to a sense of *exchange value*, through which oil and gas is exchanged for common welfare. In other words, the state makes a pact through which development is purchased at the expense of sovereignty over resources (Watts 2001). The state appears as driver of this transformation of natural wealth into social wealth and expects that a problem rooted in socio-natural relations will be solved through commoditized nature.

Furthermore, the exploitation of unconventional hydrocarbons and the resultant products are presented in relation to issues of national sovereignty. Neuquén's oil and gas is presented as helping Argentina to save important amounts of money currently spent in fuel imports and to achieve *hydrocarbon sovereignty*: *"We, the people of Neuquén, have initiated this path [of producing unconventional hydrocarbons] to recuperate hydrocarbon sovereignty and self-sufficiency. It is worth saying that the Republic of Argentina depends on 90% of gas and oil to move its industry, to produce in the fields, to move its transport and commerce and for the comfort of the households"* (Honorable Legislatura Provincia del Neuquén 2015, own translation). In this way, Neuquén is placed as a contributor to the progress of the country through its economic specialization in producing energy, in particular hydrocarbons. This is not questioned but rather understood as constituent part of the province and its identity: *"In the identity of Neuquén, in its nature, there is the energy. In mapuche,[3] nehuen means strength, energy, vigor and that is the identity of our land, rivers, mountains, people. Our identity and our nature has to be respected and our identity and our nature is indicating a clear direction"* (Honorable Legislatura Provincia del Neuquén 2013, own translation).

8.4.2 The Meanings of Environmental Protection and the Availability of Water

The need for environmental measures and controls as well as issues related to water are strongly present in the governor's speeches and in the interviews with oil industry actors.

Considering the speeches of the governor the exploitation of unconventional hydrocarbons requires not only investment but also *"ensuring extraction in an environmentally sustainable way"* (Honorable Legislatura Provincia del Neuquén 2011). The notion of "sustainability" – ambiguous as it is (O'Connor 1998) – is here associated to the idea of preserving the environment for future generations: *"We want this growth and development to be sustainable, with strict environmental control [...] in order to take care of the land, air, water, the habitat for our own and future generations"* (Honoraria Legislatura Provincia del Neuquén 2013).

In the interviews with oil industry actors, environmental measures together with environmental legislation are presented as a guarantee against any mistrust or doubts about the environmental impacts of the activity.

An IAPG member offers reassurances that the coexistence of hydrocarbons exploitation and other activities such as agriculture is possible because oil and gas activities are regulated by local, provincial and national environmental legislation, so that the industrial use of water is strictly monitored and regulated. He adds that environmental degradation or pollution could happen, but rather than from disregarding prevention measures, this would result from inevitable accidents.

Companies may also be concerned about the environment because impeachments related to environmental issues would damage their image and thereby the chances of obtaining funding from international organizations, assures a member of the IAPG.

Environmental issues also emerge in the interviews in relation to conflicts with civil society organizations. A member of the IAPG mentions two cases where exploration projects have been abandoned due to social resistance. He argues that the main cause of this social resistance is the fact that *"oil has a bad reputation"*. This bad reputation would be partly the responsibility of the industry because *"it never*

---

[3] It refers to the language of original inhabitants of the region.

took the trouble to explain to the people what they were doing". In this view, lack of knowledge would be the reason for the resentment against the industry. In conclusion, the "problem" is a product of poor public communications and not the activity itself.

The role of communication is also present in the interview with the employee of a fossil fuel company. This interviewee highlights the importance for the company to inform actors from civil society and to be in contact with the government in order to comply with current environmental regulations.

Further, water acquires an important role regarding environmental matters in unconventional hydrocarbons exploitation. In particular, the Decree 1483/12 of the province of Neuquén approves the *"norms and procedures for the exploration and exploitation of unconventional reservoirs"* in order to *"prevent mitigate and minimize environmental impacts"* (Decree 1483/12, own translation). The Decree classifies hydrocarbons as either conventional and unconventional. It suggests that the exploitation of the two types of resources differs on the quantity of water used, as hydraulic stimulation operations need significantly higher amounts of water than regular conventional operations. Further, the use of water in unconventional operations is made conditional upon the requirements of water for other purposes such as drinking water, wildlife, and agriculture. In general, the regulation centers on the procedures regarding the retrieval and disposal of used water, though also includes requirements such as the environmental license and a declaration of the chemicals used in hydraulic stimulation.

According to article 9 of the Decree 1483/12 underground waters, which are able to supply the population or irrigation, are prohibited for its use in unconventional operations. Article 10 mandates the management of flowback for its reutilization in the industry or disposal in septic tanks. Article 11 prohibits the disposal of water in surface waters and storage in open-air tanks. In its speech of 2014, the governor highlights that *"one of the most clear prohibitions [of the regulation] is the prohibition of using underground water [...]"* (Honorable Legislatura Provincia del Neuquén 2014). In this way, water becomes the main environmental concern and, to a great extent, environmental preservation acquires the meaning of water protection.

Moreover, the IAPG (2013) has created a manual with recommended practices concerning water management in unconventional hydrocarbons operations in the Neuquén Basin. The manual suggests that securing the availability and disposal of water is extremely important for the success of unconventional operations. Besides, a booklet published by the IAPG (López Anadón et al. 2014), intends to rule out a series of "myths" surrounding hydraulic fracturing. Three of five "myths" are related to water which the company counters as follows: (1) The implementation of hydraulic fracturing does not contaminate freshwater resources. (2) The acquisition of water for hydraulic fracturing does not challenge the availability of water for the population. (3) Residual waters do not constitute a problem for the environment.

These points are also present in the interviews with oil industry actors, where the role of water in the production process was continuously minimized and relativized. An IAPG member argues that water is needed mainly at the beginning of the extraction process, in order to reinforce the pressure of the deposit or to desalinate flowback. More than once, it is emphasized that the water required for secondary production is not drinking water but treated water from the same field. There may be several reasons why the companies are interested in a reduction of water use and therefore in the reuse of flowback. These are not only ecological reasons, but also economic ones (it would be cheaper for the companies to reuse return water). This would also partly solve the issue of waste water:

"There is always a cycle, a closed circle, ok? In which oil will be extracted, then water will be separated and then again pumped and so on" (IAPG member, own translation).

The employee of a fossil fuel company reaffirms this idea. Oil and gas would not compete with other water uses, as this industry would consume very little water, both in relation to the flow of the rivers in the region and in comparison to other economic activities (such as fruit farming or food-related industrial activities). The final argument to reinforce this proposition is introduced immediately afterwards: *"95% of the water of the River Negro flows into the Atlantic Ocean"* (Employee of a fossil fuel company, own translation). The very same argument appeared in a 2013 speech delivered by the governor, as well as in the justification of the Decree 1483/12.

Taken together, two interrelated issues arise. First, the availability of water becomes a central argument for the exploitation of unconventional hydrocarbons. Second, the government and industry agree in their arguments to support the access to water for unconventional operations.

8.5 Discussion and Conclusion

The above analysis of the different discursive events evidences how the state and the industry develop a discourse of inevitable necessity for fossil fuels. It is based on a perceived need to meet the social demand for fossil fuels and local/regional development. Thus, state and industry define the needs of society and present them as unproblematized facts in an attempt to legitimize the exploitation of hydrocarbons and thus manufacture consent for unconventional methods of extraction. In this way, the existing patterns of energy production and consumption are deepened and "[...] the energy-intensive, largely petroleum-dependent way of life, which has been anchored in the global north for a long time, in the infrastructure, institutions and socio-economic relations (Huber 2013), spreads with power within the middle and upper classes of emerging countries" (Wissen 2016: 361, own translation).

Further, the invocation of sustainable development and the implementation of environmental measures also play an important role in the legitimation of unconventional hydrocarbons. In fact, environmental discourse of both government and industry is produced in the context of increasing environmental conflicts and struggles not only in Neuquén in particular, but also in Latin America in general. In recent years, the region has experienced a process through which social struggles have acquired an environmental character (Leff 2006; Svampa 2012). The incorporation of issues of sustainability and environmental measures and controls in government and industry discourse could be understood as an attempt to come to terms with the general social concern for the environment.

Emerging from the analysis, the role of water is central in issues of environmental protection. In particular, both government and industry strongly argue that the development of unconventional extraction would not intercede in the availability of water for the population or other economic activities. The idea that 94% of the superficial water is "lost" because it flows into the sea promotes a sense of excess availability of water that intends to disavow struggles for both its meanings and usages. In fact, these political, legislative and technical discourses assume that water is an unlimited resource without taking into consideration ecosystem dynamics and the effects the use of water has over them. Water is transported, mixed with chemicals, used, treated, reused, and pumped into the subsurface. In this way, water is perceived as means of production and is equal to production costs. This understanding of water can be interpreted as a modern perception of water, which "can be defined as the dominant, or natural, way of knowing and relating to water, originating in western Europe and North America, and operating on a global scale by the later part of the twentieth century" (Linton 2010: 14). Without ignoring the importance of the developments in the hydrological sciences and the knowledge gained from them, the author draws attention to the way the Western scientific understanding of water has caused a deterritorialization and decontextualization of water. This leads to disregard other meanings of water as well as a decoupling of water from its social and environmental contexts (Linton 2010).

Overall, through a discourse of environmental regulation and control the state intends to ensure the extraction of unconventional oil and gas activities and therefore, reproduce specific socio-natural relations based in the commodification of nature. In so doing, it assumes a particular notion of nature as resources available for exploitation. This has political implications insofar as the solutions to environmental problems originate within particular and naturalized notions of what the environmental problems are (Forsyth 2003). Other "problems" and "solutions" emerging from alternative conceptions of the environment do not fit within the dominant paradigm of the environment. Thus, they are rejected and above all discredited.

In this sense, the denaturalization of dominant environmental discourse is essential for the recognition of alternative and multiple perceptions and notions of nature. Political ecology along with critical discourse analysis tools may contribute to this path by problematizing and deconstructing dominant environmental discourses.

References

Alimonda H (2011) La colonialidad de la naturaleza. Una aproximación a la ecología política latinoamericana. In: Alimonda H (ed) La naturaleza colonizada. Ecología política y minería en América Latina. CLACSO, Buenos Aires, pp 21–58

Amar A, Martínez R (2015) Impacto socioeconómico de YPF desde su renacionalización (Ley 26.741). Análisis del impacto fiscal de las operaciones de YPF a nivel provincial. Comisión Económica para América Latina y el Caribe. Available at: http://www.cepal.org/es/publicaciones/39399-impacto-socioeconomico-ypf-su-renacionalizacion-ley-26741-analisis-impacto. Accessed 2 Feb 2016

Bakker K (2012) Water: political, biopolitical, material. Soc Stud Sci 42(4):616–623

Bebbington A (2015) Political ecologies of resource extraction: agendas pendientes. Eur Rev Lat Am Caribb Stud 100:85–98

Bebbington A, Bury J (2013) Political ecologies of the subsoil. In: Bebbington A, Bury J (eds) Subterranean struggles: new dynamics of mining, oil, and gas in Latin America. University of Texas Press, Austin, pp 1–25

Bertinat P et al (2014) 20 Mitos y realidades del fracking. El Colectivo, Buenos Aires

Bohnsack R (2014) Rekonstruktive Sozialforschung. Einführung in qualitative Methoden. Verlag Barbara Budrich, Opladen

Brand U (2011) El papel del Estado y de las políticas públicas en los procesos de transformación. In: Miriam L, Duina M (eds) Más allá del Desarrollo. Quito. Grupo Permanente de Trabajo sobre Alternativas al Desarrollo. Ediciones Abya Yala, Quito, pp 145–147

Bridge G (2014) Resource geographies II: the resource-state nexus. Prog Hum Geogr 38(1):118–130

Bryant RL (1998) Power, knowledge and political ecology in the third world: a review. Prog Phys Geogr 22(1):79–94

Bryant R, Bailey S (2007) Third world political ecology. Routledge, Oxon

Budds J, Linton J (2014) The hydrosocial cycle: defining and mobilizing a relational dialectical approach to water. Geoforum 57:170–180

Campodónico H (2004). Reformas e inversión en la industria de hidrocarburos de América Latina. Comisión Económica para América Latina y el Caribe. Available at: www.archivo.cepal.org/pdfs/2004/S0410784.pdf. Accessed 2 Feb 2016

Coronil F (1997) The magical state. Nature, money, and modernity in Venezuela. The University of Chicago Press, Chicago/London

Fairclough N (1995) Critical discourse analysis: the critical study of language. Longman, London/New York

Fairclough N, Jessop B, Sayer A (2004) Critical realism and semiosis. In: Joseph J, Roberts JM (eds) Realism, discourse and deconstruction. Routledge, London/New York

Favaro O (2001) Estado, política y petróleo. La historia política neuquina y el rol del petróleo en el modelo de provincia, 1958–1990. Universidad Nacional de La Plata

Flick U (2014) Qualitative Sozialforschung. Eine Einführung. Rowohlt Taschenbuch Verlag, Reinbek

Forsyth T (2003) Critical political ecology. The politics of environmental science. Routledge, London/New York

Giuliani A (2013) Gas y petróleo en la economía de Neuquén. EDUCO – Universidad Nacional del Comahue, Neuquén

Gudynas E (2010) The new extractivism of the 21st century: ten urgent theses about extractivism in relation to current South American Progressivism. Americas Program Report

Instituto Argentino del Petróleo y del Gas (2011) Apache logra el primer pozo horizontal multifracturado con objetivo shale gas de la región. Available at: http://www.iapg.org.ar/noticias/201127/noticias1.htm. Accessed 23 May 2017

Instituto Argentino del Petróleo y del Gas (2013) Gestión del agua en la exploración y explotación de reservorios no convencionales en el área de influencia de la cuenca neuquina. Available at: http://www.iapg.org.ar/sectores/practicas/VF_PR_11.pdf. Accessed 24 June 2016

Joseph J, Roberts JM (2004) Introduction: realism, discourse and deconstruction. In: Joseph J, Roberts JM (eds) Realism, discourse and deconstruction. Routledge, London/New York

Köhler B (2005) Ressourcenkonflikte in Lateinamerika. Zur Politischen Ökologie der Inwertsetzung von Wasser. Journal für Entwicklungspolitik 2(XXI):21–44

La Nación (2011) YPF descubrió en Neuquén un "espectacular" yacimiento de petróleo no convencional. Available at: http://www.lanacion.com.ar/1421209-ypf-descubrio-en-neuquen-uno-de-los-yacimiento-mas-grandes-del-mundo. Accessed 30 May 2017

Latour B (2012) Nunca fuimos modernos. Ensayos de antropología simétrica. 1° edición. Siglo XXI editores, Argentina

Leff E (2006) La ecología política en América Latina. Un campo en construcción. In: Alimonda H (ed) Los tormentos de la materia. Aportes para una ecología política latinoamericana. CLACSO, Buenos Aires, pp 21–39

Linton J (2010) What is water? The history of a modern abstraction. University of British Columbia Press, Vancouver

López Anadón E et al (2014) El abecé de los Hidrocarburos en Reservorios No Convencionales. Instituto Argentino del Petróleo y del Gas, Ciudad Autónoma de Buenos Aires

Machado Aráoz H (2012) Los dolores de Nuestra América y la condición neocolonial. Extractivismo y biopolítica de la expropiación. Revista del Observatorio Social de América Latina – OSAL 13(32):51–66

Mansilla D (2007) Hidrocarburos y política energética. De la importancia estratégica al valor económico: Desregulación y privatización de los hidrocarburos en Argentina. Ediciones del CCC, Buenos Aires

Martín F, Larsimont R (2016). ¿Es posible una ecología cosmo-política? Polis (45)

Mattissek A, Pfaffenbach C, Reuber P (2013) Methoden der empirischen Humangeographie. Westermann, Braunschweig

O'Connor J (1998) Natural causes. Essays in ecological Marxism. The Guilford Press, New York/London

Organización Federal de Estados Productores de Hidrocarburos OFEPHI (2016) Pasado y presente. Available at: http://ofephi.com.ar/index.php?option=com_content&view=article&id=1&Itemid=17. Accessed 13 December 13 2016

Oszlak O (2011) Formación histórica del Estado en América Latina: elementos teórico-metodológicos para su estudio. In: Jefatura de Gabinete de Ministros de la Nación (ed) Lecturas sobre el Estado y las políticas públicas. Retomando el debate de ayer para fortalecer el actual. Jefatura de Gabinete de Ministros de la Nación, Buenos Aires, pp 115–141

Página 12, 2017. No habrá fracking en Entre Ríos. Available at: https://www.pagina12.com.ar/36346-no-habra-fracking-en-entre-rios Accessed 27 May 2017

Pérez Roig D (2016) Los dilemas de la política hidrocarburífera en la Argentina posconvertibilidad. In: di Risio D, Scandizzo H, Pérez Roig D (eds) Vaca Muerta. Construcción de una estrategia. Ediciones del Jinete Insomne, Ciudad Autónoma de Buenos Aires, pp 11–36

Przyborski A, Wohlrab-Sahr M (2014) Qualitative Sozialforschung. Ein Arbeitsbuch. Oldenbourg Wissenschaftsverlag, München

Seoane J, Taddei E, Algranati C (2013) Extractivismo, despojo y crisis climática. Herramienta/El Colectivo/GEAL, Buenos Aires

Staddon C (2009) The complicity of trees: the socionatural field of/for tree theft in Bulgaria. Slavic Review 68(1):70–94. Cambridge University Press

Staddon C, Brown J, Hayes ET (2016) Potential environmental impacts of 'fracking' in the UK. Geography 101(2):60–69

Svampa M (2011) Extractivismo neodesarrollista y movimientos sociales. ¿Un giro ecoterritorial hacia nuevas alternativas? In: Lang M, Mokrani D (eds) Más Allá del Desarrollo. Fundación Rosa Luxemburg/Abya Yala, Quito, pp 185–216

Svampa M (2012) Consenso de los commodities, giro ecoterritorial y pensamiento crítico en América Latina. OSAL Observatorio Social de América Latina XIII(32):15–38

Svampa M, Viale E (2014) Maldesarrollo. La Argentina del extractivismo y el despojo. Katz, Buenos Aires

Swyngedow E (2004) Social power and the urbanization of water. Flows of power. Oxford University Press, Oxford

Swyngedow E (2009) The political economy and political ecology of the hydro-social cycle. J Contemp Water Res Educ 142:56–60

U.S Energy Information Administration (2011) World shale gas resources: an initial assessment of 14 regions outside the United States. Available at: https://www.eia.gov/analysis/studies/world-shalegas/archive/2011/pdf/fullreport_2011.pdf. Accessed 30 May 2017

U.S Energy Information Administration (2013) Technically recoverable shale oil and shale gas resources: an assessment of 137 shale formations in 41 countries outside the United States. Available at: https://www.eia.gov/analysis/studies/worldshalegas/archive/2013/pdf/fullreport_2013.pdf. Accessed 30 May 2017

van Dijk TA (1980) Textwissenschaft. Eine interdisziplinäre Einführung. Max Niemeyer Verlag, Tübingen

Watts M (2001) Petro-violence: community, extraction, and political ecology of a mythic commodity. In: Peluso NL, Watts M (eds) Violent environments. Cornell University Press, Ithaca/London

Watts M (2004) Violent environments: petroleum conflict and the political ecology of rule in the Niger Delta, Nigeria. In: Peet R, Watts M (eds) Liberation ecologies. Environment, development, social movements. Routledge, London/New York

Wilhelm F (1997) Hydrogeographie. Westermann Schulbuchverlag, Braunschweig

Wissen M (2016) Zwischen Neo-Fossilismus und "grüner Ökonomie". Entwicklungstendenzen des globalen Energieregimes PROKLA – Zeitschrift für kritische Sozialwissenschaft 46(3):343–354

Analyzed Documents

Decree N° 1483/12, Provincia del Neuquén. 13 de agosto de 2012

Diario de Sesiones, XXXVII Período Legislativo, Reunión N°2. Honorable Legislatura Provincia del Neuquén, 1 de marzo de 2008

Diario de Sesiones, XXXVIII Período Legislativo, Reunión N°2. Honorable Legislatura Provincia del Neuquén, 1 de marzo de 2009

Diario de Sesiones, XXXIX Período Legislativo, Reunión N°2. Honorable Legislatura Provincia del Neuquén, 1 de marzo de 2010

Diario de Sesiones, XL Período Legislativo, Reunión N°2. Honorable Legislatura Provincia del Neuquén, 1 de marzo de 2011

Diario de Sesiones, XLI Período Legislativo, Reunión N°2. Honorable Legislatura Provincia del Neuquén, 1 de marzo de 2012

Diario de Sesiones, XLII Período Legislativo, Reunión N°2. Honorable Legislatura Provincia del Neuquén, 1 de marzo de 2013

Diario de Sesiones, XLIII Período Legislativo, Reunión N°2. Honorable Legislatura Provincia del Neuquén, 1 de marzo de 2014

Diario de Sesiones, XLIV Período Legislativo, Reunión N°2. Honorable Legislatura Provincia del Neuquén, 1 de marzo de 2015

Joaquín Bernáldez is a PhD fellow of the Graduate School of Socio-Ecological Research for Development at the International Center for Development and Decent Work (ICDD), a broad research network with partner universities in Africa, Asia, and Latin America. His research revolves around political ecology, extractivism, environmental conflicts, agency and qualitative methods. His PhD project focuses particularly on practices and perceptions of oil and gas extraction in the region of North Patagonia in Argentina. Bernaldez previously graduated in Anthropology at the University of Buenos Aires, Argentina. He then moved to Germany where he did his MA in Global Political Economy at the University of Kassel.

Rocío Juliana Herrera studied Political Science in Buenos Aires, Argentina. After working in public offices in Neuquén and Río Negro, she won a scholarship from the German Academic Exchange Service (DAAD) and earned a Master of Science in regional science and spatial planning at the Institute of Regional Science (IfR), Karlsruhe, Germany. She also earned her PhD at IfR, where her dissertation was "Political Ecology as an instrument for the analysis of the regulation of water uses: The case of the upper Río Negro Valley, Northern Patagonia, Argentina." She also worked as Academic Staff at IfR for the last five years. Herrera's research areas are related to regional identity, political ecology/political ecology of water, water uses and water governance, participatory planning processes, decentralization and regional integration processes (especially MERCOSUR) and qualitative empirical social research.

Chapter 9
Tight Oil and Water: Climate Change and the Extractive Waterscape of Western Siberia

Owen King

Abstract With conventional oil production declining in the Western Siberian Basin, Russia is incentivising the development of 'tight oil' reserves using hydraulic fracturing technologies. This chapter reviews the existing literature on two under-explored aspects of the unconventional hydrocarbons debate. First, that much of the research on the environmental and social implications of hydraulic fracturing for 'unconventional oil and gas' has focused substantively on shale gas. In particular, perspectives on the specific nature of tight oil and its extraction are notably scarce. Second, I argue that the increasingly apparent risks posed by the hydrological implications of climate change, extreme weather and the expansion of tight oil are worthy of much greater empirical attention.

The examples given in this chapter call attention to the specific materiality of tight oil and water, and the way in which water, nature and people mediate each other in an 'extractive waterscape.' Thus, the geophysical nature of tight oil is manifest in the intensity of production, and comes into conflict with the increasing intensity of hydrological dynamics. While this poses significant socio-ecological threats to indigenous livelihoods in Western Siberia, it is argued that water can play a key role in resistance. By placing the role of water in mediating cultural relationships with the land, at the centre of these struggles, indigenous space may be reclaimed. I conclude by highlighting three main areas for future research on the subject of tight oil extraction and water resources in the fields of the environmental sciences, physical and human geography.

Keywords Hydraulic fracturing · Unconventional hydrocarbons · Tight oil · Energy · Russia · Western Siberia · Climate change · Water · Extreme weather · Indigenous space

O. King (✉)
University of Birmingham, Birmingham, UK

© Springer International Publishing AG 2020
R. M. Buono et al. (eds.), *Regulating Water Security in Unconventional Oil and Gas*, Water Security in a New World,
https://doi.org/10.1007/978-3-030-18342-4_9

All the easy oil and gas in the world has pretty much been found. Now comes the harder
work in finding and producing oil from more challenging environments and work areas. –
William J. Cummings, Exxon-Mobil company spokesman, December 2005

As the rate of production from 'conventional' oil fields worldwide has appeared
to reach a plateau and global demand has continued to increase, energy firms and
national governments have increasingly shifted towards the development of 'unconventional' oil resources (IEA 2013, 2018). Thus we have seen an expansion of the
global oil industry from its traditional nodes, its reach extending across ever-
widening territories and into previously unfathomable strata. In the past decade,
vast reservoirs of these 'unconventional' hydrocarbons have been identified, and
huge sums have been invested in technology and infrastructure to access and process them (Farchy 2017; USDOE 2018). As these remote and hitherto economically
and technologically unviable sources of crude have been targeted, new socioecological and cultural territories have been implicated.

Consequently, much emphasis has recently been placed on research into the
environmental and social impacts of hydraulic fracturing. However, in much scholarly discussion on the extraction of what is commonly termed 'unconventional oil
and gas', the substantive emphasis has often been placed on the latter, non-liquid
form. Indeed, a brief skim of the literature on hydraulic fracturing gives the impres-
sion that the technique is almost synonymous with the extraction of shale 'gas'
specifically. Meanwhile, the use of fracking to obtain crude oil is often conflated
with conventional oil drilling and afforded little consideration. Directing attention
toward this incongruity, this chapter focuses specifically on the implications of the
shift toward deep 'tight oil' reservoirs – those located in low-porosity geologic formations at depths up to 4500 m – which have become accessible only due to the
development of multi-stage horizontal hydraulic fracturing technology. However, it
is argued here that, due to the unique nature of tight oil, the significant shift towards
this energy source poses particular environmental and social problems.

Additionally, there is an increasingly apparent intersection between the advance
of an extractive technique of which the potential environmental social impacts are
little understood; and the changing role of water in mediating these impacts in a
changing global climate. In few places is the critical nature of this emerging nexus
exemplified as it is in the Western Siberia region of Russia. Here, a landscape and a
people which – in the last century – have come to be dominated by the oil industry
are now increasingly subjected to the impacts of warming temperatures and changing hydrological regimes. With the Russian government incentivising the development of vast tight oil resources in the region, this chapter argues that the specific
nature of tight oil and water poses particular risks to the indigenous inhabitants of
Western Siberia. The region is thus considered in terms of a changing extractive
'waterscape', a concept emerging from a growing body of work that considers
water, the socioecological context within which it is contained/flows, and social
power relations as co-constitutive (e.g. Budds and Hinojosa-Valencia 2012).

This chapter continues by outlining the emergence of tight oil in Russia, and the
particular process of multi-stage horizontal hydraulic fracturing which has been

developed to exploit it. The uncertain state of knowledge on the environmental and social implication of this technological shift is then briefly reviewed. Notwithstanding much uncertainty in the latter regard, it is argued that the primary concern relates to the necessary intensity of production. Focussing on the intersection of tight oil development with the increasingly apparent consequences of climate change and extreme weather events for the Ob River Basin, the Western Siberian context is introduced. These challenges are considered in relation to the historical, political and economic context of post-communist Russia, with an emphasis on energy and environmental regulation. Here, it is argued that the particular way in which the natural resources and the environment are envisaged in Russia, combined with the recentralisation of power in recent decades, poses particular ecological and social problems for the expansion of tight oil exploitation.

Finally, the chapter focusses on anthropological studies of the experience of the Khanty, indigenous inhabitants of the Western Siberian region now known as the Khanty-Mansiisk Autonomous Okrug. The Khanty's traditional subsistence livelihoods have faced significant challenges since the discovery of vast oil reserves in the region. It is argued that the combination and intensity of impacts from tight oil, climate change and extreme weather further threaten their capacity to sustain themselves in the harsh conditions of the tundra and taiga forests. The significance of water to their relationship with the landscape is related to that of North American indigenous groups, who have productively emphasised this aspect in opposing tight oil development in the United States and Canada. In order to address these immanent problems, I conclude by highlighting some broad areas for future research across various disciplines of the environmental sciences and geography.

9.1 Oil and the Unconventional Turn in Russia

Russia's economy is highly dependent upon its hydrocarbons, with more than one-third of federal revenues coming from oil and natural gas production. It is the world's largest producer of crude oil with average production of 11.2 million barrels per day (b/d). The country exported more than 5 million b/d of crude oil and condensate in 2016, most of which (70%) went to European countries. Indeed, with more than one-third of crude oil imports into OECD Europe originating in Russia, Europe and Russia can be said to be strongly interdependent in terms of energy. Most of Russian crude oil originates from Western Siberia, where the Samotlor and Priobskoye oil fields in the Khanty-Mansiisk Autonomous Okrug[1] are the largest (EIA 2017). However, the major conventional oil fields in the region are in a state of decline, threatening Russia's ability to maintain its output and, therefore, the state's most important source of revenue (Tuzova and Qayum 2016; Farchy 2017). After a period of significant

---

[1] 'Okrug', refers to a federal administrative sub-region. The Khanty-Mansiisk Autonomous Okrug is an administrative sub-region of the Tyumen 'Oblast'.

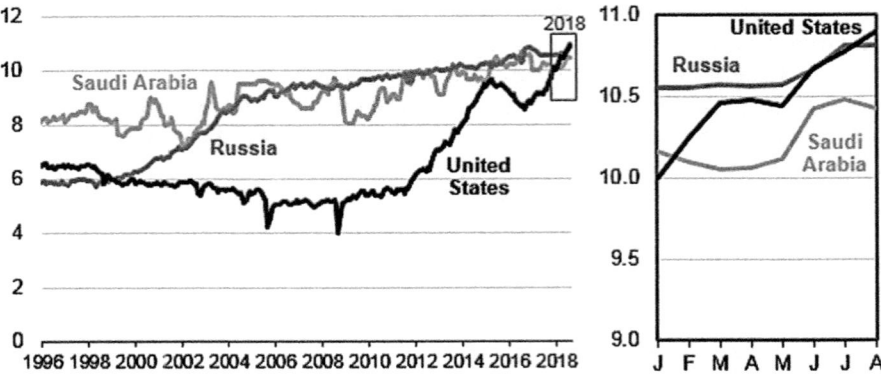

Fig. 9.1 Monthly crude oil production (million barrels per day: Jan 1996–Aug 2018). (Source: EIA 2018)

growth over the late 1990s and early 2000s, Russian oil production has displayed a notable levelling off in the past decade (EIA 2018) (see Fig. 9.1).

The burgeoning prospect of an oil production crisis in Western Siberia, and the rapid emergence of new technologies in the United States to access deeper-lying bodies of hydrocarbons has thus increasingly drawn the attention of the Russian government towards their own 'unconventional' options. One such resource, located in the geological formation known as the Bazhenov, was discovered in the 1970s, between 2000 and 3500 m beneath the oil fields of Western Siberia. This is a low-permeability, oil-saturated layer of carbonate-clay-siliceous shales, which acts as the source rock for the conventional crude reservoirs above. Between three and eight metres in thickness, the Bazhenov formation extends over a territory of more than 1 million km². This reservoir, bearing similar characteristics to that found in the Eagle Ford and Bakken shale formations in the United States and Canada, is classified as 'tight oil' and may only be obtained through multi-stage horizontal hydraulic fracturing.[2]

---

[2] It is important to clarify the nomenclature between the main conventional and unconventional oil sources, which is often mistaken in the literature. Conventional oil refers to that which is produced from wells using vertical well bores to access discrete accumulations or pools. As the strata hosting these resource bodies typically have high porosity and permeability, the extraction process requires minimal stimulation beyond drilling of the well. Unconventional oils may be referred to more specifically as 'tight oil', 'shale oil' or 'oil shale'. Shale and tight oil are conventional oils (light oils with low sulphur content) trapped in unconventional formations with extremely low porosity and permeability at depths of up to 4500 m. Shale oil and tight oil (the latter is also known as 'light tight oil' or 'LTO') reservoirs differ mainly in respect to the presence of either claystone (in the case of shale oil) or siltstone and/or mudstone (in respect to tight oil). 'Oil shale', meanwhile, is a precursor of oil called kerogen, which is trapped in rocks with low porosity and permeability, but at a much shallower depth than those containing shale oil and tight oil. Oil sands, not the subject of this chapter, generally consist of extra heavy crude oil or bitumen trapped in unconsolidated sandstone.

The process of hydraulic fracturing (hydraulic fracture stimulation, or fracking) as the key technique by which unconventional hydrocarbons are extracted has been detailed in the introductory chapter to this collection. However, the specific nature of tight oil means that the method of extraction has a number of unique aspects. Many tight oil reservoirs are relatively thin layers which cover extensive horizontal areas. Consequently, a conventional, vertically drilled well will only access a small area of the reservoir and, due to the impermeable nature of the formation, a minimal part of the resource. However, when the drilling operation can deviate from the conventional vertical plane to extend horizontally, much more of the resource becomes accessible (Speight 2016b). The in-situ oil is thus extracted using multi-stage hydraulic fracturing in horizontal wells. These tight oil wells, like shale gas wells, are characterised by very high decline rates. Production often halves in the first year, with typically 80% of the total volume being recovered within 3 years. As a result, to offset the loss of production, operators have to intensively drill new wells (IEA 2013). This pattern has been observed at major tight oil fields in the North America, with drilling becoming increasingly frantic (Maugeri 2013).

The practical application of horizontal drilling for oil production began in the early 1980s, by which time the advent of improved downhole drilling motors and the invention of other necessary supporting equipment, materials, and technologies (particularly downhole telemetry equipment) had made the technique commercially viable. The use of these techniques in conjunction with hydraulic fracturing has greatly expanded the ability of producers to profitably recover oil from low-permeability deep geologic formations (Speight 2013). It is only in recent years that commercial exploitation of tight oil has reached a significant scale, using techniques pioneered in North America (IEA 2013; Speight 2016a). As a result, the energy landscape in North America has been transformed, responding to geopolitical imperatives for the U.S. to reduce its dependence upon overseas energy markets (Henderson 2013). Tight oil is now the leading source of U.S. crude oil production (see Fig. 9.2), making up 54% of the total in

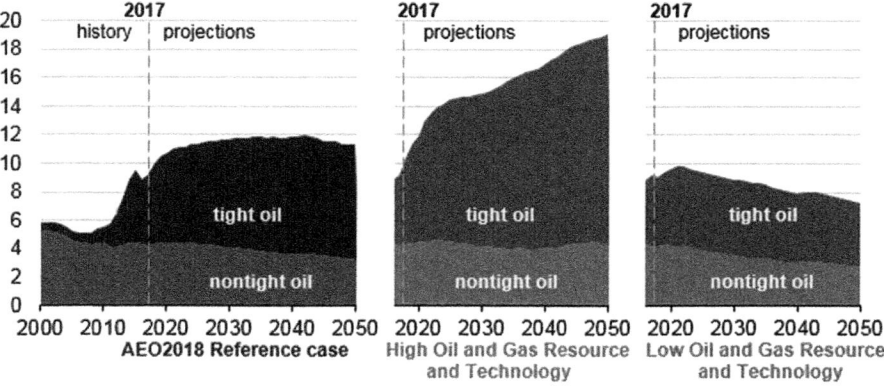

Fig. 9.2 U.S. crude oil production and projected production in three AEO2019 cases (million barrels per day, 2000–2050). (Source: EIA 2018)

2017; it is anticipated to account for nearly 70% of total U.S. production by 2050. Consequently, U.S. crude production has peaked above 10 million b/d in 2018, exceeding that of Saudi Arabia and Russia for the first time this millennium (see Fig. 9.1) (EIA 2018).

Following on from the development in the United States, governments in other regions have begun to explore the possibilities for tight oil extraction. In Russia, the adoption of unconventional recovery techniques has recently been promoted through major tax breaks. Financial incentives have been offered to Russian and international oil companies to explore tight oil resources such as in the Bazhenov shale layer, in Western Siberia. According to the Russian Ministry of Energy, total reserves in the Bazhenov amount to as much as 3758 million barrels (mb).[3] The major Russian oil firms claim that direct access to this huge reserve could stabilise the country's faltering production rates. However, a lack of both equipment – especially units for multi-stage horizontal hydraulic fracturing at such depths – and of skilled personnel, has restricted these efforts. Thus, joint ventures with global oil firms, including ExxonMobil, Statoil, Total and Royal Dutch Shell have been instigated to introduce the necessary expertise and technologies for the development of multi-stage hydraulic fracturing (Henderson 2013).

However, following the Russian annexation of Crimea and the occupation of eastern Ukraine in 2013, the United States and the European Union have imposed sanctions on the Russian government in the form of asset-freezes, visa bans, and controls on exports of energy technology (Tuzova and Qayum 2016). As a result, virtually all involvement in Russian projects by western companies has been suspended, including a joint venture between state oil firm Gazprom Neft and Royal Dutch Shell to frack for Bazhenov tight oil at the Priobskoye oil field (Farchy 2014). Nevertheless, Gazprom Neft vowed to persevere independently, and in July 2016 it was announced that a 30-stage horizontal hydraulic fracturing operation has been completed on the site. It set a new benchmark for this type of operation in Russia – the previous record being a shallower, 18-stage bore (Gazprom Neft 2016). The achievement was hailed a significant boost to hopes for Russia's future economic security.

The presence of tight oil resources, and growing technological capability and expertise in the Russian oil industry to access them, thus represent a significant moment in the history of energy development in the country. As the conventional oil resources which have sustained the Russian economy for the past 60 years become less productive, the prospects for fracking for tight oil appear to have increased. However, as discussed in the following sections, the arrival of the tight oil revolution in Western Siberia has significant environmental and cultural implications.

---

[3] Although there are huge variations in estimates stemming from assumptions about what is technically and economically recoverable, and what would be the recovery rate.

9.2 Tight Oil: Uncertainty and Intensity

As with shale gas and shale oil, the major environmental concerns around tight oil relate to the consequences of hydraulic fracturing in respect to its impact on water resources. Fracking is thought to contribute to water and land contamination, natural gas infiltration into fresh water aquifers, and the poisoning of the subsoil through intensive use of chemicals in hydraulic fluids (Jackson et al. 2013; Kappel et al. 2013). However, it has been argued that the vertical distance between shale and tight oil formations and the shallow aquifers prevents the growth of fractures that would threaten groundwater. Shale and tight oil deposits lie at depths greater than 1000 m, while freshwater aquifers are usually at depths of less than 100 m (Maugeri 2012). Little evidence of contamination of aquifers used for drinking water has been found by studies of active wells (Rodriguez and Soeder 2015).

However, much of the research suggesting the limited impacts of hydraulic fracturing on water resources is funded by energy companies and published in industry-focused reports or journals. A larger body of work has focussed on the qualitative nature of 'produced' water, including 'flowback' that results from wells drilled into high-pressure formations, which must be stored in ponds or tanks for reuse or disposal. These contributions, mostly from the environmental sciences, chemistry and geo-hydrological disciplines, have often taken hydro-geologically specific case studies situated within major shale plays in North America. Their results show the highly variable qualitative nature of the wastewater produced by the hydraulic fracturing process, and the significant challenges posed to producers in order to protect environmental and human health from their potentially toxic effects (Akob et al. 2015; Kondash et al. 2017; Orem et al. 2014).

Hydraulic fracturing is a water-intensive process. Horizontal wells have a greater rate of production, and therefore water consumption and wastewater production is higher compared to vertical wells due to an increased wellbore length exposed to the formation (Walker et al. 2017). Kondash et al. (2017) acknowledged the challenges posed by this rate of production for wastewater management. Indeed, as Cozzarelli et al. (2017) found, spillage events can have significant ecological implications. Moreover, the nature of liquid tight oil, as opposed to gas, represents a specific threat in respect to waste management and transportation of oil away from the well-site. Non-liquid gas leakages are emitted into the air, and do not pollute the immediate environment. An oil spill, however, contaminates surrounding areas for decades, rendering land and watercourses unusable for local populations (Stammler 2013). Reviewing existing research on the risks associated with shale gas and tight oil wastewater storage in the United States, Kuwayama et al. (2017) highlighted the need to seek new evidence regarding: (1) the degree of exposure to substances in wastewater through surface spills and leaching into groundwater; (2) the suitability of existing produced water and flowback storage technologies; and (3) the risks posed to ecological systems.

In a similar respect to existing research on the impacts of shale gas extraction upon water and the environment, studies on the consequences of hydraulic fracturing for tight oil is limited by the absence of a substantive historical dataset (Wheeler et al. 2015). Nevertheless, it is widely acknowledged that the technique, and its application for tight oil resources, poses significant challenges in terms of the protection of water resources from pollution. The particular intensive nature of hydraulic fracturing for tight oil production, the economic viability of which is contingent upon the sustained rapid expansion of well heads, pipelines and service roads, further exacerbates the possibility of these issues occurring (Maugeri 2013). As discussed in the following sections, for tight oil in Western Siberia, these challenges are potentially increased by existing environmental, political and cultural contexts.

9.3   The Western Siberian Basin: Changing Hydrological Regimes

Arctic and subarctic regions have unique geology, climate, hydrology, flora and fauna. The Western Siberian Basin is the largest area of unbroken flat terrain on the planet. This physiographic region covers most of the Russian territory between the Ural Mountains and the Yenisei River known as Western Siberia. The basin's rivers are strongly affected by the seasonal ice regimes and have expansive floodplains. Wetlands are common hydrologic features, including the Great Vasyugan Mire, which contains around 2% of the world's peat bogs (UNESCO 2018). Vegetation varies from mosses in the arctic tundra to grasses and boreal forests in the subarctic regions. Seasonal flooding produces shallow lakes, known as sors, which are very productive fish areas, and the region is host to a variety of birds and mammals, including migratory species (EPA 1998). These wetlands and lakes are globally significant sinks for atmospheric carbon dioxide and methane, and have received increasing attention in recent years due to the link between climate change and the release of these 'greenhouse gases' by melting permafrost.

The Ob River, upon whose banks the Siberian oil-cities of Surgut, Khanty-Mansiysk and Nefteyugansk are built, is the longest river in Russia at 5300 km and the seventh-longest in the world. One of the major river systems discharging into the Arctic Ocean, the Ob drains a total 2.9 million km^2, and its drainage basin occupies much of the Western Siberian Basin. Following the seasonal melting of snowpack in the Altai Mountains, the spring-summer period of high water begins in mid-April. In fact, levels begin to rise when the upper watercourse is still obstructed by ice; and maximum levels, which occur by May on the upper Ob, may not be reached until August further downstream (Malik et al. 2018). The drainage basin includes sphagnum marsh areas, pine woodland and many lakes and tributaries to the river, with a

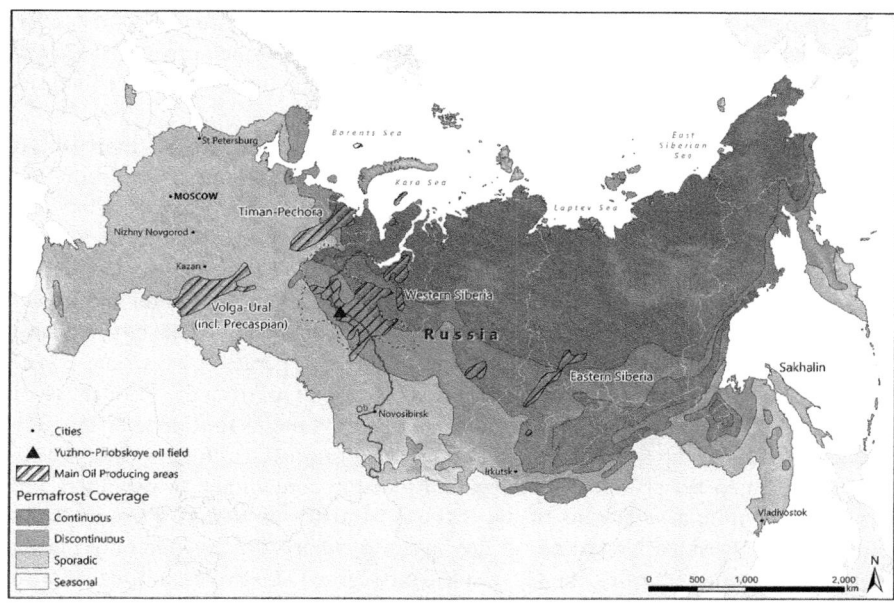

Fig. 9.3 Map of Russia with Siberian oil and gas fields and permafrost zones. (Callum Foster, UWE Bristol Cartography)

number of major fish overwintering sites. The area around the Priobskoye field, which occupies an area of 5466 km² along both banks of the Ob, is a region of permafrost transition (Fig. 9.3), with sporadic and isolated patches of permafrost to the north and south of the river (Grippa et al. 2007).

In 1998, the United States Environmental Protection Agency and Russian government conducted a joint environmental risk assessment of oil and gas activities in Russia, using the Priobskoye oil field as the demonstration site. The site was chosen due to the ecologically sensitive nature of the Ob River floodplain, beneath which the vast oil reservoir had been recently-discovered. Spills and produced water from oil extraction and transportation from the Priobskoye field, the study argued, have the potential to affect a wide area because of the annual flood and the discharge of the Ob and its tributaries. Fish concentrated in the river during the winter were assessed to be at a high risk. Spring flooding is potentially the most dangerous season, when a large section of the flood plain is inundated, rivers become active, and discharge of water intensifies. During this period, even the more elevated areas may be saturated with water, and bisected by numerous small watercourses (EPA 1998).

In recent years, significant changes have been observed to the seasonal hydrological regimes of surface and subsurface waters in the region. Increased temperatures have been accompanied by increased rainfall, extreme precipitation events, earlier and more intense snowmelts and increased instances of flooding (Gruza and Ran'kova 2009; Met Office 2011; Semenov 2011; Shiklomanov and Lammers 2009). At the same time, degradation of the permafrost layer underlying much of the

northern part of Siberia – due to increased average soil temperatures – is also causing significant hydrological and landscape changes (e.g. Goncharova et al. 2015),
increasing the incidence of flooding along rivers such as the Ob and its tributaries
(Takakura 2016; Zemtsov et al. 2014; Grippa et al. 2007).

Changing weather patterns and extreme hydrological events are increasingly
likely to coincide with incidents of pollution. In June 2015, after a wet spring and
significant snowmelt in the Altai Mountains, the Ob River overtopped its banks at
Nefteyugansk,[4] in the Khanty-Mansiysk Autonomous Okrug (region) of western
Siberia, Russia (The Siberian Times 2015a, 2015b). The river had risen to 10 m
above its normal level, and water inundated the surrounding floodplain and flowed
into parts of the city of 250,000. A few days later, 1 km outside of the city, a decay-
ing pipeline carrying oil across the floodplain ruptured. Operated by the state-owned
oil company, Rosneft, this particular pipe was conveying oil from the Priobskoye oil
field, 100 km west of Nefteyugansk, to refineries en route to European consumers.
Crude flowed into the flooded area, and continued to do so for 6 days. Local news
media published aerial imagery suggesting that the contamination extended well
beyond the public assessments of the Russian Ministry of Natural Resources and
Environment. Residents reported oil flowing into suburban areas and entering the
mains water system, turning the tap-water a dirty grey (The Siberian Times 2015b;
Luhn 2015).

With the shift towards tight oil, and the resultant expansion and increased intensity of production across this changing waterscape, the risk of coincidences such as
transpired in Nefteyugansk reoccurring is multiplied. As drilling is ramped up, so
the density and extent of extractive infrastructure increases. Water, with which many
lives and livelihoods in the Ob Basin is interwoven, is transformed into the agent of
this increased risk, conveying pollutants, ecological destruction and disease. As discussed in the following section, this additional threat is imposed upon the inhabitants of a vast Western Siberian landscape which, despite extensive environmental
regulation, has long been treated by the Russian state as a limitless sink for the
absorption of industrial waste.

9.4 Natural Resources and Environmental Regulation in Russia

The history of oil in Western Siberia begins in the 1960s, when the USSR discovered vast reserves in the region and began extracting and exporting crude to the
west. The 'Siberian Oil Rush', triggered by the 1973 Arab oil embargo on western
countries, sparked the greatest economic boom in Russia's history, as western
nations turned to their great adversary for oil. The region, historically the territory

---

[4] 'Neft' translates as 'oil' in Russian, with the city having grown out of the Siberian oil boom in the
1960s (Orttung 2016); 'yugansk' comes from the indigenous Khanty name of a nearby tributary to
the Ob.

of indigenous reindeer-herders, saw a massive expansion of oil field and infrastructure development, alongside the growth of urban centres. Ironically, the revenues actually enabled the Soviets to reach military parity with their great Cold War adversary, the United States. Meanwhile, the country was able to expand livestock production, salaries grew, it saw an influx of improved technologies, and was the precursor for an expansion of consumerism almost inconceivable during the communist era (Wenar 2015).

There has been an extensive body of literature pointing to the colonisation of the Russian Arctic, sub-Arctic and Western Siberia by state oil interests as having been accompanied by chronic environmental degradation. While it has been argued that such accounts of Soviet and post-Soviet Russia have emerged largely from the western perspective (Oldfield 2005), it is undeniable that the inefficiencies and polluting practices of the USSR's command-and-control economic system bequeathed extensive environmental problems with global effects that persist today (Bridges and Bridges 1996; Feshbach and Friendly 1992; Peterson 1993; Saiko 2001). Western Siberia has been disproportionately impacted by hydrocarbon extraction, where more than 64,000 km of pipelines have been constructed (Saiko 2001). Poor waste management, accidents and oil leaks have been commonplace, with water pollution and contamination of soils a major ecological and social threat (Balzer 2006).

In contrast to the western 'sustainable development' discourse which has become pervasive following the 1992 Rio 'Earth Summit', Russian interpretations emphasise 'natural resource management' rather than environmental regulation (Tynkkynen 2010). There is a "general concern for improving the efficiency of natural resource use and reducing the pollution-intensity of the country's industrial system with efforts to connect such ecologising intentions to wider social issues" (Oldfield 2017: 75). Such an understanding differs from the west, where environmental policy focuses on regulating the damage that human activity inflicts upon the environment. By contrast, the Russian approach has been interpreted as focussing on the economic opportunities retrievable from the environment, while linking environment damage to the impacts upon human health (Korppoo et al. 2015).

Nevertheless, following Gorbachev's reforms of the late 1980s and an environmental movement given significant impetus by post-Rio international discourse, Russia in fact inherited an extensive system of environmental regulation. Current policy is based on a set of command-and-control, economic, information, and other instruments. Most prominent is the administrative regulation of pollutants, in which a comprehensive system of environmental quality standards forms the basis for granting permits and setting fees at the federal level. The most important standards are the Maximum Allowable Concentrations (MACs), which establish maximum values for peak and average concentrations of environmental pollutants. The MACs cover over 3000 standards for soil, water and air pollutants, from which the government sets Maximum Permissible Emissions (MPEs) for stationary and mobile sources of pollution, with fees imposed for exceedances. Other regulatory measures include legal environmental liability, environmental performance ratings, and voluntary environmental management systems and corporate reporting (see IEA 2006).

However, Russian environmental policy instruments have been compromised by corruption, weakness of the government at all levels, and a shadow economy (e.g. Kotov and Nikitina 2002; Agyeman and Ogneva-Himmelberger 2009; Shvarts et al. 2016). The emission fee system lost much of its effectiveness after the 1990s as the fees failed to keep pace with rapid inflation, meaning that for the more prosperous oil companies, fees were so low as to be insignificant (IEA 2006). A general framework favouring economic development has thus compromised the work of environmental authorities and resulted in widespread non-compliance. Corporate actors are also significant for environmental policy-making, particularly in relation to industrial and energy sectors. These industries form an influential lobby which has been able to manipulate state environmental policies in their interests. Their position has been reinforced by the increasing emphasis placed by the government on economic growth through the extraction of natural resources (Korppoo et al. 2015).

Oldfield (2017) contends that the emergent system of environmental governance in Russia has been progressively undermined by a number of factors. First, a gradual shift in emphasis away from environmental protection functions and towards natural resource exploitation and use. Second, an increasingly apparent discontinuity between policy rhetoric and implementation, fomented by fiscal underfunding of enforcement and monitoring. And third, the complex and contradictory nature of policies and institutional arrangements which have been developed and repeatedly reconfigured over recent decades. One example is the dissolution of the *Ministry of Ecology* and its absorption into the *Ministry of Natural Resources and the Environment*, which has been seen as an example of the stepwise 'de-institutionalisation' of environmental policy and a discursive reframing which implies a more instrumental view (Newell and Henry 2016).

Donahoe (2009) frames this argument in terms of the relentless recentralization of power characteristic of the Putin era, giving the example of the Environmental Impact Assessment laws. Following one of a number of 'omnibus bills' pushed through parliament with minimal debate, the law now no longer mandates an independent Impact Assessment for large-scale energy, industry, and agricultural projects. Instead, only a state-level assessment of 'documentation' provided by the developer at a later stage is required. In the process, opportunities for the public to participate in the evaluation of these projects are foreclosed (ibid). As discussed in the following section, this evisceration of environmental and social protections has particularly serious implications for the indigenous people of Western Siberia, whose livelihoods stand to be transformed by the advent of a new revolution in oil extraction within their territories.

9.5   A People on the Brink?

The Khanty-Mansi Autonomous Okrug was established in 1931 on the homelands of the Khanty people and their aboriginal kin, the Mansi. The tribes historically lived on the flood plains and along tributaries of the Ob River, near to the present-day oil city of Nefteyugansk that flooded in 2015. They retained their traditional lifestyle until relatively recently, in permanent settlements of 3–10 families in yurts, with remote wooden dwellings in the forest for seasonal hunting and gathering activities. Like many indigenous groups across Siberia and the Russian north, the Khanty to the North of the Ob were also reindeer-herders (Kama 2007). Their engagements with the landscape of Western Siberia are rich and complex, and water and rivers are central to cultural attachments with their traditional territories. Settlement, ethnic and cultural patterns have tended to follow the lines of the watersheds of rivers, which play a significant role in framing spatial understandings of territory and the landscape (see Istomin and Dwyer 2009). These bonds can also be spiritual or mythical in nature, with deities and superstitions of importance. For example, each river had a sacred place occupied by the idol of a patron spirit, responsible for the protection of the wider drainage basin and the communities resident within it (Jordan 2004). However, central to Khanty's connection with the land is:

> [a] pragmatic desire to carefully tend a complex mesh of relationships linking between humans, animals, spirits and deities via a stream of physical engagements with the land itself [...] Here, the emphasis is [...] the important forces, or creative energies, that 'keep things going'. The land is the key forum – and medium – for maintaining this cultural vitality, for maintaining the supply of game animals or the birth of new human offspring. (Jordan 2004: 20)

Changes in the flows and presence of water, determined by the seasonal hydrological dynamics of the Ob River basin, have long been central to adaptive subsistence strategies employed by the Khanty people. For several millennia, the response to these factors has been the practice of seasonal mobility between summer-occupied fishing sites and wider hinterland hunting areas. As springtime temperatures thawed the deep snow-cover, leading to massive flooding and the breakup of river ice, the Khanty exploited a huge and predictable movement of fish through the drainage basin. Meanwhile, the floods dispersed game animals much more widely across the taiga, concentrating them in smaller areas, so that different seasons were marked by an uneven distribution of resources across the landscape. Falling oxygen-levels beneath winter ice would force fish to leave the river system and the Khanty to retreat to the forests. There, elk would become the major protein source; with mink, sable, fox and squirrel also being hunted for trade. The herding of domestic reindeer has provided animals for sleds as well some occasional meat reserves (Jordan 2004).

Today numbering around 22,000, many of the Khanty continue to subsist on traditional family watershed territories, settled among the low hills and marshlands along the banks and tributaries of the Ob and Irtysh Rivers. They still support themselves through herding, hunting and trapping, and fish constitutes 70% of their diet (Wiget and Balalaeva 1996). However, their lifestyles began to be disrupted in the 1930s, first by the collectivisation and dislocation campaigns of the Soviet Union, when many families were forcibly resettled for collective work in villages (Kama 2007). After Gorbachev's market restructuring in the late 1980s, the privatisation of state reindeer farms and the removal of subsidies led to dramatic declines, if not collapse, of reindeer numbers in many places across the tundra and taiga (Anderson

2006). Collectivisation and incorporation into market economies undermined evolved coping strategies of nomadism and reciprocity, which had for centuries underpinned indigenous survival in Siberia (Pluzhnikov 2009). Then, with Siberian oil rush, the 1960s began to see swathes of land appropriated by the state for the oil which lay beneath them. The birth of the most prolific oil-producing region in the world was accompanied by the fragmentation of the Khanty's hunting, gathering and herding lands by well-heads, roads, power lines and pipelines (Wiget and Balalaeva 2011).

In some localities of the Khanty-Mansiisk Autonomous Okrug, as much as a third of summer pasture has been enclosed for oil production, leading to overgrazing and soil degradation on the remaining tundra (Chance and Andreeva 1995; Forbes 1999; Henry et al. 2016; Kumpula et al. 2010). Gas flaring has also been shown to thaw frozen soils which support moss and lichen growth of the tundra ecosystems which, in turn, is an important source of reindeer food (Balzer 2006). Meanwhile, the obstruction of watercourses and the alteration of geohydrological dynamics by the construction of various types of infrastructure also affects ecological systems (Kumpula et al. 2010). One significant aspect of reindeer husbandry, for example, is the presence of biting insects that can impact on the health of herds in certain areas. Being a major determinant of the presence of mosquitoes and gadflies, changes to surface water is thus a major issue for reindeer and reindeer herders.

Pipelines, in particular, have posed a significant threat to socioecological dynamics in Western Siberian and across Russia generally. They have often been poorly constructed, poorly maintained, and do not correspond to modern, international standards, with stretches often remaining in use beyond reasonable parameters of serviceability. Indeed, there are no Russian state regulations for the safety of trunk pipeline systems. The severe climate and weather exacerbates this situation, resulting in pipeline breaks, leaks and accidents. According to Greenpeace, official statistics highlight over 10,000 oil pipeline leaks in Russia every year, with over 97% of pipeline ruptures in Russia occur because of corrosion. Many pipes in Russia are over 30 years old, while their accident-free life span is limited to 10–20 years (Greenpeace 2014). As a result, oil pollution and water contamination from pipeline leaks have become a major environmental issue in Russia. As a result, those engaged in fishing in the Ob River basin have seen their waters seriously contaminated (Saiko 2001; Chance and Andreeva 1995).

Pipelines and roads also act as barriers to reindeer, disrupting migration routes and access to watering holes. Those water bodies which remain accessible to the animals are liable to contamination by pipeline leaks, which have caused reindeer deaths (Wiget and Balalaeva 2011). As Gachechiladze and Staddon (2007) point out, pipelines can be seen as non-human 'actants' exerting a regulatory influence on processes otherwise seen as purely human. These conditioning effects becomes more profound as the intensity of oil production and the density of pipeline networks increase. Local socioecological relations in the Western Siberian waterscape have thus been reconfigured through state political-economic projects and the global oil economy. Indigenous herders and hunter-gatherers have thus been forced to spatially and temporally adapt their practices in order to circumvent particular threats in certain places and at certain times (Anderson 2006; Stammler 2013).

Stammler (2013) points to the independent lifestyle, remoteness, disconnectedness, a non-confrontational predisposition, and the egalitarian structure of nomadic and semi-nomadic pastoralists that may have inhibited the development of institutions for collective resistance to this situation. The legacy of centuries of state domination has meant that many of these groups have also internalised the idea of collective responsibility on a national level – making the idea of subverting the strategic interests of the state morally questionable. In any case, despite constitutional assurances, indigenous Siberians have no legal status, being classified simply as 'ethnic citizens' without recognition of their claims to traditional lands, or effective political or institutional representation (Wiget and Balalaeva 2011).

As the Russian economy has become increasingly dependent upon hydrocarbons, and power has become more centralised, indigenous rights conferred by law following the collapse of the Soviet Union have been eroded (Balzer 2006). In recent decades, the same legal instability that resulted in the nullification of the Environmental Impact Assessment laws has fundamentally undermined those which were intended to protect indigenous rights to the land (Donahoe 2009). While cities like Khanty-Mansiysk, Nefteyugansk and Surgut began to prosper on the proceeds of Siberian oil, the Khanty and other aboriginal communities spiralled into a critical situation. Alcoholism, AIDS and other health issues became significant problems (Wiget and Balalaeva 1996, 2011). There has since been a disproportionately high and persistent adult mortality rate and a declining fertility rate among these groups. Mortality has risen notably in comparison with the middle and later part of the 1980s, when it was at its lowest point in recent history (Bogoyavlenskiy 2010).

The fates of the Khanty and other of Siberia's indigenous peoples have mirrored those of many indigenous people across the world's resource frontiers. In the United States and Canada, the recent campaigns against the Keystone XL oil pipeline and the Dakota Access Pipeline (DAPL), has been framed around traditional and cultural ways of knowing, specifically focussing on connections to water. Whilst news coverage has centred on police confrontations at sites such as the Standing Rock Sioux' Oceti Sakowin protest camp, these efforts have in fact entailed a multitude of non-violent, co-productive activities, beyond the kinds of civil disobedience that the Khanty have avoided. These included the creation of educational spaces in which collective cultural identities around water and stewardship are reaffirmed (Roumell 2018). These common narratives have been connected with scientific and legal expertise through networks of association between tribal communities, NGOs, academia and institutional actors to contest the basis for decisions.

Thus, the 'Water is Life' movement, emerging from the Sioux' opposition to the DAPL, emphasised bodily connections to the land via water, asserting their role as 'Water Protectors' (Lane 2018). Similarly, the Black Mesa Water Coalition have productively connected Navajo and Hopi tribal culture, water and 'food sovereignty' to oppose coal mining and coal-fired power generation (Robinson-Welsh 2015). By placing water at the centre of the narratives of these campaigns, appealing to bodily, existential and normative instincts, these indigenous groups have been able to garner broader popular and political support for their causes. These efforts have provoked high-level political interventions in decision-making processes from which tribes have otherwise been marginalised.[5] However, it is notable that despite significant disruption, the U.S. administration has ultimately been able to evade opposition by allowing a legal process biased toward the interests of oil capital to take its course. Meanwhile, a growing nationalist, chauvinist discourse has continued to legitimise internal resource development.

In Russia, by contrast, an increasingly centralised, authoritarian but geopolitically marginalised regime arguably treads a narrower path of legitimacy. Popular anti-authoritarian movements have received increasing media attention in recent years, notwithstanding the government's attempts to suppress acts of resistance and freedom of expression. Moreover, should international sanctions push socioeconomic conditions beyond certain thresholds, alongside those being faced by Russia's indigenous populations in its oil-producing regions, the state may be more susceptible to concessions in order to preserve its hegemony. In such a situation, groups such as the Khanty may be able to reclaim indigenous space by connecting their own plight to metropolitan democratic movements and to the struggles of other indigenous communities worldwide. Indeed, in a similar way to the North American tribes, they may find political traction by also invoking the centrality of water to their culture and its role in mediating their relationship with their surrounding landscape.

9.6   Conclusions

The contribution of this chapter was predicated upon two converging aspects of the unconventional hydrocarbons debate that have been afforded little attention in the literature to date. First, that much of the research on the environmental and social implications of hydraulic fracturing for 'unconventional oil and gas' has focused substantively on shale gas. In particular, perspectives on the specific nature of tight oil and its extraction are notably scarce. Second, I argued that the increasingly apparent risks posed by the hydrological implications of climate change and extreme weather and the expansion of tight oil are worthy of much greater empirical attention. Such a perspective calls attention to the new frontiers of unconventional oil extraction, and the specific political, sociocultural and environmental contexts into which they are extending. The Western Siberian Basin was introduced as one such case, which exemplifies this nexus of tight oil and climate change.

The examples given in this chapter call attention to the specific materiality of [tight] oil and water, and the way in which water acts as a key mediator of the economic, ecological and cultural relationships between people and nature. Thus, the geophysical nature of tight oil and the economic imperatives of those who seek to

---

[5] Obama-DAPL.

exploit it are manifest in the intensity of drilling, hydraulic fracturing, road building, pipeline construction and other associated activities. Notwithstanding scientific uncertainties over the potential for horizontal multi-stage hydraulic fracturing to impact groundwater, this level of intensity combines with the lack of effective regulatory deterrent against the externalisation of environmental costs in Russia, increasing the probability of oil spills. Meanwhile, the changing hydrological dynamics in the Ob River Basin increases the probability of flooding events coinciding with oil spills, thus transmitting environmental risk across a greater area.

Those whose everyday subsistence livelihoods depend upon the accessibility of large areas of uncontaminated pasture, forest, and water are most exposed to these risks. The indigenous inhabitants of Western Siberia have proven remarkably resilient to decades of political, economic, cultural and legal marginalisation. In adapting their practices to accommodate the expansion of the oil industry and the fragmentation and destruction of their traditional lands, their cultural and ecological relationship with the landscape and with water has been reconfigured. In regions such as the Western Siberian Basin, with the prospects of climate change and an acceleration in the intensity of oil extraction, more people like the Khanty may be driven further towards poverty and the deterioration of their conditions of existence. However, it is argued here that water can play a key role in resistance to this threat. By placing water, its essential nature to life and its role in mediating cultural relationships with the land, at the centre of these struggles, indigenous space may be reclaimed. In turn, those with the closest connections to these waterscapes may assume a leadership role in the global movement away from unsustainable fossil fuels and carbon emissions, the effects of which threaten millions of people worldwide.

I conclude by highlighting three main areas for future research on the subject of tight oil extraction and water resources in the fields of the environmental sciences, physical and human geography. First, considering the speed at which hydraulic fracturing for unconventional hydrocarbons of all kinds is being adopted in many places across the world, the scientific knowledge-base for the impacts of this technology must be expanded as quickly as possible. The specific implications of multi-stage horizontal hydraulic fracturing for tight oil, the necessary intensity of this form of extraction, particularly in relation to groundwater contamination, requires urgent empirical attention. Second, there is an urgent need for interdisciplinary studies which combine: modelling of the hydrological conditions and the impacts of climate change; with mapping of oil pipeline density, incidence of oil spills, and the presence of tight oil resources at various stages of exploration and development; and socioeconomic and cultural factors. Third, I argue that building upon anthropological studies of indigenous peoples in regions such as Western Siberia should be made a priority in human geographic disciplines. Much potential lies in developing qualitative understandings of cultural attachments to water, and in connecting them to broader political economies and political ecologies of oil and water resource use. Participatory action research which integrates scientific and indigenous knowledges of waterscapes, such as those of Western Siberia, may help to lend greater democratic agency to marginalised groups such as the Khanty.

References

Agyeman J, Ogneva-Himmelberger Y (eds) (2009) Environmental justice and sustainability in the former Soviet Union. MIT Press, Cambridge

Akob DM, Cozzarelli IM, Dunlap DS, Rowan EL, Lorah MM (2015) Organic and inorganic composition and microbiology of produced waters from Pennsylvania shale gas wells. Appl Geochem 60:116–125

Anderson DG (2006) Is Siberian reindeer herding in crisis? Living with reindeer fifteen years after the end of state socialism. Nomad People 10(2):87–104

Balzer MM (2006) The tension between might and rights: Siberians and energy developers in post-socialist binds. Eur Asia Stud 58(4):567–588

Bogoyavlenskiy D (2010) Russia's indigenous peoples of the north: a demographic portrait at the beginning of the twenty-first century. Sibirica 9(3):91–114

Bridges O, Bridges J (1996) Losing hope: the environment and health in Russia. Avebury, Aldershot

Budds J, Hinojosa-Valencia L (2012) Restructuring and rescaling water governance in mining contexts: the co-production of waterscapes in Peru. Water Alternatives 5(1):119–137

Chance NA, Andreeva EN (1995) Sustainability, equity, and natural resource development in Northwest Siberia and Arctic Alaska. Hum Ecol 23(2):217–240

Cozzarelli IM, Skalak K, Kent D, Engle MA, Benthem A, Mumford A, Haase K, Farag A, Harper D, Nagel S (2017) Environmental signatures and effects of an oil and gas wastewater spill in the Williston Basin, North Dakota. Sci Total Environ 579:1781–1793

Donahoe B (2009) The law as a source of environmental injustice in the Russian Federation. In: Agyeman J, Ogneva-Himmelberger Y (eds) Environmental justice and sustainability in the former Soviet Union. MIT Press, Cambridge, pp 21–46

EIA (2017) Country analysis brief: Russia. U.S. Energy Information Administration

EIA (2018) The United States is now the largest global crude oil producer. Available from: https://www.eia.gov/todayinenergy/detail.php?id=37053. Accessed 28 Sept 2018

EPA (1998) Environmental risk assessments of oil and gas activities using national security and civilian data sources. United States Environmental Protection Agency

Farchy J (2014) Russian oil: between a rock and a hard place. The Financial Times, 29 October 2014

Farchy J (2017) Gazprom Neft strives to go it alone in Russian shale oil. The Financial Times, 3 January 2017

Feshbach M, Friendly A (1992) Ecocide in the USSR: health and nature under siege. Basic Books, New York

Forbes BC (1999) Land use and climate change on the Yamal Peninsula of north-west Siberia: some ecological and socio-economic implications. Polar Res 18(2):367–373

Gachechiladze M, Staddon C (2007) Towards a political ecology of oil in post-communist Georgia: the conflict over the Kulevi oil port development. J Polit Ecol 14(1):58–75

Gazprom Neft (2016) Gazprom Neft becomes the first company in Russia to undertake 30-stage multi-stage fracking. Available from: http://www.gazprom-neft.com/press-center/news/1113835/. Accessed 26 Sept 2018

Goncharova OY, Matyshak G, Bobrik A, Moskalenko N, Ponomareva O (2015) Temperature regimes of northern taiga soils in the isolated permafrost zone of Western Siberia. Eurasian Soil Sci 48(12):1329–1340

Greenpeace (2014) A brief overview of the oil spill problem in Russia. Greenpeace Russia. Available from: http://www.greenpeace.org/russia/Global/russia/report/Arctic-oil/GPRussia_Oil_spills_briefing_ENG.pdf. Accessed 26 Sept 2018

Grippa M, Mognard N, Le Toan T, Biancamaria S (2007) Observations of changes in surface water over the western Siberia lowland. Geophys Res Lett 34(15):1–5

Gruza G, Ran'kova EY (2009) Assessment of forthcoming climate changes on the territory of the Russian Federation. Russ Meteorol Hydrol 34(11):709–718

Henderson J (2013) Tight oil developments in Russia [online]. The Oxford Institute for Energy Studies. Accessed 11 June 2018

Henry LA, Nysten-Haarala S, Tulaeva S, Tysiachniouk M (2016) Corporate social responsibility and the oil industry in the Russian Arctic: global norms and neo-paternalism. Eur Asia Stud 68(8):1340–1368

IEA (2006) Optimising Russian natural gas: reform and climate policy [online]. Organisation for Economic Co-operation and Development/International Energy Agency. Accessed 15 July 2018

IEA (2013) World energy outlook 2013 [online]. International Energy Agency. Accessed 10 June 2018

IEA (2018) Oil 2018: analysis and forecasts to 2023. International Energy Agency

Istomin KV, Dwyer MJ (2009) Finding the way: a critical discussion of anthropological theories of human spatial orientation with reference to reindeer herders of northeastern Europe and western Siberia. Curr Anthropol 50(1):29–49

Jackson R, Gorody A, Mayer B, Roy J, Ryan M, Van Stempvoort D (2013) Groundwater protection and unconventional gas extraction: the critical need for field-based hydrogeological research. Groundwater 51(4):488–510

Jordan P (2004) Ethnic survival and the Siberian Khanty: on-going transformations in seasonal mobility and traditional culture. Nomad People 8(1):17–42

Kama K (2007) Spaces of indigeneity within the West Siberian oil industry: the case of Salym petroleum development [online]. Masters, University of Oxford. Available from: http://www.geog.ox.ac.uk/staff/kkama-mscthesis.pdf. Accessed 12 Sept 2018

Kappel WM, Williams JH, Szabo Z (2013) Water resources and shale gas/oil production in the Appalachian basin: critical issues and evolving developments [online]. US Geological Survey. Accessed 12 June 2017

Kondash AJ, Albright E, Vengosh A (2017) Quantity of flowback and produced waters from unconventional oil and gas exploration. Sci Total Environ 574:314–321

Korppoo A, Tynkkynen N, Hønneland G (2015) Environmental regimes and Russia's approaches to environmental and foreign policy. In: Korppoo A (ed) Russia and the politics of international environmental regimes: environmental encounters or foreign policy? Edward Elgar Publishing, Cheltenham, pp 9–22

Kotov V, Nikitina E (2002) Reorganisation of environmental policy in Russia: The decade of success and failures in implementation and perspective quests [online]. Nota di Lavoro, Fondazione Eni Enrico Mattei. Accessed 28 Sept 2018

Kumpula T, Forbes B, Stammler F (2010) Remote sensing and local knowledge of hydrocarbon exploitation: the case of Bovanenkovo, Yamal Peninsula, West Siberia, Russia. Arctic 63:165–178

Kuwayama Y, Roeshot S, Krupnick A, Richardson N, Mares J (2017) Risks and mitigation options for on-site storage of wastewater from shale gas and tight oil development. Energy Policy 101:582–593

Lane TM (2018) The frontline of refusal: indigenous women warriors of standing rock. Int J Qual Stud Educ 31(3):197–214

Luhn A (2015) Russia's Rosneft charged over pipeline leak that caused oil to come out of taps. The Guardian, 30 June 2015

Malik LK, Owen L, Micklin PP (2018) Ob River. Available from: https://www.britannica.com/place/Ob-River. Accessed 9 June 2018

Maugeri L (2012) Oil: the next revolution [online]. The Belfer Center for Science and International Affairs. Accessed 28 Sept 2018

Maugeri L (2013) The shale oil boom: a US phenomenon [online]. Harvard Kennedy School, Belfer Center for Science and International Affairs. Accessed 28 Sept 2018

Met Office (2011) Climate: observations, projections and impacts (Russia). Met Office

Newell JP, Henry LA (2016) The state of environmental protection in the Russian Federation: a review of the post-Soviet era. Eurasian Geogr Econ 57(6):779–801

Oldfield JD (2005) Russian nature: exploring the environmental consequences of societal change. Ashgate

Oldfield JD (2017) Russian nature: exploring the environmental consequences of societal change. Routledge

Orem W, Tatu C, Varonka M, Lerch H, Bates A, Engle M, Crosby L, McIntosh J (2014) Organic substances in produced and formation water from unconventional natural gas extraction in coal and shale. Int J Coal Geol 126:20–31

Orttung RW (2016) Sustaining Russia's Arctic cities: resource politics, migration, and climate change. Berghahn Books, New York

Peterson DJ (1993) Troubled lands: the legacy of Soviet environmental destruction: legacy of environmental destruction. Routledge, Boulder

Pluzhnikov NV (2009) Reindeer-herding in circumpolar regions of Russia: general issues, challenges and observations. In: Beach H, Funk D, Sillanpää L (eds) Post-Soviet transformations: politics of ethnicity and resource use in Russia. Digitala Enheten, Uppsala, pp 203–216

Robinson-Welsh AA (2015) Whose 'nature' is it in? The Navajo Generating Station and the politics of nature, space and colonialism in Northern Arizona [online]. Senior Capstone Projects. Accessed 15 July 2018

Rodriguez RS, Soeder DJ (2015) Evolving water management practices in shale oil & gas development. J Unconv Oil Gas Resour 10:18–24

Roumell EA (2018) Experience and community grassroots education: social learning at standing rock. New Dir Adult Contin Educ 2018(158):47–56

Saiko T (2001) Environmental crises: geographical case studies in Post-Socialist Eurasia. Pearson Education

Semenov V (2011) Climate-related changes in hazardous and adverse hydrological events in the Russian rivers. Russ Meteorol Hydrol 36(2):124–129

Shiklomanov A, Lammers R (2009) Record Russian river discharge in 2007 and the limits of analysis. Environ Res Lett 4(4):045015

Shvarts EA, Pakhalov AM, Knizhnikov AY (2016) Assessment of environmental responsibility of oil and gas companies in Russia: the rating method. J Clean Prod 127:143–151

Speight JG (2013) Shale gas production processes. Gulf Professional Publishing, Amsterdam

Speight JG (2016a) Deep shale oil and gas. Gulf Professional Publishing, Amsterdam

Speight JG (2016b) Handbook of hydraulic fracturing. Wiley, Hoboken

Stammler F (2013) Oil without Conflict. In: Behrends A, Reyna SP, Schlee G (eds) Crude domination: an anthropology of oil. Berghahn Books, Oxford, pp 243–270

Takakura H (2016) Limits of pastoral adaptation to permafrost regions caused by climate change among the Sakha people in the middle basin of Lena River. Pol Sci 10(3):395–403

The Siberian Times (2015a) Floods break the banks of the River Ob, with 10 metre rise in water level. The Siberian Times, 12 June 2015

The Siberian Times (2015b) Oil spill causes 'ecological disaster' close to Nefteyugansk city. The Siberian Times, 30 June 2015

Tuzova Y, Qayum F (2016) Global oil glut and sanctions: the impact on Putin's Russia. Energy Policy 90:140–151

Tynkkynen N (2010) A great ecological power in global climate policy? Framing climate change as a policy problem in Russian public discussion. Environ Polit 19(2):179–195

UNESCO (2018) The Great Vasyugan Mire. Available from: http://whc.unesco.org/en/tentativelists/5114/. Accessed 4 July 2018

USDOE (2018) Department of energy to invest $30 million to boost unconventional oil and natural gas recovery. Available from: https://www.energy.gov/articles/department-energy-invest-30-million-boost-unconventional-oil-and-natural-gas-recovery. Accessed 25 Sept 2018

Walker EL, Anderson AM, Read LK, Hogue TS (2017) Water use for hydraulic fracturing of oil and gas in the South Platte River Basin, Colorado. JAWRA J Am Water Resour Assoc 53(4):839–353

Wenar L (2015) Blood oil: tyrants, violence, and the rules that run the world. Oxford University Press, New York

Wheeler D, MacGregor M, Atherton F, Christmas K, Dalton S, Dusseault M, Gagnon G, Hayes B, MacIntosh C, Mauro I (2015) Hydraulic fracturing–Integrating public participation with an independent review of the risks and benefits. Energy Policy 85:299–308

Wiget A, Balalaeva O (1996) Black snow: oil and the Khanty of Western SIberia. Cultural Survival Quarterly Magazine, December 1996

Wiget A, Balalaeva O (2011) Khanty, people of the Taiga: surviving the 20th century. University of Alaska Press, Fairbanks

Zemtsov V, Paromov V, Kopysov S, Kouraev A, Negrul S (2014) Hydrological risks in Western Siberia under the changing climate and anthropogenic influences conditions. Int J Environ Stud 71(5):611–617

Owen King is a Research Fellow in the School of Geography, Earth and Environmental Sciences at the University of Birmingham, United Kingdom. His research relates to geographies and political ecologies of resource use, with a particular interest in environmental governance, water resources and the extractive industries. At the time this chapter was written, whilst conducting his doctoral research at the University of the West of England, Bristol, he was a member of the European Commission-funded Sustainable Water ActioN (SWAN) project, a multi-disciplinary, multi-institution collaboration on water resource issues in Europe and the United States. His PhD thesis, entitled 'Excavating the post-political: mining water and public participation in Arizona, USA', examined the politics of knowledge production and the democratization of environmental management in the United States. Owen has since been engaged in ESRC-funded post-doctoral research on European shale gas controversies.

Chapter 10
The Political Ecology of Shale Gas Exploitation in Ukraine

Olena Mitryasova, Volodymyr Pohrebennyk, and Chad Staddon

Abstract The chapter explores the main problems associated with the use of hydraulic fracturing for the production of shale gas in Ukraine. Special attention is paid to water issues. A detailed SWOT-analysis of the problem was carried out. Territorial distribution of water resources is uneven and unfortunately does not match the regions of greatest energy or mineral abundance. The smallest amount of water is found in places of concentration of powerful industrial consumers – the Donbass, Kryvorizhzhya, and the southern region of Ukraine. Areas of potential shale gas coincide with the areas of the smallest available water resources and areas with the greatest density of population. Experience elsewhere and the now well-known specifics of the hydraulic fracturing process allows us to identify potential threats to the environmental security of the country: pollution of clean groundwater and surface waters of Ukraine; utilization of large quantities of fresh water in the Yuzhivska and Oleska basins, especially in densely populated regions; degradation of large natural areas; emissions of greenhouse gases exacerbating climate change; the occurrence of seismic phenomena; and the degradation of natural landscapes.

Keywords Hydraulic fracturing · Oleska and Yuzhivska deposits · Water resources · SWOT-analysis · Political ecology

O. Mitryasova (✉)
Petro Mohyla Black Sea National University, Mykolaiv, Ukraine

V. Pohrebennyk
Lviv Polytechnic National University, Lviv, Ukraine

C. Staddon
Department of Geography and Environmental Management, University of the West of England, Bristol, UK

© Springer International Publishing AG 2020
R. M. Buono et al. (eds.), *Regulating Water Security in Unconventional Oil and Gas*, Water Security in a New World,
https://doi.org/10.1007/978-3-030-18342-4_10

10.1   Introduction: Potential Shale Gas Resources in Ukraine

Ukraine has several types of hydrocarbon fuels within its national territory that impose different ecological issues. The demand for energy in the Ukraine is high as it has high per-capita energy consumption, much higher in fact than other middle-income countries (Baker Tilly International 2012). Ukraine's population of 42 million derive their energy as follows: gas production covers about 60% of total demand, with much of the rest imported from Russia (RBC Ukraine 2015). There is also considerable reliance on coal, particularly in the politically contested eastern part of the country. With the exception of hydropower, renewable energy sources are relatively undeveloped in Ukraine. In this context, shale gas could be a relatively attractive option, both as a replacement for dirty coal and Russian natural gas and as a "transition fuel" (see introductory chapter to this volume). Thus, there has been some systemic research on available reserves and potential environmental impacts (e.g. Mykhailov et al. 2014).

The main source of shale gas is non-combustible shales and sapropelites. A peculiarity of Ukraine is that within its territory are found all types of hydrocarbon fuels in unusual abundance – a factor that led to heavy industrialization during the Soviet period.[1] Shale gas is located in the western Carpathian region, while tight oil and gas are located in the southeast of the Dniprovsk-Donetsk cavity in the eastern part of the country (Fig. 10.1). Significant accumulations of shale gas also exist in the Bovtysk cavity (Kirovohrad and Cherkasy regions) and the Volyn-Podolsk plate and other parts of Ukraine. Coal bed methane is abundant in the Donbass, which is also one of the largest coal basins in Eurasia. There are also potentially exploitable hydrocarbons underneath the Black Sea (Goshovsky et al. 2014; Soloviev 2013).

Even according to the most pessimistic prognosis, Ukraine is ranked fourth in Europe in terms of shale gas reserves after Poland, France and Norway. However, the most optimistic forecasts suggest that Ukraine may be the most productive European country (State Statistics Service of Ukraine 2017).[2] The two largest known shale gas deposits are the Oleska and Yuzhivska areas (Table 10.1), located in the western and eastern parts of the country respectively (see Fig. 10.1). The Yuzhivska shale deposit is located within the Donetsk and Kharkiv regions of Ukraine and has a total area of approximately 7886 km^2 (three times the total surface area of Luxembourg) and may have as much as 2–4 trillion m^3 of exploitable reserves. The Oleska deposit is located on the Lviv and Ivano-Frankivsk region in western Ukraine and may have as much as 2.5 trillion m^3 of exploitable shale gas (Mykhailov et al. 2014). Thus, prospective volumes of shale gas reserves in the

---

[1] As King points out in his chapter on Russian unconventional energy production (this volume), Soviet-era industrialization depended on extensive exploitation of cheap energy with little concern for efficiency or environmental impact.

[2] This information is based on generalized data on the geological structure of the region and typical indicators of shale gas and/or other rocks that contain natural gas. The real assessment of stocks can be ascertained only with detailed geological research and above all based on the results of the exploratory drilling and testing of the geological material samples (Kalynynchenko et al. 2013).

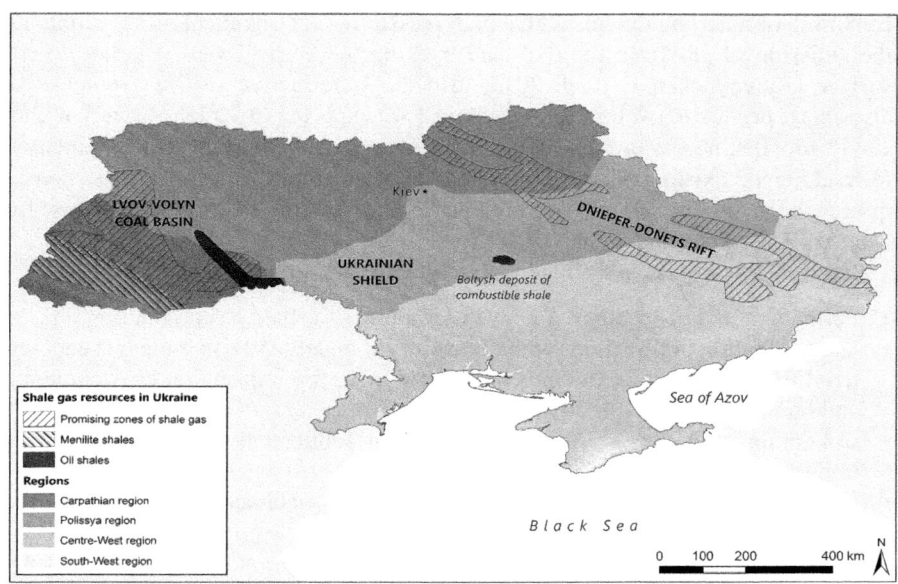

Fig. 10.1 Shale gas resources in Ukraine. (Callum Foster, UWE Bristol Cartography, with data from the Institute of Geological Sciences, NAS of Ukraine 2010)

Table 10.1 The comparative characteristic of the shale gas main basins in Ukraine (Deineko 2012)

Indicator	Yuzhivska	Olesk
Area, thousand km^2	99.82	68.61
Potential gas reserves, trillion m^3	1.36	4.22
Stocks technically suitable gas for commercial extraction, trillion m^3	0.34	0.85
Promising areas of shale gas, thousand km^2	16.12	20.32
Depth of mine, m	2950–4920	984–4920
The thickness of mine, m	8–70	394–978
The level of static pressure in the rock	Extremely high	Extremely high
Contents of organic compounds, %	4.00%	2.50%
Temperature gradient, %	1.30%	1.35%
The average concentration of gas in fractured rocks, billion. m^3/ km^2	3.08	5.79

country range from 4 to 7 trillion m^3 and considerably dwarf those of many other European countries combined.

The main stages of the shale gas extraction process are: vertical drilling to the location of the gas seam (from 1 to 5 km below the surface in the Ukrainian case); horizontal drilling of lateral mains up to 3–5 km in all directions; and hydraulic fracturing of horizontal mains with a high pressure water-chemical mix to stimulate the emission of gas from the rock and back up the vertical well structure to the surface (Kalynynchenko et al. 2013, also the Introduction to this volume). As elsewhere, production volumes for each well are expected to deplete entirely within 12–18 months, necessitating a dense arrangement of wells and also the potential re-fracking of existing wells. This technology is relatively new and the consequences of its use in Ukraine have not been extensively studied, so there is cause for caution (Kalynynchenko et al. 2013).

Environmental risks associated with hydraulic fracturing include:

- Permanent loss of locally scarce water resources to shale gas production;
- Contamination of frac fluids with industrial chemicals and contaminants encountered in situ, including radionuclides, with potential impacts on environmental and human health (NISS 2014);
- Collecting mud, reverse flow, leakage from septic tanks or vehicles during transport;
- Leaks or accidents caused by unprofessional personnel actions or related to the use of outdated technology;
- Leaks caused by seal failures (6% of wells break down at once after fracturing, and 50% within 15 years – See chapters by Miller, and Collins and Rosen in this volume);
- Leakage that occurs underground through natural or artificial cracks. Most of the liquid for fracturing remains underground (up to 80% of the injected volume). Studies show that this fluid can migrate towards natural reserves of drinking water (Kalynynchenko et al. 2013);
- In addition to problems with water quality, compounds such as benzene, ethylbenzene, toluene and n-hexane are often released in the atmosphere.

Development of new methods of hydraulic fracturing that may be less harmful to the environment (through, for example, the use of foam/clay mixtures) is underway, though it may be some years before they are brought into mainstream use.

In this chapter we adopt a "critical political ecological" approach to analyzing debates over shale gas exploration and exploitation in Ukraine. Prefacing the term "political ecology" with the appellation "critical" signals a redoubled attempt to listen to often unheard voices — an important contribution of feminist political ecologists (Rocheleau et al. 1996): those of women, ethnic minorities, the aged, etc., and of peripheral localities (Black 1990). The postulation of a "critical political ecology" also signals a deeper concern, beyond incorporating important stakeholders, with the ways in which they produce meaningful discourses about the natural world, which subsequently result in material practices of management and exploitation of the natural world. Here, debates in geography and social theory (Braun and Castree 1998; Darier 1999) about how we can move past mere constructionist theorisations of nature are enormously helpful in terms of the sophistication they have brought to scholarly attempts to balance the 'obvious' social constructedness of nature with the equally obvious non-negotiable realities of its biophysical limits (cf. Murdoch 1997). As Forsyth (2003, p.20) puts it, critical political ecology is fundamentally about the "development of an analytical approach that is biophysically grounded yet conscious of social and political constructions" of such key terms as 'environment,' 'nature' and 'natural resources'. Thus, in telling the story of recent debates over shale gas in Ukraine, we are attendant to issues of power asymmetry (especially between central government and local communities) and the use and misuse of "science" in arguing for unbridled development.

10.1.1   Ukraine's Water Resources

The potential annual resource from river waters in Ukraine is approximately 210 km^3, of which only 25% is formed entirely within Ukraine, while the rest comes into the country from Russia, Belarus and Romania. Soviet-era industrial policy meant that all large river systems were modified through dam construction for transport, energy and water supply. Ukrainian river flows are therefore highly dependent on the changing politics of upstream dam releases and vulnerable to the effects of climate change, which is making flow highly variable year on year (Institute of Geological Sciences NAS of Ukraine 2010). The country also contains considerable groundwater resources, though these are under-studied and have already been indiscriminately exploited in the past.

Total existing water withdrawal for the population and economy is approximately 15 km^3 per year though this does not account for vast system inefficiencies (leakages), evaporative loss or the requirements of the natural environment. Water demand is divided sectorally as follows:

- industry – 49.4%;
- agriculture – 25.8%;
- domestic water supply – 22.6%;
- other – 2.2% (Kovalenko 2010).

Infrastructure renewal has been poor since well before the country's independence in 1991, with the result that output per unit of water input is low and the level of water services in urban and rural areas is, in many regions, declining. Some regions, especially in the war-torn eastern part of the country are now critically short of water resources through a combination of poor infrastructure and a succession of hot, dry summers. Even parts of central Ukraine adjacent to the largest river, the Dnieper, are classified as "water scarce" (i.e. having water availability of less than 1000 m^3/person/year) (Danylko 2003).

The hydrological volatility of the Dnieper is characteristic of most Ukrainian rivers. While the Dnieper Basin receives 235 km^3 of inflow (mostly from snowmelt and groundwater) in an average year, more than 70% is either abstracted or evaporates, leaving approximately 52 km^3 (22%) to supply environmental needs along the riparian corridor to the Black Sea. Especially hot summers can mean that about 75% of the precipitation evaporates in the basin of the upper Dnieper, 87% in the basin of the middle Dnieper, and over 90% in the basin of the lower Dnieper, leaving very little outflow into the Black Sea.

Table 10.2 Regional challenges of water and land management (Kovalenko 2010)

Regions	Pressure	Main sectors affected/ challenges
1. Carpathian region	Floods from the mountain rivers; erosion; high ground water level	Tourism
		Nature protection
		Agriculture (drainage)
2. Polissya region	Floods; soil contamination; high ground water level & short dry periods	Nature protection
		Agriculture (drainage)
3. Centre-West region	Water and soil contamination & erosion	Nature protection
		Agriculture (drainage irrigation)
		Industry water supply
4. South-West region	Water scarcity; droughts floods by rising; ground water; water & soil contamination	Nature protection
		Intensive (irrigated) agriculture; Industry
		Climate change
		Tourism

Even the most water-abundant region, the Carpathian Mountains of western Ukraine, is experiencing significant water management problems. The main pressures in the region are floods, erosion, and high groundwater levels. The Polissya and centre-west zones are characterized by frequent fluvial flooding, erosive soil loss, and high groundwater levels. Only 75 km further east the picture changes markedly as one transitions into the Steppe zone, which is characterized by water scarcity, droughts, periodic flash floods and groundwater, water and soil contamination (see Table 10.2).

Territorial distribution of water resources is uneven and does not match the geography of demand. The smallest volumes of available water are co-located with concentrations of powerful industrial consumers or major cities – e.g. the Donbass, Kryvorizhzhya, central and southern regions of Ukraine. The largest water consumers are focused in the arid, heavily populated and industrialized regions. Areas of potential shale gas (Eastern and Western regions) coincide with the areas of the smallest amount of water resources (e.g. Yuzhivska deposit) and areas with the greatest density of population (Kalynynchenko et al. 2013).

10.2 SWOT Analysis of Shale Gas Development in Ukraine

In this section, a brief SWOT analysis of shale gas exploitation in Ukraine outlines the main strengths and weaknesses, as well as the possible opportunities and threats/ risks associated with shale gas exploitation in Ukraine. Rather than adopting a position on shale gas a priori, we contend that a balanced consideration of available evidence is critical to sound policy-making (cf. Dzurka and Golinko 2013; Lukin 2010, 2011; Staddon et al. 2016).

The main *strengths* (benefits) and opportunities relate to the presence of significant natural gas resources within Ukrainian sovereign territory. The identification of shale gas reserves coincides with a period in which geopolitical relations with the major hydrocarbon producer in the region, the Russian Federation, are severely strained. Having an indigenous source of energy could reduce Ukraine's dependence on imported gas from Russia and other countries and increase its energy security. Meanwhile, the possibility of exporting gas may constitute an attractive prospect to European countries, whose economies are also significantly dependent upon Russian energy supplies. European and American support for a pro-European Union Ukrainian government opens up avenues for investment in this effort, bringing the necessary international technology and expertise. This represents a considerable opportunity for economic development in Ukraine, leading to increased tax revenues and the creation of new jobs.

There are weaknesses in the economic case for shale gas in Ukraine. The territorial sovereignty of Ukraine continues to be threatened by its aggressive neighbor, Russia. Despite sanctions imposed upon the Russian Federation by the European Union, the country remains a dominant global energy producer, and this economic power has given it considerable geopolitical leverage as it seeks to extend its influence into eastern Ukraine (Yergin 1991).[3] Elsewhere in Ukraine this may create the conditions for a revanchist Ukrainian state, which may be tempted to link the development of indigenous energy with a strong and stable state. As Slavoj Zizek (1992) has pointed out, the ragged and drawn out dissolution of state socialism has sometimes combined with the unpalatability of neoliberal reform programmes to create an aporia dangerously ripe for atavistic forms of nationalism, as bloodily demonstrated in Yugoslavia and parts of the former Soviet Union. In these contexts infrastructure, including energy infrastructure, can become politically charged, as Gachechiladze and Staddon (2007) pointed out in their analysis of pipeline development in Georgia.

In general, according to economists, shale gas production can be profitable under stable high prices for traditional gas, since the process of exploration and extraction of shale gas requires a high level of ongoing capital investment due to the need for a constant increase of the number of wells.[4] However, the high cost of the extracted product, which is estimated by experts to be approximately US$350–500 per 1000 m^3 (Yakushenko and Yakovlev 2013; Yakovlev et al. 2013) compared with the lower or equal price levels for imported natural gas from Russia, as well as the large environmental losses which will accompany the shale gas extraction process, put into question the feasibility of shale gas development in Ukraine.

---

[3] Some analysts even suggest that one of the causes of the war in eastern Ukraine was a Russian desire to prevent Ukrainian production from threatening its market dominance in Europe: "In Russia's silent shale gas victory in Ukraine, the Russian-backed rebels fighting in the Donetsk and Luhansk regions ensured that at least for the near future, Ukraine's shale gas potential will not be able to challenge Gazprom's gas monopoly in the region." (Batkov 2015)

[4] Compared with conventional oil and gas which typically require huge capital investment in the start-up phase, after which operating capital becomes dominant.

Perhaps the most tangible shale gas-related threat to Ukraine comes from the potential adverse socioeconomic and environmental impacts of shale gas extraction. The necessary level of investment may require the diversion of significant government funds from other sectors. It also requires the ceding of agricultural lands, and therefore threatens existing economies, jobs and food security, particularly in rural areas of the country. Moreover, this effort propels Ukraine toward greater integration with European and global energy markets, and therefore potentially exposes the country to price variability and cycles of economic crisis. Further threats are posed by technological and economic dependency upon Europe and the United States, especially if – as seems currently to be the case – inward investment for shale gas is conducted under concession agreements with the Ukrainian government.

In place of existing agricultural lands comes the increased threat of ecological degradation, potentially affecting soil, water and air through the emission of pollutants from the processes of hydraulic fracturing, and the transportation of gas via pipelines. Table 10.3 attempts to summarise the level of risk shale gas exploitation poses to air, land and water resources. These impacts, if not well-managed, will be visited upon the human population, negatively affecting the health of those residing near to gas areas, not to mention the further implications of hydrocarbon use for people affected by climate change. The rapid production decline rates of shale gas wells necessitate the constant drilling of new wells, exacerbating and intensifying the potential environmental impacts in any given location. Meanwhile, the collocation of urban areas with some shale gas deposits increases the known risks and vulnerabilities associated with hydraulic fracturing, including contamination of groundwater, air pollution from vehicle movements and induced seismicity (see the Introductory chapter in this volume).

10.3 Foreign Investment in Shale Gas Exploitation in Ukraine to Date

At the beginning of 2013, Ukraine signed an agreement with the Anglo-Dutch energy giant Shell to explore and exploit shale gas resources in the Yuzhivska area in the northern Donetsk region. Over 10 billion USD was pledged for exploration and development activities over 50 years (Balmforth and Zhdannikov 2013). Initially the Ukrainian side was to receive 35% of the produced gas, rising to 65% in later stages of the contract. Expected profits would mean recovery of the initial investments within five years and generation of steady income from the development for its 45-year life span. Shell's interest was linked to the existence of already-established onshore gas fields, existing gas infrastructure, domestic markets (as Ukraine seeks to wean itself from reliance on Russian energy sources) and proximity to European markets.

The initial phase of the geological study undertaken by Shell involved obtaining the two-dimensional and three-dimensional seismic data and drilling 15 test wells at

Table 10.3 Estimation of potential impacts on land, air and water resources

Environment receptor		Project phase						
		Site preparation	Design, drilling, casing	Fracturing	Well completion	Production	Well abandonment	Overall across all phases
Air	Individual	Low	Moderate	Moderate	Moderate	Moderate	Low	Moderate
	Cumulative	Low	High	High	High	High	Moderate	High
Water	Individual	Low	Low	Moderate	Moderate	High	Low	Moderate
	Cumulative	Low	High	High	Moderate	High	High	High
Land	Individual	Moderate	Moderate	Moderate	Moderate	Moderate	Moderate	Moderate
	Cumulative	Moderate	Moderate	Moderate	Moderate	Moderate	Moderate	Moderate

the Yuzhivska area. Early data suggested that there could be up to 3.6 trillion m³ of exploitable shale gas reserves. However, due to escalation in the military conflict with Russia (rebels hold the southern parts of the Donetsk and Lugansk regions, within 200 kms of Yuzhivska) starting in June 2014, Shell was forced to suspend further work in the region.

In November 2013, Ukraine and US company Chevron signed a Production Sharing Agreement (PSA) for gas production at the Oleska field in western Ukraine. Over US$350 million was pledged for exploration and development of resources expected to exceed 1.5 trillion m³ (Robertson 2013). Chevron would need to drill nearly 1000 wells to achieve the specified production volumes. Given that the drilling of one well takes three months, for the given number of wells the company will need 80 modern drilling machines working round the clock for 3–4 years. While western Ukraine is not beset by the same political and military instability affecting eastern Ukraine, declining energy prices and relatively poor results from exploratory drilling here and on the Polish side of the border (this shale basin is shared by the two countries) caused Chevron to suspend operations in late 2014.

10.4   Potential Water Resource Challenges Related to Shale Gas

One of the main challenges for shale gas exploitation is associated with the use and protection of water resources. The cost of shale gas development in Ukraine has been estimated to be as high as US$8.5 million per well (see Table 10.4) (Yakovlev et al. 2013), of which the cost of providing adequate water is about US$1.2 million (14% of the total cost of shale gas extraction from a single well).

It is generally assumed that the water needed for drilling (400 m³ per conventional well) and for fracturing operations (up to 34,000 m³ per conventional well) will be accessed in one of two ways (Yakovlev et al. 2013). The first approach is to connect to existing 'mains' supply. The second is through the drilling of new wells for extraction of groundwater. However, due to the complex, time-consuming and expensive nature of exploration and extraction of new water sources, the former approach is currently the most attractive and prevalent option, as it is in other countries examined in this volume.

The short-lived nature of shale gas wells means that to maintain sufficiently high levels of extraction, it is necessary to conduct multi-stage fracturing and to build ever more horizontal wells off the main vertical well stems. This intensity of production means that shale gas extraction requires ever-larger volumes of water, which has particular implications when drawing upon mains water supplies and potentially diverting potable water from domestic, agricultural and other industrial uses.

In addition, it is necessary to take into account the fact, noted above, that Ukraine is one of the least water secure nations in Europe. For comparison, water availability in Ukraine is 1000 m3 per capita per annum; whereas in Sweden and Germany the figure is 2500 m³; in France, each person has 3500 m³ accessible to then; while in

Table 10.4 Estimation of economic costs and environmental damage associated with the extraction of shale gas (conditional deposit) (Rubel 2012)

Activity	Materials/Waste	Number	Cost (thousands of US dollars)
Construction of drilling grounds, equipment rental	Ground work	Up to 2.0 ha	2500
Drilling	Drilling of rig equipment		2000
Drilling	**Water**	**400 m³**	**200**
Drilling	Chemical substances		300
Creation of a casing	Materials of a casing	Up to 130 t	400
Creation of a casing	Cement	Up to 28 m³	50
Removal of sludge	Rocks	Up to 156 m³	100
Perforation	Explosives	Small fillers 25 g	700
Fracturing	**Water**	**Up to 34 thousands m³**	**600**
Fracturing	Chemical substances	Up to 600 m³	500
Fracturing	**Waste water**	**Up to 34 thousands m³**	**400**
Dismantling of drilling equipment	500		500
Auto transportations	6000 transfers	600	
Other expenses			2150
Total			8500

the UK, per capita water supplies are as high as 5000 m³ (Staddon 2007).[5] Consequently, almost 1300 settlements, home to more than one million people in Ukraine, live on water that is brought in by tanker trucks, often from great distances. However, due to the generally scarce nature of water resources for competing uses, variability in precipitation, and abstraction by upstream users in Russia, Romania and Belarus, water available one year is not necessarily available the next.

For the Yuzhivska area, one of the two main shale gas basins in Ukraine, the very large volumes of water required for shale gas production may have to be transferred from the distant Seversky Donets River. However, this watershed already provides drinking water for most of the population of the Kharkiv, Donetsk and Luhansk oblasts. The Ukrainian government has already invested significant sums in the construction of long-distance canals to these areas, including the 550 km Dnieper–Donbas canal and the 131 km Seversky Donets–Donbass canal. Moreover, in the event of water shortages, the smaller water regions of Kharkiv and Donetsk may not have access to a strategic reserve of groundwater.

---

[5] Though, as we see in other chapters in this volume, many European countries contain regions of water scarcity, such as the UK's east and southeastern regions.

Meanwhile, in the other shale gas area of Oleska, in the Carpathians, water reserves are also insufficient to maintain the necessary level of shale gas production, meaning that further engineered solutions will be necessary. According to preliminary projections, the consumption of water for shale gas production in the area would be about 15 million m^3 per year for the 1000 wells planned by 2030. However, all local water resources are already overallocated, with cities such as Ivano-Frankivsk experiencing chronic water shortages. Here, water used for a single shale gas well would be equivalent to the annual needs of 10,000 inhabitants (Kozlovsky 2014).

Shale gas production, which entails drilling to depths between 3.5 and 5 kms has considerable implications for the state of artesian waters. The depth of the wells means that they are virtually guaranteed to cross major groundwater bodies, with the potential to cause significant contamination of water resources. In addition, the Oleska region is characterized by a lack of options for the discharging or re-injecting of produced and wastewaters due to its proximity to water supply intakes. Ukrainian legislation prohibits the drilling of oil and gas wells and the release of wastewater in the region. In the area around Lviv, a city of 750,000 and the regional capital, the issue of transportation of wastes is a particular concern. In some countries this problem may be overcome through the construction of temporary pipelines. However, in Ukraine, there are no legislative restrictions on the transportation of fracturing fluid or its components, posing a potential environmental threat.

Drilling and hydraulic fracturing may result in the release of methane into groundwater. It also may result in the pollution of aquifers by potassium chloride, leading to salinization of water resources. During the fracking process, tonnes of hydraulic fluid are injected underground, a significant proportion of which remain in situ post-production. The exact amount of water and chemicals used depends on the permeability of rocks. Companies have tended to disavow the risks associated with chemicals used for hydraulic fracturing, claiming that these substances are often present in household detergents, cosmetics and even food, and that their consumption or inhalation is not a threat to health. However, many of these substances are either known or suspected to be dangerous. They often contain chemicals that are classified as carcinogens, allergens and hormone disruptors, including such toxic chemicals as hydrochloric acid, benzene, toluene, glutaraldehyde and 2-Butoxyethanol (Lechtenböhmer et al. 2014). Not all chemicals have verified safe exposure limits and some are known to be harmful to either or both the environment and humans (Gordalla et al. 2013) (Fig. 10.2).

In the regulations of most developed countries, which have not banned the use of fracturing, companies are required to fully disclose the composition of fracturing-liquids. In Ukraine there is no legislation requiring the provision of such information and no specific regulations on the types of chemical that can be used or which are prohibited. Thus not only is there no deterrent, in the event of an accident resulting in environmental pollution, there is no mechanism for energy companies to be held specifically accountable for harms to the environment or to human well-being.

There also exists a significant threat posed by even relatively remote pollution. For example, during the 1970s in Pervomaysk district (Kharkiv oblast) the

- Superficially active substances (SAS) – 0.096%;
- Potassium chloride – 0.06%;
- Thickener – 0.056%;
- Inhibitor of coagulation – 0.043%;
- Acid (hydrochloric or sulfuric) – 0.134%;
- Oxidant – 0.01%;
- Connecting element – 0.007%;
- A content of Ferrum regulator – 0.004%;
- Inhibitor of corrosion – 0.002%;
- Wetting agent – 0.088%

Fig. 10.2 Chemical additives present in hydraulic fluids used for fracking

'Pervomaysk Chemprom' project injected industrial wastes to a depth of more than 3 km (silvery sandstones) under slight pressure. The result of this was the pollution of the groundwater (Rymar et al. 2012).

Shale gas production in Western Ukraine could compromise conservation efforts in the Carpathian Mountains, leading to the revocation of nature protection funds and designations. Tourism and recreation industries in the region could be undermined by the potential ecological, aesthetic and social impacts of hydraulic fracturing and related activities. The development poses a potential threat to spa resorts such as Svaliava and Nemyriv and the unique Soledar salt mines, which are a major tourist attraction and also serve as a sanatorium for sufferers of asthma and other diseases. Research is urgently needed to assess the relative cost-benefits of continued operation of the above with and without shale gas development.

Another risk associated with shale gas exploitation, widely publicized, involves induced and seismic geological instabilities (see the Ehrman chapter in this volume; also Staddon et al. 2016). Buildings within urban developments in this region were not constructed to withstand the tremors that may be produced by the process. The eastern Oleska areas are already characterized by significant seismicity and tectonic activity. Hydraulic fracturing in this location could have unpredictable consequences, with potentially dangerous tremors possible. Moreover, there is a significant risk that seismic activity induced by hydraulic fracturing could itself compromise the integrity of the well casings leading to uncontrolled intrusion of fracturing liquids and gases into the adjacent strata and groundwater.

The extensive infrastructure required for shale gas production includes the construction of a large number of wells, waste-pools for used fracturing liquids, sewage, roads, and power lines. It is estimated that each repository would require no less than 300 large truck journeys to effect the final disposal of wastewater and sand. Each of these loads would travel a distance of 50–100 km for water and 200–300 km for the sand creating significant additional loading and related air pollution on the roads of both prospective shale gas regions (Kozlovsky 2014; Goshovsky et al. 2014; Lukin 2010, 2011).

10.5   Ukrainian Laws and Regulations Related to Shale Gas Production

In Ukraine the development of hydrocarbon resources is regulated by the legislative acts outlined in Table 10.5. In order to adapt Ukrainian law to meet EU regulatory standards, energy and environmental regulators need to address the following issues:

- elimination of natural gas monopolies;
- implement full competition in the internal gas market;
- enforcement against criminal access to the gas transportation networks;
- setting of economically justified tariffs for the gas market;
- measures to ensure sustainable consumption of gas and the introduction of energy-saving technologies;
- improvement of the financial and economic security of companies operating in the gas market, enhancing their profitability;
- implementation of environmental standards for the use of chemicals in the extraction of natural gas and other potential environmental impacts.

Many of the drivers for these regulatory priorities date back to the collapse of the former Soviet Union in the early 1990s and the subsequent need to dismantle the legal apparatus of the centrally-planned economy. Throughout much of the following period, Ukraine's political economic transformation was often stalled by the rise (as in other Republics of the former USSR) of oligarchs who sought to exert strong personal control over local economic affairs. After the "Maidan Uprising" (massive popular protests against the Yanukovych government) and deposition of the pro-Russian President Victor Yanukovych in 2013, the country embarked on an accelerated programme of legal reform including in the energy and environmental sectors.

After the collapse of the Soviet Union, Ukraine developed a new "Water Codex" in 1995, though as Vystavna et al. (2018) note, this Codex contained only minor changes from the Soviet era law. A key principle in current Ukrainian water law is the principle that all water bodies are the inalienable property of the Ukrainian people. This implies that the right of use of water resources does not rest with property owners, but with the Ukrainian state acting on behalf of the people. Other laws and regulations (e.g. the Ukrainian laws for "Protection of the Environment" and "On the Fundamental Principles….", Table 10.5) set out both the principles for assessing water users' abstraction charges and the maximum allowable concentration (MACs) of pollutants that can be introduced by users to abstracted water.[6] Strangely, there is no reference to any sort of ecosystems approach for setting MACs – instead the regulatory perspective is entirely human-focused. Table 10.5 also sets out

---

[6] It is an oddity of Ukrainian water quality standards that some are stricter than those used by the EC Water Framework Directive and the World Health Organisation, whilst others are less strict (Vystavna et al 2018, Table 10.2).

Table 10.5 Brief Description of Legislative Acts pertaining to hydrocarbon resources development and regulation in Ukraine (Law of Ukraine 2012)

Legislative act	Brief description
Ukraine "Mineral Resources Code"	The objectives of the Code are to regulate mining in order to promote scientific, rational, integrated use of mineral resources in the interests of present and future generations
Ukraine Law "The Basis of the Functioning of the Natural Gas Market"	This Law establishes basic legal, economic and organizational principles for the functioning of the market in natural gas. It applies to the unified national gas transportation system, trunk pipelines, gas distribution network and underground gas storage facilities. The Cabinet of Ministers and the central executive body on fuel and energy are specifically empowered to supervise and regulate the market
Ukraine Law "Pipeline Transport"	This Law determines the legal, economic and organizational principles of activities for the functioning of the pipeline transport network
Ukraine Law "Natural Monopolies"	This Law determines the legal, economic and organizational principles of state regulation of activities of subjects in relation to natural monopolies in Ukraine
Ukraine Law "Oil and Gas"	This Law determines the main legal, economic and organizational basis for the oil and gas industry of Ukraine. It governs the relations connected with the use of oil and gas subsoil, extraction, production, transportation, and storage and use of oil, gas and products of their conversion, and the utilization of these resources for the purpose of ensuring energy security of Ukraine
Ukraine Law "Energy Conservation"	This Law determines the legal, economic, social and ecological basis for energy conservation for all companies, organizations and householders in Ukraine
Ukraine Law "Production Sharing Agreements"	This Law is directed toward creating favorable conditions for investment into the exploration and production of minerals within the territory of Ukraine, its continental shelf and exclusive (sea) economic zone on the principles determined by production sharing agreements
Ukraine Law "Protection of Economic Competition"	The present Law defines legal grounds for the maintenance and protection of economic competition, for the limitation of monopolies in economic activities, and ensuring the efficient functioning of the economy of Ukraine on the basis of the development of a competitive market
Ukraine Law "Gas (methane) Coal Deposit"	This Law defines legal fundamentals for the functioning of the natural gas market in Ukraine. Founded on principles of free competition, due protection of consumers, security of supply, and integration of the states and energy industry with natural gas markets, including the creation of regional natural gas markets

(continued)

Table 10.5 (continued)

Legislative act	Brief description
Ukrainian Law "The Maximum Permissible Concentrations of Hazardous Substances in Water of Water Bodies, Used for Industrial, Drinking, Cultural and Domestic Water Use"	This law lists prescriptive norms for maximum allowable concentrations (MACs) of more than 1300 different pollutants. These MACs are sometimes higher than, and sometimes lower than, EC or WHO norms
Ukraine Law "Protection of the Environment"	The aim of the Law is the protection of nature: conservation, utilization and regeneration of natural resources; maintenance of ecological safety; prevention and mitigation of the negative effects of economic and other activity on the environment; protection of threatened and endangered species; protection of landscapes, unique territories and natural objects related to historical and cultural heritage
Ukraine Law "On the Fundamental Principles (Strategy) of Ukraine's State Environmental Policy for the Period until 2020"	The main targets of the Strategy are in the waste management sphere and include: ensuring the reduction of municipal waste in cities with populations over 250,000 at specialized and environmentally safe landfills by 2015; reduction of the whole volume of such waste by 2020, as well as a decrease of the municipal waste for the basic level of bio-degradable waste in the special conservation areas by 2020; 50% increase in capacity of waste provision, utilization and use of recyclable materials by 2020; implementation of latest solid municipal waste technologies

relevant acts and regulations related to energy development, and in these it is clear that, as with water, Soviet-era regulatory principles and perspectives are in need of revision.

Many questions remain in relation to the economic and environmental regulation of shale gas development in Ukraine. In September 2013, the Congress of the International Union of Environmental Protection (CIUEP) adopted a resolution that called on countries to suspend issuing licenses for shale gas extraction using hydraulic fracturing. It also called for bans on hydraulic fracturing in areas with drinking water resources, in areas with water shortages, in seismically unstable areas, and in protected areas ("Ecology, Law, Man" 2013). Though Ukraine is a member of the CIUEP, it has not given effect to this resolution, citing the national interest with respect to energy security in the context of the civil war in the east of the country. As we have seen above, the state, during the Yanukovych period, even lobbied for the interests of foreign energy companies such as Shell and Chevron in Ukrainian shale gas extraction. At the same time, it should be noted that the Ivano-Frankivsk Regional Council did not agree with the Government's proposed draft agreement on the distribution of revenues from hydrocarbons extracted from the Oleska region. The Parliamentary Group has urged the Ukrainian government to consider the numerous comments (138 comments) and amendments made by the special working

Fig. 10.3 Ukrainian protests against shale gas extraction. (National Ecological Center of Ukraine 2014)

group and consider them in the text of a new agreement about environmental protection and the distribution of shale gas revenues ("Ecology, Law, Man" 2013).

In the above context it is perhaps not surprising that popular support for shale gas development is low. As protests (Fig. 10.3) against shale gas in cities around the Ukraine clearly show, a growing number of citizens are just not mollified by central government reassurances regarding the safety of shale gas production. Company investors lobby for their own commercial interests, and typically do not provide local governments with transparent and clear information about the economic, social and environmental consequences of the implementation of the shale gas projects (Mìtryasova and Pohrebennyk 2017). Such projects represent a significant risk, not least because their implementation is not possible without systematic scientific support and the active consultation and participation of citizens.

In Ukraine, there have been numerous protests against shale gas extraction. Protesters are concerned about the pollution of water, air, soil, and the resultant damage to people's health. On International "Global Frackdown" Day (11th October 2014), when actions against the extraction of shale gas were undertaken in many countries of the world, there were well-attended solidarity events in Kyiv, Lviv, Poltava, Kharkiv, Zhovkva, Zolochiv and Ivano-Frankivsk. Representatives of civic organizations and eco-activists led by the National Ecological Center of Ukraine (NECU) protested in Kyiv. Whilst activists recognize the energy security challenges posed both by climate change and by the civil war in eastern Ukraine, they prefer energy demand management and alternative energy policies to the opening up of new hydrocarbon resources—especially given that so much about long-term environmental impacts of shale gas exploitation remains unclear.

10.6 Conclusions

This chapter has highlighted the high level of environmental and political risk associated with shale gas development in Ukraine and, in particular, the lack of adequate regulations and mechanisms for the protection of water resources. The development of shale gas in Ukraine requires that the state make significant investment in objective scientific research of these potential impacts. Such an effort cannot be left (as it is now) to private energy corporations, whose interests clearly lie in telling only one side of the story. What's more, climate change adaptation strategies need to recognise the requirement to reduce CO_2 production and move to a more sustainable political and economic system. It is thus very important to improve the knowledge base and develop a long-term plan for the phased transition away from hydrocarbon resources entirely.

This chapter has also identified major threats to the environmental security of Ukraine. In particular, these risks include the pollution of groundwater and surface waters by minerals, gases and hazardous chemicals used during the hydraulic fracturing process. It has also highlighted the quantitative threat to water resources posed by abstracting large quantities of potable supplies in densely populated regions. Furthermore, environmental risks associated with shale gas extraction were identified relating to the degradation of large natural areas, including the forest zone of Carpathians, due to the construction of wells, pipelines and infrastructure; emissions of greenhouse gases leading to climate change; the occurrence of seismic phenomena, including landslides and earthquakes; the removal of large areas of agricultural land from economic productivity; and the destruction of natural landscapes, flora and fauna.

References

Baker Tilly International (2012) Gas extraction in Ukraine. Добыча газа в Украине. [on-line] Available at: http://www.bakertillyukraine.com. Accessed 25 Jan 2012 (in Russian)

Balmforth R, Zhdannikov D (2013) Ukraine signs landmark $10 billion shale gas deal with Shell. Reuters, 24 January 2013. https://uk.reuters.com/article/uk-shale-ukraine-idUKB-RE90N11S20130124. Accessed 20 May 2019

Batkov S (2015) Russia's silent shale gas victory in Ukraine. Euroactiv, 2/9/2015. https://www.euractiv.com/section/energy/opinion/russia-s-silent-shale-gas-victory-in-ukraine/

Black R (1990) Regional political ecology in theory and practice: a case study from Northern Portugal. Trans Inst Br Geogr New Ser 15:35–47

Braun B, Castree N (1998) Remaking reality: nature at the Millennium. Routledge, London

Danylko VK (2003) Water resources of Ukraine: ecological statistic analysis. Herald SAU 1:254–260. (in Ukrainian)

Darier E (ed) (1999) Discourses of the environment. Blackwell, London

Deineko VV (2012) Shale gas: environmental aspects of the production (World experience for Ukraine, analytical evaluations). Reg Econ 4:98–108. (in Ukrainian)

Deposits of shale gas in Ukraine (2010) Institute of Geological Sciences NAS of Ukraine. – Access mode: http://nuina.net/ymperator-benzokolonky-prodolzhaet-vojnu-za-slancevyj-haz/

Dzurka GF, Golinko II (2013) Prospects of development and analysis of modern technology to extract shale gas in Ukraine. Ecol Plus 6(39):2–5

Ecology, Law, Man (2013) Вимагаємо мораторрію на видобуток сланцевого газу в Україні! [on-line] Available at: http://epl.org.ua/announces/vymahaiemo-moratoriiu-na-vydobutok-slantsevoho-hazu-v-ukraini-2/. Accessed 31 Jan 2013 (in Ukrainian)

Energy balance of Ukraine/State Statistics Service of Ukraine (2017) [Electronic resource]. – Access mode: access: http://www.ukrstat.gov.ua/

Forsyth T (2003) Critical political ecology: the politics of environmental science. Routledge, London/New York

Gachechiladze M, Staddon C (2007) Towards a political ecology of oil in post-communist Georgia: the conflict over the Kulevi Oil Port Development. J Polit Ecol 14:58–75

Gordalla BC, Ewers U, Frimmel FH (2013) Hydraulic fracturing: a toxicological threat for ground-water and drinking-water? Environ Earth Sci 70(8):3875–3893

Goshovsky SV, Krasnozhon MD, Liuta NH, Vasylenko AP, Kostenko MM (2014) Mineral resources base of Ukraine. Article 1. About the need for changes in the national program for the development of mineral resource base of Ukraine for the period until 2030. Miner Recourses 4:4–7. (in Ukrainian)

Kalynynchenko AV, Kopishynska OP, Kopishynsky AV (2013) Environmental risks of shale gas in the gas-bearing areas of Ukraine. Bull Poltava State Agrarian Acad 2:127–131. (in Ukrainian)

Kovalenko P (2010) Water resources of Ukraine: state and prospects of use. – Access mode: http://www.riob.org/IMG/pdf/P-_Kovalenko_Prezentaciya1.pdf

Kozlovsky SV (2014) Status and trends of shale gas in the World. Prospects for Ukraine: economic and ecological aspects, collection of scientific papers VNAU. Ser Econ Sci 2:49–60. (in Ukrainian)

Law of Ukraine «On Approval of the State Program development of the of mineral resource base of Ukraine for the period until 2030» № 4731-VI from 17.05.2012 (in Ukrainian)

Lechtenböhmer S, Altmann M, Capito S (2014) Impacts of shale gas and shale oil extraction on the environment and on human health. European Parliament, Brussels, 2011. – 86 p

Lukin A (2010) Shale gas and the prospects of its production in Ukraine. Article 1. The current state of the problem of shale gas (In the light of the experience of development of its resources in the United States). Geolog 3:17–33 (in Russian)

Lukin O (2011) Gas resources in Ukraine: current state and prospects of development. Bull NAS Ukraine 5:40–48. (in Ukrainian)

Mitryasova O, Pohrebennyk V (2017) Environmental aspects of the extraction of shale gas in Ukraine, work first international internet conference. Applied Science and technology research: Academy of Sciences of Ukraine, p 3 (in Ukrainian)

Murdoch J (1997) Inhuman/non-human/human: actor-network theory and the prospects for a nondualistic and symmetrical perspective on nature and society. Environ Plann D Soc Space 15(6):731–756

Mykhailov VA et al (2014) Unconventional sources of hydrocarbons in Ukraine: monograph. In 8 books. Book 8. The Theoretical Substantiation of the Unconventional Hydrocarbons of Sedimentary Rocks of Ukraine/National Joint-Stock Company «Naftogas of Ukraine», etc. – K.: Nika-Center, 280 p (in Ukrainian)

National Ecological Center of Ukraine (2014) Ukraine is on the brink. Extraction of unconventional gas Україна на розрив. Видобуток нетрадиційного газу [on-line] Available at: http://necu.org.ua/ukraina-na-rozryv/. Accessed 11 June 2014 (in Ukrainian)

NISS (2014) The danger of shale gas production in Ukraine – Access mode: http://www.niss.gov.ua/content/articles/files/slanets-19b15.pdf   http://anvictory.org/blog/2013/04/18/skaz-o-tom-kak-shell-na-ukraine-slancevyj-gaz-dobyval/

RBC – UKRAINE (2015) Газ собственной добычи покрывает 60% потребностей населения Украины. [on-line] Available at: https://www.rbc.ua/rus/news/gaz-sobstvennoy-dobychi-pokryvaet-potrebnostey-1432046272.html. Accessed 15 May 2015 (in Russian)

Robertson H (2013) Chevron signs Ukraine shale exploration deal. Petroleum Economist, 31 October 2013. https://www.petroleum-economist.com/articles/upstream/exploration-production/2013/chevronsigns-ukraine-shale-exploration-deal. Accessed 20 May 2019

Rocheleau D, Thomas-Slayter B, Wangari E (eds) (1996) Feminist political ecology: global issues and local experiences. Routledge, London

Rubel OE (2012) Economic and environmental basis and institutional preconditions for making decisions about prospects for shale gas extraction, scientific works of DonNTU. Ser Econ 2:72–77

Rymar MV, Kraevska AS, Dulin IS (2012) Environmental safety of extraction of shale gas in Ukraine. Reg Econ 4:109–114. (in Ukrainian)

Soloviev VO (2013) Unconventional sources of hydrocarbons. Geography 15–16:55–59. (in Ukrainian)

Staddon C (2007) Managing Europe's water: 21st century challenges. Ashgate Press, Farnham

Staddon C, Brown J, Hayes E (2016) Potential environmental impacts of 'fracking' in the UK. Geography 101(Part 2, 5):–14

Ukrainians Protest Against Shale Gas Extraction on the Fracturing Technology. – Access mode from site National Ecological Center of Ukraine: http://necu.org.ua/proty-vydobutku-slantsevoho-hazu-za-tehnolohiyeyu-hidrorozryvu-protestuyut-ukrayintsi. (in Ukrainian)

Vystavna Y, Cherkashyna M, van der Valk MR (2018) Water laws of Georgia, Moldova and Ukraine: current problems and integration with EU legislation. Water Int 43(3):424–435. https://doi.org/10.1080/02508060.2018.1447897

Yakovlev EO, Lyalko VI, Azimov OT, Kostiuchenko Yu V (2013) Development of Deposits of Shale Gas (SHALE GAS): a New Ecological and Technogenic Threats and Risks/Centre for Aerospace Research Land NASU: http://www.niss.gov.ua/public/File/2012_table/1126_prez1.pdf. (in Ukrainian)

Yakushenko L, Yakovlev E (2013) Prospects shale gas in Ukraine: Environmental aspects/department of environmental and technical safety/ – Access mode: http://www.niss.gov.ua/content/articles/files/slanets-19b15.pdf. (in Ukrainian)

Yergin D (1991) The prize: the epic quest for oil, money and power. Touchstone Books, New York

Zizek S (1992) Eastern Europe's republics of Gilead. In: Mouffe C (ed) Dimensions of radical democracy: pluralism, citizenship, community. Verso, London, pp 193–210

Olena Mitryasova is a Professor and Head of the Ecology and Environmental Management Department at Petro Mohyla Black Sea National University in Ukraine. She is the author of more than 250 scientific and methodological publications, including monographs, manuals in general chemistry, organic chemistry, environmental chemistry and the history of chemistry. Her research interests include problems of higher education pedagogy; curricula, methods, forms, and programs of natural science, environmental education, and education for sustainable development; environmental assessment and water resources management; environmental monitoring of water objects; water security research (especially at the urban scale); and water-related postgraduate programs, including opportunities for mobility. She holds a diploma with honors in organic chemistry from Mechnikov Odessa National University, and a PhD from the Institute of Pedagogics and Psychology of Professional Education of NAPS of Ukraine (Kiev).

Volodymyr Pohrebennyk is a professor in the Ecological Safety and Nature Protection Activity Department of National University (Lviv Polytechnic) in Lviv, Ukraine. He previously served as a professor of Engineering at the Karpenko Physico-Mechanical Institute of the NAS of Ukraine, the Higher Attestation Commission of Ukraine, and a senior researcher for the Higher Attestation Commission of the U.S.S.R. He also served as the Scientific Secretary of the Academic Council of Physico-Mechanical Institute of NAS of Ukraine (1989-2007); a member of the academic councils at the Institute of Electrodynamics of NAS of Ukraine (2012-2014). He is the author of more than 550 scientific and methodological papers, including six monographs, thirteen book chapters in books, three manuals, four textbooks, two preprints, sixteen methodical materials, and 27 patents. He holds a diploma in electrical engineering from Lviv Polytechnic Institute, Lviv, Ukraine, and a PhD in information-measuring systems from the All-Union Research Institute for Optical and Physical Measurements, Moscow, Higher Attestation Commission of U.S.S.R.

Chad Staddon is Professor of Resource Economics and Policy in the Department for Geography & Environmental Management at the University of the West of England, Bristol. Chad's research revolves around the social, political and economic issues related to sustainable resource management. Current projects focus on the historical geography of urban water systems around the world, water-energy trade-offs in unconventional oil and gas operations, and economic policy for resilient urban water services. He received his PhD in geography from the University of Kentucky in 1996 for research on the political economy of water (mis)management in post-communist Bulgaria.

Part III
What Comes Next? Disposing of Water from Hydraulic Fracturing

Chapter 11
Disposal of Water for Hydraulic Fracturing: Case Study on the U.S.

Romany Webb and Katherine R. Zodrow

Abstract In 2012, the U.S. oil and gas industry produced approximately 3.4×10^9 cubic meters (m^3) of water, equivalent to 9.1×10^6 m^3 per day and greater than six times the amount of water treated by the City of Houston, Texas. This "produced water" consists of drilling or completion fluids that exit a well shortly after it is brought into production, along with water occurring naturally in the rock formation that exits with the oil and/or gas. Produced water can be contaminated by hydrocarbons, metals, radioactive material, and salts, which can make recycling and disposal difficult. In this chapter, we will discuss two aspects of produced water handling—regulation and technology—specifically focusing on five U.S. regions—the Permian, Eagle Ford, Bakken, Marcellus, and Niobrara. We will explore various disposal practices used in each region and consider how the regulatory framework influences those practices. The focus will be on regulations in six states – Texas, North Dakota, Pennsylvania, Ohio, Colorado, and Wyoming – with jurisdiction over the above regions. Just as the regions have remarkably different geology, and therefore different quality of produced water, these six states also have different regulatory frameworks. To illustrate these differences, we undertake a detailed exploration of the regulations in Texas and Pennsylvania and compare other states' regulations where appropriate. The analysis highlights the complexity of produced water regulation, treatment, and disposal within the United States.

Keywords Waste disposal · Injection wells · Produced water · Recycling · Frac fluid · Texas · Pennsylvania

R. Webb (✉)
Climate Law Fellow, Sabin Center for Climate Change Law, Columbia Law School,
New York City, NY, USA
e-mail: rwebb@law.columbia.edu

K. R. Zodrow
James A. Baker III Institute for Public Policy's Center for Energy Studies,
Rice University, Houston, TX, USA

Environmental Engineering, Montana Technological University, Butte, MT, USA

© Springer International Publishing AG 2020   221
R. M. Buono et al. (eds.), *Regulating Water Security in Unconventional Oil and Gas*, Water Security in a New World,
https://doi.org/10.1007/978-3-030-18342-4_11

11.1 Introduction

Oil and gas development uses large amounts of water, both in the initial drilling of a well (e.g., to clean and lubricate the drill-bit), and during subsequent well completion processes (e.g., hydraulic fracturing). A portion of the water injected during well drilling and completion returns to the surface, as does water occurring naturally in the rock formation (together "produced water"). The amount of these return flows varies between geological formations, ranging from just 10% of injected volumes in the Marcellus to over 100% in the Barnett (EPA 2015, p. 4–3). The produced water is often contaminated—containing oils, solids, salts, metals, hydrocarbons, and naturally occurring radioactive materials ("NORM")—which makes its treatment difficult. Perhaps for this reason, most produced water in the U.S. is currently disposed of through underground injection, with little or no treatment. Underground injection can have serious environmental impacts and results in produced water, particularly flowback water, being permanently removed from the hydrological cycle. This, along with the use of freshwater for hydraulic fracturing operations, may contribute to water shortages, particularly in arid areas.

Rather than disposing of produced water, oil and gas operators could reuse it. This has dual benefits for operators, reducing their need to source freshwater and dispose of produced water. Despite these benefits, however, recycling is limited in many areas. This is likely due to economic factors, including the cost of treating produced water for reuse. Unless and until the economics change, regulatory intervention may be needed to encourage recycling. At a minimum, it is important that regulations not prevent or hinder recycling. This may occur where, for example, recycling operations are subject to overly burdensome and/or complex permitting requirements. However, care must be taken to ensure that any change in those requirements does not undermine environmental protections.

Several major oil and gas producing states, including Pennsylvania and Texas, have recently streamlined the permitting of recycling operations (PDEP 2012; RRC 2016). Texas has also sought to encourage recycling by providing tax incentives therefor (Texas Tax Code § 151.355(7)). Recycling could be further encouraged through other regulatory changes, such as restrictions on produced water disposal. This chapter discusses the regulatory framework for disposal in six oil and gas producing U.S. states, namely Texas, North Dakota, Pennsylvania, Ohio, Colorado, and Wyoming, with a particular focus on the regulations in Texas and Pennsylvania. At the time of writing, regulations in all six states allowed produced water to be disposed of through underground injection and surface water discharge. Five of the states' regulations also permitted wastewater disposal on land. The wide range of disposal options has likely hindered adoption of produced water recycling.

11.2 Water Production in Oil and Gas Operations

Oil and gas in both conventional and unconventional reservoirs coexist with water, and water exits a producing well along with the targeted oil and gas. Unless otherwise specified, the produced water regulations discussed here do not apply only to wells that are treated using hydraulic fracturing, but rather all wells that produce oil and gas. As the well ages, the water-to-oil ratio ("WOR") and/or the water-to-gas ratio ("WGR") increases. It was estimated that the U.S. national average WOR in 2012 was 9.2 cubic meters ("m^3") of water per m^3 of oil (Veil 2015). Therefore, the main fluid exiting oil wells is, in fact, water, and 3.4×10^9 m^3 of water was produced in the U.S. in 2012 (Veil 2015). While water is coproduced in conventional oil and gas and other types of unconventional wells, the hydraulic fracturing process changes some of the characteristics of that water, notably at the start of production. Specifically, fracturing fluid contributes to flowback water, which is more similar chemically to the fracturing fluid than the reservoir water. Flowback water, therefore, may contain corrosion or scale inhibitors, disinfectants, friction reducers, acids, or surfactants that are not naturally present in the formation. However, over time, produced water composition more closely resembles the formation water.

Onshore generation of produced water has increased since the early 2000s (Fig. 11.1a). This increase coincides with an increase in oil and gas activity due to exploitation of unconventional sources, including shale formations, oil sands, and coal bed methane. However, the increase in water production in more recent years is not as high as one may expect given the rise in oil and gas production (Fig. 11.1b), perhaps indicating that younger unconventional wells do not tend to produce as much water as older conventional wells. In 2012, Texas was the biggest generator of

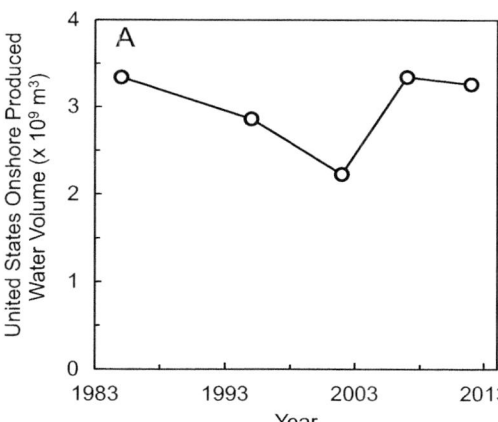

B) Produced Water Volumes (x 10^9 m^3)

State	Year 2007	Year 2012
Texas	1.173	1.182
Wyoming	0.374	0.346
Colorado	0.061	0.057
North Dakota	0.021	0.046
Pennsylvania	0.001	0.005
Ohio	0.001	0.001

Fig. 11.1 Onshore produced water volumes. (**a**) Trends in onshore produced water production in the United States from 1985 to 2012. (**b**) Produced water volume estimates for the six states described in this chapter. Data from API (1988), Veil et al. (2004), Clark and Veil (2009), and Veil (2015)

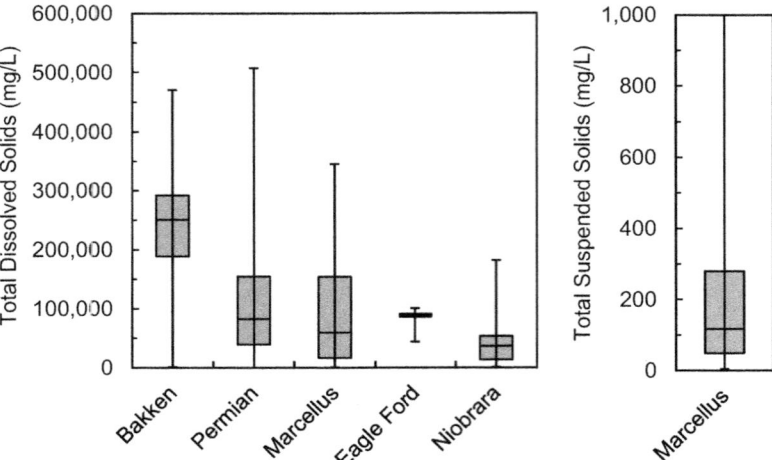

Fig. 11.2 Total dissolved solids (TDS) and total suspended solids (TSS) concentrations in selected basins and formations in the United States. In these box plots, the bottom of the box represents the first quartile and the top represents the third quartile. The horizontal line in the box represents the second quartile or median. The error bars show the spread of the data (minimum to maximum value). Data from Blondes et al. (2016)

produced water, generating 1.2×10^9 m^3, 36% of the produced water generated onshore (Veil 2015). Behind Texas, top water producing states include California, Oklahoma, Wyoming and Kansas.

The composition of produced water impacts the management method and treatment necessary (Igunno and Chen 2014). Although many different water parameters are considered when determining if water must be treated prior to disposal, reuse in enhanced oil recovery ("EOR"), or other potential uses, a few parameters will be discussed here. It is important to note that the concentration of each of these parameters varies (Fig. 11.2) widely depending on the formation and the well (Fakhru'l-Razi et al. 2009; Blondes et al. 2016). For example, in oil and gas wells, produced water may have a pH[1] as low as 3.1 or as high as 10. Likewise, the total suspended solids ("TSS")[2]—including clays, sand, precipitated salts, and bacteria—can range from 1.2 to 1000 mg/L. A value of interest for water recycling, especially if it is to be reused for irrigation, livestock watering, or released into a freshwater body, is the number of total dissolved solids ("TDS"). These dissolved solids include salts, such as sodium chloride, and metals, and many dissolved solids can be costly to remove. In produced water, TDS may range from 2600 to 360,000 mg/L (Fakhru'l-Razi et al. 2009; Blondes et al. 2016). For reference,

---

[1] A water's pH is an indicator of its acidity. Water with a lower pH value is acidic, while water with a higher pH value is basic. Neutral pH is 7.0. For reference, lemon juice has a pH of about 2, while an ammonia solution has a pH of about 11.

[2] TSS can be expressed using the unit "mg/L" referring to milligrams per liter. Therefore, there may be as much as 1000 mg or 1 g of suspended solid particles in 1 liter of produced water.

seawater has about 32,000 mg/L TDS—so, some produced water has greater than ten times the salt concentration of seawater. Both TSS and TDS concentrations, like all water quality parameters, vary between and within formations, as shown in Fig. 11.2. Finally, water from oil producing wells contains between 2 and 565 mg/L of oil, which must be removed prior to surface discharge to protect aquatic organisms. The wide variation in produced water composition contributes to regional variation in management strategies, state regulations and, likely, the treatment technologies employed.

11.3   Regulatory Framework Governing Produced Water

Despite its potentially dangerous nature, produced water is not subject to federal hazardous waste regulations adopted under the Resource Conservation and Recovery Act ("RCRA") (42 U.S.C. § 6901 et seq.). The RCRA aims to, among other things, assure "that hazardous waste management practices are conducted in a manner which protects human health and the environment" (42 U.S.C. § 6902(a) (4)). Hazardous waste is defined in section 2(5) of the RCRA (42 U.S.C. § 6903(5)) as:

> solid waste[3], or a combination of solid wastes, which because of its quantity, concentration, or physical, chemical, or infectious characteristics may –
> A) cause, or significantly contribute to an increase in mortality or an increase in serious irreversible, or incapacitating reversible, illness; or
> B) pose a substantial present or potential hazard to human health or the environment when improperly treated, stored, transported, or disposed of, or otherwise managed.

Certain wastes with these characteristics are, however, exempt from regulation as hazardous wastes. In October 1980, Congress provided a conditional exemption for certain wastes from oil and gas exploration and production ("E&P Waste"), pending review of their adverse effects (Pub. L. 96-482, October 1, 1980, 94 Stat. 2334). The U.S. Environmental Protection Agency ("EPA") conducted the review, which determined that regulation of E&P Waste as a hazardous waste is "not warranted," because existing state regulatory programs are "generally adequate" for controlling such waste, and additional federal controls would be uneconomical (EPA 1987). Thus, E&P Waste remains exempt from the hazardous waste regulations.

The exemption for E&P Waste covers "drilling fluids, produced waters, and other wastes associated with the exploration, development or production of crude oil or natural gas" (42 U.S.C. § 6921(b)(2)(A); 40 C.F.R. §261.4(6)).[4] These wastes are, however, only exempt from regulation under Subtitle C of the RCRA (i.e., the provi-

---

[3] Section 2(27) of the RCRA (42 U.S.C. § 6903(27)) defines "solid waste" to mean any "discarded material, including solid, liquid, semisolid, or contained gaseous material resulting from industrial, commercial, mining, and agricultural operations, and from community activities."

[4] The term "other wastes" encompasses waste material intrinsically derived from primary field operations associated with oil and gas exploration, development, or production, such as materials produced from a well in conjunction with oil or gas (EPA 2002).

sions dealing with hazardous wastes). RCRA Subtitle D, dealing with non-hazardous waste, continues to apply. The subtitle confers primary authority for regulating non-hazardous wastes on the states. The six oil and gas producing states examined in this chapter have each adopted their own regulations governing the management of produced water and other E&P Waste. Such waste is also subject to regulation under several federal statutes, including the Safe Drinking Water Act (42 U.S.C. § 300f et seq.) and Federal Water Pollution Control Act (commonly known as the "Clean Water Act") (33 U.S.C. § 1251 et seq.). Further information on the state and federal regulations is provided in Sects. 11.5 and 11.6.

11.4   Handling of Produced Water in the U.S.

In 2007, it was estimated that 96% of produced water (onshore and offshore) in the U.S. was disposed of via injection. Approximately 58% was injected into producing formations for EOR (Clark & Veil 2009). Approximately 38% of injected water was placed in non-producing formations. The remaining ~4% was discharged to surface waters. In 2012, injection was still the preferred disposal method in the U.S., with 45% of produced water injected for EOR, 39% sent to disposal wells, and 7% sent to off-site commercial disposal wells (Veil 2015).

Notable differences in management practices in the states are observed (Fig. 11.3). All states in our case study, except for Pennsylvania, manage their produced water primarily through injection for either disposal or EOR. (In Fig. 11.3a, data was only gathered for these two options.) In 2012, Pennsylvania, notably, allocated >85% of its produced water for beneficial reuse, mainly as fluid for oil and gas operations, such as completions (Fig. 11.3b). This shift occurred because of economics and geography, and is discussed in detail below. Colorado and Ohio also show modest beneficial reuse rates, 12% and 5%, respectively (Veil 2015). Produced water reuse in Colorado, like Pennsylvania, is primarily for use in oil and gas operations. Reuse of produced water in this manner—for example, as part of the fracturing fluid—is attractive because the dissolved solids do not need to be removed. Although no reuse in Texas was reported in the study depicted in Fig. 11.2, Nicot (2013) reported produced water reuse and recycling in Texas in 2012, stating that 5% of hydraulic fracturing makeup water in the Barnett Shale and in East Texas was sourced from reused or recycled water. This number was as high as 20% in the Anadarko Basin in North Texas. Several industry reports suggest reuse has increased in Texas in the past few years, as discussed in Sect. 11.6.2.

11.5   Regulation of Produced Water Disposal

As shown in Fig. 11.3, underground injection is the primary means of disposing produced water in the six oil and gas producing states examined in this chapter, except for Pennsylvania. During the early development of the Marcellus shale, large

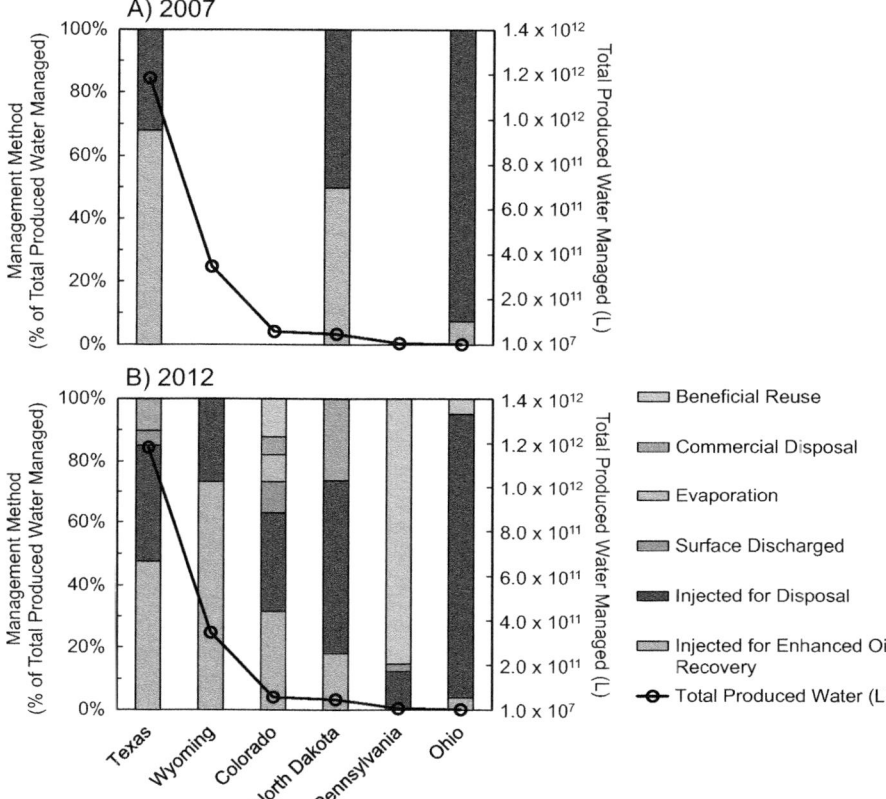

Fig. 11.3 Estimated volumes of produced water and management methods in Texas, North Dakota, Pennsylvania, Ohio, Colorado, and Wyoming in (**a**) 2007 and (**b**) 2012. Data from Clark and Veil (2009), and Veil (2015). Although it is likely that beneficial reuse, commercial disposal, evaporation, and surface discharge were produced water management strategies employed in 2007, data on volumes managed using these strategies was not collected in the referenced study. Management strategies were self-reported to the authors of these two studies, and numbers were not reported if the data are not shown

amounts of produced water were sent to sewage treatment plants ("publicly owned treatment works" or "POTWs"). However, conventional sewage treatment plants are unequipped to remove the large amounts of dissolved solids (i.e. salts) present in produced water. This practice, therefore, led to degradation of the water into which these treatment plants discharged their effluent. Seeking to minimize the potential for water contamination, oil and gas producers in other states often inject produced water into disposal wells. However, geologic conditions in Pennsylvania are such that the state has few sites suitable for injection. Oil and gas operators are, therefore, often forced to truck produced water to neighboring states, such as Ohio, for injection. The high cost of trucking has led some to pursue alternative practices. This section focuses on common disposal practices—underground injection, surface discharge, and land application—with the regulations governing each summarized in Table 11.1.

11.5.1 Disposal Via Underground Injection

Primary regulatory authority over underground injection rests with EPA. Through its Underground Injection Control ("UIC") Program, established under the Safe Drinking Water Act (42 U.S.C. § 300f et seq.), EPA regulates the injection of produced water into non-producing formations. The UIC Program does not regulate injection into producing formations for EOR (EPA 2016a). Such injection is generally considered part of the production process and may be regulated as such by the state in which it occurs.

EPA's UIC Program aims to prevent the contamination of underground sources of drinking water due to fluid injection (EPA 2016b). Wells used for injection are divided into six classes, based on the type of fluid they accept, with those accepting produced water falling into Class II (EPA 2016b). Federal regulations provide for the permitting of Class II wells. Existing wells are permitted by rule, meaning that the operator generally does not have to obtain an individual permit, unless specifically required to do so by EPA (40 C.F.R. § 144.21(a)). An individual permit must be obtained for any new well (40 C.F.R. § 144.31(a)).[5] Permits are issued by the EPA or, in states that have assumed primary responsibility for underground injection, the relevant state agency. The permit holder must comply with minimum standards relating to well construction and operation (40 C.F.R. pt. 144, 146). These include ensuring that the well is situated outside any formation containing underground sources of drinking water (40 C.F.R. § 144.22(a)) and cased and cemented to prevent the movement of waste into drinking water (40 C.F.R. § 146.22(b)–(e)).

By establishing permitting and other requirements for Class II wells, the UIC Program may affect the pace at which new wells are constructed. Limited availability of wells could increase the costs of disposal via underground injection and thereby encourage greater produced water recycling. That has been the experience of Pennsylvania which, as of February 2015, had just nine active Class II wells (EPA 2015, p. 8–69). Other key oil and gas producing states have a much larger number of wells, however. There are approximately 36,000 disposal wells nationwide, primarily in the west and south (EPA 2016a). Nearly one-quarter of the wells are in Texas, which had 8100 active disposals wells as of July 2015 (RRC). Due to the widespread availability of disposal wells, underground injection is typically inexpensive, often costing less than 6.3 USD/m^3 (Cook et al. 2015, p. 57). This is likely less than the cost of recycling.

---

[5] The Director may issue a permit on an area basis, rather than for each well individually, in certain circumstances (40 C.F.R. § 144.33).

Table 11.1 Regulation of produced water disposal in major oil and gas producing states

State	Regulation of produced water disposal methods		
	Underground injection	Treatment and discharge	Land application
Texas	*Authorized* with permit from the RRC	*Authorized* at a CWT that holds an NPDES permit issued by the EPA	*Authorized* subject to permitting by the RRC. Permit not required to dispose of:
		Authorized at a permitted POTW if the produced water originates from conventional oil and gas extraction facilities or coal-bed methane extraction facilities	Low-chloride water-based drilling fluids on-site through land-farming;
			Dewatered low-chloride water-based drilling and other fluids through on-site burial; and
			Spent completion and work-over fluids through burial
North Dakota	*Authorized* with permit from the North Dakota industrial commission	*Authorized* at a CWT that holds an NPDES permit issued by the EPA	*Not authorized*
		Authorized at a permitted POTW if the produced water originates from conventional oil and gas extraction facilities or coal-bed methane extraction facilities	
Pennsylvania	*Authorized* with permit from the EPA	*Authorized* at a CWT that holds an NPDES permit issued by the PDEP	Land application *authorized* subject to permitting by the PDEP
		Authorized at a permitted POTW if the produced water originates from conventional oil and gas extraction facilities or coal-bed methane extraction facilities	Disposal in pits *not authorized*

(continued)

Table 11.1 (continued)

State	Regulation of produced water disposal methods		
	Underground injection	Treatment and discharge	Land application
Ohio	*Authorized* with permit from the Ohio Department of Natural Resources	*Authorized* at a CWT that holds an NPDES permit issued by the Ohio Environmental Protection Agency	Road-spreading of brine produced from a vertical well *authorized* subject to permitting by the local authority
		Authorized at a permitted POTW if the produced water originates from conventional oil and gas extraction facilities or coal-bed methane extraction facilities	Road-spreading *not authorized* for:
			Brine produced from a horizontal well;
			Fluids from the drilling of a well;
			Flowback from the stimulation of a well; and
			Other fluids used to treat a well
			Disposal in pits *not authorized* for any fluids
Colorado	*Authorized* with permit from the Colorado oil and gas conservation commission	*Authorized* at a CWT that holds an NPDES permit issued by the Colorado water quality control commission	Land application *authorized* only for:
			Water-based bentonitic drilling fluids;
		Authorized at a permitted POTW if the produced water originates from conventional oil and gas extraction facilities or coal-bed methane extraction facilities	Produced water (outside sensitive areas only); and
			Drilling fluids and oily waste
			Disposal in pits *authorized* only for water-based bentonic drilling fluids and produced water
Wyoming	*Authorized* subject to permitting by the Wyoming oil and gas conservation commission (for non-commercial wells) or Wyoming Department of Environmental Quality (for commercial wells)	*Authorized* at a CWT that holds an NPDES permit issued by the Department of Environmental Quality	*Authorized* subject to permitting by the oil and gas conservation commission (for disposal through road spreading or in pits) or Department of Environmental Quality (for disposal through land farming)
		Authorized at a permitted POTW if the produced water originates from conventional oil and gas extraction facilities or coal-bed methane extraction facilities	

11.5.2 Discharge to Surface Waters

In addition to underground injection, oil and gas producers also have other options for disposing of produced water, including discharging it to surface waters. Any such discharge must be permitted under the Clean Water Act (33 U.S.C. § 1251 et seq.). Section 2(a) of the Act (33 U.S.C. § 1311(a)) prohibits the "discharge of any pollutant from any point source" without a permit. A "point source" is "any discernible, confined, and discrete conveyance from which pollutants are discharged" (40 C.F.R. § 122.2). This includes the discharge of waste by oil and gas producers into surface water bodies.[6]

Section 2 of the Clean Water Act (33 U.S.C. § 1311) established a National Pollution Discharge Elimination System ("NPDES") Program, under which EPA or an authorized state agency may permit the discharge of waste to surface water. While most waste must be treated prior to discharge, there is an exemption for certain classes of oil and gas waste (40 C.F.R. § 435.32).[7] These include produced water generated from onshore facilities located west of the 98th meridian[8] with a use in agriculture or wildlife propagation (40 C.F.R. §§ 435.50, 435.52) and wastewater from facilities producing 1.6 m³ or less of crude oil per day (40 C.F.R. § 435.60). No other oil and gas waste may be discharged without treatment.

Treatment can occur at private facilities known as centralized waste treatment facilities ("CWTs"). CWTs may be authorized, by permit, to treat and discharge produced water and/or other oil and gas waste. Such waste was, in the past, also treated and discharged by POTWs. However, as those facilities are typically designed to treat municipal wastewater with low pollutant concentrations, their treatment processes may be inadequate for highly polluted oil and gas waste. In the late 2000s, the Monogahela River in western Pennsylvania was polluted by inadequately treated oil and gas waste, discharged from a POTW. In response, the Pennsylvania Department of Environmental Protection ("PDEP") adopted regulations restricting the discharge of "wastewater resulting from fracturing, production, field exploration, drilling, or well completion of natural gas wells" (formerly 25 Pa. Code § 95.10(3)). The regulations prohibited discharges from POTWs unless the wastewater was first treated at a CWT. Because CWTs perform additional treatment processes, not undertaken by POTWs, this likely increased the costs faced by oil and gas operators. These surface water discharge costs, together with the limited

---

[6] Uncontaminated storm water discharges associated with oil and gas construction and field operation activities are exempt from the permitting requirements in the Clean Water Act (Kundis Craig 2013).

[7] Permits issued under the NPDES Program include limits on the maximum concentration of pollutants in the discharge, which are set based on the available treatment technologies, as well as the desired quality of the receiving water. Procedures for establishing those limits are set out in regulations adopted under the Clean Water Act (40 C.F.R. Pt. 131).

[8] The 98th meridian runs through North Dakota, South Dakota, Nebraska, Kansas, Oklahoma, and Texas.

availability of other disposal methods (e.g., underground injection), may have contributed to the high rate of recycling in Pennsylvania.

Following Pennsylvania's lead, EPA has adopted its own regulations with respect to the treatment of produced water by POTWs, which apply nationwide (40 C.F.R. pt. 435, subpt. C). The regulations establish a "zero discharge" requirement, which prevents POTWs accepting any waste from onshore facilities[9] used in the extraction of unconventional oil and gas, defined as oil and gas produced from a shale or other tight formation (40 C.F.R. §§ 435.30, 435.33). POTWs can accept waste from conventional oil and gas extraction facilities and coal-bed methane extraction facilities. Such waste must not, however, contain pollutants that will "pass through"[10] or cause "interference"[11] with the operations of the POTW (40 C.F.R. § 430.5(a)(1)). A POTW receiving such waste must specify pollutant limits, which translate the general prohibition on pass-through and interference into site-specific limitations, based on the POTW's capabilities (40 C.F.R. §§ 403.5(c), 403.8(f)(4)). All persons delivering waste to the POTW must comply with those standards.

11.5.3 Land Application

Produced water can also be disposed of on land, though this is less common than both underground injection and discharge to surface waters. There are no federal regulations governing land disposal. The practice is, however, generally regulated by the states. Regulations in five of the six oil and gas producing states examined in this chapter allow some land disposal of oil and gas waste. Three of those states— Texas, Ohio, and Colorado—restrict the types of waste that can be disposed of on land. Texas permits only low-chloride water-based drilling fluids to be disposed of through land-farming (i.e., where the waste is mixed with or applied to soil in such a manner that it will not migrate to other areas). Certain drilling and other fluids can, however, be disposed of through burial in Texas. In Colorado and Ohio, road-spreading is permitted for certain wastes that meet pollutant concentration limits.

Most states allow produced water and certain other oil and gas waste to be disposed of in earthen impoundments or pits. In Texas, for example, produced water may be disposed of in a pit with a permit from the Railroad Commission ("RRC"). The RRC may only issue a permit if it determines that the disposal will not result in

---

[9] "Onshore facilities" are those located landward of the inner boundary of the territorial sea (40 C.F.R. § 435.30).

[10] Clean Water Act regulations define "pass-through" as occurring where a pollutant is not removed through treatment at the POTW (40 C.F.R. § 403.3(p)).

[11] Clean Water Act regulations define "interference" as occurring where a pollutant inhibits or disrupts the POTW, its treatment processes or operations, or its sludge processes, use, or disposal, resulting in a violation of the POTW's NPDES permit, or certain statutory provisions (40 C.F.R. § 403.3(k)).

the waste of oil, gas, or geothermal resources or the pollution of surface or ground water. The permit will specify minimum requirements for pit construction and operation, designed to protect water resources. Many other states have similar requirements in their regulations. Some states, such as Colorado, also impose requirements aimed at minimizing air pollution from disposal pits. Open-air pits, where oil and waste is left to evaporate, often emit volatile organic compounds ("VOCs") which are harmful to human health and contribute to ground-level ozone formation.[12] Seeking to reduce emissions, the Colorado Air Quality Control Commission has prohibited the disposal of VOC-containing waste through evaporation, unless the so-called "RACT" (reasonably available control technology) standard is met. In broad terms, RACT reflects the degree of emissions reduction that can be achieved through application of control technology that is found to be reasonably available, considering technological and economic feasibility. Thus, compliance with the standard may require changes to pit design and/or the installation of emissions controls, thereby increasing the cost of disposal.

11.6 Regulation of Produced Water Recycling

Most produced water reuse is for EOR, though some is reused as makeup water for hydraulic fracturing fluid. These two uses require relatively minimal treatment, as there is little need for salt removal. Some states, noting the potential benefit of reusing produced water for industrial purposes, have encouraged oil and gas operators to do so through streamlined regulation.

11.6.1 Available and Emerging Recycling Technologies

Many technologies for the treatment of produced water and other oily or saline wastewaters are currently in use, and several more are under development. Ultimately, the choice of technology depends upon the quality of the produced water (discussed in Sect. 11.2), the intended fate of the treated water, the scale of treatment (i.e. the size of the plant), the duration of treatment, and the cost. In most cases, several processes will be used to meet the water quality requirements of the treated water. Produced water treatment can be broadly grouped into two categories: (1) removal of oil and solids prior to injection for disposal, EOR, or other industrial uses and (2) removal of dissolved solids and potentially toxic compounds for non-industrial beneficial reuse. Disposal via injection and EOR require less treatment than non-industrial reuse. In some cases, the quality of the produced water is such that beneficial reuse for irrigation of salt tolerant crops or livestock watering is achievable without dissolved solids removal (Fipps 2016; Higgens et al. 2016);

---

[12]VOCs include benzene, ethylbenzene, toluene, and xylene.

however, most non-industrial beneficial reuse applications would require some degree of salt removal. Thus, produced water treatment desalination is presently an active area of research (Fakhru'l-Razi et al. 2009; Igunno and Chen 2014; Hayes and Arthur 2004).

Most produced water treatment in the U.S. removes primarily oil and suspended solids (Stewart and Arnold 2011). Treatment technologies that remove oil and solids can be grouped roughly into three different categories: (1) gravity separation, (2) gas flotation, and (3) filtration. Both gravity separation and gas flotation take advantage of the density[13] (or specific gravity[14]) of oil droplets and suspended solids. The settling velocity, v, of a suspended particle (either oil or solid) can be estimated using Stokes Law. Specifically, given the density of a particle, ρ_p, the particle radius, R, the density and viscosity of the fluid, ρ_f and μ, and the gravitational force, g, one can predict the settling velocity of the particle per the following equation:

$$v = \frac{2\left(\rho_p - \rho_f\right)gR^2}{9\mu}$$

Thus, the settling velocity of the particle, v, will increase with the particle size, R, or the gravitational force, g. The settling velocity also increases if the difference between the density of the particle, ρ_p, and the density of water, ρ_f, increases. Thus, larger, more dense particles sink more quickly. In addition to being applied to solid particles, this equation may also be applied to oil droplets, which have a density less than that of water. The resulting negative velocity value indicates that these particles will rise to the surface rather than settle to the bottom. This theory outlines the basic principle of separation between oil, solids, and water using gravity, where oil naturally rises to the top, and solids settle to the bottom. Skimmer tanks and API separators operate on this principle. These separators work well with relatively large particles. Smaller particles (that is, particles with a smaller R) have a slower settling (or rising) velocity. Thus, to separate these particles, some assistance may be needed. This assistance may come in the form of increased settling forces (that is, increased g) imparted by a hydrocyclone or centrifuge, or induced coalescence. Coalescers are built so that particles hit an object (for example, a flat or corrugated plate), accumulate there, and are bombarded by other particles. When the particles hit each other, they coalesce into larger particles that can be separated using gravity. Enhanced coalescence may take advantage of chemical additives to induce precipitation or a filter.

Oil droplets may also be removed using gas flotation. In gas flotation, bubbles are forced through the water. Because oil droplets stick to the bubbles, the oil is

---

[13] The density of a material is its mass divided by its volume. For example, water has a density of approximately 1 g/cm³.

[14] The specific gravity of a material is its density divided by the density of a reference material. The reference material often used for solids and liquids is water.

carried up by the bubbles to the top of the reactor. The foam on the top of the reactor (created by bubbles and oil) is skimmed off the surface. The efficiency of this process may be enhanced by using coagulants, polyelectrolytes, or demulsifiers to destabilize the particles suspended in the water, increasing their ability to stick to each other.

If, for a use like EOR, further particle removal is needed, filtration may be employed. Media filters use sand, anthracite coal, or nutshells. The suspended particles intercept the media particles as the produced water is forced through the media. Periodically, the media must be cleaned to remove the accumulated particles. In some cases, polymeric or ceramic ultrafiltration or even nanofiltration membranes may be used to filter water (Ashaghi et al. 2007). Membranes offer an advantage over media filtration because they can offer a high degree of particle removal with less chemical addition, they occupy a smaller footprint, and they have relatively low energy cost.

Although the technologies described above effectively remove oil and suspended solids, further treatment is required to remove dissolved solids and degrade or remove potentially harmful organic contaminants. Although these treatment processes are not widespread in the field, they are currently under development for produced water treatment applications, and they would be required if that water were to be used in non-industrial applications. (An extreme example of this would be treated produced water used for drinking water.) Dissolved solids removal is accomplished using desalination. Desalination approaches include evaporation, multi-stage flash, multi-effect distillation and mechanical vapor recompression. Most assisted evaporation technologies are very energy intensive while solar evaporation requires large land area. Some membrane processes, including forward osmosis (Coday et al. 2014) and membrane distillation (Duong et al. 2015), show promise for treating high concentration TDS produced water because they use a concentration and temperature gradient, respectively, rather than a pressure gradient to purify water. Processes such as reverse osmosis, which require very high pressures to treat high salinity water, will be limited in their application for produced water treatment (Shaffer et al. 2013), although some have used lower-pressure nanofiltration to remove divalent ions, including calcium and sulfate. Several technologies are being studied to remove potentially harmful organic chemicals from produced water. Oxidation and photocatalysis can degrade these chemicals. Chemicals may also be adsorbed onto carbon-based adsorbents, organoclays, polymers, or zeolites (Fakhru'l-Razi et al. 2009).

Removal of dissolved solids, including potentially harmful contaminants, will be necessary for non-industrial beneficial reuse applications. The applicability of these advanced produced water treatment technologies will vary by location. Desalination of produced water will likely prove more expensive than conventional treatment of freshwater and will thus probably only be applicable in regions prone to water shortages.

11.6.2   Trends in Recycling

Recycling of produced water in Pennsylvania has grown enormously in the past decade (Fig. 11.3). The primary driver for recycling in Pennsylvania is financial and driven by disposal costs, as discussed previously. However, the incentive for produced water recycling in Texas is quite different. Texas, is a more arid state than Pennsylvania, especially in the western region where the Permian Basin and Eagle Ford Shale are located. To decrease water demand for hydraulic fracturing, the Texas Legislature has encouraged operators to tap unconventional sources of water, including brackish groundwater and recycled produced water. Recycling additionally reduces water transportation costs, decreases traffic on rural roads, and thus reduces noise and wear and tear on roads. Thus, companies have a financial and regulatory incentive for water reuse (and water reuse is good for publicity). Chesapeake claims to reuse 870 m^3 per well in the Barnett Shale (Mantell 2011). In the Haynesville Shale in East Texas, where produced water is high in total dissolved solids, the company prefers to reuse the lower salinity drilling wastewater. They state that reusing produced water reduces the overall cost of operations. Fasken Oil and Ranch and Apache Corporation both limit the amount of freshwater they withdraw for hydraulic fracturing, targeting instead brackish groundwater and recycled produced water (Midland Reporter Telegram 2015). A representative of the Apache Corp. said the company treated 1.6 million m^3 of produced water in 2014, enough to fill 80,000 trucks (Boyd 2015). Although widely-available produced water recycling data is limited, the consensus from the industry and officials is that regulations (discussed below) are facilitating reuse of produced water and helping to lessen the industry's impact on water demand.

11.6.3   Regulatory Framework for Recycling

Produced water recycling is assumed to be legal throughout the U.S., though many states have yet adopted regulations with respect to the practice (Richardson et al. 2013). Of the six oil and gas producing states examined in this chapter, for example, Wyoming has no regulations governing recycling. The regulations in other states are summarized in Table 11.2 below. As indicated there, most require recycling operations to be permitted, typically by the state oil and gas regulator. The permitting requirements are intended to enhance state oversight of recycling to ensure that it is conducted safely and does not endanger public health or the environment. They may, however, have the unintended consequence of discouraging recycling by leading to burdensome and/or time-consuming reviews. Recognizing this, a number of states have recently taken steps to streamline the permitting process. One example is Pennsylvania, wherein regulations require all recycling operations to be permitted by the PDEP.[15] In 2012, the PDEP issued a general permit authorizing the recycling

---

[15] The PDEP is authorized to issue general permits under 25 PA. CODE § 287.612.

Table 11.2 Regulation of produced water recycling in major oil and gas producing states

State	Regulation of produced water recycling	
	Non-commercial recycling	Commercial recycling
Texas	*Authorized* without a permit if the recycled fluid will be used as make-up water in a fracking fluid treatment or as another type of oil field fluid. In all other circumstances, a permit is required from the RRC	*Authorized* at a commercial recycling facility that has been permitted by the RRC
North Dakota	*Authorized* with a permit from the North Dakota industrial commission	
Pennsylvania	*Authorized* with a permit from the PDEP. A general permit has been issued for the recycling of oil and gas liquid waste to develop or fracture a well. Persons wishing to operate under the general permit must obtain a registration from the PDEP	
Ohio	*Authorized* with a permit from the Ohio Department of Natural Resources	
Colorado	*Authorized* without a permit if recycling occurs at the well site. Recycling may occur off-site, at a non-commercial centralized waste management facility, that holds a permit from the Colorado oil and gas conservation commission	*Authorized* at a facility that has been registered with the Department of Public Health and Environment
Wyoming	No state regulations	

of oil and gas liquid waste for re-use in developing or fracturing a well (PDEP 2012).[16] Oil and gas producers recycling waste for use in future operations do not, therefore, have to be permitted on an individual basis and need only register with the PDEP under the general permit (PDEP 2012, p. 2).

In half of the six oil and gas producing states, regulations differentiate between commercial and non-commercial recycling operations, with less stringent requirements applied to the latter. Texas, for example, has established a simplified permitting process for non-commercial operations. Until 2013, Texas regulations required all recycling facilities to be permitted. Although this requirement continues to apply to commercial facilities, in March 2013, state regulations were amended to allow certain non-commercial recycling without a permit. Under the amended regulations, a permit is not required for the recycling of flowback fluid at a drilling site if the recycled fluid will be used "as make-up water for a hydraulic fracturing fluid treatment(s), or as another type of oilfield fluid to be used in the wellbore of an oil, gas, geothermal, or service well" (16 Texas Administrative Code § 3.8(d)(7)(B)).

These and other similar policies should, in theory, encourage increased recycling of produced water by lowering the costs faced by oil and gas operators. Their practical effect is, however, difficult to assess as most operators do not report on the extent to which they recycle. While there is some anecdotal evidence that recycling is increasing, in many areas, the bulk of produced water is simply disposed of. This is

---

[16] Oil and gas liquid waste is defined to include "liquid wastes from the drilling, development and operation of oil and gas wells and transmission facilities" (PDEP 2012, p. 2).

likely due to economic factors, with studies finding that recycling is generally more expensive than disposal, particularly through underground injection (Cook et al. 2015). Operators are, therefore, unlikely to recycle wastewater absent regulatory mandates or other incentives.

To our knowledge, no state has mandated recycling. Of the six oil and gas producing states examined in this chapter, only Texas has actively sought to encourage recycling through tax incentives. Texas legislation exempts "tangible personal property specifically used to process, reuse, or recycle wastewater that will be used in fracturing work performed at an oil and gas well" from state sales, excise, and use taxes (Texas Tax Code § 151.355(7)). The Texas Legislature has also considered providing tax credits to oil and gas producers who use recycled wastewater and/or other alternatives to fresh water in their operations (H.B. 4021, 84th Legislature, Regular Session (2015)).

Texas has considered imposing restrictions on produced water disposal to encourage increased recycling. A bill introduced in the Texas Legislature in 2013 would have prohibited the disposal of produced water from wells subject to hydraulic fracturing "unless [it] is incapable of being treated to a degree that would allow [it] to be: (1) used to perform a hydraulic fracturing treatment on another oil or gas well; (2) used for another beneficial purpose; or (3) discharged into or adjacent to water in the state" (H.B. 2992, 83rd Legislature, Regular Session (2013)). Another bill, also introduced in 2013, would have imposed a fee 0.06 USD/m^3 "on oil and gas waste disposed of by injection in a commercial well" (H.B. 379, 83rd Legislature, Regular Session (2013)). Neither bill passed.

11.7 Conclusion

Increasing produced water recycling will minimize the impact of future oil and gas operations on water resources. While there is currently some recycling of produced water for EOR and industrial uses in the U.S., this and other reuse remains fairly limited in most oil and gas regions, likely due to the cost and complexity of treating produced water. Produced water is often contaminated with oil, solids, salts, metals, and hydrocarbons which must be removed or substantially reduced prior to reuse in oil and gas and/or other applications. The cost of treatment may discourage recycling if other financial incentives—such as a relatively high cost of disposal—are absent. Recycling rates may also be impacted by the regulatory framework governing produced water disposal. In Pennsylvania, for example, regulatory restrictions on surface discharge have led to increased recycling by oil and gas operators. Recycling is less common in other states, likely due to the widespread availability of disposal wells for underground injection and a permissive regulatory framework.

There have recently been a few recycling success stories in Texas. This is likely due, at least in part, to changes in the regulation of recycling. The changes removed regulatory barriers to recycling by streamlining the permitting process. Texas' experience thus suggests that states wishing to increase recycling should take steps to simplify their regulatory frameworks. Care should, however, be taken to ensure that any simplification does not compromise environmental protections. The experience of Pennsylvania, where water resources were contaminated by improperly treated produced water, highlights the need for careful oversight of produced water handling.

Pennsylvania's experience also suggests that restrictions on disposal may encourage increased recycling. While in Pennsylvania the restrictions are largely a result of geology, which limits the sites suitable for underground injection, other states could achieve similar results through regulatory action. States could, for example, adopt regulations limiting the amount or type of produced water that may be disposed of through underground injection, surface discharge, or land application. Such regulatory action seems unlikely, however, particularly in major oil and gas producing states. In those states, restricting produced water disposal could have economic impacts, leading to a slowdown in oil and gas production (i.e., due to the higher cost of recycling). This is also likely to discourage the tightening of federal disposal regulations, for example, to treat oil and gas waste as hazardous under the RCRA. In the absence of regulation, recycling is likely to remain limited, at least for the foreseeable future.

References

American Petroleum Institute (API) (1988) Production waste survey. American Petroleum Institute (API), Washington, DC

Ashaghi SK, Ebrahimi M, Czermak P (2007) Ceramic ultra- and nanofiltration membranes for oilfield produced water treatment: a mini review. Open Environ Sci 1:1–8

Blondes MS, Gans KD, Rowan EL, Thordsen JJ, Reidy ME, Engle MA, Kharaka YK, Thomas B (2016) U.S. Geological survey national produced waters geochemical database v2.2 (PROVISIONAL). Available at: https://energy.usgs.gov/EnvironmentalAspects/EnvironmentalAspectsofEnergyProductionandUse/ProducedWaters.aspx#3822349-data. Accessed 13 Jan 2016

Boyd D (2015) Water management: recycling, treating practices enable operators to conserve freshwater supply resources. The American Oil and Gas Reporter. Available at: http://www.aogr.com/magazine/cover-story/recycling-treating-practices-enable-operators-to-conserve-freshwater-supply

Clark CE, Veil JA (2009) Produced water volumes and management practices in the United States (No. ANL/EVS/R-09-1). Argonne National Laboratory (ANL), Washington, DC

Coday BD, Xu P, Beaudry EG, Herron J, Lampi K, Hancock NT, Cath TY (2014) The sweet spot of forward osmosis: treatment of produced water, drilling wastewater, and other complex and difficult liquid streams. Desalination 333(1):23–35

Cook M, Huber K, Webber M (2015) Who regulates it? Water policy and hydraulic fracturing in Texas. Texas Water J 6(1):45–63

Duong HC, Chivas AR, Nelemans B, Duke M, Gray S, Cath TY, Nghiem LD (2015) Treatment of RO brine from CSG produced water by spiral-wound air gap membrane distillation—a pilot study. Desalination 366:121–129

EPA (1987) Management of waste from exploration, development and production of crude oil, natural gas, and geothermal energy: EPA530-SW-88-003. Available at: https://nepis.epa.gov/Exe/ZyPURL.cgi?Dockey=9100WI1P.TXT

EPA (2002) Exemption of oil and gas exploration and production wastes from federal hazardous waste regulations: EPA530-K-01-004. Available at: http://www.epa.gov/epawaste/nonhaz/industrial/special/oil/oil-gas.pdf

EPA (2015) Assessment of the potential impact of hydraulic fracturing for oil and gas on drinking water resources. Available at: http://ofmpub.epa.gov/eims/eimscomm.getfile?p_download_id=523539

EPA (2016a) Class II oil and gas related injection wells. Available at: https://www.epa.gov/uic/class-ii-oil-and-gas-related-injection-wells. Accessed 22 Jan 2017

EPA (2016b) General information about injection wells. Available at: https://www.epa.gov/uic/general-information-about-injection-wells. Accessed 22 Jan 2017

Fakhru'l-Razi A, Pendashteh A, Abdullah LC, Biak DR, Madaeni SS, Abidin ZZ (2009) Review of technologies for oil and gas produced water treatment. J Hazard Mater 170(2–3):530–551

Federal Water Pollution Control Act; 33 U.S.C. § 1251 et seq

Fipps G (2016) Irrigation water quality standards and salinity management. Texas A&M University, AgriLife Extension. Available at: http://soiltesting.tamu.edu/publications/B-1667.pdf. Accessed 28 Jan 2017

Hayes T, Arthur D (2004) Overview of emerging produced water treatment technologies. In: The 11th annual international petroleum environmental conference, Albuquerque, New Mexico, USA

Higgens SF, Agouridis CT, Gumbert AA (2016) Drinking water quality guidelines for Cattle. University of Kentucky College of Agriculture, Cooperative Extension Service. Available at: http://www2.ca.uky.edu/agcomm/pubs/id/id170/id170.pdf. Accessed 28 Jan 2017

Igunno ET, Chen GZ (2014) Produced water treatment technologies. Int J Low Carbon Technol Adv Access 9:157–177

Kundis Craig R (2013) Hydraulic fracturing (fracking), federalism, and the water-energy Nexus. Idaho Law Review 49:241–264

Mantell ME (2011) Produced water reuse and recycling challenges and opportunities across major Shale plays. Available at: www.epa.gove/sites/production/files/documents/09_Mantell_-Reuse_508.pdf

Midland Reporter Telegram (2015) Permian basin drillers lead the way in water recycling push. Available at: http://www.mrt.com/business/energy/article/Permian-Basin-drillers-lead-the-way-in-water-7402392.php

Nicot J (2013) Hydraulic fracturing and water resources: a Texas study. Gulf Coast Assoc Geol Soc Trans 63:359–368

Pennsylvania Department of Environmental Protection (PDEP) (2012) General permit WMGR123 processing and beneficial use of oil and gas waste. Available at: http://files.dep.state.pa.us/Waste/Bureau%20of%20Waste%20Management/WasteMgtPortalFiles/SolidWaste/Residual_Waste/GP/WMGR123.pdf

Railroad Commission of Texas (RRC) (2016) Recycling. Available at: http://www.rrc.state.tx.us/oil-gas/applications-and-permits/environmental-permit-types-information/recycling/. Accessed Jan 23 2017

Resource Conservation and Recovery Act; 42 U.S.C. § 6901 et seq

Richardson N, Gottlieb M, Krupnick, A, & Wiseman, H (2013) The state of state shale gas regulation. Available at: http://www.rff.org/files/sharepoint/WorkImages/Download/RFF-Rpt-StateofStateRegs_Report.pdf

Safe Drinking Water Act; 42 U.S.C. § 399f et seq

Shaffer DL, Arias Chavez LH, Ben-Sasson M, Romero-Vargas Castrillón S, Yip NY, Elimelech M (2013) Desalination and reuse of high-salinity shale gas produced water: drivers, technologies, and future directions. Environ Sci Technol 47(17):9569–9583

Stewart M, Arnold K (2011) Produced water treatment field manual. Gulf Professional Publishing, Waltham

Texas Tax Code

Veil J (2015) US produced water volumes and management practices in 2012. Report prepared for the Ground Water Protection Council

Veil J, Puder MG, Elcock D, Redweik RJ Jr (2004) A white paper describing produced water from production of crude oil, natural gas, and coal bed methane. Available at: http://www.veilenvironmental.com/publications/pw/ProducedWatersWP0401.pdf

Romany Webb is an Associate Research Scholar at Columbia Law School and Senior Fellow at the Sabin Center for Climate Change Law. Romany's research focuses on the use of law to address climate change and promote clean energy development. Prior to joining Columbia, Romany completed a fellowship at the University of Texas at Austin, where she researched legal options for minimizing the climate and other environmental impacts of energy development. The fellowship followed several years practicing energy and environmental law in Sydney, Australia. Romany holds a Master of Laws, with a certificate of specialization in environmental law, from the University of California, Berkeley. She also holds a Bachelor of Laws and Bachelor of Commerce (Economics), awarded with first class honors, from the University of New South Wales in Australia.

Katherine R. Zodrow is an Assistant Professor of Environmental Engineering at Montana Technological University and a nonresident scholar in the Center for Energy Studies at Rice University's Baker Institute for Public Policy. Her work has appeared in several academic journals, including Environmental Science & Technology and the Proceedings of the National Academy of Sciences. Zodrow's interests include water and wastewater treatment systems, the water-energy nexus, interactions between microorganisms and engineered systems, and nanotechnology. Previously, she was a postdoctoral researcher at Rice University with a joint appointment in the Center for Energy Studies and the Department of Civil and Environmental Engineering. Zodrow completed her Ph.D. in chemical and environmental engineering at Yale University as a National Science Foundation Graduate Research Fellow. She also holds a B.S. in civil engineering and an M.S. in environmental engineering, both from Rice University.

Chapter 12
Regulating the Disposal of Produced Waters from Unconventional Oil and Gas Activities in Australia

Tina Soliman Hunter and David Campin

Abstract Production of unconventional petroleum resources in Australia comprises the exploration for and extraction of shale gas and coal seam gas (CSG, also known as coalbed methane). This chapter examines the issues associated with produced water from CSG and shale gas extraction, which differ greatly in both content and regulation. In examining the regulation of produced water from the extraction of CSG, only the Queensland jurisdiction will be assessed, since it is the only jurisdiction where production is occurring. Due to a moratorium on shale gas exploration and extraction in the Northern Territory, the regulation of produced water from shale gas exploration and production in Western Australia and South Australia is considered, with a particular focus on Western Australia given the advanced development of shale gas exploration in that state. This chapter provides an overview of unconventional petroleum resources (UPR) in Australia, and the regulation of UPR exploration and production in Queensland, Western Australia, and South Australia. It considers issues relating to produced water from both shale gas and CSG production and analyses the legal and environmental issues related to produced water in shale gas and CSG activities.

Keywords Australia · Unconventional petroleum resources · Coal seam gas · Dewatering · Shale gas

T. S. Hunter (✉)
Centre for Energy Law, University of Aberdeen, Aberdeen, UK
e-mail: thunter@abdn.ac.uk

D. Campin
The University of Queensland, Brisbane, Australia

© Springer International Publishing AG 2020
R. M. Buono et al. (eds.), *Regulating Water Security in Unconventional Oil and Gas*, Water Security in a New World,
https://doi.org/10.1007/978-3-030-18342-4_12

12.1 Introduction

Production of unconventional petroleum resources (UPR) in Australia comprises the exploration for and extraction of shale gas and coal seam gas (CSG, also known as coalbed methane). It does not cover oil shale or underground coal gasification, which are produced using in-situ heating or combustion processes. The regulation of these onshore petroleum activities is undertaken by the individual states and the Northern Territory as a consequence of the constitutional division of powers in Australia at the time of Federation. Each state has developed their onshore petroleum at different stages, with some states—such as Western Australia and South Australia—having highly sophisticated, objective-based regulatory frameworks that have been developed as a result of petroleum activities occurring for the last half-century. For other states, such as Queensland, New South Wales and Victoria, where approximately 80% of the Australian population resides, there have been small amounts of onshore conventional petroleum development.

CSG has been under development in Queensland since the mid-2000s. It has been fully commercialised by three exporting consortia (Gladstone LNG Project (Santos), Queensland Curtis LNG Project (QGC, a subsidiary of BG), and Australia Pacific LNG Project (Origin) and Arrow Energy supplying the domestic market and the export market, with first gas exported in 2015. The Queensland CSG industry is the first significant commercialisation of CSG outside of the United States, and has evoked much concern from the public, community and landowners. The regulation of CSG activities has been at the heart of community concerns, with Queensland's adaptive management approach to environmental regulation of CSG extraction meaning that there have been thousands of regulatory amendments over the last 10 years as the impact and consequences of CSG extraction has been felt by landowners and communities. At the heart of such concerns have been the use of water and the disposal of produced water from CSG wells. As a result of these concerns, the CSG resources of New South Wales have been subject to a lengthy moratorium (2011–2016) for development. Consequently, it is only since late 2016 that CSG exploration in New South Wales has recommenced. In Victoria, a permanent ban on CSG development was legislated (*Resources Amendment Legislation (Fracking Ban) Act 2017*, Victoria), making it the first jurisdiction to permanently ban CSG extraction.

Unlike CSG, shale gas development in Australia remains in the exploratory phase, with shale gas exploratory wells drilled numbering fewer than 100 and no commercially producing unconventional gas wells drilled to date. This low rate of exploration is a result of a number of factors, including the location of the shale resources (many in remote desert areas), the climate (much drilling can only be undertaken in the six-month dry season), public pressures and community concerns, and governments' apprehension in approving drill programs for fear of community backlash. However, there has been some exploratory drilling undertaken, particularly in Western Australia, where the onshore petroleum framework is particularly advanced and comprehensive due to conventional petroleum activities occurring for decades. While the Northern Territory contains UPR, the current (as of October 2017) moratorium and previous public concerns mean that little exploration activity has occurred. To date, there is no exploration allowed while the Northern Territory government considers the future of shale gas exploration. The Northern Territory government has commissioned the *Scientific Inquiry into Hydraulic Fracturing in the Northern Territory*.[1]

Given the distribution and type of unconventional petroleum activities in Australia, this chapter will examine the issues associated with produced water from CSG and shale gas extraction, which differ greatly in both content and regulation. In examining the regulation of produced water from the extraction of CSG, only the Queensland jurisdiction will be assessed, since it is the only jurisdiction where production is occurring. Similarly, because there is a moratorium on shale gas exploration and extraction in the Northern Territory, the regulation of produced water from shale gas exploration and production in Western Australia and South Australia will be considered in this chapter, with a particular focus on Western Australia given the advanced development of shale gas exploration in that state. This chapter will first provide an overview of UPR in Australia, and the regulation of UPR exploration and production in Queensland, Western Australia, and South Australia. It will then examine concerns relating to produced water from both shale gas and CSG production. Finally, it will analyse the legal and environmental issues related to produced water in shale gas and CSG activities.

12.2 Overview of Unconventional Oil and Gas Resources in Australia

As a geologically old and complex continent spanning over 3.8 billion years of the Earth's geological history, Australia contains almost all known rock types (Johnson 2009). As a result of this complex geology, Australia's UPR are geographically separated. Shale gas resources generally dominate the western and central regions of Australia, while coal seam gas resources dominate the east coast (Hunter 2012, 55–58). This physical division in the location of UPR broadly follows the geological division of Australia, where the western and central areas are dominated by Precambrian geology while the eastern third is dominated by the newer Cambrian and Phanerozoic geology, with an abundance of tertiary geology (Johnson 2009, 33). To date, only terrestrial unconventional gas resources in Australia have been identified (Hunter 2012, 58). The location of Australia's unconventional gas resources can be seen in Fig. 12.1.

---

[1] The Interim Report, released in July 2017, can be found at https://frackinginquiry.nt.gov.au/interim-report.

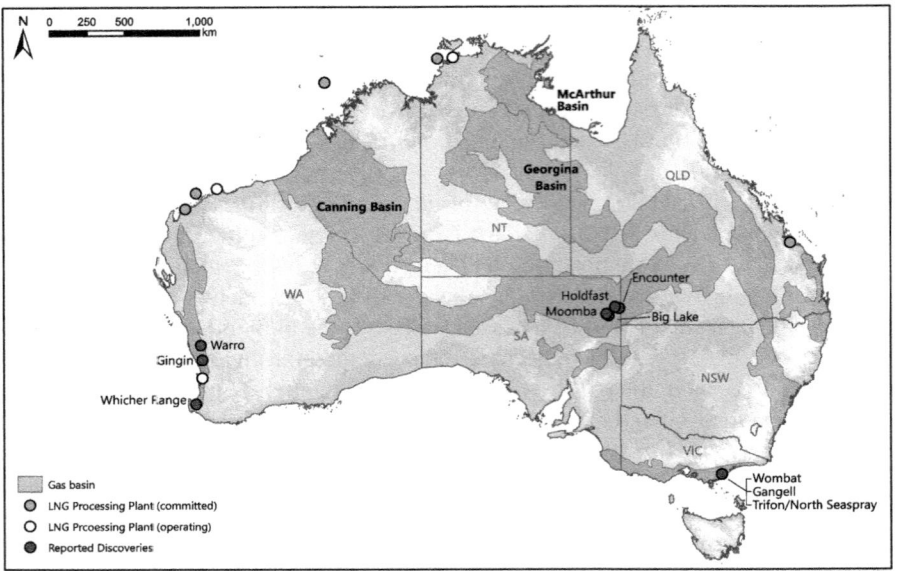

Fig. 12.1 Overview of gas basins in Australia (Callum Foster, UWE, Bristol Cartography)

12.2.1 Coal Seam Gas Resources in Queensland

Australia is endowed with massive coal resources, possessing 6% of the world's black coal and 25% of recoverable brown coal (Geoscience Australia 2016). CSG reserves in Australia are largely confined to the east coast of Australia, with a small amount in Western Australia's Perth Basin. The primary CSG activity has been confined to Queensland as a result of the quality and type of coal, where gas production has occurred since the late 1990s, beginning with small-scale commercial exploitation of the methane gas from coal seams in 1996. Since the mid-1990s commercial production of CSG has increased, initially providing gas for Queensland electricity. Large-scale development is also targeted for export of liquefied natural gas (LNG) to Asian energy markets on long-term forward contracts (Hunter 2013, 83–84).

The pioneering development of CSG occurred in the Surat and Bowen Basins over an area of around 270,000km², with additional area for pipeline corridors and LNG processing and transport facilities on Curtis Island. This development has occurred since the mid-2000s, and has involved four individual consortia, with a capital investment exceeding US$60 billion. Australia's annual CSG production has increased from 4 Pj (petajoules) in 1996 to 1199 Pj in 2016 (Qld Government 2017), with all production coming from the Bowen and Surat Basins in Queensland. The export of CSG as LNG has occurred since January 2015, exclusively to Asian markets. This export of CSG from Queensland marks the first time that unconventional gas has been exported from Australia, and the first time that UPR have been developed for the specific purpose of underpinning the LNG export market (Towler et al. 2016, 249). The total contracted volume of CSG for LNG export is approximately 24 MT/a (million tonnes per annum), dropping to approximately 20 MT/a in 2033 (Geoscience Australia 2016, 11). To date, approximately 9000 wells have been drilled, with intensive drilling of approximately 1500–2000 wells per annum to be drilled through until 2030 to recover the CSG (Geoscience Australia 2016, 12–14).

12.2.2   Shale Gas Resources in Central and Western Australia

There are four major geological basins where shale gas resources exist in Australia and, with the exception of the small, scarcely explored Maryborough Basin in coastal Queensland, shale gas resources in Australia are located in Western Australia, the Northern Territory, and South Australia. Estimates of shale gas resources in Australia places the reserves at 437 Tcf (trillion cubic feet), the seventh largest in the world in terms of country reserves (US Energy Information Administration 2013, 10).

Western Australia has demonstrably large strategic gas reserves. The U.S. Energy Information Administration has estimated the presence of shale gas reserves of 268 Tcf, almost double that of offshore conventional gas reserves (WA Department of Mines and Petroleum 2014, 13). The Perth Basin is the smallest of Western Australia's shale gas-bearing basins, however, due to its proximity to Perth and favorable climatic conditions, it is perhaps the best explored. The large Canning Basin in Western Australia has deep, Ordovician-age marine shales that are roughly correlative with the Bakken shale in the United States (US Energy Information Administration 2013, III-1). Buru Energy, an Australian exploration and production company, holds significant exploration permits in the Canning Basin. In 2010, Mitsubishi agreed to fund an A\$152.4 million exploration and development program to earn a 50% interest in Buru's permits. The two companies executed a State Agreement with the Government of Western Australia in 2012 (*Natural Gas (Canning Basin Joint Venture) Agreement Act 2013,* Western Australia), which extends for 25 years, with an option of a further 25 years. The agreement enables appraisal work undertaken to relieve the exploration permits from their existing relinquishment obligations and to enable exploration work to be credited against adjacent exploration permits that are not covered under the State Agreement. In addition, under the terms of the State Agreement, the Western Australia Department of State Development will take the lead agency role in the development of a LNG facility in the Pilbara as well as a domestic gas pipeline from the Canning Basin in order to secure domestic energy supplies in the future. The primary role of the State Agreement and the exploration for and development of shale gas is to secure long-term accessible domestic supplies of gas for Western Australia.

In the Northern Territory, the Beetaloo Basin and the Georgina Basin have reported oil and gas shows in shale exploration wells (US Energy Information Administration 2013). If proved commercial, these two shale petroleum basins would become some of the oldest producing hydrocarbon source rocks in the world. Aside from the now depleted Mereenie and Palm Valley conventional petroleum fields, the Northern Territory utilizes gas from the Blacktip gas field for domestic energy supplies (Ahmad and Munson (compilers) 2013, 2:4). As such, any development of shale gas in the Northern Territory (if the moratorium is lifted) will be for export purposes only. The Amadeus Basin, located in central-southern Northern Territory, contains producing conventional oil and gas fields (Mereenie and Palm Valley), and is one of the most prospective onshore areas in the Northern Territory for UPR (Munson et al. 2013, 62). The Georgina Basin, covering an area of 330,000 km², is located in central-eastern Northern Territory and extends into western Queensland (Ahmad and Munson (compilers) 2013). The basin is also one of the most prospective for UPR, although exploration is still in a frontier stage and, given the limited amount of seismic and geological data available, no estimate is available for potential shale gas resources in the Northern Territory section of the Georgina Basin (Ahmad and Munson (compilers) 2013, 62). The Beetaloo Sub-Basin has attracted a considerable amount of exploration activity, probably since it is a significant subsurface depositional center within the McArthur Basin (Ahmad and Munson (compilers) 2013, 21).

South Australia has been Australia's leading onshore gas producer and a pioneer in the exploration for and development of shale gas (US Energy Information Administration 2013, III-8). The shale gas formations in South Australia are confined to the Cooper Basin, which partly extends into southern Northern Territory and western Queensland, although the majority of the field lies in South Australia (US Energy Information Administration 2013, III-8). To date, several exploratory shale gas wells have been drilled, with Beach Energy's Encounter 1 well the first shale gas exploration well drilled in the Cooper Basin in 2010. Unlike other basins in Australia, much of the shale gas in the Cooper Basin is located below operational conventional gas fields (Cook et al. 2013, 1, 15–16, 85). Santos, a major operator in the Cooper Basin, estimates the potential range of net recoverable gas from under existing conventional petroleum licenses to be 15–125 Tcf (Santos 2012). Of the six shale basins assessed, the Cooper Basin, with its existing gas processing and transportation infrastructure, will provide the first commercial source of shale hydrocarbons. Santos, Beach Energy, and Senex Energy continue to explore the Cooper Basin shale reservoirs, expecting to find huge commercial reserves of gas, most likely under existing conventional gas reserves. Given the existing conventional petroleum activities, the extent of associated infrastructure for delivery of gas to east coast markets, and the sophisticated regulatory framework for onshore petroleum activities, it is highly likely that shale gas from the Cooper Basin will be the first shale gas to market in Australia.

12.3 Regulatory Framework for Unconventional Oil and Gas Activities

The regulation of UPR in Australia is complex as a result of the pre-existing Australian colonies at the time of the formation of the Federation (1 January 1901 under the *Commonwealth of Australia Act 1900,* Australia). Australia comprises six states (New South Wales, Queensland, Western Australia, South Australia, Victoria, and Tasmania), each with their own political and legal system, as well as two self-governing territories, the Northern Territory and the Australian Capital Territory. Given that Tasmania and the Australian Capital Territory do not hold UPR, they will not be considered here.

Onshore petroleum activities are regulated by the states and territories, as there is no enumerated power for the Commonwealth to regulate petroleum and mineral activities under the Australian Constitution. The only authority under which the Commonwealth could regulate the extraction of UPR is § 51(i) of the Constitution (Interstate and overseas trade and commerce), or § 51(xx) of the Constitution (Corporations power), (*Australian Constitution § 51).* It is unlikely to invoke either of these sections as it would be highly controversial and a sign of 'aggression' towards the states or Northern Territory. In contrast, each Australian state has the capacity to regulate all other activities not enumerated in the Australian Constitution for the "peace, welfare and good government" of that state, for example, as set out in § 5 of *Constitution of Queensland 2001.* All onshore petroleum activities, be they conventional or unconventional, are regulated under the relevant petroleum legislation in each state and the Northern Territory with the exception of Victoria, where CSG activities are regulated under the *Mineral Resources (Sustainable Development) Act 1990* (Victoria).

Exploration for and extraction of shale gas in Australia is generally governed by the main petroleum act in each jurisdiction. Shale gas is a gaseous form of hydrocarbons, and therefore falls under the definition of petroleum in the various acts. Petroleum is defined in each jurisdiction.[2] Similarly, the regulation of environmental issues relating to UPR (including the management of produced water) is largely a matter for individual states under the ambit of the states' consti-

---

[2] The definition of "petroleum" in each jurisdiction may be found in the following legal authorities: Petroleum (Onshore) Act 1991 (NSW) § 3; Petroleum (Onshore) Act 1991 (NSW) § 3; Petroleum Act (Northern Territory) § 5; Petroleum Act 1923 (Queensland) § 2 and Petroleum and Gas (Production and Safety) Act 2004 (Queensland) § 10(1); Petroleum and Geothermal Energy Act 2000 (South Australia.) § 4; Petroleum Act 1998 (Victoria) § 6 (defined under "conventional petroleum") and Mineral Resources (Sustainable Development) Act 1990 (Victoria) § 5 (defined under "unconventional petroleum"); and Petroleum and Geothermal Energy Resources Act 1967 (Western Australia) § 5.

tutional plenary power to make laws for the 'peace, welfare and good government' of the state.

Although the Commonwealth does not have an enumerated power to make laws with respect to environmental matters, there are sections of the Australian Constitution that grant the Commonwealth the capacity to regulate environmental management of petroleum activities. In particular, the interpretation of the *Trade and Commerce Power* (§ 51(i) of the Constitution) by the High Court in *Murphyores* (*Murphyores Inc. Pty Ltd. v. Commonwealth* (1976) 136 CLR 1 (Austl.), displaying various opinions by the justices, concluded that the Minister is within her/his power to prohibit activities of a company exporting certain items due to the commerce and trade powers in the Constitution. Further, in interpreting the *Corporations Power* (§ 51(xx) of the Constitution) in the *Work Choices* case (*NSW v. Commonwealth* [2006] HCA 52, the High Court held by a majority of 5 to 2 that changes to the Workplace Relations Act were valid, thus enabling the Commonwealth to enact a comprehensive regime of industrial relations law and substantially widening the scope of the corporations power. These landmark decisions signified a shift in the distribution of power from the states to the federal parliament (*Australian Constitution* § 51(xx)). However, it is under the External Affairs power (§ 51 (xxix) of the Constitution) that the Commonwealth Government has a substantial ambit to regulate environmental matters (*Australian Constitution* § 51(xxix), declaring the Australian government to "have power to make laws for the peace, order, and good government of the Commonwealth with respect to … external affairs") and enabling the Commonwealth to make laws with respect to environmental treaties and conventions to which Australia is a signatory.

The primary legislative tool enacted to regulate environmental impacts at the Commonwealth level is the *Environmental Protection and Biodiversity Conservation Act 1999* (Australia) (EPBCA). As the EPBCA is Commonwealth legislation, it does not apply to all petroleum activities. Rather it only applies where the activity falls into an area where referral for assessment is required under Chap. 2, EPBCA. Day-to-day environmental management falls under the ambit of state and territorial law. Therefore, while the EPBCA is not core environmental legislation, it is nonetheless important and needs to be considered when examining environmental regulation of petroleum activities. Under the EPBCA, development of UPR (an action) will require approval from the Commonwealth Environment Minister if it will have, or is likely to have, a significant impact on a matter of national environmental significance (§§ 11, 130), where an action is defined broadly to include a project, a development, an undertaking, an activity or a series of activities, or an alteration of any of these things (§ 523). A "significant impact" is defined as:

> An impact which is important, notable, or of consequence, having regard to its context or intensity. Whether or not an action is likely to have significant impact depends on the sensitivity, value and quality of the environment which is impacted, and on the intensity, duration, magnitude and geographic extent of the impacts.[3]

---

[3] www.environment.gov.au/epbc/about/glossary.html#significant

The matters of national environmental significance comprise: listed threatened species and ecological communities; migratory species protected under international agreements; Ramsar wetlands of international importance; the Commonwealth Marine Environment; world heritage properties; the Great Barrier Reef Marine Park; nuclear actions; and a water resource in relation to coal seam gas development and large coal mining development (water trigger: EPBCA ch 2).

To determine whether an action requires approval by the Commonwealth Minister, the project is referred (by the project proponent or a third party) to relevant onshore environmental authority in the state or territory if the project includes one of the matters of national environmental significance. The state minister considers the application for referral to determine whether it requires further assessment on the grounds of posing a significant risk to a matter of national environmental significance (a controlled action). If it is deemed not to pose a significant risk (an uncontrolled action), the project is referred to the state minister for assessment. Where an action is deemed to be a controlled action, it is referred to the Commonwealth Minister and assessed under the EPBCA. If approved, the action will also be assessed under the appropriate state and Northern Territory environmental legislation.

The 'water trigger' was implemented by reforms to the EPBCA in 2013 to address community concerns regarding CSG activities in Queensland. It is important to note that the water trigger addresses the use and consumption of water in relation to CSG activities and does not apply to shale gas activities. This is a significant weakness of the water trigger, given the vast shale gas resources in Australia, the location of most of those resources in areas of low rainfall, and the amount of water required to hydraulically fracture shale gas resources to enable the production. Given the volume of water required to fracture a well and the low rainfall in many areas where shale gas occurs, it is logical that the water trigger as a matter of national environmental significance should also apply to shale gas projects. As shale gas projects are developed in the future, the matter of national environmental significance should be expanded to apply to water use for shale gas development.

Although the Commonwealth does not have constitutional capacity to regulate shale gas and CSG activities, Commonwealth energy and resources Ministers have nonetheless utilized the Standing Council of Energy and Resources, a subcommittee of the Council of Australian Governments, to address issues relating to UPR development. The *Harmonised Regulatory Framework for Natural Gas From Coal Seams* (the "Framework") has been established to address community concerns. The Framework covers well integrity, water use and disposal, hydraulic fracturing, and the use and disclosure of chemicals used in operations. Although the Framework is called a 'regulatory framework' it is non-binding on the states since the Commonwealth has no power to regulate these activities. Rather, it is an overview of issues that regulators should consider when developing coal seam gas resources and guidance on developing requisite regulatory tools for CSG. Like the water trigger under the EPBCA, the Framework applies to CSG only, although it recognizes that the Framework may have partial applicability to shale gas development (Australia Standing Council on Energy and Resources 2013, 9). Some of the principles outlined in the Framework are relevant and should be expanded to incorporate shale gas activities, however, produced water management for shale relates principally to flowback, as on-going water production is likely to be non-existent, given the dry nature of the shales. Water management for shale focuses much more on short-term sourcing water for hydraulic fracturing rather than long-term disposal.

12.3.1 Regulation of Produced Waters from CSG Activities

The use and production of water in CSG activities in Queensland are primarily regulated under the *Petroleum and Gas (Production and Safety) Act 2004* (Queensland) (PGPSA). Generally, water use for agriculture and other activities requires a water entitlement under the *National Water Initiative* (NWI), which is an intergovernmental agreement between Australian states/territories and the Commonwealth under the Council of Australian Governments (CoAG) regarding the use and licensing of water. The NWI provides a blueprint for the reform of water use in Australia, regulated under the *Water Act 2000* (Queensland). However, article 34 of the NWI provides a mining and petroleum exclusion, with all parties agreeing that there may be special circumstances facing the minerals and petroleum sector that require 'specific management arrangements outside the scope of th[e] agreement'. The capacity for unfettered water use is outlined in Section 185(1) of the PGPSA, wherein the 'holder of a petroleum tenure may take or interfere with the underground water in the area of the tenure if the taking or interference happens during the course of, or results from, the carrying out of another authorised activity for the tenure.' Further, section 185(3) specifies that there is no limit to the volume of water that may be taken as part of the underground water rights, and this water taken is known as 'associated water' (§ 185 (4)), produced from the dewatering of the coal measures. However, for the purposes of this paper, the associated water will be known generically as produced water. The impact of unfettered water rights is considerable, and there is industry recognition that there are localised effects due to the interconnection between coal seams and aquifers in which aquifers may be depleted by 6–20 m, significantly impacting landholders (Qld Office of Groundwater Impact Assessment 2016).

The disposal of water and fluids from CSG wells is regulated primarily under the *Waste Reduction and Recycling Act 2011* (Queensland). Fluid produced from CSG wells is saline/briny and unable to be used for agricultural purposes without treatment. In addition, as outlined above, there is a high volume of produced water. Therefore, the fluids are processed so they can be used for agricultural purposes and additional uses given the exceptionally high volume. The regulation of produced water from CSG for beneficial use is discussed in Sect. 12.5.

12.3.2 Regulation of Produced Water from Shale Gas Activities

As stated previously, there has been little activity regarding shale gas in Australia, with the majority of activities confined to Western Australia and South Australia.

The development of shale gas resources in this region is largely characterized by the use of principle or objective-based regulation, particularly by Western Australia and South Australia. In Western Australia, although there has been no production of shale gas for commercial use, interest in UPR is high due to the dominance of gas as the primary energy source in that state. The prospective nature of the Canning Basin and increasing community concern drove the Western Australia Department of Mines and Petroleum to develop a robust and comprehensive objective-based regulatory framework to effectively regulate future unconventional gas activities, as well as new environmental and resource regulations based on the principles of (1) minimizing harm to the environment from petroleum activities by identifying and reducing the risks and (2) managing the environmental effects of the petroleum activity.

In order to implement these principles, the Western Australia Department of Mines and Petroleum implemented the *Petroleum and Geothermal Energy Resources (Environment) Regulations 2012* (Western Australia), which requires the operator of a petroleum activity to have an approved environmental plan in place prior to a petroleum activity being undertaken. The objective of the 2012 Regulations is to ensure that any petroleum activity carried out in Western Australia occurs in a manner consistent with the principles of ecologically sustainable development and in accordance with an environmental plan that demonstrates that environmental impacts and risks associated with the activity (including produced water) will be made 'as low as reasonably practicable' (*Petroleum and Geothermal Energy Resources (Environment) Regulations 2012* (Western Australia) r 3). In order to achieve this risk reduction, the environment plan is required to have (1) appropriate environmental performance objectives and standards and (2) appropriate measurable criteria to determine whether the objectives and standards have been met (*Petroleum and Geothermal Energy Resources (Environment) Regulations 2012* (Western Australia), r 11). The operator is required to:

- identify the risks in the specific environment in which they are undertaking the activity;
- identify the impact of those activities;
- assess the identified risks and impacts; and
- then formulate a plan to reduce those risks as low as reasonably possible.

The contents of environmental plans are set out in Part 2 of the 2012 Regulations rr 13–17, and further clarified in the *Guidelines for the Preparation and Submission of an Environment Plan*. Given the principle-based nature of the 2012 Regulations, operators have flexibility in preparing a program to address produced water issues. In preparing the plan, the document must comply with provisions of the 2012 Regulations that stipulate the requirements for the plan, but the method in which they comply is entirely up to the operator. Complementing the requirements for an environmental plan under Part 2 of the 2012 Regulations, Part 4 outlines the environmental requirements relating to emissions and discharges, including the monitoring and reporting requirements for such emissions and discharges.

In South Australia, the Department of State Development regulates petroleum activities. The principal act regulating environmental aspects of onshore petroleum activities is the *Petroleum and Geothermal Energy Act 2000* (South Australia) (PGEA) and the associated *Petroleum and Geothermal Energy Regulations 2000* (South Australia) (PGEA Regulations). Unlike other onshore jurisdictions, the principles of environmental management are embedded in the PGEA, and no shale gas activities can be undertaken unless there is an approved statement of environmental objectives (*Petroleum and Geothermal Energy Act 2000* (South Australia), § 96.). An environment impact report is required for low impact or medium impact activities (*Petroleum and Geothermal Energy Act 2000* (South Australia), §§ 98–99.), while a separate report is required for high impact activities (*Petroleum and Geothermal Energy Act 2000* (South Australia), § 100.). The statement of environmental objectives must be prepared in accordance with PGEA Regulations that address the natural, cultural, social and economic aspects of the area, locality or region where the petroleum activity occurs.

To further capture and address possible community concerns regarding shale gas development, in 2010 the South Australian Government set up the *Roundtable for Unconventional Gas Projects in South Australia* (now known as the *Roundtable for Oil and Gas*) (the Roundtable) to assist in developing the burgeoning unconventional gas industry. The Roundtable comprises industry, government, universities, academics, media, and key individuals and takes a holistic approach to the regulation of unconventional petroleum activities. One of the critical areas of consideration for the Roundtable is water, particularly the use of water, the monitoring of water for use and contamination, and the reuse and ultimately disposal of produced water from shale gas activities.

12.4   Unconventional Oil and Gas Activities and Produced Waters – What Is the Issue?

The produced water that arises from unconventional petroleum activities varies as a result of a combination of technique, geology, and type of unconventional reservoir. The produced water from shale gas differs markedly from that of CSG, primarily as a result of the geology, but also the techniques used.

In order to produce gas from coal seams (CSG), there is a need to depressurise them (Jakubowski et al. 2014, 133) through removing water from the coal measure, allowing the gas to desorb from the coal cleats and flow to the well. The resulting fluid is pumped or flows to the surface, where the raw gas and water are separated. This process is known as dewatering. It typically produces large volumes of produced water that contain high levels of sodicity and salinity and require gathering, storage, and treatment (Jakubowski et al. 2014, 134–5). The coal measures of the Surat Basin, where much of the CSG production occurs, are part of a large interconnected aquifer system that is located within the 1.7 Mkm² Great Artesian Basin (Towler et al. 2016, 263). These aquifers provide water to 35 towns and numerous farms for both stock and crops (Towler et al. 2016, 263), and have provided a source of community discontent as a result of rapid dewatering. The dewatering of coal seams to produce CSG produces approximately 120 GL/y (gigalitres/year) (Towler et al. 2016, 263), although estimates vary (Tan et al. 2015, 684). These large volumes of produced water resulting from CSG extraction have provided many challenges for project proponents, as there is a difficulty in processing and disposing of such high volumes of water. Santos GLNG has developed an integrated water management plan that involves the use of a number of techniques, including irrigation and aquifer reinjection after treatment, primarily by reverse osmosis (Jakubowski et al. 2014, 134). This is known as the beneficial use of produced water, with its use contingent on the water not negatively impacting forest, pasture, and crop production (Jakubowski et al. 2014, 134). One of the major issues associated with CSG production is that, due to the interconnectivity of the Surat Basin, the dewatering of the coal measures adversely affects nearby aquifers, which have been used for decades to undertake agricultural activities. The 2016 *Surat Basin Underground Water Impact Report* has identified long-term affected areas, denoted by the lowering of bores of more than five metres as a result of dewatering (Qld Office of Groundwater Impact Assessment 2016, xii). In 2016, 459 bores were identified as long-term affected.

In order to produce oil and gas from shale formations, it is necessary to undertake hydraulic fracturing. It is well established that hydraulic fracturing is undertaken by forcing water, combined with necessary chemicals and sand, at high pressure into shale formations in order force the shale to fracture. The water that has

It is important to realise that most of the produced water arising from CSG originates from dewatering and is not the result of hydraulic fracturing. To illustrate the composition of produced water from CSG, Table 12.1 identifies the composition of produced water samples collected between July and September 2012 from 150 CSG production wells in three separate fields (designated A, B, and C) located at the eastern end of the Surat Basin. Water samples were obtained from a locally defined region of the Basin in order to identify the relationships between a large set of water quality parameters within small geographical proximity. Fields A and C were relatively close to each other (≈1 km) whereas field B ranged from approximately 9 km distance in some locations from fields A and C, up to 36 km in other locations. The samples were from both the Juandah and the Taroom Coal Measures and ranged in maximum depth of 574 m for field A, 640 m for field B and 498 m for field C.

Table 12.1 Range and mean values (in brackets) for water quality parameters – Surat Basin associated water (dissolved species in terms of mg/L)

Parameter	A field (54 wells)	B field (73 wells)	C field (23 wells)
EC (mS/cm)	3850–13,300 (7084)	3630–9410 (6743)	5150–17,200 (9765)
SAR (ratio)	69.6–177 (120)	86.4–163 (121)	62–156 (112)
pH (pH units)	7.92–8.89 (8.47)	7.94–8.76 (8.43)	7.83–8.63 (8.30)
Total suspended solids	5–7560 (435)	6–1520 (272)	7–265 (72)
Total dissolved solids	2940–7600 (4444)	2190–5790 (4046)	3050–10,200 (5655)
Hardness	276–1620	12–80 (30)	12–482 (102)
Bicarbonate alkalinity	5–2030 (1039)	470–1540 (960)	108–1350 (714)
Bromide	1.93–12.7 (6.0)	2.82–11.7 (5.6)	2.75–16.6 (8.9)
Chloride	471–4390 (1595)	875–2930 (1579)	823–5910 (2938)
Sulfate	1–18 (4.4)	1–48 (8.0)	1–5 (3.3)
Calcium	2–55 (9.0)	3–19 (7.0)	3–137 (26.2)
Magnesium	1–16 (3.8)	1–8 (3.1)	1–34 (8.8)
Sodium	909–2700 (1487)	786–2010 (1452)	1130–3700(2098)
Potassium	4–14 (6.9)	3–10 (6.1)	5–20 (10.6)
Aluminium	0.02–40.9 (2.9)	0.01–17.8 (2.1)	0.01–2.08 (0.65)
Manganese	0.002–0.54 (0.06)	0–3.59 (0.11)	0–0.13 (0.04)
Strontium	0.65–9.03 (2.30)	0.7–4.5 (2.12)	0.99–20.2 (5.52)
Boron	0.22–0.54 (0.34)	0.24–0.68 (0.41)	0.17–0.61 (0.31)
Iron	0.16–45.1 (5.12)	0.09–35.1 (8.73)	0.3–6.16 (2.44)
Silica	13.1–19.6 (16.9)	13.9–23.1 (17.7)	13.8–19.2 (16.5)
Fluoride	0.8–3.2 (2.2)	1–3.3 (2.0)	0.4–2.7 (1.6)
Barium	0.53–4.39 (1.37)	0.38–2.32 (1.18)	0.62–9.38 (2.73)
Strontium	0.65–9.03 (2.30)	0.7–4.5 (2.12)	0.99–20.20 (5.52)
Total organic carbon	1–71 (19)	1–138 (25)	4–138 (27)

Source: (Table 2, Rebello et al. (2017))

been utilized to fracture the geology will flow back from the formation. This water is often referred to flowback water. For purposes of this study, all water produced as a result of unconventional activities will be referred to as produced water.

The composition of produced water from shale gas is a combination of the original fluid injected down-hole under pressure during the fracturing activity, combined with the compounds and contaminants present in the shale formation being fractured. Given that shale formations are fluvial and marine depositional environments, those compounds can vary, and may include salts, organic materials, ions, and naturally occurring radioactive materials such as barium, radium and strontium. Therefore, the produced water from shale is variable, and it can bring to the surface compounds that are normally buried deep in the geology.

12.5   Disposal of Produced Water from Unconventional Oil and Gas Activities

12.5.1   Coal Seam Gas

The processing of produced water in Queensland is regulated under § 111A of the PGPSA, which was inserted in amendments in 2012 in response to community concerns. The produced water is treated to remove salts and other chemicals, often utilizing reverse osmosis, and then disposed of. The resultant treated water is seen as a 'beneficial use', since under the *Waste Reduction and Recycling Act 2011* (Queensland), CSG companies are required to identify beneficial uses for produced water, such as irrigation, town water supplies, environmental flows, and aquifer recharge. Given that CSG production in Queensland occurs in an area of water stress, innovative disposal options for appropriate treatment have been developed, as required under the PGPSA. Such innovative treatments include aquaculture, potable aquifer replenishment, coal washing at existing coal facilities, irrigation, and feedlot watering.

12.5.1.1   Legal Processes for Use of Produced Water

The relatively rapid development of the CSG sector in Queensland after 2002 led to various impacts on the environment which were not particularly well recognized under existing regulatory frameworks where petroleum activities were seen as confined to a few wells and associated infrastructure. Extensive use of land, many wells, compressor stations and a need for many access roads and pipelines was novel to this activity. Interventions and action were required to develop new policies and regulations to address some of the unique aspects of CSG development, such as addressing the burgeoning water and brine storage requirements (Queensland Minister for Climate Change and Sustainability The Honourable Kate Jones 11 June, 2010). As distinct from the common use of discharge to surface waters or irrigation with the lower salinity CSG produced water, like that often found in the USA (Hamawand et al. 2013), the higher salinity Queensland CSG required a number of policies to address disposal options. Irrigation and other surface applications in the extensive CSG basins in the Rockies and elsewhere in North America are subject to an annual replenishment of snow melt into the soil column, allowing for sequential amendment of soil ion balance (Veil et al. 2004, US Environmental Protection Agency 2010), whereas Queensland's climate across the CSG basins is arid with intermittent rain (QGC 2016a).

Following a policy debate, the Queensland Government resolved a strategy hierarchy for managing the use of CSG produced water under the *Environmental Protection Act 1994* (Qld Department of Environment and Heritage Protection 2012), which is as follows:

Priority 1 – CSG water is used for a purpose that is beneficial to one or more of the follow-
ing: the environment, existing or new water users, and existing or new water-dependent
industries.

Priority 2 – After feasible beneficial use options have been considered, treating and dispos-
ing CSG water in a way that firstly avoids, and then minimises and mitigates, impacts
on environmental values.

This policy also addresses the management of saline wastes with similar hierarchy, but Priority 1 must be demonstrated to be non-feasible before Priority 2 can be
adopted:

Priority 1 – Brine or salt residues are treated to create useable products wherever
feasible.

Priority 2 – After assessing the feasibility of treating the brine or solid salt residues to cre-
ate useable and saleable products, disposing of the brine and salt residues in accor-
dance with strict standards that protect the environment.

CSG produced water in Queensland is initially a waste, but its legal status can be
reviewed under the *Waste Reduction and Recycling Act 2011* in order to approve its
use as a resource for a purpose other than disposal. If an operator has exhausted
Priority 1 above, the department can approve a discharge provided any potential
environmental risks identified are appropriately avoided, minimized, and
mitigated.

The Queensland Government has adopted some aspects of a performance-based
regime (Qld Department of Environment and Heritage Protection 2016) to address
the management of CSG produced water and saline waste for individual operators
with a requirement to develop a suite of management criteria which must address
the following elements: *the quantity and quality of the water used, treated, stored or*
disposed of; protection of the environmental values affected by each relevant CSG
activity; and, *the disposal of waste—including, for example, salt—generated from*
the management of the water. These elements seek to supersede the previous policy
tool of *Beneficial Use Approvals* that included a general approval (Qld Department
of Environment and Heritage Protection 2014b) and a specific CSG produced water
irrigation approval (Qld Department of Environment and Heritage Protection
2014a). These beneficial use approvals are due to lapse in 2019. The management
criteria are to be documented and addressed in the annual report to the department
noting whether CSG produced water has been effectively managed and if not what
and when actions are to be taken to resolve the issue.

Unlike North America where each petroleum producing state may be home to
hundreds of operators, UPR development in Australia is dominated by few major
players, each with many hundreds or thousands of wells. The water management
strategies of those companies have largely taken an aggregation approach with central collection and treatment rather than satellite systems. Consequently the treatment facilities are substantial and designed (QGC 2016a) for high initial flow rates
that may not continue through the life of the basin. CSG produced water production
of 49 and 53 GL/year (2015 & 2016, gigaliters per year) in Queensland[4] is directed
(APPEA 2015) as follows:

---

[4] https://data.Queensland.gov.au/dataset/petroleum-gas-production-and-reserve-statistics/
resource/9746212a-e0c6-484d-95ad-b2be1c46027d

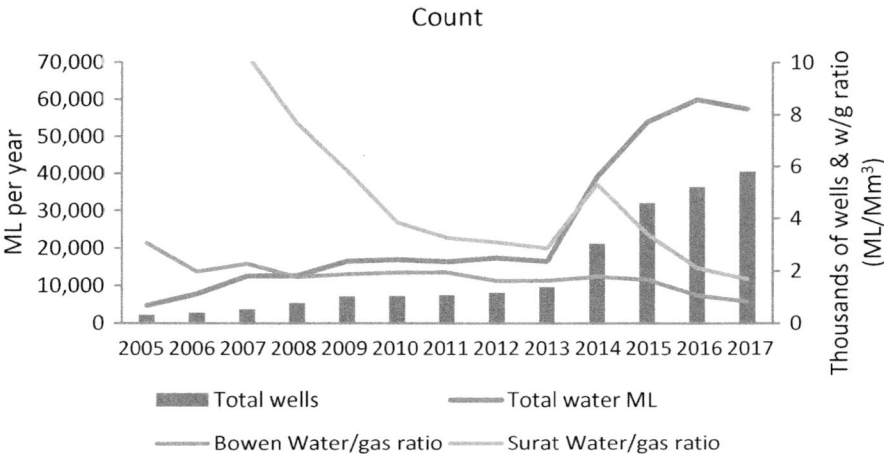

Fig. 12.2 Relationship of CSG water production to number of CSG wells. (Data from https://data. Queensland.gov.au/dataset/petroleum-gas-production-and-reserve-statistics/resource/9746212a-e0c6-484d-95ad-b2be1c46027d)

- 97% is treated and desalinated
- 59% is made available for agricultural purposes
- 24% is reinjected into underground aquifers
- 14% is used for industrial purposes such as mining, roads and construction
- 3% remains as brine or salt

Produced water flow is reasonably correlated to productive CSG well numbers, as shown in Fig. 12.2.

Each of the major players has adopted flexible strategies for managing their produced water, thus allowing for interruption of flows to any particular endpoint such as could arise through limiting irrigation under high rainfall conditions or other circumstances for other pathways.

Long-term management of the salt generated as a by-product of water treatment is one of the most intractable problems of waste treatment with CSG production in Queensland. It is estimated that around 4.5 tons of salt is produced per 1ML of produced water (Arrow Energy 2013). At that rate, annual production in 2016 is estimated at around 200,000 tons for the Queensland basins. The carbonate/chloride ion dominance limits the economic prospect of commercial options (QGC 2016b). The Surat Basin is located in the Murray-Darling catchment, which is subject to oversight from the Australian Commonwealth, including the cross-border level of salinity in the Murray River. Section 25(2) *Water Act 2007* (Australia), specifies salinity targets for both place and percentage of time. Given the high evaporation rates in the Surat and Bowen basins, evaporation/storage dams have provided an interim solution prior to commitment to advanced treatment technologies such as thermal crystalisers (Arrow Energy 2013; QGC 2016a).

12.5.1.2 Treatment

A number of review papers examining CSG water treatment options have been prepared since CSG operations reached a scale of significance (Millar et al. 2016; Hamawand et al. 2013; Rebello et al. 2016; Towler et al. 2016). Due to the policy of aggregation of produced water streams prior to treatment, salinity variability tend to be evened out. Similar facilities have been developed for each of the major players, with the use of high technology treatment trains variously incorporating major aggregation dams, pressure ultrafiltration, ion exchange, multi-stage reverse osmosis and brine concentration (Carter 2015). Further amendment of the permeate by blending raw CSG to approximate end user quality requirements is commonplace.

12.5.1.3 Irrigation

Soils of the Surat and Bowen basins have been developed for agricultural purposes for more than a century. Although subject to occasional extended drought conditions the soils respond to good agronomic practices and increased application of water. They are, however, susceptible to salinity and sodicity issues if excessive water application is allowed in the absence of appropriate drainage.

The Surat and Bowen basins produce typically high salinity carbonate/chloride-dominated CSG waters requiring ionic amendment prior to irrigation. Alternative approaches to expensive water treatment trains may also allow for irrigation through land amendment irrigation where agricultural chemicals such as sulphur bentonite and gypsum can be applied to address sodicity changes and soil structure brought about by the CSG water. However, under the particular trial conditions, upper soil column leaching was insufficient to manage root zone salinity (Bennett et al. 2016). Use of soil amendments rather than water treatment, however, can lead to scaling problems in the irrigation equipment due to high carbonate concentrations (Dale 2015).

12.5.1.4 Aquifer Injection

In the U.S., the oil and gas sector has made extensive use of underground injection to manage various waste streams. The Underground Injection Control[5] program, administered by the Environment Protection Agency, provides for different purposes, including:

- Class I industrial and municipal waste disposal wells
- Class II oil and gas injection wells
- Class V wells that inject non-hazardous fluids into or above underground sources of drinking water

---

[5] https://www.epa.gov/uic/general-information-about-injection-wells

Class II wells receive a significant quantity of produced water with minimal or no treatment to assist oil recovery through reservoir flooding (enhanced oil recovery). Class I wells are built to a significantly higher level of protection (multiple barriers) and may receive hazardous waste. Class V wells are those that may convey water for aquifer remediation and a number of other non-hazardous uses. In Queensland, the lithology of the Surat Basin does not have any significant extent of suitable receiving formations for Class I type wastes. However, municipal sourcing of groundwater in the Surat Basin for potable use has led to active aquifer recharge activities in the basin area.

12.5.1.5 Brine Injection

Brine injection may be permitted in Queensland where the operator can demonstrate (1) isolation of the target zone from any aquifers that are contiguous with water source aquifers and (2) that the injection fluid shows inconsequential reactivity with the injection zone formation fluids. Formations demonstrating such character are not common in Queensland, particularly in the area of the CSG resources. However, Santos's Fairview Arcadia area has access to the Timbury Hills Formation and this formation has received brine since 2006 (Santos GLNG Project 2013). Environmental approval conditions for fluid injection are similar to US EPA Class I wells, and include the need for an overlying aquitard, production casing fully cemented into the injection zone, an injection tube passing through an annulus packer, injection pressures below the formation fracture gradient, the well annulus containing an inert fluid and fluid level monitoring in the annulus (Qld Department of Environment and Heritage Protection 2017). Permit conditions for brine injection include conditions that assess:

- the volumes and rates of injection;
- a description of the nature of the fluid, source and management;
- a demonstration of inconsequential reactivity with native fluids, characteristics of the receiving formation;
- description of the surface and subsurface hydrology of the area;
- description of the wells and any impacts on environmental values; and
- risk assessment and mitigation measures, monitoring and performance verification measures

12.5.1.6 Aquifer Recharge

Aquifer injection to supplement existing resources for potable purposes has been regarded as a beneficial use from the first recognition of reuse of CSG produced water by the Queensland Government. Pressures on local aquifers providing water resources for towns in the Surat Basin have provided a framework and an incentive for negotiation between Darling Downs municipalities, such as Roma, and the CSG sector, which offers high quality treated water. The supply aquifer providing Roma's water has been drawn upon for over 100 years and the pressure head has dropped over 80 m. Following early trials, recharge rates of over 20 ML/day are maintained (Santos GLNG Project 2013). Conditions applied to aquifer recharge have similarity in respect to well construction and operations aspects and to brine injection conditions, discussed above, with the addition of specific contaminant limits for oxygen, electrical conductivity, total dissolved solids, and pH (Qld Department of Environment and Heritage Protection 2015).

12.5.1.7 Disposal to Surface-Flowing Waters (Streams and Rivers)

Release of CSG produced water is tightly regulated with specified release points, volume, rate and contaminant limits. Conditions for release include the following:

- erosion, sediment and flooding prevention
- notification to the regulator in the event of out-of-specification discharge water with associated risk assessment
- flow measuring devices, monitoring, record keeping and good access to the discharge point

Over and above these discharge criteria, the operator is required to implement a prescriptive receiving environment monitoring program (Qld Department of Environment and Heritage Protection 2015) that is to be maintained by a qualified person. The program must address:

- a description of the ecological character of the receiving environment including consideration of seasonality
- a description of environmental values to be protected and water quality objectives to be achieved
- the hydrology and geomorphology of the receiving environment
- impact assessment
- administrative matters.

12.5.2   Disposal of Flowback Fluids

Flowback is specifically the term given to returning hydraulic fracturing fluid rather than produced waters. With Queensland CSG basins, hydraulic fracturing is not common, having been undertaken in just 6% of producing gas wells, but this may rise to 40% as the industry develops (URS Australia 2014). In Queensland, neither oil-based nor synthetic drilling muds are permitted in carrying out of petroleum activities, and polycyclic aromatic hydrocarbons are not permitted in hydraulic fracturing. Queensland does not specify flowback management requirements but does stipulate monitoring of water bores within a timeframe extending to that period finishing after 150% of the injected fluid volume has returned to the surface. This may be several months after the hydraulic fracturing activity if the well is left shut-in. Permit conditions require a detailed prospective environmental and health risk assessment of potential reactions of the various additives used through the hydraulic fracturing process (Santos GLNG Project 2016). NSW does not stipulate treatment processes but does require the operator to report the flowback management strategy to the regulator (NSW Department of Trade and Investment 2012).

12.5.3 Information Disclosure Requirements for Wastewater

Public reporting of impacts from produced water disposal or water quality is not required under relevant Acts in any states of Australia, however, operators are required to report annually on produced water management activities to the regulator under permit conditions. The regulator may choose to publish such material or make it available under provisions of a public register. The operators may also report water quality data publicly. Santos, QGC and AGL all maintain detailed open GIS water portals.[6] In Queensland the operators are required to report performance against water management criteria but, with very little specificity in permits, the information included in annual reports are variable, either a brief yes/no or an extensive description of performance (APLNG 2017; Santos Toga 2017).

12.6 Conclusion

This chapter has outlined the regulation of produced water and flow-back fluid in Australia. It identified UPRs in Australia and discusses the regulation of those activities at both Commonwealth and state/territory level. In undertaking this discussion, it identifies that shale gas activities occur in central and Western Australia, whilst commercial production of CSG is presently occurring in Queensland. It focuses on the management of CSG produced water, given that this is the only UPR commercial development at present. It concludes by outlining how the produced water is regulated, with a particular emphasis on Queensland.

References

Ahmad M, Munson TJ (compilers) (2013) Geology and mineral resources of the Northern Territory. Northern Territory Geological Survey, Special Publication 5
APLNG (2017) Annual return for environmental authority EPPG00885313 (Spring Gully)
APPEA (2015) Reference document: coal seam gas and water volumes
Arrow Energy (2013) Coal seam gas water management strategy

---

[6] See http://www.santoswaterportal.com.au/,http://watermap.qgc.com.au/& https://www.agl.com.au/about-agl/how-we-source-energy/natural-gas/water-portal)

Australia Standing Council on Energy and Resources (2013) The National harmonised regulatory framework for natural gas from coal seams 2013. http://www.coagenergycouncil.gov.au/publications/national-harmonised-regulatory-framework-natural-gas-coal-seams

Bennett JML, Marchuk A, Raine SR et al (2016) Managing land application of coal seam water: a field study of land amendment irrigation using saline-sodic and alkaline water on a Red Vertisol. J Environ Manag 184:178–185. http://www.sciencedirect.com/science/article/pii/S0301479716307496

Carter A (2015) GE Kenya WTP Queensland. J Aust Water Assoc 42(Supplement 2015):28–29

Cook P, Beck V, Brereton D et al. (2013) Engineering energy: unconventional gas production: a study of shale gas in Australia (Report for the Australian Council of Learned Academies). https://www.acola.org.au/PDF/SAF06FINAL/Final%20Report%20Engineering%20Energy%20June%202013.pdf

Dale G (2015) Considerations and challenges in sustainable use of coal seam gas water for irrigated agriculture. Water J Aust Water Assoc Suppl 42:18–23

Geoscience Australia (2016) Coal seam, shale and tight gas in Australia: resources assessment and operation overview 2016, Upstream Petroleum Resources Working Group Report to COAG Energy Council

Hamawand I, Yusaf T, Hamawand SG (2013) Coal seam gas and associated water: a review paper. Renew Sust Energ Rev 22:550–560. http://www.sciencedirect.com/science/article/pii/S1364032113001329

Hunter T (2012) Australia's unconventional gas resources. Committee for Economic Development of Australia, Melbourne, VIC

Hunter T (2013) Rising demands for Australian gas exports in the Asian century: implications for Japan's energy security. In: Proceedings the Australia Japan dialogue: energy security: challenges and opportunities, pp 83–84

Jakubowski R, Haws N, Ellerbroek D et al (2014) Development of a management tool to support the beneficial use of treated coal seam gas water for irrigation in Eastern Australia. Mine Water Environ 33(2):133–145. https://doi.org/10.1007/s10230-013-0246-4

Johnson D (2009) The geology of Australia, 2nd edn. Cambridge University Press, Cambridge/New York

Millar GJ, Couperthwaite SJ, Moodliar CD (2016) Strategies for the management and treatment of coal seam gas associated water. Renew Sust Energ Rev 57:669–691. http://www.sciencedirect.com/science/article/pii/S1364032115014707

Munson TJ, Ahmad M, Dunster JN (2013) Chapter 39: Carpentaria basin. In: Ahmad M and Munson TJ (compilers). Geology and mineral resources of the Northern Territory. Northern Territory Geological Survey, Special Publication 5

NSW Department of Trade and Investment, Regional Infrastructure and Services (2012) Code of practice for coal seam gas fracture stimulation activities, New South Wales Government

QGC (2016a) EPBC referral 2008/4398 approval condition 49i Stage 3 water monitoring and management plan – annual report

QGC (2016b) QGC Stage 3 water monitoring management plan 2015 annual report

Qld Department of Environment and Heritage Protection (2012) Coal seam gas water management policy 2012, Queensland Government

Qld Department of Environment and Heritage Protection (2014a) General beneficial use approval—Irrigation of associated water (including coal seam gas water), Queensland Government

Qld Department of Environment and Heritage Protection (2014b) General beneficial use approval associated water (including coal seam gas water), Queensland Government

Qld Department of Environment and Heritage Protection (2015) Spring gully environmental authority EPPG00885313, Queensland Government. https://environment.ehp.qld.gov.au/env-authorities/pdf/eppg00885313.pdf

Qld Department of Environment and Heritage Protection (2016) CSG water management: Measurable criteria, Queensland Government

Qld Department of Environment and Heritage Protection (2017) Fairview arcadia project area environmental authority EPPG00928713, Queensland Government

Qld Government (2017) Petroleum and gas production statistics (December 2016). https://data.qld.gov.au/dataset/petroleum-gas-production-and-reserve-statistics/resource/9746212a-e0c6-484d-95ad-b2be1c46027d

Qld Office of Groundwater Impact Assessment (2016) Underground water impact report for the surat cumulative management area 2016

Queensland Minister for Climate Change and Sustainability The Honourable Kate Jones. 11 June, 2010. Gas companies compelled to change water storage practices

Rebello CA, Couperthwaite SJ, Millar GJ et al (2016) Understanding coal seam gas associated water, regulations and strategies for treatment. J Unconv Oil Gas Resour 13:32–43. http://www.sciencedirect.com/science/article/pii/S221339761500052X

Rebello CA, Couperthwaite SJ, Millar GJ et al (2017) Coal seam water quality and the impact upon management strategies. J Pet Sci Eng 150:323–333. http://www.sciencedirect.com/science/article/pii/S0920410516313006

Santos (2012) Cooper basin unconventional gas opportunities & commercialisation. https://www.santos.com/media/2134/121112_eabu_cooper_basin_unconventional_gas_opportunities_and_commercialisation.pdf

Santos GLNG Project. (2013) Santos GLNG Project CSG water monitoring and management plan summary plan – Stage 2 Revision 2

Santos GLNG Project (2016) Santos GLNG upstream hydraulic fracturing risk assessment compendium of assessed fluid systems

Santos Toga (2017) Annual return for environmental authority EPPG00928713 (Fairview)

Tan P-L, George D, Comino M (2015) Cumulative risk management, coal seam gas, sustainable water, and agriculture in Australia. Int J Water Resour Dev 31(4):682–700. https://doi.org/10.1080/07900627.2014.994593

Towler B, Firouzi M, Underschultz J et al (2016) An overview of the coal seam gas developments in Queensland. J Nat Gas Sci Eng 31:249–271. http://www.sciencedirect.com/science/article/pii/S1875510016300853

URS Australia (2014) Hydraulic fracturing ('fraccing') techniques, including reporting requirements and governance arrangements

US Energy Information Administration (2013) Technically recoverable shale oil and shale gas resources: an assessment of 137 shale formations in 41 countries outside the United States

US Environmental Protection Agency (2010) Coalbed methane extraction: detailed study report. EPA, Washington

Veil JA, Puder MG, Elcock D et al (2004) A white paper describing produced water from production of crude oil, natural gas, and coal bed methane. US Department of Energy, Washington, DC

WA Department of Mines and Petroleum (2014) Petroleum and geothermal explorer's guide. http://dmp.wa.gov.au/Documents/Petroleum/PD-RES-PUB-100D.pdf

Tina Soliman Hunter specializes in energy and petroleum law and is the Director of the Aberdeen University Centre for Energy Law (AUCEL). She teaches and researches in the areas of national and international petroleum law, unconventional petroleum regulation, Arctic petroleum law and governance, international investment protection in the energy sector, and resources law and policy. She has received academic qualifications in marine sediments, geology, political science, applied science, and law, completing her PhD at the University of Bergen, Norway. Hunter has been a visiting scholar at the University of Bergen, University of Oslo's Norweigian Institute for Sjørett (NIFS), Murdoch University, Bond University and The University of Texas at Austin. Her expertise has been sought by many national and state/regional governments, where she has been engaged to analyze petroleum laws, draft legislation and advise on technical, policy and governance issues relating to conventional and unconventional petroleum. Hunter has written over 50 peer-reviewed articles and 20 peer-reviewed book chapters. She is the author and editor of several books, including the *Handbook of Shale Gas Law and Policy*, *Regulation of the Upstream Petroleum Sector*, and *Petroleum Law in Australia*. She is the co-leader of the project on the *Scientific Regulation of*

Energy Installations in the Arctic (SciBAR Installations Project) and undertakes interdisciplinary research with engineers, scientists and geologists. She is currently researching and writing on oil spills in the Euro-Barents Arctic Region and heads the Consortium of Researchers & Experts in North & Arctic Marine Ecosystems Oil Contamination (CRENAME), a research initiative between the University of Aberdeen, Tomsk State University, Northern Arctic Federal University and the Murmansk Biological Research Institute.

David Campin is an engineer and ecologist who completed his PhD in 2018. His PhD thesis examined the structure, form and effectiveness of hydraulic fracturing regulations for unconventional resource jurisdictions across the world. He has studied conventional and unconventional resource basins across the U.S. and Australia as part of his research and presented elements of his findings to conferences in the U.S., Scandinavia, Europe and Australia. He commenced working as a biologist in a natural resources company, progressed through extremely diverse environmental management roles dealing with land-based, freshwater and marine environments. As an engineer he worked as a regulator in the environmental policy and compliance arena, largely dealing with heavy industry and unconventional resource development. He has authored and co-authored over twenty peer reviewed papers and government publications on a range of issues including unconventional resource risks and regulation, process safety, ultra-trace chemistry, toxicology and the application of ecological factors to land planning. Campin has worked for an Australian state environmental protection agency, and with the U.S. Environmental Protection Agency and the U.K. Environment Agency. He represented Australia and New Zealand on the Environmental Protection and Safety Panel of the International Ocean Discovery Program. He won a Queensland International Fellowship in 2010.

Chapter 13
Unconventional Oil and Gas: Interactions with and Implications for Groundwater

Brett A. Miller

Abstract The actual and potential impacts of the "shale revolution" on groundwater supplies are subject to intense scholarly debate in scientific, legal, and policy domains. Unconventional development of shale gas through the dynamic combination of hydraulic fracturing and horizontal drilling will continue as a fundamental component of energy policy in the United States, particularly with regards to notions of energy independence and security. At a regional level, the water-related risks associated with hydraulic fracturing operations include impacts on water quality and quantity. This chapter examines the potential implications of hydraulic fracturing operations for groundwater drinking supplies through direct, indirect, and natural contamination pathways, including subsurface migration of methane, accidental surface spills, leak-off implicating fracturing fluids, well-casing integrity, and water table interactions with produced water. These effects are controversial because the best available scientific research is often contradictory, offering both support and opposition to establishing a causal relationship between contamination pathways and hydraulic fracturing. Regulatory uncertainty and challenges in establishing legal causation further contribute to the difficulties associated with detecting, monitoring, and assigning liability for groundwater contamination. This chapter examines the science behind the conduits that could impact drinking water supplies and analyzes regulatory regimes that monitor groundwater interactions with unconventional oil and gas development.

Keywords Groundwater · Contamination pathways · Subsurface · Threats · Drinking water · Causation

B. A. Miller (✉)
Water Asset Management, LLC, New York, NY, USA

© Springer International Publishing AG 2020
R. M. Buono et al. (eds.), *Regulating Water Security in Unconventional Oil and Gas*, Water Security in a New World,
https://doi.org/10.1007/978-3-030-18342-4_13

13.1   Introduction

Remarkable technological advancements in hydraulic fracturing techniques and horizontal drilling continue to allow oil and natural gas companies to extract more energy resources from unconventional tight shale plays at economically viable production costs (Gold 2014). Hydraulic fracturing operations have an enormous positive economic impact on oil and gas production and job creation; however, the industry has also drawn strong opposition from stakeholders concerned about negative effects such as air and water pollution (Cunningham 2014). At the intersection of recent legal developments and prevailing scientific research, this chapter explores important distinctions that underscore the relationship between hydraulic fracturing and groundwater contamination among various theatres of contention.

Unconventional oil and gas exploration and production present a number of potential threats to groundwater sources, including contamination by fracture fluids, methane migration, produced water, and flowback. I first examine these threats and then survey the direct, indirect, and naturally occurring conduits and pathways that may facilitate aquifer contamination, including shallow fracturing operations, natural hydraulic connectivity, leakoff, inadequate well-casing, unlined wastewater pits, operator error, and accidental surface spills. I then review current legal, regulatory, and scientific developments in three U.S. states, representing the nation's major shale plays in the Permian Basin, Eagle Ford Shale, Marcellus Shale, Niobrara Shale, and Bakken Formation (See Fig. 5.3, Chap. 5). Ultimately, much of the tension around this issue stems from the scientific, legal, and regulatory uncertainty that characterizes the relationship between groundwater and unconventional production.

With understandings and explanations as diverse and numerous as the potential contamination pathways, any baseline consensus regarding how and to what extent hydraulic fracturing interacts with groundwater is elusive across disciplines. For example, the standard of legal causation required to assign liability in a judicial context may be heightened when compared with the statistical standards underlying scientific correlation. The distinct nuances between the legal and scientific standards may be further skewed when the debate enters the public realm.

As a practical matter, the volume of fracture operations on a regional level suggests that even minute risks could be consequential if aggregated. In 2014, the Railroad Commission of Texas[1] ("RRC") permitted more than 8800 Permian Basin wells (RRC 2014). In research quantifying water use for hydraulic fracturing of unconventional horizontal wells, Scanlon et al. (2017) found that median water use in 2015 increased to 184,000 barrels per well (~23.7 acre-feet/well) in the Permian Basin. Nevertheless, water for irrigation (91%) and municipal (6%) purposes in the Permian Basin region of west Texas combine for 97% of total water use. The

---

[1] Despite its name, the Railroad Commission regulates the oil and gas industry in Texas. The agency was originally created to oversee the rail industry but, despite the transfer of those functions to other state agencies, legislation to change the name of the agency overseeing oil and gas has failed to pass, and the agency retains its misleading title.

controversy is exacerbated, however, because the disposal of wastewater from hydraulic fracturing permanently removes it from the hydrologic cycle.

Effectively and appropriately regulating oil and gas can be challenging because developments in the industry may outpace legislative endeavors, enforcement regimes, and scientific consensus. Divergence between federal and state law further contributes to the expansive scope of these challenges. In a 2004 report, the Environmental Protection Agency ("EPA") exempted hydraulic fracturing from the Safe Drinking Water Act, which found that the process of hydraulic fracturing posed no discernible threat to underground drinking water (2004 EPA Report). The 2004 EPA Report cited dilution, adsorption, and biodegradation as attenuation factors that would effectively reduce contaminant concentrations before reaching shallow domestic groundwater wells. States like Texas and Pennsylvania have robust regulations prohibiting the contamination of groundwater by oil and gas activities (see, e.g., 16 Tex. Admin. Code §§3-5; 58 Pa.C.S. §§3201-3274). According to the U.S. Energy Information Administration (2011), shale plays exist within the jurisdictions of at least 30 states. The vast number of wells in a particular state, however, may limit the ability of administrative agencies to effectuate monitoring and enforcement regimes. In Texas, Pennsylvania, and Wyoming, plaintiffs have faced difficulty establishing the heightened standard of causation as it relates to establishing legal liability for potential groundwater contamination.

It cannot be said with certainty that the process of hydraulic fracturing causes groundwater contamination—nor can it be said that hydraulic fracturing does not cause groundwater contamination. Scientists, lawyers, policy-makers, industry advocates, and environmentalists should therefore exercise caution before making blanket statements associated with hydraulic fracturing because factors that drive this debate are nuanced, vary based on geological characteristics across shale plays, and are best analyzed on a case-by-case basis. Although the risk that any particular production operation will contaminate groundwater is small, the possibility of human error means there exists a baseline risk of groundwater contamination (Spence 2014).

13.2 Potential Threats to Groundwater from Fracturing Operations

Public health concerns attributed to fracturing operations stem from the potential for groundwater contamination due to fracturing fluids, natural formation waters, and stray gases (Jackson et al. 2013). The debate surrounding hydraulic fracturing and groundwater contamination has focused on distinct categories of interactions that could impact groundwater quality (Vengosh et al. 2013): (1) stray gas migration to shallow groundwater aquifers (Osborn et al. 2011), (2) possible connectivity between deep shale formations and shallow aquifers (Warner et al. 2012), and (3) potential contamination from fracture fluids, flowback waters, and produced brines containing toxic substances during drilling, transport, and disposal (Gregory et al. 2011). Research studies near hydraulic fracturing activities have indicated both elevated levels of methane (Osborn et al. 2011) and stray thermogenic natural gas in groundwater samples (Darrah et al. 2014). Other studies, however, have found the natural occurrence of dissolved gases in areas that are not close to active hydraulic fracturing activities (Siegel et al. 2015), as well as the presence of methane contamination in groundwater before drilling activities (Molofsky et al. 2013).

The majority of horizontal wells in Texas's Permian Basin and Eagle Ford shale are drilled at depths greater than 10,000 feet below the surface. In the Delaware Basin, one of the Permian's sub-basins, the horizontal wells producing from the Bone Spring and Wolfcamp formations are drilled at depths between 8,200–14,150 feet. In comparison to the shale layer, the groundwater table is typically much closer to the surface (within a few hundred feet of the surface). Freshwater aquifers in these Texas shale regions are typically found at stratigraphic intervals that are much shallower than 4,000–5,000 feet. For instance, aquifers in the Permian Basin, such as the Pecos River Valley and Rustler aquifers, are found much closer to the surface at depths of 50–400 feet. This suggests that oil and gas production zones are separated from drinking water supplies by, approximately, a minimum of 7,000–7,500 feet.

Interactions between groundwater and surface oil and gas activities present additional risks given the natural hydrological connectivity between surface water and groundwater. In the Bakken (Lauer et al. 2016) and Marcellus (Warner et al. 2013) formations, groundwater contamination has been attributed to surface spills, as researchers have documented the presence of inorganic constituents and heavy metals in drinking water supplies. Deficient baseline groundwater quality data may limit the scope of regulatory oversight and may even hinder the task of establishing legal causation to assign liability. In the Eagle Ford, scientists call for more extensive investigations into groundwater quality (Hildenbrand et al. 2017), including, but not limited to, the quantification of noble gas analyses and dissolved compound-specific hydrocarbon isotopes as a means to differentiate between anthropogenic pathways attributed to unconventional oil and gas activities and naturally occurring geogenic and biogenic pathways (Darrah et al. 2014). Furthermore, injection wells provide another conduit for fracture fluids and wastewater to contaminate drinking water supplies (Jackson et al. 2015).

Although it did not characterize the severity or frequency of the impact on drinking water sources, a 2016 report by the EPA found evidence of drinking water contamination from hydraulic fracturing activities (2016 EPA Report). The 2016 EPA Report examined various stages in the fracturing process, including water acquisition, chemical mixing to prepare fracture fluids, injecting fluids to create fractures in the production zone, collecting produced water after injection, and wastewater disposal. In particular, the 2016 EPA Report highlighted several notable conditions or pathways that could result in drinking water contamination, including injection of fracture fluids into wells with inadequate mechanical integrity (thereby allowing contaminants to move to groundwater resources), injection of fracture fluids directly into groundwater resources, and disposal of wastewater into unlined pits resulting in contamination of groundwater resources.

Operators, too, have an economic incentive to prevent interactions between hydrocarbon recovery and groundwater. Unconventional drilling and production operations are expensive as Permian Basin operators may fracture the stacked shale intervals upwards of 28 times along their horizontal wellbores that can extend lengths up to 7100 feet (Oxy Permian 2017). If the fracture operation exceeds past the production zone, the entire well may become "watered out" if water migrates into the wellbore (Hyne 1996), which—from an operator's perspective—would make the well uneconomical. Given the intensive capital and operating expenses associated with fracture operations, operators have a serious incentive not to contaminate the well in ways that will inhibit recouping costs or limit revenues (Deweese 2010). Standard industry practice of unconventional oil and gas production does not involve blatant or intentional contamination of groundwater. On the other hand, operator error and human-derived accidents do present a direct pathway for groundwater contamination. Regardless of how minimal a risk that a certain drilling site poses to groundwater contamination, Spence (2014) explains that "the possibility of human error means that the risk of groundwater contamination is not zero."

13.2.1   Physical Substances That Threaten Groundwater Supplies

13.2.1.1   Frac Fluids

Although frac fluids are primarily comprised of water and sand, at least 0.5% of mixture includes toxic chemicals and their constituents. Companies use a variety of different formulas and range of contaminants for their confidential and proprietary "frac fluids," which may include benzene, ethylbenzene, toluene, boric acid, monoethanolamine, xylene, diesel-range organics, methanol, formaldehyde, hydrochloric acid, and ammonium bisulfite, among others (Merrill and Schizer 2013).

Studies have tended to find little to no evidence of frac fluids contaminating groundwater supplies. Moniz et al. (2011) who later became Secretary of Energy in the Obama Administration, stated that despite concern that fracturing of shale formations could penetrate shallow freshwater zones and contaminate them with fracturing fluid—there is "no evidence" of the migration of fracture fluids into shallow freshwater zones. Merrill and Schizer (2013, p. 184) concluded that the "paucity of confirmed incidents of water contamination from the underground migration of fracturing fluid provides power evidence that the risk is small," given the more than two million fracture operations in the past 60 years. However, the authors called for additional study, given the continued expansion of fracturing operations.

Geological considerations suggest that the risk of groundwater contamination is remote from a practical perspective, as fracturing in low-permeability shale intervals takes place at production zones that are 5,000–10,000 feet (1–2 miles) below the surface, whereas the water table is typically only 500–1,000 feet below the surface. From a legal perspective, this challenges landowners' ability to establish causation in lawsuits that may allege contamination of their groundwater wells (King et al. 2012). Because fracturing of shale formations takes place far below the surface, the toxic chemicals in fracture fluids would have to migrate upward against the massive weight of rock and soil pressing down on the layer of shale being fractured (Merrill and Schizer 2013). To establish this direct pathway, significant quantities of contaminants would have to migrate upwards over a mile from the shale seam in order to potentially pollute the overlying groundwater aquifer. In a comprehensive study analyzing over 15,000 fracture operations in the Barnett and Marcellus formations, Fisher (2010) suggested that the results demonstrate that hydraulic fractures are not migrating far enough to reach groundwater supplies, emphasizing it would be highly unlikely that cracks produced in horizontal drilling operations so far below the surface would produce permeable fissures extending upward thousands of feet.

The EPA investigated potential groundwater contamination in 2008 as a response to complaints in Pavillion, Wyoming, regarding the taste and odor of drinking water from domestic wells. The study was the first and only time that the Comprehensive Environmental Response and Liability Act (CERCLA) has been invoked to investigate the groundwater-hydraulic fracturing relationship. Much controversy stemmed from the EPA's response of installing monitoring wells, and the subsequent draft report released by the EPA in 2011 that attributed unconventional drilling activities as the source of groundwater contamination (2011 EPA Draft Report). DiGiulio[2] and Jackson (2016) further documented injection of stimulation fluids in groundwater and evaluated the impact to groundwater as a result of acid stimulation and hydraulic fracturing, specifically the potential upward migration of contaminants to depths of current groundwater use for domestic drinking water supplies. DiGiulio and Jackson (2016) found that "inorganic and organic geochemical anomalies in the [monitoring wells] appeared to be attributable to production well stimulation."

13.2.1.2 Produced Water: Flowback and Formation Water

During the drilling process, the initial water produced from the well is primarily *flowback water* from fluids injected for hydraulic fracturing (Scanlon et al. 2017). Chemical additives that may be included in flowback water during the fracturing process include viscosifiers, descaling agents, anticorrosive compounds, lubricants, and pH stabilizers (Vidic et al. 2013). Eventually, the amount of water from the formation itself increases. Formation brines naturally occur in water that accumulates in shale deposits irrespective of the drilling process. Even though it does not contain toxic fracturing chemicals, produced formation water has natural contaminants (including salt, other organic compounds, silt, clay, oil, grease, and naturally occurring radioactive material). Flowback and formation water are collectively known as *produced water*.

Among the most notable of public health concerns is the potential for drinking water contamination from produced water and the natural migration of formation

---

[2] Digiulio was the lead scientist on 2011 EPA Draft Report.

brines (Warner et al. 2012; Chapman et al. 2012). Operator adherence to best-management practices regarding these substances is crucial because oil and gas wells in shale plays generally produce more water than oil (Deweese 2010). In 2012, operators generated nearly 10 times more produced water than oil from combined unconventional and conventional reservoirs in the U.S (Veil 2015).

Operators must secure this fluid when it rises to the surface so that it does not seep into the water table and drinking water supply. Along with man-made chemicals that comprise the minority of produced wastewater, naturally occurring brine water contains varying levels of salts, heavy metals, and radioactive elements. This naturally occurring formation water may also contain elevated concentrations of chloride, bromide, sodium, and sulfate (Warner et al. 2013). Meanwhile, elevated levels of chloride and bromide, as well as chloride/bromide mass ratio, in groundwater samples may indicate contamination from anthropogenic origins (Hildenbrand et al. 2015). Such origins of contamination could be from oil and gas activity (Hudak 2010), as a result of formation water commingling with groundwater supplies (Warner et al. 2013). In research that analyzed 550 groundwater samples from the Barnett Shale region of Texas, Hildenbrand et al. (2015) found 97 of the total 550 groundwater well samples had chloride/bromide ratios indicating contamination from oilfield brine formation water.

Hildenbrand et al. (2015) further detected elevated levels of 10 different metals and the presence of 19 different chemical compounds, including benzene, toluene, ethyl-benzene, and xylene (BTEX), although the authors suggest that the data do not necessarily identify unconventional oil and gas activities as the source of contamination. In a study that quantified the inorganic and organic chemical composition of produced water samples from Pennsylvania shale gas wells, Akob et al. (2015) detected volatile organic compounds ("VOCs") in four samples, including benzene, toluene, and tetrachloroethylene; but noted that the source is unclear because VOCs can occur both naturally and from industrial activity. Kondash et al. (2017) found that wastewater coming from hydraulically fractured unconventional oil and gas wells is mostly comprised of naturally occurring brines, rather than man-made fracture fluids. The study suggests that chemical-laden fracture fluids account for only 4–8% of wastewater during the productive lifespan of hydraulically fractured wells in major U.S. shale basins. More than 92% of the cumulative flowback and produced water is derived from naturally occurring brines that are extracted with the oil and gas.

Researchers have been successful in developing innovative techniques that reuse the produced brines and flowback from fracturing operations as an alternate water source for unconventional oil and gas production (Burnett et al. 2015). However, legal questions exist in Texas regarding the ownership of produced water, specifically whether recycled water belongs to the surface or mineral estate. The Texas Supreme Court in *Edwards Aquifer Authority v. Day* (2012) held that the owner of the surface estate also owns groundwater in place beneath her land as a vested property right subject to protection under the Texas Constitution. Because the mineral estate's right to use water is usufructuary for fracturing operations, mineral interest owners in Texas will likely be prevented from benefitting from the sale of treated produced water (Hosey and Lotay 2017).

13.2.1.3 Methane and Natural Gas

Natural gas itself, primarily methane, constitutes an additional threat of groundwater contamination. Contamination may arise from methane migration along natural fractures unrelated to drilling activities. Methane naturally migrates upwards and gets trapped in shallow porous formations. This presents regulatory challenges because even if researchers and administrative agencies identify contamination events, they often struggle to differentiate between geogenic and anthropogenic contamination pathways. Perhaps methane migration may have always existed—as reports from the 1800s describe gas bubbles in water wells, streams, and fields after heavy rains (USGS 2000). The presence of methane in groundwater before shale drilling activities complicates efforts to establish legal liability. For instance, a 2006 USGS survey of 47 West Virginia counties reported the presence of methane in 131 out of 170 residential water wells (USGS 2006). Boyer et al. (2012) found methane in 78% of water samples taken in Pennsylvania before shale drilling began, further noting historical evidence of natural methane migration dating back to the late 1700s.

Investigations near hydraulic fracturing operations have identified both elevated levels of methane (Osborn et al. 2011) and stray thermogenic natural gas in groundwater samples (Darrah et al. 2014; Jackson et al. 2013). The possibility that methane could migrate up into aquifers from the fractured shale seam through pre-existing, natural fissures in the overlying rock, or even through fissures created or enlarged by fracturing—compounds the risks associated with hydraulic fracturing (Merrill and Schizer 2013). Many oil and gas operators have initiated the practice of testing groundwater wells before commencing drilling activities in certain locations. In fact, some state management regimes require pre-drilling water sampling, such as in Pennsylvania's "rebuttable presumption" statute discussed below in Section III(b).

In the Marcellus Shale, Osborn et al. (2011) documented higher methane concentrations and less negative $\delta^{13}C$ isotopic signatures for methane, which were consistent with a natural gas source, in domestic water wells that were less than 1 km from shale gas wells. The research sampled well water before and after fracturing and found no evidence of groundwater contamination by frac fluids or wastewater, but did find evidence that levels of thermogenic methane were higher in shallow groundwater aquifers near natural gas production wells than elsewhere in the same aquifers (despite not establishing a cause and effect relationship) (Osborn et al. 2011). Additional research by Jackson et al. (2013) documented significantly higher concentrations of methane in the drinking water of homes near shale gas wells compared to homes farther away. Molofsky et al. (2013), however, noted the potential correlation between the presence of methane in groundwater and topography, rather than shale gas production. Yet in 2014, the Pennsylvania Department of Environmental Protection ("Pennsylvania DEP") released a list of more than 200 examples of hydraulic fracturing related water-supply contamination cases.

As depicted in the documentary *Gasland* (2010), groundwater contamination in Dimrock Township, Pennsylvania, sparked a national debate on the relationship between hydraulic fracturing and drinking water quality. In 2009, residents of Dimock Township sued Cabot Oil & Gas ("Cabot") after the company's hydraulic fracturing activities allegedly contaminated their groundwater supplies. Almost all plaintiffs, 42 out of 44, reached confidential settlements with Cabot in order to dismiss their respective lawsuits (DeKok 2016). Cabot maintains that the methane contamination occurred naturally near the water wells and was not caused by the company's drilling operations, pointing to affidavits from longtime Dimock residents that note the presence of methane in water supplies dating back to the 1950s (McNeal Affidavit 2010).

In December 2010, Pennsylvania DEP entered a Consent Order and Settlement Agreement finding that 18 drinking water supply wells in Dimock Township were contaminated by methane as a result of natural gas drilling activities operated by Cabot (In re Dimrock/Springville 2010). In March 2016, the two plaintiffs who did not settle their 2009 groundwater contamination lawsuit against Cabot received a $4.24 M jury verdict in the U.S. District Court for the Middle District of Pennsylvania on their causes of action for private nuisance and negligence (Cabot Jury Verdict 2016). In March 2017, however, a federal judge reversed the jury verdict because of insufficient evidence on the issues of causation and damages (Cabot Mem. Op. 2017). After almost a decade of litigation, the final two plaintiffs settled their lawsuit with Cabot in September 2017 (Hurdle 2017).

In Texas's Barnett Shale, the saga between Range Resources Corporation's ("Range") hydraulic fracturing operations and Steve Lipsky's methane-rich water well underscores the contentious nature of the debate. In 2010, the EPA initially maintained that methane in Lipsky's water may have been caused by Range's nearby natural gas wells, but the agency withdrew its complaint in 2012 after the RRC cleared the company of charges and found that Range did not contaminate Lipsky's groundwater (Park 2012). Range's lawyers noted that natural gas, predominantly methane, is naturally present in the Trinity Aquifer area (Solomon 2015). Lipsky filed a petition seeking $6.5 M in damages for the alleged groundwater contamination. Range then moved to dismiss all claims and filed a counterclaim for defamation. At issue in the defamation suit was a video that Lipsky created of himself setting water on fire coming out of his well. The trial court dismissed all of Lipsky's claims against Range but found that Lipsky was igniting gas connected to the water well vent, not the water line (*Lipsky v. Range* 2011). In 2015, the Texas Supreme Court ruled that Range could proceed with its defamation suit because Lipsky's accusations of groundwater contamination were harmful to Range's capabilities as a natural gas producer (In re Lipsky 2015).

13.2.2 Conduits and Methods of Contamination

13.2.2.1 Limited Vertical Separation in Shallow Fracturing Operations

In general, the large vertical separation between shallow drinking water wells and shale gas formations means that the potential for shallow groundwater contamination is often dismissed (Warner et al. 2012). However, when hydraulic fracturing occurs at shallow intervals (i.e. when vertical separation is minimal) potential problems exist because chemicals could migrate into overlying drinking water supplies regardless of issues with well integrity (Jackson et al. 2015). In the U.S., depths of hydraulic fracturing and horizontal drilling operations range from 100 feet (30 m) to more than 15,840 feet (5000 m) (Jackson et al. 2015). The risk of fractures propagating from deep shale formations to reach overlying aquifers is generally mitigated by the fact that a majority of oil and gas production occurs several miles underground. However, drilling wells at shallow intervals intensifies the risk of groundwater contamination given the uncertainty in determining adequate vertical separation between the depth of hydraulic fracturing and the overlying surface aquifer. From a regulatory perspective, it is notable that no state currently has restrictions that place an upper limit on how shallow depth at which a horizontal well may be hydraulically fractured (Vaidyanathan 2016).

Jackson et al. (2014, 2015) maintain that the lack of vertical separation between hydraulically fractured wells and drinking water in certain instances will increase potential hydraulic connectivity and the likelihood of groundwater contamination. With regards to adequate vertical separation in various U.S. shale plays, Davies et al. (2012) found that the greatest upward propagations in the Marcellus, Woodford, and Eagle Ford shales were 536, 588, and 556 m (~1800 to 1900 ft.), respectively. Even shallow fractures that do not reach all the way to an aquifer, however, can still link formations acting as a conduit through natural faults, fissures, or other pathways that characterize regional geology.

In the first comprehensive analysis of hydraulic fracturing depths in relation to water use, Jackson et al. (2015) noted that out of 44,000 hydraulic fracturing observations in the U.S. between 2010–2013, more than 80% were at least a mile underground. Conversely, the fact that 16% (n = 6896) of these wells were hydraulically fractured at depths shallower than one mile may increase the potential for contamination events considering the limited vertical separation in particular instances. In documenting more than 20,000 (n = 20,267) wells in Texas, Jackson et al. (2015) observed 541 cases of hydraulically fractured wells within 1000 feet of the surface. In Oklahoma, high-volume fracturing is occurring as shallow as 2800 feet below the surface (Soraghan 2017). Despite the relative limited number, the existence of fractured wells at shallow intervals may increase the potential for groundwater contamination in comparison to horizontal wells fractured at depths greater than one mile underground.

The lack of vertical separation between fractured intervals and the overlying aquifer may be an influential factor causing the alleged groundwater contamination in a natural gas field in Pavillion, Wyoming. Well stimulation records indicate that fracturing occurred in drinking water sources as shallow as 1060 feet (323 m), and acid stimulation occurred at depths as shallow as 699 feet (213 m) below ground surface (2011 EPA Report). Domestic water wells in the Pavillion area often extend to depths of at least 750 feet (229 m). The alleged contamination could be a product of shallow drilling and the limited vertical separation between the water table and drilling activities (Jackson et al. 2015). To ensure adequate vertical separation between fracked wells and drinking water aquifers, many state rules mitigate the risk by enforcing casing and cementing requirements, rather than through spatial or distance limitations on vertical separation between protected groundwater and shallow wells.

13.2.2.2   Natural Hydraulic Connectivity

Warner et al. (2012) presented geochemical evidence to indicate the existence of natural hydraulic connections as migration pathways between deep underlying gas formations and overlying shallow drinking water aquifers. Although their study did not report any evidence of groundwater contamination from drilling activities, these naturally occurring pathways could theoretically function as the conduit between drilling activities and overlying water resources. The extent of potential contamination via this particular conduit may be limited as the occurrence of mixing relationships between shallow groundwater and deep formation brine water did not correlate with the location of shale-gas wells (Warner et al. 2012).

Natural hydraulic connectivity from biogenic sources further emphasizes the uncertainty that underscores regulatory decisions associated with protecting groundwater resources (Warner et al. 2012). Additional evidence of natural hydraulic connectivity between shallow groundwater and brines from deep formations is suggested through the presence of inorganic elements attributed to biogenic sources using noble gas geochemistry to understand migration processes (Darrah et al. 2015) and through geochemical analyses of bedrock permeability (Llewellyn 2014).

13.2.2.3   Leakoff

DiGiulio and Jackson (2016) concluded that migration of fracture fluids into the Pavillion Field drinking water supplies likely occurred during fracture propagation and subsequent leakoff. Leakoff is the loss of fluid into a formation in or near the target stratum. Leakoff increased in complex fracture networks as a result of the lithologic variation present in Wyoming's Wind River shale formation (DiGiulio and Jackson 2016). From a contamination perspective, leakoff may remove or divert much or most of the fracturing fluid from the produced wastewater even for moderate-sized induced fractures (DiGiulio and Jackson 2016). DiGuilio and Jackson (2016) further explained that the loss of zonal isolation during well stimulation results in the migration of stimulation fracture fluids into water-bearing units. Kondash et al. (2017) documented the presence of fracture fluids in produced and flowback water in their comprehensive analysis of drilling activities in the Bakken, Marcellus, Barnett, Eagle Ford, and Niobrara formations. They found that most of the fracturing fluids injected into wells do not return to the surface but are instead retained in the deep shale formations (Kondash et al. 2017). Transport of fluids through micro-scale annular fissures in unconventional gas wells was noted by Burton et al. (2016), indicating a significant risk if extrapolated.

13.2.2.4   Inadequate Mechanical Integrity and Cracked Well Casing

The drilling process itself may create additional conduits for gas to contaminate overlying aquifers as the wellbore passes through shallower formations if the well lacks proper steel and cement casing (Yoxtheimer 2012). However, groundwater contamination from inadequate casing is not unique to hydraulic fracturing, as this method of contamination may occur in vertical drilling operations with or without hydraulic fracturing. State regulatory regimes rely on technical standards for casing requirements to protect groundwater from contamination, requiring field-specific minimum surface casing depths or that surface casing be set some distance below the deepest protected groundwater (Jackson et al. 2015). The 2016 EPA Report outlines various pathways for potential movement of fluids in the wellbore, including leakage from the casing and tubing into a permeable formation, migration between the drill pipe and outside of an uncemented wellbore, migration between casing and cement, migration through inadequate cementing, and migration between cement and the formation.

Moniz et al. (2011) noted evidence of natural gas migration into freshwater zones that most likely resulted from sub-standard well completion practices (i.e., cracks in well-casing). DiGiulio and Jackson (2016) explain that surface casing in production wells is the primary line of defense to protect groundwater during oil and gas extraction from conventional and unconventional resources. In terms of responsibility, liability, safety, and precautionary standards, a diverse array of operators, ranging from major corporations to "wildcatters," are producing oil and gas throughout the U.S. Some companies embrace environmental stewardship and social responsibility while emphasizing the long-term benefits of a respectful landowner-operator relationship, whereas others have a general disregard for the environment and fail to error on the side of caution. Safeguards by other companies may be influenced by tight profit margins in an era of decreasing oil prices.

Despite regulations that focus on casing requirements, the scientific literature and legal conflicts are replete with examples of inadequate mechanical integrity. DiGiulio and Jackson (2016) found evidence of casing failure at multiple production wells in the Pavillion Field following well stimulation. Operator error has been documented by research that found stimulation fluids injected less than five meters below an interval lacking cement outside of the production casing, a scenario that could indicate potential entry into the annular space (DiGiulio and Jackson 2016).

13.2.2.5   Disposing of Wastewater and Contaminants in Unlined Pits

Disposal of drilling fluids (i.e., well completion practices) into unlined pits represents an indirect conduit to groundwater through seepage into shallow aquifers. When improper, these completion practices create "legacy" issues from a legal perspective when operators move on from particular wells or when companies endure financial hardship. The EPA has documented the existence of production fluids in unlined pits in the Pavillion Field, including flowback, condensate, and produced water (Folger et al. 2012). The 2011 EPA Report attributed groundwater contamination in the surficial Quaternary unconsolidated alluvium to the existence of numerous unlined pits used for disposal of diesel-oil-based drilling mud and production fluids. 2-Butoxyethanol and lower concentrations of BTEX compounds (benzene, toluene, ethylbenzne, xylenes) have also been detected in shallow domestic wells located in the vicinity of unlined pits used for the disposal of production fluids (DiGiulio and Jackson 2016). Breaches in flowback wastewater pits can occur from overflow or faulty linings, and may introduce dichloromethane and other species into the surface environment (Vengosh et al. 2014).

The lateral extent of groundwater contamination attributed to disposal in unlined pits is magnified when uncontaminated water overlies portions of a contaminated plume via a process known as "plume diving" (DiGiulio and Jackson 2016). Low molecular weight organic acids, such as lactate, formate, acetate, and propionate are anaerobic degradation products associated with hydrocarbon contamination in groundwater (DiGiulio and Jackson 2016). Formate and acetate are degradation products of dichloromethane (methylene chloride) occurring from both natural and anthropogenic sources. Acetate has been detected in produced water contained in storage pits on numerous occasions (Orem et al. 2014) (Cluff et al. 2014). In the Marcellus Shale, acetate has been detected in impoundment pits used to contain flowback water (Mohan et al. 2013). The potential for toxicity in wastes that are present in storage impoundments underscores the importance of proper barriers and best-management practices to protect groundwater.

Permits are generally required for storage and disposal of oil and gas waste (16 TEX. ADMIN. CODE §3.8(d)(1)–(2)). Texas requires that fluid recycling pits be lined, routinely monitored, and designed to prevent any migration of materials from the pit into adjacent groundwater (16 TEX. ADMIN. CODE §3.8(d)(4)(G)(i)-(v)). A permit to use an unlined pit, other than an emergency saltwater storage pit for the disposal of oil field brines, may only be issued if the RRC determines that the applicant has conclusively shown that the pit cannot cause pollution of groundwater (16 TEX. ADMIN. CODE §3.8(d)(6)(A)).

13.2.2.6 Accidental Surface Spills and Seepage into the Water Table

With respect to natural hydrologic connectivity between groundwater and surface water, chemicals used in fracturing operations could be spilled before or after the drilling process, eventually seeping into the water table (Wiseman 2013). In some regions, hydrogeologic factors outside the control of operators may magnify connections between surface water and aquifers, making certain aquifers more susceptible to contamination. In the Bakken and Marcellus formations, groundwater contamination has been attributed to surface spills as researchers have documented the presence of inorganic constituents and heavy metals in water supplies (Lauer et al. 2016; Warner et al. 2013). The EPA found that the frequency and typical causes of surface spills remain unclear (2012 EPA Report). Shallow water wells may be vulnerable to contamination from surface sources, such as mishandled produced water or fluid spills during stimulation and completion processes (Hildenbrand et al. 2015).

The increased use of trucks and other traffic that accompany the transportation of toxic chemicals to the well location for use in fracturing presents legitimate risks. Accidents involving trucks are particularly dangerous with regards to groundwater contamination because they transport biohazard chemicals in concentrated form. Although this transportation risk of surface spills is ubiquitous throughout all industrial and commercial activities, fracture operations increase the total volume of toxic chemicals on a regional level (Merrill and Schizer 2013).

13.2.2.7   Frac Hits: Vertical Wells Impacted by Horizontal Drilling

In particular shale play regions, the combination of older vertical wells constructed without groundwater protections and high-volume fracturing along long horizontal wellbores could increase the probability for "frac hits" resulting in potential groundwater contamination (Soraghan 2017). Frac hits occur when the hydraulic treatment from one well communicates with another well (McEwen 2018). In a 2017 study, the Oklahoma Energy Producers Alliance estimated that more than 400 frac hits occurred in just one Oklahoma county (OEPA Study 2017). Although modern cementing techniques are designed to protect groundwater—older vertical wells may have been drilled and plugged well before the current cementing regulations were introduced. For example, small producers in Oklahoma allege that hundreds of their older vertical wells have been flooded by high-pressure fracturing of horizontal wells (Soraghan 2017). Oklahoma's spacing and pooling provisions allow horizontal wells to be fractured within 600 feet of older vertical wells (OKLA. ADMIN. CODE 165:10–1-21).

Although the Oklahoma Corporation Commission claims there is no proof of groundwater contamination, several producers claim there could be an impact on freshwater (Soraghan 2017). Aside from the potential for groundwater contamination in certain areas, operators have an economic incentive to limit frac hits as the consequences may result in a loss of production depending on the extent of mechanical, physical, and/or chemical damage to the wells (McEwen 2018). With more than 100,000 drilling locations in the Permian Basin, the industry is actively sharing data to mitigate frac hits, with companies like Schlumberger amassing case studies in the Eagle Ford, Barnett, and Bakken shales (McEwen 2018).

13.3   Regulatory Regimes Addressing the Groundwater-Hydraulic Fracturing Nexus

Legal scholars emphasize the need for both best-practice regulations and legal liability regimes to protect groundwater resources (Merrill and Schizer 2013). Oil and gas activities are primarily regulated by state-level administrative agencies rather than the federal government. Best-practice regulation monitors surface spills, vertical well leaks, blowouts, and disposal of produced water and flowback, among many other technical aspects of exploration and production, and offer consistency and predictability for corporations that make significant investments in drilling operations. Nevertheless, regulatory oversight must be backstopped by liability and enforcement regimes, or the regulations will be ineffective. Best-practice regulations may be ineffective for novel risks associated with unconventional oil and gas activities with respect to the underlying scientific and legal uncertainty.

Many states with active shale plays require disclosure of the names and Chemical Abstract Service ("CAS") numbers of the chemicals used in hydraulic fracturing, although exemptions exist regarding the disclosure of trade secrets (Jackson et al. 2014). For example, the Texas Administrative Code maintains hydraulic fracturing disclosure requirements, in which operators must complete the chemical disclosure registry form providing information on the hydraulic fracturing treatment before well completion. The required information includes the total volume of water used, each additive used, each chemical ingredient used, and concentrations and CAS number for each chemical ingredient listed (16 TEX. ADMIN. CODE §3.29). The information submitted to the RRC on the chemical disclosure form is considered public, unless the information is entitled to trade secret protection under Chapter 552 of the Texas Government Code. The FracFocus website provides a publicly available chemical disclosure registry for wells that have been hydraulically fractured in the U.S.

Property owners seeking redress in court for alleged groundwater contamination from hydraulic fracturing activities have filed numerous causes of action with varying degrees of success, including gross negligence, negligence per se, common law negligence, subsurface tress, and private nuisance (Goldman 2013; Watson 2017). Groundwater contamination regulations specific to several states are described below.

13.3.1 Texas

The Texas Administrative Code makes it unlawful for a person conducting oil and gas activities to cause or allow pollution of surface or subsurface water in the state (16 TEX. ADMIN. CODE §3.8(b)). Pollution is defined as an alteration of the water that renders the water harmful, detrimental, or injurious to humans, vegetation, property or public health, safety, or welfare (16 TEX. ADMIN. CODE §3.8(a)(28)). Violations of this rule may result in criminal penalties under the Texas Natural Resources Code or other law (16 TEX. ADMIN. CODE §3.8(h); TEX. NAT. RES. CODE §91.002). In any well drilled for oil and gas, hydrocarbon resource fluid must be confined in its original stratum until it can be produced and utilized without waste, such that each stratum must be adequately protected from infiltrating waters (16 TEX. ADMIN. CODE §3.7). The RRC requires each stratum to be cased off and protected (16 TEX. ADMIN. CODE §3.7). Operators that encounter gas-bearing stratums when drilling oil or gas wells must confine the gas to its original stratum and protect the gas-bearing strata from infiltrating water (TEX. NAT. RES. CODE §91.016).

In response to the City of Denton's 2014 ordinance prohibiting fracking operations within city limits, the Texas legislature in 2015 passed an express preemption statute providing that oil and gas operations are subject to the exclusive jurisdiction of the state (TEX. NAT. RES. CODE §91.0523). Regulatory authority over drilling activities in Texas is deferential to the RRC. Specifically, the RRC and Texas Commission on Environmental Quality ("TCEQ") have adopted a memorandum of understanding ("MOU") regarding the division of jurisdiction between the agencies over wastes associated with exploration, development, and production of oil and gas (16 TEX. ADMIN. CODE §§3.30, 3.8(i)). Under the MOU and Texas Water Code, the RRC must submit a written notice to the TCEQ of any documented cases of groundwater contamination that may affect a drinking water well (16 TEX. ADMIN. CODE §3.30(e)(7)(A)). In the Permian Basin, the Middle Pecos Groundwater Conservation District maintains additional rules that prohibit the pollution or degradation of groundwater quality by means of saltwater or other deleterious matter admitted from another stratum, or by activities which cause pollutants to enter the groundwater, whether natural or manmade (Rule 14.3 Middle Pecos GCD).

The RRC rules provide clear standards for well casings in order to protect groundwater. Casing must be securely anchored for well control in order to isolate and seal off all usable-quality water zones to effectively prevent contamination or harm (16 TEX. ADMIN. CODE §3.13(a)(1)). Operators must set and cement sufficient surface casing to protect all usable-quality water strata as defined by the Groundwater Advisory Unit of the RRC's Oil and Gas Division and prevent upward migration of deeper formation fluids into protected water (16 TEX. ADMIN. CODE §3.13(b)(1)(A)-(B)). Pursuant to the MOU, the RRC defines the base usable-quality water as generally less than 3000 mg/L total dissolved solids (TDS), but may include higher TDS if identified as a water source for desalination (16 TEX. ADMIN. CODE §§3.13(a)(2)(P), 3.30(e)(7)(B)(i)). The RRC establishes field-specific requirements for surface casing depths that must be set below the base of usable quality water, and with regards to groundwater protection requirements (TEX. NAT. RES. CODE §§91.011, 91.1015). As part of its permit application review process, the RRC has previously found that facilities must be designed and operated to prevent ground and surface water pollution (RRC Oil & Gas Docket 2013). In this particular case, the RRC denied the permit because leaching of contaminated material down into the surface outcrop of the aquifer recharge zone was a "real possibility" since freshwater is located close to the surface (RRC Oil & Gas Docket 2008).

If oil and gas wastes cause the pollution of groundwater, the RRC may use the oil and gas cleanup fund for site investigation, environmental assessment, and cleanup if the responsible person has failed to or is unknown (TEX. NAT. RES. CODE §91.113(a)). In addition to any lease forfeiture provided by law, violations of prohibitions on groundwater pollution will subject operators to penalties of not more than $10,000 when the rule pertains to the prevention of pollution, and the applicable maximum penalty may be assessed for each and every day of violation and each and every act of violation (16 Tex. Admin. Code §3.8(h); Tex. Nat. Res. Code §85.381(a)(1)-(b)).

Lawsuits alleging groundwater contamination often reach settlement agreements (e.g., *Scoma v. Chesapeake Energy* (2010); *Mitchell v. Encana Oil & Gas* (2010)), or are dismissed by courts for insufficient evidence among other reasons (e.g., *Harris v. Devon Energy* (2010); *Beck v. ConocoPhillips* (2011)). Some are still pending, such as the case of a Texas rancher who alleges groundwater contamination in addition to other claims after a poorly sealed natural gas well allowed methane to build up in the rancher's water well and ultimately resulted in an explosion (*Murray v. EOG Resources, et al* (2015)).

13.3.2   Pennsylvania

The Marcellus Formation is a shale play that stretches across most of Pennsylvania and West Virginia, and shale gas extraction from the low-permeability Marcellus generally requires hydraulic fracturing. Natural gas production from the Marcellus Region has surged since 2008, increasing from less than 5000 million cubic feet/day to almost 20,000 million cubic feet/day in 2017 (EIA 2017 Marcellus Report). The U.S. Energy Information Administration estimates that the Marcellus Formation has more than 141 trillion cubic feet of recoverable natural gas (EIA 2012 Report).

In 2012, the Pennsylvania General Assembly recodified the state's Oil and Gas Act to include language that explicitly targeted the development of unconventional natural gas drilling operations that use hydraulic fracturing (58 Pa.C.S. §§ 3201–3274). Operators are required to complete a fracturing fluid chemical disclosure form and post the information to a publicly available registry (i.e., FracFocus) (Pa. HB 1950, 58 Pa.C.S. § 3222.1). Operators applying for an unconventional well drilling permit are required to send notice to all surface landowners and water suppliers that are within 3000 feet of the vertical well bore (Pa. HB 1950, 58 Pa.C.S. § 3222.1 (b)(2)).

Pennsylvania's "rebuttable presumption" statute presumes that, unless rebutted by an established defense, well operators are responsible for groundwater pollution if (1) the water supply is within 2500 feet of the unconventional vertical wellbore and (2) the pollution occurred within 12 months of the latter of completion, drilling, stimulation, or alteration of the unconventional well (58 Pa.C.S. § 3218(c)(2)). To rebut the presumption of responsibility for groundwater pollution, operators must affirmatively prove that either the pollution existed prior to drilling activities as determined by a predrilling survey, or that the landowner refused to allow the operator access to conduct the predrilling survey, or that the pollution occurred as the result of a cause other than drilling activity (58 Pa.C.S. § 3218(d)).

Pennsylvania statutes require well operators who pollute a public or private water supply to restore or replace the affected water supply with an alternate source of adequate quantity and quality that meets the standards set forth in the Pennsylvania

Safe Drinking Water Act (58 Pa.C.S. § 3218(a)). Landowners that suffer pollution of a water supply as a result of drilling activities may request an investigation, in which the Pennsylvania DEP must investigate within 10 days and make its determination within 45 days (58 Pa.C.S. § 3218(b)).

To protect groundwater sources, operators are obligated to control and dispose of brines and produced water in compliance with the Clean Streams Law (58 Pa.C.S. § 3217(a)). Well-casing regulations are designed to prevent migration of methane gas or fluids into groundwater sources, such that strings of casing are required to be permanently cemented in each well drilled through the fresh water-bearing strata to depths required by Pennsylvania DEP (58 Pa.C.S. § 3217(b)). The Pennsylvania DEP also publishes an updated list of water supply determination letters that record when a private water supply was impacted by unconventional oil and gas activities, including any case that results in confirmed water supply contamination from hydraulic fracturing or an increase in constituents above background conditions (Pennsylvania DEP 2014).

13.3.3   Wyoming

Advanced unconventional drilling techniques have increased statewide production in Wyoming from the Niobrara Shale, as well as from the Powder River Basin, Green River Formation, and Wind River Formation (EIA 2016 Wyoming Profile). Wyoming accounts for 2–3% of U.S. crude oil production and is one of the top 10 natural gas-producing states in the nation (EIA 2016 Wyoming Profile). Wyoming has more producing oil and gas leases on federal lands than any other state (2015 BLM Report).

Per statewide rules set forth by the Wyoming Oil and Gas Conservation Commission ("WOGCC"), all operators in Wyoming are required to submit a groundwater baseline sample, analysis, and monitoring plan with their respective drilling permit applications (WOGCC Rules Ch. 3, §46). The groundwater monitoring program consists of initial baseline water sampling and testing, as well as subsequent sampling and testing after setting the production casing. The initial and subsequent sampling and testing must include standard water physicochemical variables (temperature, specific conductance, total dissolved solids, pH, etc.), and must check for the presence of dissolved gasses (methane, ethane, propane), total petroleum hydrocarbons (TPH), and BTEX compounds (benzene, toluene, ethylbenzene, and xylenes), among others (WOGCC Rules Ch. 3, §46(h)).

Surface casing is required to reach a depth that is below all known or reasonably estimated utilizable groundwater. Within a minimum of one-quarter mile radius, operators must set surface casing at a minimum of 100–120 feet below the depth of any Wyoming Office of State Engineer- permitted water supply wells designated for domestic, stock water, irrigation or municipal use, and the casing must be cemented to the surface (WOGCC Rules Ch. 3, §22(a)). Any freshwater flows detected during drilling must be recorded and reported to WOGCC on the next business day. If cement is not circulated to the surface during the primary drilling operation, operators must perform supplemental cementing to ensure that the annular space from the casing shoe to the surface is filled with cement (WOGCC Rules Ch. 3, §22(a)).

In the litigation arena, Pavillion-area farmers sued Encana Corporation ("Encana") for negligently reworking a well, alleging that groundwater contamination resulted when Encana failed to test the water supply for petroleum products after a 2003 settlement agreement (EE News 2017). The U.S. District Court for the District of Wyoming denied Encana's motion for summary judgment, which argued a statute of limitations defense, instead allowing the lawsuit to proceed to trial in late 2017 (*Locker v. Encana* (2014)). Plaintiffs reached a settlement agreement with Encana in January 2018 (Detrick 2018).

13.4   Conclusion

Absent human-induced error, and when conducted with adequate separation between the aquifer and shale interval, hydraulic fracturing, in and of itself, likely does not elevate the risk of groundwater contamination. Nevertheless, the proliferation of unconventional oil and gas production across U.S. shale plays invariably presents more opportunities for operator error by orders of magnitude and produces more wastewater that must be accounted for.

There exist common sense steps based on best-practices that jurisdictions can take when seeking to regulate the interaction between groundwater and unconventional oil and gas. For instance, jurisdictions should implement aquifer-specific restrictions that limit the depth for how shallow a horizontal well can be hydraulically fractured, thereby insuring adequate separation between the shale interval and groundwater table. Further, jurisdictions should require companies to administer groundwater sampling surveys before drilling similar to the "rebuttable presumption" statute in Pennsylvania, a measure that will protect both those landowners seeking redress for groundwater contamination and those oil and gas operators defending fraudulent allegations.

Described as "so secret, occult, and concealed that an attempt to administer any set of legal rules [involves] hopeless uncertainty" (*Houston & T.C. Ry. Co. v. East* (1904)), the Texas Supreme Court's early venture into the realm of groundwater jurisprudence foreshadowed the regulatory challenges underscoring the intersection of groundwater and the current shale revolution. Science is on the cusp of understanding the dynamics of groundwater contamination in localized scenarios, both from natural and anthropogenic sources, but law and regulation maintain the tendency to lag behind. Although the current regulatory atmosphere incorporates better understandings of the geological, economic, legal, and public policy challenges, divisive rhetoric championed by environmental advocates, energy industry lobbyists, and agenda-driven research impedes the evolving relationship between science and law by compounding certain degrees of uncertainty that already pervade the complex groundwater-hydraulic fracturing nexus.

References

Akob DM, Cozzarelli IM, Dunlap DS, Rowan EL, Lorah MM (2015) Organic and inorganic composition and microbiology of produced waters from Pennsylvania shale gas wells. Appl Geochem 60:116–125

Beck v. ConocoPhillips. Co., No. 2011–484 (123rd Dist. Ct. Panola County, Tex., Dec. 1, 2011) (case dismissed with prejudice on May 27, 2015)

Boyer EW, Swistock BR, Clark J, Madden M, Rizzo DE (March 2012) Center for rural Pennsylvania, The impact of Marcellus gas drilling on rural drinking water supplies 1–26., from http://www.rural.palegislature.us/documents/reports/Marcellus_and_drinking_water_2011_rev.pdf

Burnett DB, Platt FM, Vavra CE (2015) Achieving water quality required for fracturing gas shales: cost effective analytic and treatment technologies', SPE-173717-MS, Society of Petroleum Engineers. Presented at Society of Petroleum Engineers International Symposium in Oilfield Chemistry (April 13–15, 2015). https://doi.org/10.2118/173717-MS

Burton TG, Rifai HS, Hildenbrand ZL, Carlton DD Jr, Fontenot BE, Schug KA (2016) Elucidating hydraulic fracturing impacts on groundwater quality using a regional geospatial statistical modeling approach. Sci Total Environ 545–546:114–126

Cabot Oil & Gas (20 July 2010) Affidavit of Norma McNeal. Viewed 15 October 2017., from http://www.cabotog.com/pdfs/Tab1.pdf

Chapman EC, Capo RC, Stewart BW, Kirby CS, Hammack RW, Schroeder KT, Edenborn HM (2012) Geochemical and strontium isotope characterization of produced waters from Marcellus shale natural gas extraction. Environ Sci Technol 46(6):3545–3553

Cluff MA, Hartsock A, MacRae JD, Carter K, Mouser PJ (2014) Temporal changes in microbial ecology and geochemistry in produced water from hydraulically fractured Marcellus shale gas wells. Environ Sci Technol 48(11):6508–6517

Commonwealth of Pennsylvania Department of Environmental Protection (15 December 2010) In the matter of Cabot Oil & Gas Corporation Dimock and Springville Townships. Available at: http://files.dep.state.pa.us/OilGas/BOGM/BOGMPortalFiles/OilGasReports/Determination_Letters/EAST/CO258482-1_Redacted.pdf

Cunningham N (16 July 2014) As fracking expands, so does opposition – even in Texas. In: Oil price. Viewed 10 June 2017, from https://oilprice.com/Energy/Energy-General/As-Fracking-Expands-So-Does-Opposition-Even-In-Texas.html#. (Discussing fracking opposition movement)

Darrah TH, Vengosh A, Jackson RB, Warner NR, Poreda RJ (2014) Noble gases identify the mechanisms of fugitive gas contamination in drinking-water wells overlying the Marcellus and Barnett shales. Proc Natl Acad Sci 111:14076–14081

Darrah TH, Jackson RB, Vengosh A, Warner NR, Whyte CJ, Walsh TB, Kondash AJ, Poreda RJ (2015) The evolution of Devonian hydrocarbon gases in shallow aquifers of the northern Appalachian Basin: insights from integrating nobel gas and hydrocarbon geochemistry. Geochim Cosmochim Acta 170:321–355

Davies RJ, Mathias SA, Moss J, Hustoft S, Newport L (2012) Hydraulic fractures: how far can they go? Mar Petrol Geol 37:1–6

DeKok D, Reuters (10 March 2016) Pennsylvania families win $4.2 million damages in fracking lawsuit., from http://www.reuters.com/article/us-pennsylvania-fracking-idUSKCN0WC2I8

Detrick A (24 January 2018) Gillette news record. 'Pavillion couple, Encana settle lawsuit over tainted water allegation', from http://www.gillettenewsrecord.com/news/wyoming/article_3590d791-3f5b-562f-aec3-cb62921cda98.html

Deweese W (2010) Fracturing misconceptions: a history of effective state regulation, groundwater protection, and the ill-conceived FRAC Act. Okla J Law Technol 6(1):49–81

DiGiulio DC, Jackson RB (2016) Impact to underground sources of drinking water and domestic wells from production well stimulation and completion practices in the Pavillion, Wyoming, field. Environ Sci Technol 50(8):4524–4536

DiGiulio DC, Wilkin RT, Miller C, Oberley G (December 2011) U.S. Environmental Protection Agency. Investigation of Ground Water Contamination Near Pavillion, Wyoming – Draft Report. EPA 600/R-00/000 from http://www2.epa.gov/region8/draft-investigation-ground-water-contamination-near-pavillion-wyoming. ("2011 EPA draft report")

Edwards Aquifer Authority v. Day, 369 S.W.2d 814 (Tex. 2012)

Environment & Energy News (4 January 2017) Judge: Pavillion fracking contamination case can go to trial. Available at: http://www.wyofile.com/judge-pavillion-fracking-contamination-case-can-go-to-trial/. ("EE News 2017")

Fisher K (July 2010) Data confirm safety of well fracturing. American Oil & Gas Reporter, from http://www.bfenvironmental.com/pdfs/inducedfracturingoilreporter.pdf

Folger P, Tiemann M, Bearden DM (26 January 2012) The draft report of groundwater contamination Near Pavillion, WY: main findings and stakeholder responses, congressional research service, R42327, from http://wyofile.com/wp-content/uploads/2012/01/R42327-2.pdf

FracFocus chemical disclosure registry. Available at: https://fracfocus.org/

Gasland (2010) New Video Group. Available at: http://www.gaslandthemovie.com/

Gold R (2014) The boom: how fracking ignited the American energy revolution and changed the world. Simon & Schuster, New York

Goldman M (2013) A survey of typical claims and key defenses asserted in recent hydraulic fracturing litigation. Texas A&M Law Rev 1:305–334

Gregory KB, Vidic RD, Dzombak DA (2011) Water management challenges associated with the production of shale gas by hydraulic fracturing. Elements 7(3):181–186

Harris v. Devon Energy Production Co., L.P., No. 4:10-cv-00708 (E.D. Tex., December 22, 2010)

Hildenbrand ZL, Carlton DD Jr, Fontenot BE, Meik JM, Walton JL, Taylor JT, Thacker JB, Korlie S, Shelor CP, Henderson D, Kadjo AF, Roelke CE, Hudak PF, Burton T, Rifai HS, Schug KA (2015) Comprehensive analysis of groundwater quality in the Barnett shale region. Environ Sci Technol 49:8254–8262

Hildenbrand ZL, Carlton DD Jr, Meik JM, Taylor JT, Fontenot BE, Walton JL, Henderson D, Thacker JB, Korlie S, Whyte CJ, Hudak PF, Schug KA (2017) A reconnaissance analysis of groundwater quality in the eagle ford shale region reveals two distinct bromide/chloride populations. Sci Total Environ 575:672–680

Hosey PE, Lotay JS (14 April 2017) Quench my thirst: water rights in the context of water treatment technologies, 1–16. Presented at 43rd annual Ernest E. Smith Oil, Gas and Mineral Law Institute, Houston, Texas

Houston & T.C. Ry. Co. v. East, 66 L.R.A. 738 (Tex. 1904)

Hudak PF (2010) Solutes and potential sources in a portion of the trinity aquifer, Texas, USA. Carbonates Evaporites 25(1):15–20

Hurdle J (2017) StateImpact NPR. Last two Dimock families settle lawsuit with Cabot over water, September 26, 2017, from https://stateimpact.npr.org/pennsylvania/2017/09/26/last-two-dimock-families-settle-lawsuit-with-cabot-over-water/

Hyne NJ (1996) Dictionary of petroleum exploration, drilling, and production 563

In re Steven Lipsky, Decided 24 April 2015, No. 13–0928, Supreme Court of Texas

Jackson RB, Vengosh A, Darrah TH, Warner NR, Down A, Poreda RJ, Osborn SG, Zhao K, Karr JD (2013) Increased stray gas abundance in a subset of drinking water wells near Marcellus shale gas extraction. Proc Natl Acad Sci 110(28):11250–11255

Jackson RB, Vengosh A, Carey WJ, Davies RJ, Darrah TH, O'Sullivan F, Petron G (2014) The environmental costs and benefits of fracking. Annu Rev Environ Resour 39:327–362

Jackson RB, Lowry ER, Pickle A, Kang M, DiGiulio D, Zhao K (2015) The depths of hydraulic fracturing and accompanying water use across the United States. Environ Sci Tech 49:8969–8976

Jury Verdict, Ely v. Cabot Oil & Gas Corp., No. 3:09-cv-02284 (M.D. Pa 2016)

Kerr RA (2010) Natural gas from shale bursts onto the scene. Science 328(5986):1624–1626

King JC, Bryan JL, Clark M (2012) Factual causation: the missing link in hydraulic fracture—groundwater contamination litigation. Duke Environ Law Policy Forum 22:341–360

Kondash AJ, Albright E, Vengosh A (2017) Quantity of flowback and produced waters from unconventional oil and gas exploration. Sci Total Environ 574:314–321

Lauer NE, Harkness JS, Vengosh A (2016) Brine spills associated with unconventional oil development in North Dakota. Environ Sci Technol 50(10):5389–5397

Llewellyn GT (2014) Evidence and mechanisms for Appalachian Basin brine migration into shallow aquifers in NE Pennsylvania, USA. Hydrogeology 22:1055–1066

Locker et al. v. Encana Oil & Gas (USA), Inc., (Case Number 1:14-cv-00131) (Federal District Court Wyoming, Casper) (filed July 2, 2014)

McEwen M (20 February 2018) Midland Reporter-Telegram. Frac Hits growing issue as infill drilling, Frac stages rise, from https://www.mrt.com/business/oil/article/Frac-hits-growing-issue-as-infill-drilling-12617883.php

Memorandum Opinion, Ely v. Cabot Oil & Gas Corp., No. 3:09-cv-2284 (M.D. Pa March 31, 2017)

Merrill TW, Schizer DM (2013) The shale and gas revolution, hydraulic fracturing, and water contamination: a regulatory strategy. Minn Law Rev 98:145–264

Middle Pecos Groundwater Conservation District, Rule 14.3, effective July 19, 2016., from http://www.middlepecosgcd.org/pdf/rules/2014/07-19-2016_rules.pdf

Mitchell v. Encana Oil & Gas (USA), No. 3:10-cv-02555 (N.D. Tex., December 15, 2010)

Mohan AM, Hartsock A, Hammack RW, Vidic RD, Gregory KB (2013) Microbial communities in flowback water impoundments from hydraulic fracturing for recovery of shale gas. FEMS Microbiol Ecol 86(3):567–580

Molofsky LJ, Connor JA, Wylie AS, Wagner T, Farhat SK (2013) Evaluation of methane sources in groundwater in northeastern Pennsylvania. Ground Water 51:333–349

Moniz EJ, Jacoby HD, Meggs AJM (2011) The future of natural gas: an interdisciplinary MIT study. Viewed 30 October 2017, from http://www.mit.edu/~jparsons/publications/NaturalGas_Report_Final.pdf

Murray v. EOG Resources, et al, No. DC-15-08865 (95th Dist. Ct., Dallas County, Tex., August 6, 2015)

Oklahoma Energy Producers Alliance (14 September 2017) Are vertical wells impacted by horizontal drilling? A study of Kingfisher County, from https://www.eenews.net/assets/2017/10/27/document_pm_07.pdf ("OEPA 2017 Study)

Orem W, Tatu C, Varonka M, Lerch H, Bates A, Engle M, Crosby L, McIntosh J (2014) Organic substances in produced and formation water from unconventional natural gas extraction in coal and shale. Int J Coal Geol 126:20–31

Osborn SG, Vengosh A, Warner NR, Jackson RB (2011) Methane contamination of drinking water accompanying gas-well drilling and hydraulic fracturing. Proc Natl Acad Sci 108(20):8172–8176

Oxy Permian (10 March 2017) Driving value in the Permian. Viewed 5 October 2017, from http://www.oxy.com/investors/Documents/PermianBasin%20InvestorDayMarch2017.pdf

Park M (20 March 2012) Texas tribune. EPA withdraws order against natural gas driller, from https://www.texastribune.org/2012/03/30/epa-withdraws-order-against-range-resources/

Pennsylvania Department of Environmental Protection (2014) Water supply determination letters, from http://files.dep.state.pa.us/OilGas/BOGM/BOGMPortalFiles/OilGasReports/Determination_Letters/Regional_Determination_Letters.pdf ("Pennsylvania DEP 2014")

Railroad Commission of Texas. Data online research queries, from http://www.rrc.state.tx.us/about-us/resource-center/research/online-research-queries/ (operators obtained only twenty-six (26) drilling permits in the Eagle Ford Shale in 2008. In 2012 and 2013, the RRC issued a total of 4,143 and 4,416 Eagle Ford drilling permits, almost exclusively for horizontal wells)

Railroad Commission of Texas Oil & Gas Docket (21 November 2008) No. 09-0257041. Show Cause Hearing & Contingent Application of WEC, Inc., Examiners' report & proposal for decision, pp 19–20 ("RRC Oil & Gas Docket 2008")

Railroad Commission of Texas Oil & Gas Docket (4 November 2013) Nos. 01-0281775, 01-0281868. Application of Kenmare Invs., Examiners' report & proposal for decision, pp 5–6 ("RRC Oil & Gas Docket 2013")

Scanlon BR, Reedy RC, Male F, Walsh M (2017) Water issues related to transitioning from conventional to unconventional oil production in the Permian Basin. Environ Sci Technol 51(18):10903–10912

Siegel DI, Azzolina NA, Smith BJ, Perry AE, Bothun RL (2015) Methane concentrations in water wells unrelated to proximity to existing oil and gas wells in northeastern Pennsylvania. Environ Sci Technol 49(7):4106–4112

Scoma v. Chesapeake Energy Corp., No. 3:10-cv-01385 (N.D. Tex., July 15, 2010)

Solomon D (28 April 2015) Texas Monthly. The Texas Supreme Court rules that a Fracking Company's defamation suit against a guy who claims his tap water is on fire can proceed, from https://www.texasmonthly.com/the-daily-post/the-texas-supreme-court-rules-that-a-fracking-companys-defamation-suit-against-a-guy-who-claims-his-tap-water-is-on-fire-can-proceed/

Soraghan M (1 November 2017) EENews. Now it's oilmen who say fracking could harm groundwater, from https://www.eenews.net/stories/1060065209

Spence DB (2014) The political economy of local vetoes. Texas Law Rev 93:351–413

StateImpact (2012) Tap water torches: how faulty gas drilling can lead to methane migration. Viewed 9 June 2017, from https://stateimpact.npr.org/pennsylvania/tag/methane-migration/. (Quoting Penn State University geologist Dave Yoxtheimer)

Steven and Shyla Lipsky v. Range Resources Corp. et al (14 July 2011) Cause No. CV11-0798 in the 43rd District Court, Parker County, Texas

U.S. Department of the Interior, Bureau of Land Management. Number of producing leases on federal lands (Updated October 29, 2015). Viewed on October 21, 2017, from https://www.eia.gov/state/analysis.php?sid=WY. ("BLM 2015 report")

U.S. Energy Information Administration (June 2012) Annual energy outlook 2012 with projections to 2035. Viewed 15 October 2017, from https://www.eia.gov/outlooks/aeo/pdf/0383(2012).pdf. ("EIA 2012 report")

U.S. Energy Information Administration (June 2016) Lower 48 states shale plays., from https://www.eia.gov/maps/images/shale_gas_lower48.pdf

U.S. Energy Information Administration. Wyoming state profile and energy estimates (Updated December 15, 2016). Viewed on October 21, 2017, from https://www.eia.gov/state/analysis.php?sid=WY#40. ("EIA 2016 Wyoming profile")

U.S. Energy Information Administration ("EIA") (July 2017) Marcellus region drilling productivity report. Viewed 15 October 2017, from https://www.eia.gov/petroleum/drilling/pdf/marcellus.pdf. ("EIA 2017 Marcellus report")

U.S. Environmental Protection Agency (2012) Study of the potential impacts of hydraulic fracturing on drinking water resources: progress report 1:31–32., from http://www.epa.gov/hfstudy/pdfs/hf-report20121214.pdf. ("2012 EPA report")

U.S. Environmental Protection Agency, Office of Water, Office of Ground Water and Drinking Water (June 2004) Evaluation of impacts to underground source of drinking water by hydraulic fracturing of coalbed methane reservoirs, (4606M), EPA 816-R-04-003 ("2004 EPA report")

U.S. Environmental Protection Agency, Washington, DC (2016) Hydraulic fracturing for oil and gas: impacts from the hydraulic fracturing water cycle on drinking water resources in the United States (final report). EPA/600/R-16/236F. ("2016 EPA report")

U.S. Geological Survey (2000) Coal-bed methane: potential and concerns. Viewed 9 June 2017, from https://pubs.usgs.gov/fs/fs123-00/fs123-00.pdf

U.S. Geological Survey (2006) Methane in West Virginia ground water. Viewed 9 June 2017, from http://pubs.usgs/gov/fs/2006/3011/pdf/FactSheet2006_3011.pdf

Vaidyanathan G (4 April 2016) Fracking can contaminate drinking water. Climate Wire, Scientific American, from https://www.scientificamerican.com/article/fracking-can-contaminate-drinking-water/

Veil J (April 2015) Groundwater Protection Council, U.S. Produced Water Volumes and Management Practices in 2012, from http://www.gwpc.org/sites/default/files/Produced%20Water%20Report%202014-GWPC_0.pdf

Vengosh A, Warner NR, Jackson R, Darrah T (2013) The effects of shale gas exploration and hydraulic fracturing on the quality of water resources in the United States. Proc Earth Planet Sci 7:863–866

Vengosh A, Jackson RB, Warner N, Darrah TH, Kondash A (2014) A critical review of the risks to water resources from unconventional shale gas development and hydraulic fracturing in the United States. Environ Sci Technol 48(15):8334–8348

Vidic RD, Brantley SL, Vandenbossche JM, Yotheimer D, Abad JD (2013) Impact of shale gas development on regional water quality. Science 340(6134):1235009

Warner NR, Jackson RB, Darrah TH, Osborn SG, Down A, Zhao K, White A, Vengosh A (2012) Geochemical evidence for possible natural migration of Marcellus formation brine to shallow aquifers in Pennsylvania. Proc Natl Acad Sci 109(30):11961–11966

Warner NR, Christie CA, Jackson RB, Vengosh A (2013) Impacts of shale gas wastewater disposal on water quality in Western Pennsylvania. Environ Sci Technol 47(20):11849–11857

Watson B (21 December 2017) Hydraulic fracturing tort litigation summary. University of Dayton, from   https://www.udayton.edu/directory/law/documents/watson/blake_watson_hydraulic_ fracturing_primer.pdf

Wiseman HJ (2013) Risk and response in fracturing policy. Univ Colo Law Rev 84(3):729–818

Wyoming Oil & Gas Conservation Commission Rules Ch. 3, §§ 22(a), 46 (2014)

58 Pa.C.S. §§ 3201-3274, 3217(a)-(b), 3218(a)-(d), 3222.1, 3222.1(b)(2) (Pa. HB 1950) (2012)

16 Tex. Admin. Code. §§ 3.13(a)(1), 3.13(b)(1)(A)-(B), 3.13(a)(2)(P), 3.29, 3.30 ("Memorandum of Understanding"), 3.30(e)(7)(A), 3.30(e)(7)(B)(i)), 3.7, 3.8(a)(28), 3.8(b), 3.8(d)(1)-(2), 3.8(d)(4)(G)(i)-(v), 3.8(d)(6)(A), 3.8(h), 3.8(i)

Tex. Nat. Res. Code §§ 85.381(a)(1)-(b), 91.002, 91.011, 91.016, 91.0523, 91.1015, 91.113(a)

165 Okla. Admin. Code §10–1-21

Brett A. Miller is an associate for the water investment firm Water Asset Management, LLC. Mr. Miller clerked at the Texas Oil & Gas Association in Austin, Texas, and won the 2016 Hartrick Scholar Writing Competition Award from the Institute for Energy Law (Center for American and International Law). He also received a 2015–16 ConocoPhillips Energy Institute Fellowship at the Texas A&M Energy Institute, served as an articles editor on the Texas A&M Law Review, and interned at the international private equity firm, TPG Global, LLC. Miller has published articles in both law reviews and peer-reviewed science journals, and he received the 2016 best publication award from the Louisiana Association of Professional Biologists. He holds a J.D. (*Magna Cum Laude*) from Texas A&M University School of Law, a M.S. from Louisiana State University with a concentration in freshwater ecology, and a B.S. in biology and history from Rhodes College.

Chapter 14
Overview of Oil and Gas Wastewater Injection Induced Seismicity in Hydrocarbon Regions in the United States, Canada, and Europe

Monika U. Ehrman

Abstract There has been a tremendous increase in earthquake activity in traditionally non-seismically active areas, such as the American states, such as Kansas, Ohio, Oklahoma, and Texas and the Canadian provinces of Alberta and British Columbia. In Europe, the United Kingdom and The Netherlands experienced seismicity events that were associated with oil and gas development. Studies ensued and are in process to determine the correlations between oil and gas activity and seismicity. Recently, many researchers have identified a correlation between seismic activity and certain oil and gas operations, such as wastewater fluid injection, which is a common practice used to dispose of wastewater generated during oil and gas operations. Oil and gas companies, state regulatory agencies, and local and state governments are proceeding surely, but cautiously, given that most of this activity is occurring in areas with a strong and economically vested interest in petroleum production. This chapter reviews the geologic mechanism, scientific studies, applicable U.S. federal environmental legislation, state and provincial overview of agency work, in addition to a brief review of anthropogenic seismic events in the U.S., Canada, and Europe.

Keywords Induced seismicity · Earthquakes · Fluid injection · Wastewater · Texas · Oklahoma · Canada

M. U. Ehrman (✉)
University of Oklahoma School of Law, Norman, OK, USA
e-mail: mehrman@ou.edu

© Springer International Publishing AG 2020
R. M. Buono et al. (eds.), *Regulating Water Security in Unconventional Oil and Gas*, Water Security in a New World,
https://doi.org/10.1007/978-3-030-18342-4_14

14.1   Introduction

This chapter[1] reviews the geologic mechanism, scientific studies, applicable U.S. federal environmental legislation, state and provincial overview of agency work, in addition to a brief review of anthropogenic seismic events in the U.S., Canada, and Europe.

14.2   Seismology Background

Seismology is the study of earthquakes and waves in the earth, which may be generated both naturally and artificially (Gubbins 1990). Seismicity refers to the "geographic and historical distribution of earthquakes" (U.S. Geological Survey n.d.).

An earthquake generally occurs from the buildup and release of stress within the tectonic plates that make up the earth's lithosphere, which is the solid, outer part of the earth and includes the brittle upper portion of the mantle and crust (Schick 2002). Originating in the 1950s and developing over two decades, the plate tectonics theory evolved out of Alfred Wegener's continental drift theory, first proposed in 1912. Plate tectonics theorizes that Earth's outer shell is divided into several tectonic plates—comprised of both continental and oceanic crust—that glide over the mantle—the rocky inner layer above the core. These rigid plates move independently and their boundaries form ridges, trenches, or transform faults (Gubbins 1990). Although Wegener did not have an explanation for how continents could move around the planet, scientists now explain this movement using plate tectonics, which is considered geology's unifying theory.

Unlike puzzle pieces, the plates do not neatly connect with each other. Instead, they are part of a dynamic geologic process whereby they push up, slide against, and move away from each other. These movements result in varying terrestrial and planetary effects, such as earthquakes, but also include the creation of ocean floor, mountain ranges, and rift formations. On a larger geologic time scale, plate tectonics is responsible for the movement of the continents. The supercontinents Rodinia and Pangaea, which existed nearly one billion and 300 million years ago respectively, formed from the movement of the tectonic plates and have since been rifted apart by those same forces to form the current plate structure.

The release of stored stress energy "associated with rapid movement on active faults" causes most earthquakes (Hurong 2015). Although smaller micro-earthquakes rupture faults for only a small fraction of a second, the duration of very large earthquakes is measured in minutes.

Earthquake seismologists record seismic waves generated by earthquakes to understand the geometry and motion of Earth's internal structure. These waves are

---

[1] Sections 14.2 through 14.7 of this chapter include some information and analysis published previously in Ehrman 2017 (internal citations are omitted).

generated at either a natural source, such as an earthquake, or artificial source, such as an explosion (Stein and Wysession 2003). Although "the term 'earthquake' describes a sudden shaking of the ground," geoscientists usually employ the term "to describe the 'source' of seismic waves, which is nearly always sudden shear slip on a fault within the Earth" (Beroza and Kanamori 2015). These resulting waves travel through the earth and may be recorded by a ground receiver. Strong waves may be felt by people or may affect surface structures and are accordingly referred to as *felt* earthquakes. The receivers record ground motion when waves pass and collect various other information about a wave's origin and receiver arrival time. This data set allows for calculations of wave velocity and resulting properties of the medium through which the wave travels. In fact, petrophysicists employ similar data to understand and model subsurface oil and gas formations.

14.3 Induced Seismicity

Induced seismicity is earthquake activity caused by anthropogenic activities, including "fluid injection for waste disposal and secondary recovery of oil, geothermal energy production, oil and gas extraction, reservoir impoundment, mining and quarrying" (Cypser and Davis 1994). It is often identified by increased seismic activity over historical levels. Thus, areas that experience "a certain level of seismic activity" before the artificial activity begins are likely to continue experiencing seismic activity (Cypser and Davis 1994). But, if seismicity increases after the onset of the human activity, induced seismicity may be the culprit. Further, if the seismic activity returns to historical levels after the artificial activity stops, it suggests the likelihood that the increase was due to induced seismicity.

Many scientific studies are underway regarding the possible mechanisms of induced seismicity. The term *mechanism* is preferable to *cause* as there is not a single cause of induced seismicity. Rather, induced seismicity likely occurs due to a complex system of subsurface stresses, fluid pressures, and fracture and faulting geology.

Subsurface rock formations contain porous spaces and fractures. Fluids may be present in these rock pores and fractures, causing an outward pressure termed *pore pressure*. This pore pressure counterbalances the weight of the rock and its interstitial forces, resulting from tectonic forces. When pore pressures are low, especially compared to the stresses caused by the overlying strata, seismic activity results when imbalances of natural in situ earth stresses occur. When pore pressures increase, it takes less of this imbalance to trigger an earthquake, and seismicity accelerates. This type of failure is termed *shear failure*. Injecting fluids into the subsurface artificially increases pore pressures, which can cause certain faults and fractures to slip, thereby releasing stored stress energy. Notably, not only can subsurface fluid injection induce seismicity, fluid extraction can also cause subsidence or slippage along planes of weakness in the earth.

Geoscientists have long been aware of induced seismicity by various human activities impacting the surface or subsurface. Such major activities include mining, water impoundment like dams and hydroelectric projects, waste disposal, and geothermal activities. Numerous studies observing and analyzing these activities "bear evidence to the presence of critically stressed rocks in the earth's crust, wherein small stress changes induced by human activity trigger earthquakes" (Chen and Talwani 2001).

14.4   Induced Seismicity and Oil and Gas Operations

Scientists previously observed that fluid injection could trigger earthquakes. In disposal wells, seismic activity resulted after fluid injection caused shock waves or fluids to "release strain on a preexisting fault" (Fairley 2012). This high-pressure fluid squeezes into and pushes apart a planar fault, "freeing adjacent rock formations to slide past one another" (Fairley 2012). The surmised phenomenon is often attributed to the injected fluid increasing pore pressure around a fault plane—or *lubricating the fault*—making it easier for a slip to occur. Given a relatively sudden increase in seismic activity in predominantly American oil and gas regions, scientists concluded that more research must be done on the relationship between wastewater reinjection and seismicity, and hydraulic fracturing and seismicity.

Since the 1960s, seismologist recognized that there may be an association between wastewater injection and seismicity (Healy et al. 1968; Frolich et al. 2016). It was only recently that scientists first began researching the various series of earthquakes in oil and gas producing areas. In 2008, ten felt earthquakes occurred near an injection well at the Dallas–Fort Worth (DFW) Airport (Frolich et al. 2016). Earthquakes followed in various midwestern regions, including Oklahoma (Keranen et al. 2014; Frolich et al. 2016), Arkansas (Horton 2012; Frolich et al. 2016), Kansas (Nolte et al. 2017), and Ohio (Ellsworth 2013). Due to the string of suspect earthquakes in shale gas regions, the U.S. Geological Survey established a project to study induced seismicity (Fountain 2011).

Some of these studies concluded that the hydraulic fracturing process was likely triggering the earthquakes (Elsworth et al. 2016). The hydraulic fracturing involves the injection of high volumes of water with relatively small quantities of sand or proppant and chemical into the subsurface shale gas formations, creating microearthquakes (Friberg et al. 2014).

Conversely, researchers studying the "Jones swarm" of earthquakes in Oklahoma noted that "four high-rate disposal wells in southeast Oklahoma City probably induced a group of earthquakes …, which accounted for 20% of the seismicity in the central and eastern United States between 2008 and 2013" (Branson-Potts 2014). Researchers from Cornell University and the University of Colorado surmised that the activity was a result of "a few highly active disposal wells, where wastewater from drilling operations—including hydraulic fracturing—is forced into deep geological formations for storage" (Branson-Potts 2014).

The National Research Council, the arm of the National Academy of Sciences which conducts research, found it was not able to "accurately predict the magnitude or occurrence of such events due to the lack of comprehensive data on complex natural rock systems and the lack of validated predictive models" even though the general mechanisms of induced seismicity were well understood (National Research Council 2013). However, by 2017 and 2018, scientific research in Kansas, Oklahoma, and Texas linked induced seismicity in those states with certain wastewater injection activities (Rubinstein et al. 2018; Walsh and Zoback 2015; Hincks et al. 2018).

In Oklahoma, scientists recently asserted that the rise in seismicity since 2009 is due to wastewater injection and that there is a strong correlation to the depth of injection (Hincks et al. 2018). Although the Oklahoma Corporation Commission requires operators to prove injected wastewater does not encounter geologic basement lithologies, the question remains over whether there is a time lapse between injection and a seismic event. Interestingly, in a larger regional study of the Central and Eastern United States, there was "no significant correlation between proximity to basement and earthquake occurrence" (Hincks et al. 2018).

As additional data is collected and further studies performed, scientists are likely to make similar conclusions and reach a general scientific consensus about the causes of oil and gas induced seismicity. Presently, the two major theories appear to be wastewater injection and disposal, and hydraulic fracturing as triggers for seismic activity. The focus of this chapter is on wastewater injection as a trigger mechanism, and not hydraulic fracturing.

14.5 Wastewater Injection Disposal

Many scientists accept that wastewater injection is capable of inducing seismic activity. During oil and gas operations, water injection primarily occurs as a disposal mechanism for wastewater generated by production and hydraulic fracturing. During the production process, exploration and production companies drill through the subsurface, targeting hydrocarbon-rich formations. These formations also contain salt water—essentially the brine from an ancient sea. Production companies cannot dispose of this non-potable salt water in public facilities or as effluent into a stream or other body of water because it often mixes with the produced hydrocarbons and various other minerals, chemicals, and sediments. Once the hydrocarbons and accompanying fluids flow through the production wellhead, the hydrocarbons separate from the salt water, and the salt water must be disposed of, often in deep disposal wells. Private companies and sometimes the oil and gas operator itself will operate a disposal well, which are usually depleted oil and gas wellbores. Wastewater is injected into the depleted geologic formation that formerly held oil and gas.

In addition to injection volume, other factors influence the probability of seismicity near wastewater disposal operations. For example, plate tectonics can dictate whether seismic activity will occur and in what magnitude. In Oklahoma, the plates are squeezing the region from east to west, which results in most earthquakes occurring along a northwest-southeast oriented fault. Further, a propensity for wastewater injection seismicity may be highly correlated to a region's geology. The Arbuckle formation underlies much of Oklahoma. Its porosity and geologic features allow for absorption of huge volumes of water, making it a good target for wastewater disposal. Unfortunately, the formation sits on ancient and brittle basement rocks that can fracture along major faults under stress (Witze 2015). Thus, the deeper the injection depth, the greater the likelihood that the injected fluid will make its way into a seismogenic fault zone that is prone to forming earthquakes (Hincks et al. 2018; Witze 2015). The resulting earthquakes range in magnitude depending on the geologic structure framework and regional in situ tectonic stress.

At present, there are more than 30,000 injection wells permitted for the disposal of wastewater generated by oil and gas operations in the United States. But of those wells, only a very small number is suspected of inducing seismicity. Indeed, one recent report linked an estimate of nine such wells to induced seismic events. Although seismic events over the past few years likely have increased that number, even now, the fraction remains small. Nevertheless, in the last few years, geologists suspect that injection disposal induced hundreds of seismic events, though many were not felt events (Frohlich et al. 2016).

14.6   Major Applicable U.S. Federal Legislation

The Safe Drinking Water Act (SDWA) "regulates contaminants in drinking water supplied by public water systems and requires the [U.S. Environmental Protection Agency] (EPA) to set national drinking water regulations that incorporate enforceable maximum contaminant levels or treatment techniques" (Farber and Carlson 2013). Specifically, the SDWA works to prevent the release of toxic contaminants in water from underground sources, such as landfills and underground injection wells. The Underground Injection Control (UIC) regulations affect those wells where fluid is injected subsurface into geologic formations. Injected fluids typically include wastewater such as brine and chemical-mixed water.

The UIC program protects underground sources of drinking water from endangerment by setting minimum quality requirements for injection wells. Therefore, injection requires authorization under either general rules or specific permits. "Injection well owners and operators may not site, construct, operate, maintain, convert, plug, or abandon wells or conduct any other injection activity that endangers underground sources of drinking water" (Environmental Protection Agency, *General information about injection wells* n.d.). The UIC program seeks to ensure that either (1) injected fluids stay within the well and the intended injection zone or (2) fluids that are directly or indirectly injected into an underground source of drinking water do not cause a public water system to violate drinking water standards or otherwise adversely affect public health.

The EPA organizes injection wells into six classes, ranging from Class I to VI. A specific set of technical requirements and regulation applies to each well class. Class II injection wells are used to inject fluids associated with oil and gas production. Under the Class II classification, wells are categorized as (1) disposal wells, (2) enhanced recovery wells, or (3) hydrocarbon storage wells. There are approximately 180,000 Class II wells in operation in the country, about 80% of which are enhanced recovery wells.

Under the SDWA, the individual states, federally recognized tribes and U.S. territories have the option of requesting primary enforcement authority—*primacy*—for Class II wells (Environmental Protection Agency, *Class II oil and gas related injection wells* n.d.). Primacy establishes that the state, tribe, or territory oversees the UIC program within its jurisdiction (Environmental Protection Agency, *Primary enforcement authority for the underground injection control program* n.d.). A majority of the states have primacy (Environmental Protection Agency, *Class II oil and gas related injection wells* n.d.). States must meet EPA's minimum requirements for UIC programs under Section 1422. Disposal wells require permits that entail owners or operators meet all applicable requirements, including strict construction and conversion standards and regular testing and inspection. Section 1425 provides that states must demonstrate that their existing standards are effective in preventing endangerment of underground sources of drinking water. "These programs must include requirements for (1) permitting, (2) inspection, (3) monitoring, (4) record-keeping, and (5) reporting" (Environmental Protection Agency, *Class II oil and gas related injection wells* n.d.; Environmental Protection Agency, *Primary enforcement authority for the underground injection control program* n.d.).

From an induced seismicity perspective, concerned parties may seek to utilize the UIC to regulate oil and gas operator activity with respect to wastewater injection and hydraulic fracturing operations to curb or prevent seismic activity. However, in the sweeping Energy Policy Act of 2005, Congress exempted hydraulic fracturing—provided there is no use of diesel fuel—from the SDWA. Hydraulic fracturing is therefore "excluded from the definition of underground injection" and not subject to UIC regulation (Environmental Protection Agency, *Class II oil and gas related injection wells* n.d.).[2] Although some operators used to mix diesel fuel in the injected slurry during the hydraulic fracturing process, today most operators prohibit the injection of diesel fuel. The UIC program is thus not likely to apply to suspected seismic activity possibly resulting from hydraulic fracturing; it is, however, likely to arise in the wastewater disposal context.

---

[2] Citing 42 U.S.C. § 300 h (Safe Drinking Water Act § 1421(d)(1)(B))).

14.7 Exemplar U.S. State Regulation

14.7.1 Oklahoma

Oklahoma is the troubled heart of induced seismic activity. In 2014, the state experienced 585 magnitude 3-plus earthquakes, a five-fold increase from 2013. It now has the unfortunate distinction of being the most seismically active state in the United States. Scientists have observed a relationship between produced water disposal from oil and gas production operations and triggered seismic activity. With over 4200 disposal wells in the state—3600 actively used—wastewater injection volumes have doubled in 6 years, from 800 million barrels in 2009 to 1.5 billion barrels in 2014.

In January 2011, "small earthquakes of magnitude 2.9 and lower were allegedly induced by hydraulic fracturing activities," (Nicholson et al. 2016) while wastewater disposal injection was the suspected cause of the November 2011 magnitude 5.7 earthquake—the largest recorded in Oklahoma. A destructive earthquake in the vicinity of Cushing, Oklahoma—home to one of the largest oil storage hubs in the world and the pricing location for West Texas Intermediate crude oil prices—could have global financial consequences.

Scientists from state and federal institutions began studying the activity to determine causes and correlations. An increase in oil and gas development activity leads to an increase in wastewater production. Thus, operators bear the burden of disposing of greater volumes of water, often at higher pressures, in the same decades-old Class II UIC wells. Even though Oklahoma Class II UIC wells fall under the state permitting purvey, traditionally Oklahoma did not consider seismicity risk during its permitting process. Rather, its consideration focused on risks related to underground sources of drinking water. Therefore, regulators and state officials faced difficulty determining a clear connection between wastewater disposal operations and seismicity. This difficulty was "exacerbated in part by the vast number of UIC wells and earthquakes in the area." (Sundstrom 2015) Finally on April 21, 2015, the Oklahoma Geological Survey (OGS) "determined that the majority of recent earthquakes in central and north-central Oklahoma [were] very likely triggered by produced water disposal" (Sundstrom 2015). The OGS "issued a public statement that rates and geographical patterns of seismicity observed in the state 'are very unlikely to represent a naturally occurring rate change and process'" (Andrews and Holland 2015). State geologists Richard Andrews and Austin Holland first concluded that the primary source for the suspected induced seismicity was from the oil and gas wastewater injection activities (Andrews and Holland 2015).

The identification of a likely source of induced seismicity—wastewater disposal—allowed regulators and legislators to establish regulations governing operations. Adopting an approach supportive of the oil and gas sector, a large and dominant industry in Oklahoma, Governor Mary Fallin maintained the state's position that the Oklahoma Corporation Commission (OCC or the Corporation Commission), which regulates state oil and gas operations, retains exclusive authority over oil and gas operations in the state. However, with swift execution in September 2014, the Governor "directed the Oklahoma Secretary of Energy and Environment to assemble the Coordinating Council on Seismic Activity." (Office of the Oklahoma Secretary of Energy & Environment 2016) The council's "primary responsibility is to work cooperatively to develop solutions, identify gaps in resources[,] and coordinate efforts among state agencies, researchers and the state's oil and gas industry" (Office of the Oklahoma Secretary of Energy & Environment 2016). In January 2016, Governor Fallin further approved a $1.38 million transfer of state emergency funds to support earthquake research by certain state agencies, including the OGS. State agencies will use this funding to increase seismic monitoring in the state and hire additional geoscientists.

From the regulatory perspective, the Corporation Commission has done much to address seismic activity, while continuing oil and gas operations in the state. The OCC, an independent agency with three statewide elected commissioners, is "statutorily granted exclusive jurisdiction over the conservation of oil and gas and Class II UIC wells" (Sundstrom 2015). And although it has legal authority "to take extraordinary measures in the interest of public safety, without notice and hearing," the OCC "normally operates under its general authority to permit oil and gas and UIC well operations" (Sundstrom 2015). Following the state legislature, the Corporation Commission, too, "disavowed a moratorium on injection operations" (Nicholson et al. 2016).

Recently, the OCC instituted several state regulations pertaining to wastewater disposal. Some of these regulations include the large-scale regional reduction in oil and gas wastewater disposal within an approximate 5000 square mile radius in Western Oklahoma. This reduction affects over 200 disposal wells in the Arbuckle formation, identified as a formation predisposed to seismic activity. The OCC also ordered certain injection well operators to reduce wastewater disposal volumes on five wells operating within ten miles of the center of earthquake activity near Edmond, Oklahoma, a prosperous suburb north of Oklahoma City that suffered an earthquake in January 2016.

The Corporation Commission has also been working with its sister agency, the Oklahoma Geological Survey, to identify faults in the state. The OGS disclosed a preliminary map of known faults. Realizing the importance of identifying the state's faulting system, the OGS began compiling a fault database with voluntary contributions from the Oklahoma Independent Petroleum Association, the state's largest oil and gas industry association.

14.7.2 Texas

Texas is the largest energy producer in the United States. And like Oklahoma, Texas faces considerable challenges balancing citizen and property concerns with the interests of a robust oil and gas sector. Texas is taking a slightly different path than its northern neighbor, Oklahoma, perhaps due to the fact that its earthquakes have not been as severe or frequent as Oklahoma's. Residents in the Barnett shale area of north Texas complained of earthquakes as early as 2006. But, the Railroad Commission of Texas (the RRC or Railroad Commission) first denied any correlation between oil and gas operations and seismic activity. However, in recent years, and after several studies conducted by scientific and academic institutions, the RRC has moved forward with some actions relating to induced seismic activity.

In 2014, the Railroad Commission amended its rules concerning wastewater disposal. Beginning November 17, 2014, "disposal well operators must research US Geological Survey data for a history of earthquakes within 100 square miles of a proposed well site before applying for a permit" (CBS DFW 2014). The Commission also has the ability to modify or rescind a permit if it determines that the well may be contributing to seismic activity. Confident that the new measures did not substantially increase the cost of operations, the RRC estimated that the new rules "would cost companies an additional $300" (Atkin 2014). The Commission also hired seismologist Craig Pearson, who advised a newly-formed Texas House of Representatives' Subcommittee on Seismic Activity that "regulations would help make sure injected wastewater does [not] migrate onto inactive fault lines and cause man-made quakes" (Atkin 2014). Though Pearson noted that "most of the earthquakes occurring in Texas are too small to be felt," some scientific groups warned that the accumulation of fracturing and wastewater injection activities may result in stronger seismic movement (Atkin 2014).

But Texas falls short of Oklahoma's overall acceptance regarding oil and gas wastewater induced seismicity. The Railroad Commission stated that there was not yet a clear link to oil and gas activity despite a recent study by Southern Methodist University (SMU) seismologists in Dallas. The SMU team, also consisting of The University of Texas at Austin and the U.S. Geological Survey, studied the Azle-Reno earthquakes and concluded that wastewater disposal wells represented "the most likely cause of recent seismicity" (Hornbach et al. 2015). Studies examining past seismic activity indicates induced earthquakes may have been distributed across various geographic regions of Texas over the last 90 years (Frolich et al. 2016).

But even given the Texas regulator's doubts, the Texas legislature created the TexNet Seismic Monitoring Program, to be overseen by The University of Texas. The legislature approved the program last year with $4.5 million, including the creation of an Integrated Seismicity Research Center housed at The University of Texas Bureau of Economic Geology. Twenty-two permanent seismograph stations will be installed throughout the state, in addition to 36 temporary seismometers to deploy in areas of scientific interest. Given the increase of seismicity in the country's largest oil and gas producing state, Texas legislators and regulators may have to implement additional protective efforts.

14.8   Canadian Provinces and Wastewater Induced Seismicity

Unlike the United States, where many affected states were initially reluctant to identify a relationship between oil and gas wastewater injection and induced seismicity, the hydrocarbon-rich Canadian provinces of Alberta and British Columbia were quicker to make the identification. However, it is the hydraulic fracturing process, as opposed to wastewater injection, that appears to be the cause of much of the induced seismic events.

14.8.1   Alberta

Alberta is the heart of Canada's energy industry. It is the largest producer of oil and natural gas in the country. Due to its geological origins, much of past recorded seismic activity occurred near the spine of the Canadian Rocky Mountains and "occur within the thrust-fault systems associated with the ancient mountain-building processes that created the Rocky Mountains" (Alberta Energy Regulator and Alberta Geological Survey n.d.).

Recently, seismic activity beginning in December 2013 has occurred in Northwest Alberta, near Fox Creek. Seismologists working with the provincial geological agency, the Alberta Geological Survey, have studied the activity and determined that the injection of fluid during the hydraulic fracturing process is triggering earthquakes (Schultz et al. 2018).

Wastewater injection induced seismicity also occurs in Alberta. Scientists from the Alberta Geological Survey and the University Alberta studied seismic events near the Cordel disposal well, which is located at the western edge of Western Canadian Sedimentary Basin, near the Alberta-British Columbia border. The study concluded that those seismic events correlated with monthly injection activities. The scientists further concluded that the time lag between injection and seismic events is explained as "the time required for the diffusion of the pore pressure front to nearly critically stressed faults and the subsequent buildup to stimulate slip" (Schultz et al. 2014).

Presently, the Alberta scientific and regulatory agencies are in the process of conducting further studies, with the belief that it is the hydraulic fracturing process itself, as opposed to wastewater injection, that is causing the majority of oil and gas operation induced seismic events in the province. The conclusion is drawn from the statistically insignificant seismic activity associated with disposal operations. Noting that that Western Canadian Sedimentary Basin ("WCSB") "is an extensive basin containing petroleum resource in North America, with more than 2500 active wastewater injection wells operating during our study period," Schultz, Stern, and Gu note that there appears to be only "two cases of documented, injection-related seismicity in the WCSB" (Schultz et al. 2014).

14.8.2 British Columbia

Like its sister agency, the British Columbia (B.C.) oil and gas regulator, the British Columbia Oil and Gas Commission ("BCOGC"), identified two causes of induced seismicity—hydraulic fracturing and wastewater injection (BCOGC, *Induced seismicity* n.d.) British Columbia suffered from earthquake activity in the northeast area of the province, which is in the general region of the Alberta earthquakes. The BCOGC undertook several studies to monitor and analyze seismic activity.

To regulate disposal activity and mitigate the seismic effects of disposal, the BCOGC limits injection rates to control subsurface formation pressure. It also has the ability to terminate disposal activity (BCOGC, *Induced seismicity* n.d.)

As in Alberta, the majority of induced seismic activity in British Columbia appears to arise out of the hydraulic fracturing operation and not wastewater disposal. The BCOGC instigated two studies on seismic activity in the province: (1) Report of Observed Seismicity in the Horn River Basin (2012) and (2) Report of Induced Seismicity in the Montney Trend (2014) (BCOGC, *What's being done* n.d.).

The study on the Montney shale "found that during the study period 231 seismic events in that play were attributed to oil and gas operations – 38 induced by wastewater disposal and 193 by hydraulic fracturing operations" (BCOGC, *What's being done* n.d.). It further found that "[n]one of the recorded events resulted in any injuries, property damage or loss of wellbore containment outside of the formation and only 11 were felt at the surface" (BCOGC, *What's being done* n.d.). Likewise, the study on the Horn River basin "found that seismic events observed within remote and isolated areas of the Horn River Basin between 2009 and 2011 were caused by hydraulic fracturing operations, including 38 events recorded by Natural Resources Canada" (BCOGC, *What's being done* n.d.). And "[n]one of th[o]se events resulted in any injuries, property damage or loss of wellbore containment and only one was felt at the surface" (BCOGC, *What's being done* n.d.).

14.9 Certain European Oil and Gas Induced Seismicity Events

As in Canada, the U.K. likely has more effect from the hydraulic fracturing operation rather than induced seismicity. A multi-stage hydraulic fracturing operation occurred in Lancashire in the Namurian Bowland shale in Spring 2011. The British Geological Society reported seismic activity less than two kilometers from the wellsite. Upon recommencement of the operation almost 2 months later, another earthquake event occurred, and operational activities were suspended.

The recent spate of earthquake activity in The Netherlands' Groningen Field appears associated with subsurface reservoir pressure and pressure differential when the gas reservoir is drilled and produced and not related to wastewater disposal operations (Van Eck et al. 2006). Joint efforts were undertaken by the Dutch

Petroleum Company and the government to attempt to manage the seismic risk (Ellsworth et al. 2015). But the Dutch government recently announced that it would halt production of the Groningen field by 2030 to limit seismic hazard (Deutsche Welle 2018).

14.10 Conclusion

Currently, the literature points to oil and gas wastewater injection induced seismicity limited mainly to the North American continent—and the United States, in particular. Although anthropogenic induced seismicity occurs globally, the challenge to comprehend and mitigate the effects of wastewater injection appears to be limited to the producing American states, such as Kansas, Oklahoma, and Texas.

Regulatory agencies are working with geological state and federal agencies to compile faulting maps of the subsurface in an effort to understand the complex geologic structures and predict the effect of volumetric and pressure differentials on a system borne of an ancient marine past. Oil and gas operators and technology companies also have an opportunity to reduce seismicity risk by reducing wastewater injection via treatment methodologies. Further research and development may also allow for the hydraulic fracturing operation to be conducted using another medium instead of water. Further research is essential for the continued and safe operation of petroleum development.

References

Alberta Energy Regulator, & Alberta Geological Survey (Can.) (n.d.) Alberta's earthquakes. Retrieved from http://ags.aer.ca/albertas-earthquakes.htm
Andrews RD, Holland A (2015) Statement on Oklahoma seismicity: Oklahoma Geological Survey. Retrieved from http://wichita.ogs.ou.edu/documents/OGS_Statement-Earthquakes-4-21-15.pdf
Atkin E (2014, August 27) Texas proposes tougher rules on fracking wastewater after earthquakes surge. ThinkProgress. Retrieved from http://thinkprogress.org/climate/2014/08/27/3476207/texas-earthquake-rules-fracking/
Beroza GC, Kanamori H (2015) Earthquake seismology: an introduction and overview. In: Schubert G (ed) Treatise on geophysics, 2nd edn. Elsevier, Amsterdam, pp 1–50
Branson-Potts H (2014, July 3). Study links Oklahoma earthquake swarm with fracking operations. Los Angeles Times. Retrieved from http://www.latimes.com/science/sciencenow/la-sci-sn-oklahoma-earthquakes-fracking-science-20140703-story.html
British Columbia Oil & Gas Commission (2012, August) Investigation of observed seismicity in the Horn River Basin. Retrieved from https://www.bcogc.ca/node/8046/download
British Columbia Oil & Gas Commission (Can.) (n.d.) What's being done. Retrieved from https://www.bcogc.ca/public-zone/seismicity/whats-being-done
British Columbia Oil & Gas Commission (Can.) (n.d.) Induced seismicity. Retrieved from https://www.bcogc.ca/public-zone/seismicity/induced-seismicity

British Columbia Oil and Gas Commission (2014, December) Investigation of observed seismicity in the Montney Trend. Retrieved from https://www.bcogc.ca/node/12291/download

CBS DFW (2014, October 28) Texas amends waste disposal rules for fracking. CBS DFW. Retrieved from http://dfw.cbslocal.com/2014/10/28/texas-amends-waste-disposal-rules-for-fracking/

Chen L, Talwani P (2001) Mechanism of initial seismicity following impoundment of the Monticello Reservoir, South Carolina. Bull Seismol Soc Am 91(6):1582–1594. https://doi.org/10.1785/0120000293

Cypser DA, Davis SD (1994) Liability for induced earthquakes. J Environ Law Litig 9(2):551–589

Deutsche Welle (2018, March 29) Netherlands to shut Europe's biggest gas field to limit quake risk. Retrieved from http://www.dw.com/en/netherlands-to-shut-europes-biggest-gas-field-to-limit-quake-risk/a-43190065

Ehrman MU (2017) Earthquakes in the oilpatch: the regulatory and legal issues arising out of oil and gas operation induced seismicity. Ga State Law Rev 33(3):609–657. http://readingroom.law.gsu.edu/gsulr/vol33/iss3/2

Ellsworth W (2013) Injection-induced earthquakes. Science 341(6142):1225942-1–1225942-7. https://doi.org/10.1126/science

Ellsworth W, Llenos A, McGarr A, Michael A, Rubinstein J, Mueller C et al (2015) Increasing seismicity in the U.S. midcontinent: implications for earthquake hazard. Lead Edge 34(6):618–626. https://doi.org/10.1190/tle34060618.1

Elsworth D, Spiers CJ, Niemeijer AR (2016) Understanding induced seismicity. Science 354(6318):1380–1381. https://doi.org/10.1126/science.aal2584

Environmental Protection Agency (U.S.) (n.d.) Primary enforcement authority for the underground injection control program. Retrieved from https://www.epa.gov/uic/primary-enforcement-authority-underground-injection-control-program

Environmental Protection Agency (U.S.) (n.d.) General information about injection wells. Retrieved from https://www.epa.gov/uic/general-information-about-injection-wells

Environmental Protection Agency (U.S.) (n.d.) Class II oil and gas related injection wells. Retrieved from https://www.epa.gov/uic/class-ii-oil-and-gas-related-injection-wells

Fairley P (2012, January 20) Fracking quakes shake the shale gas industry. MIT Technology Review. Retrieved from http://www.technologyreview.com/news/426653/fracking-quakes-shake-the-shale-gas-industry/

Farber DA, Carlson AE (2013) Cases and materials on environmental law (9th edn). West

Fountain H (2011, December 12) Add quakes to rumblings over gas rush. New York Times. Retrieved from http://www.nytimes.com/2011/12/13/science/some-blame-hydraulic-fracturing-for-earthquake-epidemic.html?_r=0

Friberg P, Besana-Ostman G, Dricker I (2014) Characterization of an earthquake sequence triggered by hydraulic fracturing in Harrison County, Ohio. Seismol Res Lett 85(6):1295–1307. https://doi.org/10.1785/0220140127

Frohlich C, Deshon H, Stump B, Hayward C, Hornbach M, Walter J (2016) A historical review of induced earthquakes in Texas. Seismol Res Lett 87(4):1022–1038. https://doi.org/10.1785/0220160016

Gubbins D (1990) Seismology and plate tectonics. University of Cambridge, Cambridge

Healy JH, Rubey WW, Griggs DT, Raleigh CB (1968) The Denver earthquakes. Science 161(3848):1301–1310

Hincks T, Aspinall W, Cooke R, Gernon T (2018) Oklahoma's induced seismicity strongly linked to wastewater injection depth. Science 359(6381):1251–1255

Hornbach M, Deshon H, Ellsworth W, Stump B, Hayward C, Frohlich C et al (2015) Causal factors for seismicity near Azle, Texas. Nat Commun 6(6728):1–11. https://www.nature.com/articles/ncomms7728

Horton S (2012) Disposal of hydrofracking waste fluid by injection into subsurface aquifers triggers earthquake swarm in Central Arkansas with potential for damaging earthquake. Seismol Res Lett 83(2):250–260

Hurong D (2015) Influence of fault asymmetric dislocation on the gravity changes. Geod Geodyn 5(3):1–7. https://doi.org/10.3724/SP.J.1246.2014.03001

Keranen KM, Weingarten M, Abers GA, Bekins BA, Ge S (2014) Sharp increase in central Oklahoma seismicity since 2008 induced by massive wastewater injection. Science 345(6195):448–451

National Research Council (2013) Induced seismicity potential in energy technologies. http://www.nap.edu/

Nicholson B et al (2016) Proceedings from the 67th Institute of Energy Law: what's shaking: seismic activity and unconventional oil and gas activity

Nolte KA, Tsoflias GP, Bidgoli TS, Watney WL (2017) Shear-wave anisotropy reveals pore fluid pressure-induced seismicity in the U.S. midcontinent. Sci Adv 3(12):E1700443. https://doi.org/10.1126/sciadv.1700443

Office of the Oklahoma Secretary of Energy & Environment (U.S.) (2016) Earthquakes in Oklahoma. Retrieved from http://earthquakes.ok.gov/

Rubinstein JL, Ellsworth WL, Dougherty SL (2018) The 2013–2016 induced earthquakes in Harper and Sumner counties, southern Kansas. Bull Seismol Soc Am 108(2):674–689. https://doi.org/10.1785/0120170209

Schick R (2002) The little book of earthquakes and volcanoes. Springer, New York

Schultz R, Stern V, Gu YJ (2014) An investigation of seismicity clustered near the Cordel Field, West Central Alberta, and its relation to a nearby disposal well. J Geophys Res Solid Earth 119(4):3410–3423. https://doi.org/10.1002/2013JB010836

Schultz R, Atkinson G, Eaton DW, Gu YJ, Kao H (2018) Hydraulic fracturing volume is associated with induced earthquake productivity in the Duvernay play. Science 359(6373):304–308. https://doi.org/10.1126/science.aao0159

Stein S, Wysession M (2003) An introduction to seismology, earthquakes, and earth structure. Blackwell Pub, Malden

Sundstrom CD (2015) Oklahoma regulators implement evolving regulatory directives in response to earthquakes. Am Bar Assoc Trend, vol 46, p 5. Retrieved from https://www.americanbar.org/groups/environment_energy_resources/publications/trends/2014-2015/july-august-2015/oklahoma_regulators_implement_evolving_regulatory_directives_response_earthquakes.html

U.S. Geological Survey (n.d.) Earthquake glossary: seismology. Retrieved from http://earthquake.usgs.gov/learn/glossary/?term=seismology

van Eck T, Goutbeek F, Haak H, Dost B (2006) Seismic hazard due to small-magnitude, shallow-source induced earthquakes in The Netherlands. Eng Geol 87(1):105–121. https://doi.org/10.1016/j.enggeo.2006.06.005

Walsh F, Zoback M (2015) Oklahoma's recent earthquakes and saltwater disposal. Sci Adv 1(5):1–9. https://doi.org/10.1126/sciadv.1500195

Witze A (2015) Artificial quakes shake Oklahoma: earthquakes linked to oil and gas operations prompt further research into human-induced seismic hazards. Nature 520(7548):418–419. www.nature.com

Monika U. Ehrman is Professor of Law and Faculty Director of the *Oil & Gas, Natural Resources, and Energy Center* (ONE C) at the University of Oklahoma College of Law, where she leads the oil and gas, natural resources, and energy program. Her scholarly interests are in the area of oil and gas real property issues, the intersection between law and petroleum technology, and energy policy. Prior to teaching, she served as general counsel of a privately held oil and gas company in Dallas; senior counsel with Pioneer Natural Resources; and associate attorney at Locke Lord LLP. Her practice experience includes oil and gas litigation and energy transactional work. Before law school, Ehrman worked as a petroleum engineer in the upstream, midstream, and pipeline sectors of the energy industry. In addition to her experience with the technical aspects of the industry, she also worked as an analyst in the areas of commodity risk management and energy trading. She holds a B.Sc. in petroleum engineering from the University of Alberta; a J.D. from SMU Dedman School of Law; and a LL.M. from Yale Law School.

Part IV
Regulatory Regimes and Issues: Regional Perspectives

Chapter 15
Hydraulic Fracturing in Canada: Regulation by Moratorium or Specialized Agencies in Landscapes of Aboriginal and Treaty Rights

Deborah Curran

Abstract Over the past decade hydraulic fracturing activities have rapidly transformed the landscape in some regions of Canada, with both public and private sector drives to expand the oil and gas industry taking precedence over long term water stewardship. Within the Canadian federation, provincial governments have devolved responsibility for both water management and the regulation of unconventional oil and gas. Provinces tend to issue renewable short-term water licences for fracking activities under regulatory review processes that are separate from normal water licensing processes. This separation of regulatory function, the sheer volume of water use involved, and the scale of fracking has resulted in conflicts over water use and unregulated storage and use of water for the industry. In addition, Indigenous communities' aboriginal and treaty rights to water-based activities, such as fishing, are threatened by the extent of fracking activity and lack of hydrological data. These same First Nations are advocating for region- and watershed-wide water strategies to create objectives for long-term land and water planning that address the impacts of fracking and establish collaborative management structures for decision-making.

Keywords Water · Hydraulic fracturing · Canada · Law · Regulation · Aboriginal rights · Indigenous law · Indigenous governance

Thanks to Kelly Firth for careful research on the regulatory regime in BC. Special thanks to Kara Shaw, Lana Lowe, Ben Parfitt, and Oliver Brandes for ongoing discussions and information.

D. Curran (✉)
University of Victoria, Victoria, BC, Canada
e-mail: dlc@uvic.ca

15.1   Introduction

In 2012, the Ministry of Forests, Lands and Natural Resource Operations in British Columbia (BC), Canada issued the first "regular" water licence for fracking activities that would retain some priority in the prior allocation system used in water law in BC (most oil and gas activities rely on short term water use approvals of up to 24 months). Just a few months later, the licensee, Nexen Inc., withdrew approximately one third of the water (53 cm) from North Tsea Lake during low flow conditions that were only 15% of the flow level at which the licensee was required to cease taking water, the zero withdrawal threshold in the licence (Fort Nelson First Nation 2015). The five-year licence authorized water diversions from this shallow lake in what is known in oil and gas circles as the Horn River Basin of up to 60,000 cubic metres per day, and 2.5 million cubic metres per year, between April and October. The licence also prohibited water withdrawals when flow levels in the nearby Tsea River had decreased to less than 0.351 cubic metres per second, and required monitoring and reporting of actual use.

At the time of this remarkable diversion, the licence was the subject of an appeal to the Environmental Appeal Board (EAB) by the Fort Nelson First Nation (FNFN) on whose traditional territory the water withdrawal was occurring and with whom the provincial regulator had consulted during the three-year application process. The First Nation argued that the Assistant Regional Water Manager (the Manager) relied on incomplete and inadequate information when making the decision, which placed the Tsea River watershed and First Nation's treaty rights at risk. The FNFN also asserted that the provincial government failed in its assessment of the potential impact of the water licence on those rights.

The panel members of the EAB found that the licence was fundamentally flawed in concept and operation, and, almost uniquely, ordered that the Manager's decision be reversed. Although the licence application had stated that water withdrawals were grounded in a precautionary approach, the panel criticized both the hydrological and ecological information on which the licence was based as the licence permitted a water diversion approach "that is not supported by scientific precedent, appropriate modelling, or adequate field data", and that the withdrawal parameters "are arbitrary and have no basis in scientific theory or hydrometric modeling" (para 337). The panel members also found that the Manager had relied on "incorrect, inadequate, and mistaken factual information and modelling results" as the basis for his conclusion that water diversions under the licence would have no significant impacts on fish, riparian wildlife and the riparian environment (para 338). Finally, the panel members also found that the consultation process with the FNFN was seriously flawed, in part because the Ministry failed to explain the process and the proponent's role in that process to the FNFN. The Manager considered inaccurate and irrelevant information, and did not consider important information, about the FNFN's exercise of its treaty rights in the Tsea Lakes watershed (para 490).

While startling in effect, this decision is about more than the impact of one water licence on the FNFN's treaty rights in the Tsea Lakes region in northeastern BC. The context is one where gas exploration and operation activities have transformed the FNFN's traditional territory over the past decade, a territory that includes three of the four major gas plays in BC. Between just 2006 and 2012, the FNFN saw an increase in square kilometres of well pads from 0.23 to 31.9, the extension of roads from 1673 to 11,287 km, and the expansion of the network of seismic lines from 91,490 to 168,970 km (O'Shannassey 2012; BCOGC 2014). The FNFN have repeatedly and in many fora called for strategic cumulative effects assessment of gas activities in their territory, which they also identify as the upstream impacts of the burgeoning liquid natural gas industry in the province (Garvie et al. 2014/15).

This one appeal also encapsulates the primary challenges with current regulation of hydraulic fracturing in Canada, and especially the impacts on water. It raises issues of how water licensing is implicated in the cumulative impacts of one industry, and challenges the legal priority that water use for industrial activities could obtain over other rights-based activities in a watershed. The appeal highlights the lack of scientific information upon which the Government of BC bases water allocation decisions in some regions, and the inherent uncertainty in making predictions about impacts on the environment. It underscores the disconnect between requirements for monitoring water use and regulatory enforcement. Finally, it addresses, at length, the provincial government's responsibility to meaningfully consider the impacts of proposed activities on aboriginal and treaty rights under its constitutional duties to consult and accommodate First Nations. In short, the water licence appeal in *Fort Nelson First Nation* highlights the constellation of issues facing the public interest in safeguarding water quality and quantity under old principles of water law in the face of a new regulatory regime for hydraulic fracturing. Indeed, the primary fracking storyline in North America is about its impact on water quality (Olive and Delshad 2017).

The swift development of shale gas extraction operations in Canada over the past 10 years has spawned diverse regulatory approaches across the country, from moratoria prohibiting unconventional gas exploration to extensive production facilitated by a specialized regulator. Quickly changing industrial practices overlying a landscape for which there is little scientific baseline has meant that regulatory approvals are proceeding without a thorough understanding of environmental impact:

> [T]he rapid expansion of shale gas development in Canada over the past decade has occurred without a corresponding investment in monitoring and research addressing the impacts on the environment, public health, and communities. The primary concerns are the degradation of the quality of groundwater and surface water (including the safe disposal of large volumes of wastewater); the risk of increased greenhouse gas (GHG) emissions (including fugitive methane emissions during and after production), thus exacerbating anthropogenic climate change; disruptive effects on communities and land; and adverse effects on human health... The phrase *environmental impacts from shale gas development* masks many regional differences that are essential to understanding these impacts. (Council of Canadian Academies 2014 at xii)

The provinces of BC and Alberta already have extensive shale gas exploitation, with notable reserves in Quebec, New Brunswick, Nova Scotia and the Territories. The potential exists in Canada for tens of thousands of hydraulically fractured horizontal wells. The purpose of this chapter is to explore the onshore regulatory regimes for fracking in Canada, with a focus on the impacts on water in the province of BC. Like in other countries, the swift expansion of this industry and evolution of techniques over the past decade has meant that environmental regulations lag behind impacts across the landscape where there are significant data gaps and no requirements for cumulative effects assessment. These impacts occur in a context where the Canadian constitution acknowledges and affirms aboriginal and treaty rights, which First Nations are using frequently to force broader watershed planning and assessment of impacts.

15.2 Context: Fracking and Water Use in Canada

Canada is one of the top producers and exporters of natural gas in the world, producing some 150 billion cubic metres per year (CIA 2015). Eight provinces and territories in Canada produce natural gas, however, BC and Alberta, Canada's westernmost provinces, produce 75% (NEB 2013). While estimates vary widely for the amount of shale gas reserves in Canada, ranging from 1000 trillion cubic feet (NEB 2009) to three times that amount for Alberta alone (Rokosh et al. 2012), it is clear that Canada's shale gas reserves are significantly larger than its conventional marketable gas reserves (NEB 2011). The vast majority of shale gas production occurs in the northeast of BC, Canada's westernmost province, with smaller production occurring in Alberta and Saskatchewan. Marketable shale reserves are also located in Quebec, New Brunswick and Nova Scotia (Map 15.1; Council 2014).

It is important to note the unique regional contexts that make shale gas production possible in different areas of Canada. The extensive shale gas production in BC and Alberta mostly takes place in a landscape that is thinly populated and where the provincial government asserts ownership and regulatory jurisdiction. However, this landscape is the traditional territory of many Indigenous peoples who continue to exercise their treaty rights to hunt and trap across that territory. It also transpires on some private farmland. Conversely, shale gas reserves in eastern Canada underlie populous areas where residents own the land and have a greater reliance on groundwater. For example, in Quebec the Utica Shale underlies the St Lawrence River between Montreal and Quebec City, which is an agricultural area where over two million people reside.

The expansion of the shale gas industry in Canada occurs in a water law and management context that is one of the last frontiers of environmental regulation where decisions are based on insufficient science about ecosystem conditions. The knowledge gaps include basic information such as how much water is available, and how much water is actually being used (Curran and Brandes 2012; Curran 2014; Nowlan 2012). Fundamentally, provincial water law regimes in Canada still reflect their colonial origins that emphasized settlement of the land for development, agriculture, and mining. Although literature on the impacts of hydraulic fracturing on

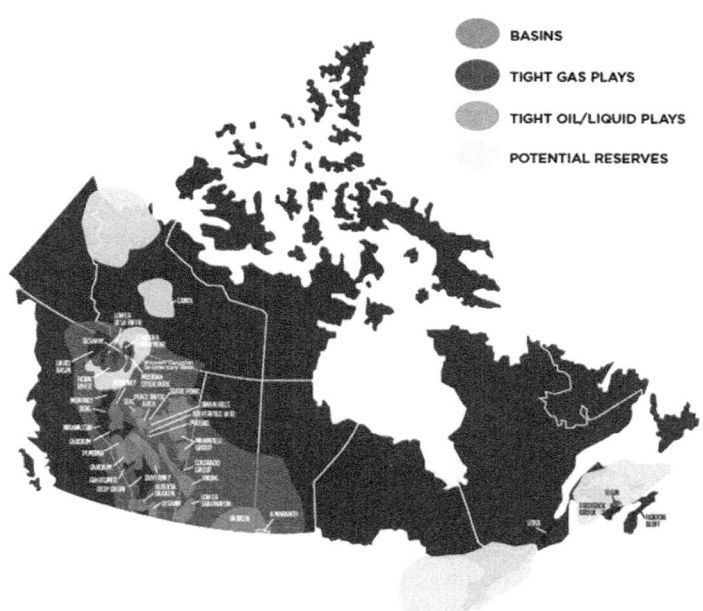

Map 15.1 Map of unconventional oil and gas plays in Canada. Reproduced with permission from Canadian Water Network's 2015 report, Water and Hydraulic Fracturing, page 20: http://cwn-rce.ca/report/2015-water-and-hydraulic-fracturing-report/

groundwater has increased significantly in the past 5 years, data is still limited and do not support confident conclusions (Council 2014).

This context for water and hydraulic fracturing in Canada - a rapidly expanding industry where insufficient environmental data supports decision-making and where regulatory jurisdiction rests with provincial governments who have taken very different approaches to risk management – has resulted in divergent regulatory regimes for fracking. In the provinces that are producing the most shale gas, regulatory authority is concentrated with specialized decision makers who have the dual authority for permitting all gas activities, which include water withdrawals and pollution permitting, as well as industry monitoring and enforcement. This dual function is increasingly raising conflict of interest concerns, and the public and First Nations are directly challenging regulators for the damage that gas activities have caused to land and waters.

15.3   Regulatory Oversight

The geographic, geologic, hydrological and economic diversity in Canada has spawned several different regulatory approaches to water and hydraulic fracturing. Sub-national provincial and territorial governments are responsible for regulating and managing both water and gas, except where a federal interest is paramount. This results in four different types of water law regimes, each with its own characteristics, and three primary regulatory methods for overseeing fracking operations. With the majority of lands in Canada in public ownership, as much as 95% in the westernmost province of British Columbia, the provincial governments have extensive discretion to shape how proponents use public land, and the standards to which they adhere. The primary burden underlying this wide authority are constitutionally acknowledged aboriginal and treaty rights that, since 1982, place largely procedural limitations on provincial licensing and permitting activities. First Nations using these procedural requirements of consultation and accommodation of aboriginal and treaty rights, as well as declarations of Indigenous law, are most directly challenging the risks posed to water by hydraulic fracturing.

15.3.1 Federal and Provincial Jurisdiction

Canada is a federation of ten provinces and three territories in the north, with the Constitution Act 1867 allocating jurisdiction between the federal and provincial governments to make laws relating to natural resources and property. Indigenous peoples and governments do not factor into this original constitutional apportionment, however, it is clear that the Indigenous laws and governance systems that preceded colonization continue in many forms (Borrows 2002; Napoleon and Friedland 2016; Napoleon 2015). Likewise, the word 'water' does not appear in Canada's constitutional documents but courts have presumed that its ownership and regulatory jurisdiction flowed pursuant to section 109 of the *Constitution Act* that grants ownership of land to most provinces (Burrard Power 1910; Kennett 1991; Saunders 1988). The *Natural Resources Transfer Agreements* granted essentially the same rights to water, oil and gas to the prairie provinces of Alberta, Saskatchewan and Manitoba in the 1930s (Percy 1988). Provincial assertion of ownership over land carries with it provincial ownership and regulation of water and non-renewable resources, subject to widespread claims by First Nations for aboriginal and treaty rights and aboriginal title (see Sect. 15.3.4). Provinces gain specific regulatory authority over non-renewable resources under section 92A. In practice, Canada's constitution establishes a shared and multijurisdictional responsibility for water, like for health and agriculture, where overlapping authorities may require action from several governments (Brandes and Curran 2017). Conversely, authority for non-renewable natural resources resides exclusively with the provinces.

While the federal government has a small ownership interest in water on federal lands and works, such as Indian reserves and national parks (Constitution Act 1982 at s. 108), its regulatory jurisdiction touches on water management in several areas. The federal government has direct regulatory responsibility for inland fisheries, navigation and shipping, federal works and undertakings (Constitution Act 1982 at s. 91). Constitutional authority plus interpretations by courts have extended federal authority to interprovincial waterways and flows (s. 91; Citizens, 1880), and environmental assessment (Friends of the Old Man River 1992). The federal government has almost no involvement in day-to-day decisions about permitting decisions that affect water quality and quantity, however it may influence provincial regulation where provincial decisions may affect its authority (Curran 2015). This is most evident when dealing with provincial decisions that have an impact on fish, which the federal government regulates directly. Federal interests in interprovincial trade and commerce relate to natural gas pipelines such that when pipelines cross provincial boundaries they may be subject to approval processes by the National Energy Board (National Energy Board Act 1985).

In addition to ownership, several heads of constitutional authority delegate provincial legislative authority for water. These include property and civil rights in a province, local works and undertakings, municipal institutions, and generally all matters of a local or private nature (Constitution Act 1982, s 92). Like exploration and development of non-renewable gas resources, provinces have virtually complete regulatory authority over domestic water management and are responsible for permitting the use of water under licence.

15.3.2   Water Regulation

Canada's approach to water law in general, and water allocation in particular, varies significantly from province to province. Most provinces assert ownership and jurisdiction over water (Curran 2017a), with the original water entitlements received from Britain in the form of riparian rights. Provinces began immediately modifying riparian rights in response to settlement, hydrogeology, and climatic conditions. Over the past 150 years four distinct approaches to water regulation have evolved in Canada (Brandes and Curran 2017).

(i) *Modified or Regulated riparian* – Riparian rights continue as a central water law tenet in Ontario and the Atlantic provinces. However, legislated licensing and permitting supplement riparian rules, requiring all uses over a certain volume to have provincial approval.

(ii) *Prior allocation* – Western Canadian provinces adopted the prior allocation system and purported to do away with riparian rights. It provides security of water entitlement on a first-in-time, first-in-right basis. Different from the prior appropriation rules in the United States, Canadian prior allocation vests priority of water entitlement upon obtaining a licence from the provincial government.

(iii) *Civil law* – The province of Quebec relies on a civil law tradition, whereas the rest of Canada uses a common law approach. The civil law incorporated riparian rights, which still exist in a modified form, with additional legislation providing for more environmental protection and modern management.

(iv) *Authority or Administrative management* – The territories – Yukon, Northwest Territories and Nunavut – delegate water management to natural resources boards or administrative bodies. This structure is established in the land claims agreements between the federal government and First Nations, and operationalized by federal and territorial legislation. The province of Saskatchewan also uses a modified version of this approach, however its delegation to the Water Security Agency does not result from government-to-government agreements between First Nations and senior government (The Water Security Agency Act 2005).

All provinces now regulate the use of groundwater in some way, with BC mandating groundwater licensing only in 2016 (Groundwater Protection Regulation 2016). There are widespread concerns that water laws do not adequately protect groundwater in Canada (Council of Canadian Academies 2009, 2014; Nowlan 2005).

Two characteristics of Canadian water law distinguish it from other water licensing and management regimes, for example from those in the United States. Water entitlements in Canada are not viewed as property rights (Brandes and Nowlan 2009; Curran 2014; Percy 1988). They are a right to use a specified amount of water under enumerated conditions subject to provincial regulation. Although they may behave as property rights because many licences are perpetual and there are few avenues for their amendment, provincial governments actually have extensive authority to make regulations and orders about water use under different conditions. Second, some water laws generally preclude compensation for harms cause by a provincial government changing the rules for water use, or making specific orders that affect only some water licences (Water Sustainability Act 2016 at s. 121; Curran 2017a). Analogies, therefore, can be drawn between water and gas regulation in that provincial governments have a high degree of discretion and regulatory flexibility to craft management regimes.

15.3.3 Gas Regulation

Similar to water, each province regulates gas activities differently, leading to three distinct approaches: regulation by moratorium, provincial government ministry regulation and specialized regulator. It is fair to say that as one moves from east to west, provincial governments become more comfortable with oil and gas activities and with more actively facilitating the industry (Canadian Water Network 2015). Sub-surface rights are generally reserved to the "Crown" (provincial government) and do not belong to private landowners.

(i) *Regulation by Moratorium* – Several provinces have enacted moratoria on hydraulic fracturing. Nova Scotia prohibits the use of high volume hydraulic fracturing for shale gas extraction (Nova Scotia n.d.). New Brunswick imposed a moratorium in 2014 and reaffirmed its commitment to the moratorium in 2016 citing the need for more information on environmental impacts and a process for consulting with First Nations (New Brunswick 2016; Prohibition Against Hydraulic Fracturing Regulation 2015). Until late 2016 Quebec had a moratorium on exploration for oil and gas, and continues with a moratorium in some specific locations, such as Anticosti Island, due to their ecological and cultural sensitivity (Government of Quebec 2017). In supporting the local government on Anticosti Island to seek UNESCO designation as a world heritage site, the Quebec provincial government paid over 40 million Canadian dollars in compensation to existing oil and gas permit holders.

(ii) *Provincial Government Ministry Regulation* – Several provinces rely on shared regulation between multiple provincial government ministries. In Quebec, the Ministry of Energy and Natural Resources takes the lead on permitting gas activities under the new *Petroleum Resources Act* (Bill 106 2016). The Quebec government issued its first permits for hydraulic fracturing in 2016 (Quebec Oil and Gas Association 2016), although no drilling may occur in the St Lawrence River (*An act to limit oil and gas activities* 2011). A member of the National Assembly of Quebec introduced a private members bill in 2017 intending to prohibit hydraulic fracturing and the chemical stimulation of wells (National Assembly of Quebec 2017).

(iii) *Specialized Regulator* – The two westernmost provinces in Canada, BC and Alberta, have specialized regulators dedicated to the oil and gas industry. Intended as "one stop shops" for permitting and oversight, the BC Oil and Gas Commission (OGC) and the Alberta Energy Regulator (AER) are responsible for a broad range of permits and approvals for oil and gas activities. They also generate regulations for the industry and undertake inspection and enforcement action. Distinct from provincial ministry oversight, both the OGC and AER exercise authority over water management – from permitting to disposal - as it pertains to hydraulic fracturing.

The broad authority for regulating water quality and quantity and gas exploration and development sits squarely within provincial jurisdiction. Federal authority has little impact on most management and approval decisions at the provincial level. However, the third level of government in Canada, First Nations, have unextinguished aboriginal and treaty rights across the Canadian landscape that impose substantive and procedural obligations on regulators and proponents. As seen in the case study of BC below, First Nations may be the most effective force advocating for reforms to hydraulic fracturing regimes in Canada.

15.3.4 Aboriginal Rights and Title

Since 1982, section 35 of the *Constitution Act, 1982*, has "acknowledged and affirmed aboriginal and treaty rights". Courts have interpreted the purpose of section 35 as the "reconciliation of the preexistence of Aboriginal societies with the sovereignty of the Crown" through negotiated settlements and court judgements (Van der Peet 1996; Delgamuukw 1997). As a process and not a final legal remedy, the day-to-day expression of reconciliation is twofold. First, Indigenous individuals have the right to activities that are integral to their cultures, such as hunting, fishing, trapping and ceremonies. Many of these activities rely on a physical environment where healthy ecological processes support flourishing species. Second, the federal and provincial governments have a duty to consult and accommodate First Nations on the potential impacts of proposed industrial activities (Haida Nation 2004; Mikisew Cree First Nation 2005). The provincial and federal governments must justify potential development that may infringe or deny Aboriginal or treaty rights (Sparrow 1990; Tsilhqot'in 2014).

Most of the contemporary jurisprudence on section 35 addresses whether or not the Crown has fulfilled it procedural duty to consult and accommodate, and accepts the governments' justification of infringement of Aboriginal and treaty rights (Ritchie 2013; Vermette 2011; Christie 2006). Courts will rarely direct specific consultation and accommodation procedures, nor will they give substantive direction on acceptable impacts to a First Nation's traditional territory. With few substantive remedies or limitations on Crown approvals in traditional territories resulting from this procedural requirement, overarching provincial jurisdiction for lands and water continues, except in a few pockets (Curran 2017b), and development of natural resources continues apace (Grassy Narrows First Nation 2014).

Therefore, the constitutional acknowledgement of treaty and aboriginal rights in Canada leads to extensive paper trails between senior governments seeking input from affected First Nations on proposed industrial activities. For example, the BC OGC consulted with the FNFN on 615 applications in the Horn River Basin alone in 2009/2010 (BCOGC 2010). Consultation and accommodation do not often result in watershed- or territory-based planning or cumulative effects assessment. The Canadian constitutional structure concentrates jurisdiction for both water and gas regulation at the provincial level. There is little federal oversight, except where commercial species of fish may be affected, or where federal regulation, such as for pipelines, is implicated. Aboriginal and treaty rights are a burden on government authority, however, as the case study of BC underscores, the rapidly growing industry for unconventional gas is outpacing provincial regulation. First Nations are at the forefront of challenging cumulative effects, particularly related to the quality and quantity of water.

15.4   Case Study: British Columbia

British Columbia is the largest producer of unconventional gas in Canada, with production starting in 2005 and accounting for 80% of total gas production in the province in 2015 (Government of Canada 2017). The Montney Basin alone, with 3100 active gas wells, produces upwards of 3.4 billion cubic feet per day, which makes it "the most important gas producing horizon in Canada", contributing approximately 67% of BC's monthly raw gas volume (Government of Canada 2017). As an example of the dominance of unconventional gas production, in the Horn River Basin 85% of all new wells target shale gas, an increase from 3.4% in

2006, which amounted to 501 new wells in 2013 (BCOGC 2010, 2014). While the percentage of the landscape in northeast BC subject to direct oil and gas activities is just 2.14% (BCOGC 2014), the fragmentation caused by seismic lines and roads creates an extensive network of disturbance in what was formerly largely uninhabited landscape or wilderness to the western eye but well-used traditional territory of the Indigenous communities of the northeast (Shackelford et al. 2017; Garvie et al. 2014/15).

As highlighted in the *Fort Nelson First Nation* appeal, the regulatory regime in BC is characterized by decision-making absent sufficient scientific or watershed information, a piecemeal project-by-project approach to considering aboriginal rights, and, until recently, permitting of only a portion of the hydrological system as the province did not require groundwater licensing. Although the province has paid closer attention to water and fracking in the past 2 years through the development of new strategies and law, the lack of oversight of the single regulator, the OGC, and lack of public confidence in the connection between monitoring and enforcement, means there is still considerable skepticism about the ability of the regulatory regime to uphold the public interest. The backdrop for this regulation is a 2012 Natural Gas Strategy that underscored the industry's role in the BC economy and the government's intent to significantly increase natural gas production and exports using liquid natural gas technologies (BC Ministry of Energy and Mines 2012).

The provincial government created the OGC, a Crown corporation, in 1998 as a single regulator for upstream oil and gas operations in BC with the explicit mandate to streamline regulation to grow the industry. Following a report from the Canadian Association for Petroleum Producers pointing out the complexity of oil and gas regulation in the province, the BC government created this single-window, one stop regulatory regime in just 5 months following the province and industry representatives signing a memorandum of understanding, the purpose of which was to "make British Columbia one of the most attractive places in North America for oil and gas investment" (Rankin et al. 2000). Between 1998 and 1999 the OGC reported a 60% increase in applications (BCOGC 2000).

The OGC is responsible for approvals as well as enforcement of oil and gas activities pursuant to the *Oil and Gas Activities Act* (OGAA) while the Ministry of Energy, Mines and Petroleum Resources retains authority for granting subsurface oil and gas tenures, which include the right to explore or produce natural gas, under the *Petroleum Natural Gas Act*. Although some gas activities that meet specified thresholds require an environment assessment certificate (Reviewable Projects Regulation 2002; Environmental Assessment Act 2002), the OGC evaluates and permits the majority of gas activities. It regulates, with considerable discretion, surface land use primarily through the Environmental Protection and Management Regulation that directs permit holder action and also establishes procedures for site restoration (2010). The Regulation establishes objectives for each listed value, such as water, riparian areas and wildlife, which include specific operating criteria. For example, the water objectives state that wellsites, facility areas, road right of way and pipeline corridors may not be located within 100 m of where water is diverted by a waterworks or water supply well unless adverse effects can be mitigated or the person proposing the works operates the waterworks or water supply well (at s. 4). This wiggle room – discretion to the OGC to determine that the gas activity will not have or mitigation of adverse effects – is also present in the performance-based standards in the Regulation. Causing material adverse impacts to waterworks or a water supply well is prohibited except where: it is not practicable to comply; the adverse effect is minimized; and the person gives 72 h notice to the owner and provides them with an alternate supply of water (at s. 9). Finally, the provincial government may establish, by regulation, areas where gas activities may not operate, for example within a designated watershed or on top of an identified aquifer, unless the operating areas will not have a "material adverse effect on the quality and quantity of water and the natural timing of water flow" (at s. 4).

Two trends in the regulatory regime have emerged since the establishment of the OGC – the concentration of authority over environmental permitting at the OGC and weak public oversight. Over time the provincial government has delegated authority for most environmental permitting to the OGC for all oil and gas activities. The OGC has replaced the normal regulator, such as the Ministries of Forests, or Environment, or Forests, Lands and Natural Resources Operations and Rural Development. This has created a two-tier environmental management regime in BC where cutting permits and water licences for gas activities go through the OGC and virtually all other activities through regular ministry channels. The OGC now has control of permitting in the areas of forestry, waste management, including for storage of hazardous waste, heritage conservation, leases and licences for the use of provincial land, water diversions, and changes in and about streams for all applications dealing with oil and gas activities. The provincial government has delegated to the OGC the "discretion, function and duty" of an officer under specified enactments, such as the *Water Sustainability Act*, which also carries the responsibility to enforce the relevant provisions of those enactments (OGAA 2008 at s. 8).

This two-tier approach to the environment brings up structural issues about the interaction between OGC permitting for water and parallel but separate permitting under the *Water Sustainability Act* for other activities on the same landscape. It also raises unanswered questions about the priority for different types of uses, for example long-term permitting for drinking water versus short-term use approvals for fracking in the same watershed. On the other hand and perversely, the OGC short-term permits generally require better monitoring and reporting than do normal water licences in BC. Permit conditions for fracking activities typically include prohibitions on diversions where the waterbody flow volume or depth is less than a prescribed amount, prohibitions on drawdowns of more than 0.1 m, mandatory reporting every six months where withdrawals exceed a specific volume per day (i.e. 500 cubic metres), and implementation of hydrological and limnological monitoring if contemplated in the application for the water licence.

The second trend relates to public oversight. In contrast to other areas of environmental regulation where interested parties can appeal to the Environmental Appeal Board, there is more limited scope for review of OGC decisions, which narrows public oversight of OGC approvals and evolving industry practice. The administrative review process is limited to permit holders and applicants, with landowners having a right of appeal only for a determination (a decision) that was made without due regard to landowner submissions or to a consultation report submitted under the OGAA (OGAA 2008 at 70–72). Unlike earlier iterations of the regime, other interested parties do not have a right of review or appeal, or access to a dispute resolution process. Since 2008, the OGC also has had broad authority to establish regulations for the industry, a status it shares with the provincial cabinet, the usual delegate of regulation-making authority in Canada. The OGC may make regulations respecting consultation, notice, permits, security, recovery of expenses, and the carrying out of oil and gas activities (OGAA 2008 at Part 9, Div 2), while cabinet retains authority for orphan sites, administrative penalties and environmental protection (OGAA 2008 at 95 and 105). Therefore, even if BC has world class regulations in terms of technical standards for environmental protection, given this concentration of authority in a non-elected statutory regulator and the significant discretion the OGC has in making permitting decisions, the regime is weak in application absent consistent monitoring and public reporting, and public confidence in enforcement and oversight of the regulator.

Most recently this lack of oversight came to light when critics discovered over 100 unlicensed and unregulated dams and ponds constructed to store water for fracking (Parfitt 2017). Located on both public and private land, sometimes without even a land lease in place for the use of public land, this water infrastructure behaves, in some cases, like a private water market for the fracking industry. Private landowners store water and then sell it to companies engaged in fracking activities outside of public regulation. In many cases companies engaged in fracking filled their storage structures before applying for a water licence (Parfitt 2017). The provincial Environmental Assessment Office ordered one company to hold no more than 10% of the water storage capacity, which amounts to dewatering the facility, in two of its dams after the company applied for an exemption from the requirement for Environmental Assessment Certificate under the provincial *Environmental Assessment Act* (Progress Energy 2017; Environmental Assessment Office 2017).

Such a specialized industry-specific regulator leads to a fundamental conflict of interest: the OGC is responsible for both environmental protection and growing the industry. It facilitates development activities, as well as monitors and enforces permit conditions and regulations. Between 2007 and 2015 the OGC carried out between 4000 and 6500 site inspections annually, which resulted in a maximum in any one year of 31 warnings, 36 compliance orders and, since 2014, 16 administrative penalties totaling $179,000 (the maximum penalty for any one infraction is $500,000) (BCOGC Compliance and Enforcement, various dates). The BC Auditor General has identified the weaknesses in this type of dual function regulatory structure – industry facilitation and environmental protection – as contributing to unacceptable environmental and public health risks in the mining sector (Office of the Auditor General of BC 2016). The Auditor General concluded that the role of promoting an industry comes into conflict with regulatory, compliance and enforcement functions.

Three recent provincial government initiatives attempt to address some of the weaknesses in the water-energy regulatory regime. First, the provincial government enacted a new water law in 2014 and brought it into force in 2016. The *Water Sustainability Act* attempts to address some of the gaps in decision-making for environmental health by introducing groundwater regulation for the first time and requiring that decision makers consider environmental flows when making licensing decisions (at ss. 6 & 15; Groundwater Protection Regulation 2016). However, there are no regulatory standards for environmental flows, and the Act is silent on the interaction between water licensing decisions and aboriginal rights to water (Curran 2017a; Brandes et al. 2015).

Second, in 2015 the OGC introduced area-based analysis as a method for addressing cumulative impacts when assessing applications (BCOGC n.d.). The intent is to consider broad landscape impacts on specific values, such as riparian and wildlife, when making approval decisions.

Finally, the Province of BC released the non-regulatory Northeast Water Strategy in 2015. Acknowledging that the northeast will be a centre of economic growth in the next two decades, this high level multi-government, multi-ministry and broad stakeholder strategy purports to set out "…a proactive, long-term approach for the responsible use and stewardship of water resources in the region" (at 2). The five areas identified for action in the Strategy mirror the weaknesses with the regulation of fracking as it affects water:

- Enhance information to support decision-making;
- Strengthen the regulatory regime;
- Coordinate and streamline decision-making processes;
- Enhance monitoring and reporting; and
- Build a water stewardship ethic.

The Strategy also reveals the discord between provincial and Indigenous governance and management interests in the region. In Appendix C of the Strategy, the Treaty 8 First Nations (referring to those first nations of Saskatchewan, Alberta, Northwest Territories and northeastern BC included in an 1899 treaty with the Crown) clearly set out their position and interests on their water rights, which reads in part (at 28–31):

> In accordance with their custom and traditional modes of governance, First Nations communities assert that, as the indigenous peoples of northeastern B.C., they have and continue to exercise Aboriginal and Treaty rights over water management within their traditional territories and they accept their responsibility for keeping water healthy and abundant as traditional stewards of these watersheds… [T]he First Nations must be equal partners with participating provincial, and federal Crown agencies consistent with government-to-government relationships respecting the design and implementation of water management plans… and the development of a joint-decision making approach to consultation.

***

> The Treaty established an ongoing Crown obligation to secure a continued supply of game and fish for the support and subsistence of the First Nations; and this implies ongoing Crown duties to protect fish and wildlife populations and to protect and safeguard their habitat, including water resources within lands not taken up. But that is not the end of the

story. First Nation rights to occupy and use Crown lands and resources to support their way-of-life and *livelihood, have been interpreted to include rights which are incidental to rights-based practices. The right of access to safe drinking water within their traditional territories, and their legitimate expectation that the Crown will manage waters so as to maintain reasonable access to safe drinking water by Treaty 8 peoples is one aspect of the duties owed to First Nation peoples by the Crown.* (emphasis in original)

First Nations are developing their own processes, such as generating baseline data, to counteract continued cumulative pressure from industry in their traditional territories. After drumming the provincial government out of the BC First Nations Liquid Natural Gas Summit in 2014 while inviting industry to stay (Hume 2014), the Fort Nelson First Nation has developed a water monitoring program, and is working on a water management plan that is based on its own Indigenous laws (Fort Nelson First Nation n.d.). Following the federal government release of a recovery strategy under the federal *Species At Risk Act* in 2012 for boreal caribou, the majority of which exist in Fort Nelson First Nation territory, the Nation released its own Medzih Action Plan (Boreal Caribou Recovery Plan) in 2017 (Fort Nelson First Nation 2017). The Plan cites lack of meaningful protection of the provincial landscape due to expanding industrial activities and its attendant seismic lines, roads and drill sites that fragment caribou habitat.

Similarly, the Blueberry River First Nation filed a civil action in 2015 for declaratory and injunctive relief for infringement of treaty rights due to the cumulative impacts of industrial activity in their traditional territory (Blueberry River First Nation 2015; Brend 2017). Citing destruction and loss of access to key hunting, fishing, and trapping sites, diminution in the abundance and health of wildlife, and impacts to water quality and quantity, the plaintiffs are seeking declarations from the court that the provincial government may not continue to authorize activities that breach their treaty rights. Notably, the statement of claim focuses on the failure of the provincial government to obtain sufficient information to: understand the potential cumulative impacts; assess, monitor and manage cumulative impacts; and manage the pace, scale and location of activities within the Blueberry River First Nation's traditional territory. The rapid increase in industrial activity in the Blueberry River First Nation's traditional territory is shown in Map 15.2.

Although local governments have no jurisdiction to ban fracking or its impacts, some are taking steps to protect drinking water. One example is the Quebec village of Ristigouche-Sud-Est that enacted a bylaw based on the precautionary principle limiting industrial well drilling within 2 km of domestic wells, effectively creating a no-drill zone, and prohibited the introduction of chemical substances into the soil (Page 2018). The company that held a drilling permit sued the municipality for $1.5 million in damages claiming that the village had enacted an illegal bylaw for the purpose of stopping the project (Kassam 2018). In rejecting the company's argument, the judge of the Superior Court of Quebec found that absent a provincial scheme for protecting water sources, the municipality had acted in the public interest of its citizens (*Gastem v Municipalite de Ristigouche-Sud-Est* 2018).

In conclusion, and personifying the lack of public confidence in the regulation of hydraulic fracturing in BC, in the fall of 2017 a coalition of 17 organizations issued

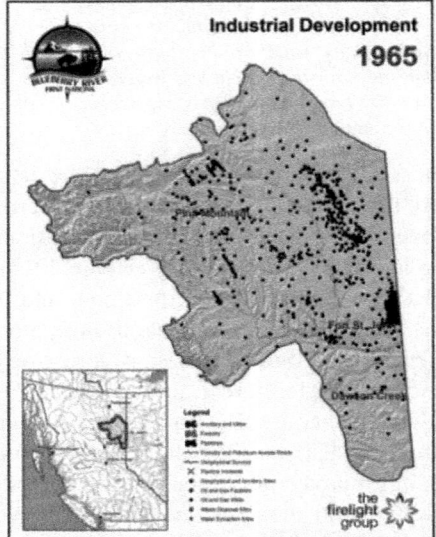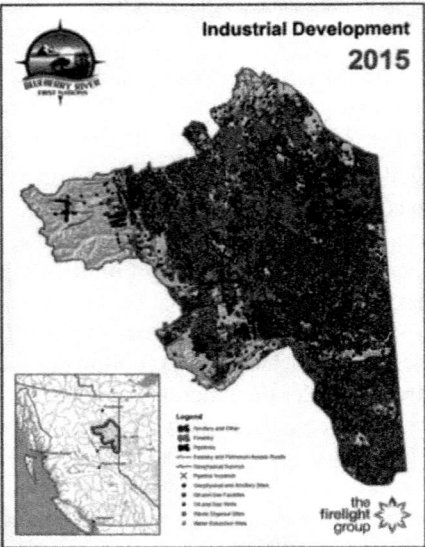

Map 15.2 Extent of industrial development in Blueberry River First Nations traditional territory, 1965 and 2015. Reproduced with permission from the Blueberry River First Nation

a call for a public inquiry. Motivated by concerns about water, GHG emissions, seismic activity and inadequate public process, particularly for First Nations, the call focuses on the regulatory apparatus for fracking in BC: "…whether or not provincial agencies adequately oversee fracking operations, ensuring that companies comply with existing laws and regulations, safeguard public health, and protect the environment" (Canadian Centre for Policy Alternatives 2017). The concerns of these environmental, First Nations, health and public policy organizations in BC encapsulate the current challenges with regulating a fast-growing industry that has significant capital available to it in an operating landscape that is largely unknown by western science. Water is the focus for the critique of fracking regulation, governance and impacts.

15.5 Conclusion: Challenges with Fracking in Canada

Several chapters in this volume outline the generic risks that hydraulic fracturing poses to the natural environment and human communities. The provincial regulation of fracking and water in Canada presents heightened challenges at the watershed scale, with regulatory authority conflicted in the most productive provinces and reconciliation with Indigenous rights failing. The regulatory apparatus for hydraulic fracturing has not caught up with the rapid pace of expansion of the industry such that exploration and production activities have taken precedent over long term ecological health and Indigenous rights in the provinces with the largest reserves (Moore et al. 2015).

At the watershed scale, there is no environmental baseline in the areas where fracking is occurring. Shale gas development is proceeding without publicly available hydrological and other environmental data that would assist regulators and the public to weight the risks posed by fracking (Council 2014). Likewise, there are no comprehensive frameworks for cumulative effects assessment or landscape-level planning. The OGC in BC is beginning to address systemic impacts with its area-based analysis, however has chosen an internal methodology rather than as part of a broader public conversation about the extent of gas activities in the province. Quebec also formed the Strategic Environmental Assessment Committee on Shale Gas in 2011 (Council 2014).

At the regulatory scale, public perception is that the regulatory structure is representing industry interests much more effectively than public interests. Lack of baseline information, meaningful monitoring and consistent enforcement by regulators does not adequately account for public risks. This view is bolstered by the expose on unlicensed dams and the OGC's recent release of a 4 year old report on gas migration or leakage from wells (Lavoie 2017), some of it contaminating water, where the regulator admits that it does not have an "accurate understanding of the total number of wells" involved (BCOGC 2013). Academics are also turning to evaluating regulatory effectiveness and finding, in at least the province of Saskatchewan, that "regulation of fracking is comparable to governments that have minimally responded to the new risks presented by the rapid expansion of fracking" (Carter and Eaton 2016).

These weaknesses of the regulatory apparatus for shale gas regimes inevitably mean heightened risks to the fundamental conditions of life such as drinking water, foodsystems (Pothukuchi et al. 2017) and Indigenous foodscapes (Turner et al. 2013). The City of Dawson Creek in the northeast of BC has mapped the oil and gas infrastructure in their drinking water watershed, and a pilot study has shown benzene levels, a chemical used in fracking, in pregnant women in the same region to be 3.5 times higher the general population (Caron-Beaudoin et al. 2018). The Supreme Court of Canada recently ruled that individuals whose land is contaminated by regulated oil and gas activities cannot sue the Alberta Energy Regulator for damages (Ernst 2017). Landowner Jessica Ernst sued the Alberta Energy Regulator, provincial government and an energy company in 2007 for damages for aquifer contamination near her home that she asserts is caused by shallow well fracking. The court ruled that Ms. Ernst could sue the government, but not the regulator because of immunity clauses in provincial legislation. The court did not rule on the primary issue before it, whether a legislature can prohibit a constitutional damages claim against an agency of the government. These risks, all occurring in rural and wilderness areas of western Canada, will increase if hydraulic fracturing occurs in more populous areas of eastern Canada.

For Indigenous communities in BC and Alberta, the extent of impacts to landscape and water, particularly perceived chemical contamination, has meant that they no longer feel safe drinking the water in their traditional territories or eating some game. Indigenous peoples and scholars note that Indigenous-Crown reconciliation must involve treating the environment differently and reconciliation with the earth. This includes first protecting the ecological integrity of watersheds (Borrows 2019), as well as sharing in the development of natural resources as a key part of local economies (LaForme and Truesdale 2013).

In a comprehensive review of shale gas development in 2014, the Council of Canadian Academies, an independent science-based organization, identified five qualities that an effective framework for managing the risks with this industry requires. In addition to two elements dealing with technology and management systems to control risks, the Council focused on transparent systems of accountability (at xix and 191):

(iii) An effective regulatory system. Rules to govern the development of shale gas must be based on appropriate science-driven, outcome-based regulations with strong performance monitoring, inspection, and enforcement.

(iv) Regional planning. To address cumulative impacts, drilling and development plans must reflect local and regional environmental conditions, including existing land uses and environmental risks. Some areas may not be suitable for development with current technology, whereas others may require specific management measures.

(v) Engagement of local citizens and stakeholders. Public engagement is necessary not only to inform local residents of development, but to receive their input on what values need to be protected, to reflect their concerns, and to earn their trust. Environmental data should be transparent and available to all stakeholders.

The Council goes on to note that government and company reassurances of safety and technical prowess will not win public acceptance and social licence for hydraulic fracturing. What is needed are "go-slow" approaches to generating the needed baseline data, and transparent and credible monitoring of environmental impacts. Such an approach will likely have impacts on the rate of expansion of the industry, however those impacts are necessary to address the currently unacceptable public health and environmental risks posed by shale gas development.

References

An Act to limit oil and gas activities, SQ 2011, c 13

BC Ministry of Energy and Mines (2012) British Columbia's natural gas strategy. http://www.gov. bc.ca/ener/popt/down/natural_gas_strategy.pdf. Accessed 16 Dec 2017

BC Oil and Gas Commission (2010) Compliance and enforcement activity report for 2010/2011. https://www.bcogc.ca/node/6107/download. Accessed 17 Dec 2017

BC Oil and Gas Commission (2014) 2013–2014 compliance and enforcement activity summary. https://www.bcogc.ca/node/12466/download. Accessed 17 Dec 2017

BC Oil and Gas Commission (2000) Information release. https://www.bcogc.ca/node/5634/download. Accessed 17 Dec 2017

BC Oil and Gas Commission (2010) Horn River Basin status report 2010. BC Oil and Gas Commission.   http://www.bcogc.ca/node/5930/download?documentID=1015&type=.pdf. Accessed 16 Dec 2017

BC Oil and Gas Commission (2012) Compliance and enforcement activity summary. https://www. bcogc.ca/node/11247/download. Accessed 17 Dec 2017

BC Oil and Gas Commission (2013) Gas migration preliminary investigation report – December 2013. https://www.bcogc.ca/node/14620/download. Accessed 18 Dec 2017

BC Oil and Gas Commission (2014) Oil and gas land use in Northeast British Columbia. BC Oil and Gas Commission. http://www.bcogc.ca/node/12908/download. Accessed 16 Dec 2017

BC Oil and Gas Commission (2015) Compliance and enforcement summary 2015. https://www.bcogc.ca/node/13558/download. Accessed 17 Dec 2017

BC Oil and Gas Commission (n.d.) Area-based analysis (ABA). http://www.bcogc.ca/public-zone/area-based-analysis-aba. Accessed 17 Dec 2017

Bill 106, An Act to implement the 2030 Energy Policy and to amend various legislative provisions, 1st Sess. 41th Leg, Quebec, 2016

Blueberry River First Nation v. British Columbia (2015) Vancouver Registry S-151727 Notice of Civil Claim. http://www.ratcliff.com/sites/default/files/news_articles/2015-03-03%20Notice%20of%20Civil%20Claim.PDF. Accessed 16 Dec 2017

Borrows J (2002) Recovering Canada: the resurgence of indigenous law. University of Toronto Press, Toronto

Borrows J (2019) Earth-bound: indigenous laws and environmental reconciliation. In: M Asch, J Borrows, J Tully (eds) Resurgence and reconciliation: indigenous-settler relations and earth teachings. University of Toronto Press, Toronto, pp 49–81

Brandes O, Curran D (2017) Changing currents: a case study in the evolution of water law in Western Canada. In: Renzetti S, Dupont D (eds) Water policy and governance in Canada. Springer, New York, pp 45–67

Brandes OM, Nowlan L (2009) Wading into uncertain waters: using markets to transfer water rights in Canada—possibilities and pitfalls. J Environ Law Pract 19(3):267–287

Brandes OM, Carr-Wilson S, Curran D, Simms R (2015) Awash with opportunity: ensuring the sustainability of British Columbia's new water law. POLIS Project on Ecological Governance, University of Victoria, Victoria, Canada. http://poliswaterproject.org/awashwithopportunity. Accessed 10 Dec 2017

Brend Y (2017) Collision between indigenous hunting and oil development rights set for legal showdown in B.C. court. CBC News June 10 2017. http://www.cbc.ca/news/canada/british-columbia/bc-treaty-rights-law-blueberry-river-first-nation-oil-and-gas-development-court-1.4151779. Accessed 16 Dec 2017

Burrard Power Co. v. R. (1910), [1911] A.C. 87 (Canada P.C)

Canadian Centre for Policy Alternatives (2017) Public inquiry needed to properly investigate deep social and environmental harms of fracking, coalition says. https://www.policyalternatives.ca/newsroom/news-releases/public-inquiry-needed-properly-investigate-deep-social-and-environmental. Accessed 16 Dec 2017

Caron-Beaudoin E, Valter N, Chevrier J, Ayotte P, Frohlich K, Verner M-A (2018) Gestational exposure to volatile organic compounds (VOCs) in Northeastern British Columbia, Canada: a pilot study. Environ Int 110:131–138. https://doi.org/10.1016/j.envint.2017.10.022

Carter AV, Eaton EM (2016) Subnational responses to fracking in Canada: explaining Saskatchewan's "wild west" regulatory approach. Rev Policy Res 33(4):393–419

Canadian Water Network (2015) Water and hydraulic fracturing: where knowledge can best support decisions in Canada waterloo. ON, Canadian Water Network. http://cwn-rce.ca/wp-content/uploads/2015/10/CWN-2015-Water-and-Hydraulic-Fracturing-Report.pdf

Christie G (2006) Developing case law: the future of consultation and accommodation. UBC Law Rev 39:139

CIA (Central Intelligence Agency) (2015) The world Factbook. Country comparison: natural gas – production. Retrieved February 2013, from https://www.cia.gov/library/publications/the-world-factbook/rankorder/2249rank.html?countryName=Canada&countryCode=ca&regionCode=noa&rank=5#ca

Citizens and the Queen Insurance Cos of Canada v Parsons 4 SCR 215

Constitution Act, 1867 (U.K.), 30 & 31 Vict., c. 3, s. 91 and 92, reprinted in RSC 1985, App. II, No. 5

Constitution Act, 1982, being Schedule B to the *Canada Act 1982* (UK), 1982, c 11

Council of Canadian Academies. 2009. The Sustainable Management of Groundwater in Canada – Expert Panel on Groundwater. Ottawa: Council of Canadian Academies

Council of Canadian Academies (2014) Environmental impacts of shale gas extraction in Canada. Council of Canadian Academies, Ottawa

Curran D (2014) British Columbia's water sustainability act – a new approach to adaptive management and no compensation regulation. The University of Calgary Faculty of Law Blog on Developments in Alberta Law. https://ablawg.ca/2014/05/28/british-columbias-water-sustainability-act-a-new-approach-to-adaptive-management-and-no-compensation-regulation/. Accessed 17 Dec 2017

Curran D (2015) Water law as watershed endeavour: federal inactivity as an opportunity for local initiative. J Environ Law Pract 28(1):53–88

Curran D, Brandes OM (2012) When the water dries up: lessons from the failure of water entitlements in Canada, the US and Australia—Workshop Discussion Paper. University of Victoria, Canada. http://poliswaterproject.org/publication/478

Curran D (2017a) Leaks in the system: environmental flows, aboriginal rights and the modernization imperative for water law in British Columbia. UBC Law Rev 50(2):233–291

Curran D (2017b) 'Legalizing' the great bear rainforest agreements: colonial adaptations towards reconciliation and conservation. McGill Law J 62(3):813–860

Delgamuukw v BC, [1997] 3 SCR 1010, [1997] SCJ No 108

Environmental Assessment Act, S.B.C. 2002, c. 43

Environmental Assessment Office (2017) In the matter of the environmental assessment Act S.B.C. 2002, c.43 (the act) and progress energy non-compliance with section 8.1 of the act – order under section 34(1). https://projects.eao.gov.bc.ca/api/document/59fb5731dc09b600192 19a81/fetch. Accessed 16 Dec 2017

Environmental Protection and Management Regulation. B.C. Reg. 200/2010

Ernst v Alberta Energy Regulator 2017 SCC 1

Fort Nelson First Nation v. Assistant Regional Water Manager (2015) BC EAB decision, 2012-WAT-013(c), 3 September 2015

Fort Nelson First Nation (2017) Medzih action plan: Fort Nelson first boreal Caribou recovery plan. http://www.fortnelsonfirstnation.org/uploads/1/4/6/8/14681966/2017-sept-29_fnfn_medzih_action_plan_final_medres.pdf. Accessed 15 Dec 2017

Fort Nelson First Nation (n.d.) Liard and Horn River Basin water monitoring. http://lands.fnnation.ca/project/liard-horn-river-basin-water-monitoring. Accessed 16 Dec 2017

Friends of the Old Man River Society v Canada (Minister of Transport) [1992] 1 SCR 3

Garvie KH, Lowe L, Shaw K (2014) Shale gas development in Fort Nelson first nation territory: potential regional impacts of the LNG boom. BC Stud 184:45–72

Gastem, Inc. v *Municipalite de Ristigouche-Partie-Sud-Est* 28 fevrier 2018 Cour Superiere du Quebec No. 105-17-000384-132

Government of Quebec (2009) An act to affirm the collective nature of water resources and provide for increased water resource protection, CQLR c C-6.2. http://canlii.ca/t/52b2f

Government of Quebec (2017) Île d'Anticosti – Le Gouvernement du Québec soustrait définitivement le territoire de l'île d'Anticosti à l'exploration pétrolière et gazière: http://www.fil-information.gouv.qc.ca/Pages/Article.aspx?idArticle=2507281358&lang=en

Grassy Narrows First Nation v Ontario (Natural Resources) 2014 SCC 48

Groundwater Protection Regulation. B.C. Reg. 39/2016

Haida Nation v BC, 2004 SCC 73

Hume M (2014) First nations' LNG fight takes wing on an eagle's feather. Globe and Mail, April 20 2014 online: https://www.theglobeandmail.com/news/british-columbia/first-nations-lng-fight-takes-wing-on-an-eagles-feather/article18073092/

Kassam A (2018) Tiny Canada town defeats oil firm in fight over drinking water. The Guardian, March 3 2018 online: https://www.theguardian.com/world/2018/mar/03/canada-oil-drilling-town-lawsuit-ristigouche-sud-est

Kennett SA (1991) Managing interjurisdictional waters in Canada: a constitutional analysis. Canadian Institute of Resources Law, Calgary

LaForme HHS, Truesdale C (2013) Section 25 of the charter; section 35 of the constitution act, 1982: aboriginal and treaty rights – 30 years of recognition and affirmation. Supreme Court Law Rev 2(687):62

Lavoie J (2017) B.C. coughs up fracking report four years late and only after it was leaked to journalist. DeSmog Canada. https://www.bcogc.ca/node/14620/download. Accessed 18 December 2017

Mikisew Cree First Nation v Canada (Minister of Canadian Heritage) 2005 SCC 69

Moore M-L, Castleden H, Shaw K (2015) Regional snapshot. Building capacity for building trust: key challenges for water governance in relation to hydraulic fracturing. Canadian Water Network, Waterloo

Napoleon V (2015) Tsilhqot'in law of consent. UBC Law Rev 48:3 873–3 901

Napoleon V, Friedland H (2016) An inside job: engaging with indigenous legal traditions through stories. McGill L J 61:4 725–4 754

National Assembly of Quebec. Bill 990 – an act to prohibit hydraulic fracturing and chemical stimulation of wells throughout Quebec. https://www.google.ca/url?sa=t&rct=j&q=&esrc=s&source=web&cd=1&ved=0ahUKEwjNtZ2VopLYAhUG6GMKHamVA4oQFgguMAA&url=http%3A%2F%2Fwww.assnat.qc.ca%2FMedia%2FProcess.aspx%3FMediaId%3DANQ. Vigie.Bll.DocumentGenerique_130605en%26process%3DDefault%26token%3DZyMoxNwUn8ikQ%2BTRKYwPCjWrKwg%2BvIv9rjij7p3xLGTZDmLVSmJLoqe%2FvG7%2FYWzz&usg=AOvVaw129_P2IsdXdlxUatbzS0AX. Accessed 17 Dec 2017

National Energy Board Act. R.S.C. (1985) C. N-7

Natural Resources Transfer Agreements, Constitution Act, 1930, 20-21 Geo. V c 26 (uk), Schedule 2 s 1

NEB (National Energy Board) (2009) Energy brief – understanding Canadian shale gas. National Energy Board, Calgary

NEB (National Energy Board) (2011) Canadian energy overview 2010. National Energy Board, Calgary

NEB (National Energy Board) (2013) Marketable natural gas production in Canada. Retrieved February 2013, from http://www.neb-one.gc.ca/clfnsi/rnrgynfmtn/sttstc/mrktblntrlgsprdctn/mrktblntrlgsprdctn-eng.html

New Brunswick (2016) News release: moratorium on hydraulic fracturing to continue indefinitely. http://www2.gnb.ca/content/gnb/en/news/news_release.2016.05.0462.html. Accessed 17 Dec 2017

Nova Scotia (n.d.) Hydraulic fracturing review. https://energy.novascotia.ca/oil-and-gas/onshore/hydraulic-fracturing-review. Accessed 17 Dec 2017

Nowlan L (2005) Buried treasure: groundwater permitting and pricing in Canada. Walter and Duncan Gordon Foundation. https://qspace.library.queensu.ca/bitstream/handle/1974/8509/Buried%20Treasure_Groundwater%20Permitting%20and%20Pricing%20In%20Canada.pdf?sequence=1&isAllowed=y. Accessed 17 Dec 2017

Nowlan L (2012) CPR for Canadian rivers – law to conserve, protect, and restore environmental flows in Canada. Environ Law Pract 23:203–252

O'Shannassey J (2012) Fort Nelson first nation's action: moving towards a water management plan. Presentation to keepers of the water IV conference. http://lands.fnnation.ca/moving-toward-water-management-plan. Accessed 16 Dec 2017

Office of the Auditor General of British Columbia (2016) An audit of compliance and enforcement of the mining sector. Office of the Auditor General of British Columbia, Victoria

Oil and Gas Activities Act, SBC 2008, c 36

Olive A, Delshad AB (2017) Fracking and framing: a comparative analysis of media coverage of hydraulic fracturing in Canadian and US newspapers. Environ Commun 11:784–799. https://doi.org/10.1080/17524032.2016.1275734

Page J (2018) Small Quebec village, sued for trying to protect its drinking water, wins legal battle. February 28, 2018 CBC, online: http://www.cbc.ca/news/canada/montreal/small-quebec-village-sued-for-trying-to-protect-its-drinking-water-wins-legal-battle-1.4555774

Parfitt B 2017. A dam big problem: regulatory breakdown as fracking companies in BC's northeast build dozens of unauthorized dams. Vancouver: Canadian Centre for Policy Alternatives. http://www.policynote.ca/dam-big-problem/. Accessed 16 Dec 2017

Percy DR (1988) The framework of water rights legislation in Canada. Canadian Institute of Resources Law, Calgary

Petroleum Natural Gas Act, RSBC 1996, c 361

Pothukuchi K, Arrowsmith M, Lyon N (2017) Hydraulic fracturing: a review of implications for food systems planning. https://doi.org/10.1177/0885412217733991

Progress Energy (2017) Letter from Jarred Anstett to Theresa Morris July 20 2017. https://projects.eao.gov.bc.ca/api/document/5994c0758ee539001822a2c8/fetch. Accessed 16 Dec 2017

Prohibition Against Hydraulic Fracturing Regulation – Oil and Natural Gas Act. New Brunswick Regulation 2015–28

Quebec Oil and Gas Association (2016) Quebec government gives the first authorization certificate for fracking in Quebec. http://www.apgq-qoga.com/en/2016/06/21/quebec-government-gives-the-first-authorization-certificate-for-fracking-in-quebec/. Accessed 17 Dec 2017

R v Van der Peet, [1996] 2 SCR 507, 137 DLR (4th) 289

R. v Sparrow, [1990] 1 SCR 1075, 70 DLR (4th) 385

Rankin M, Carpenter S, Burchmore P, Jones C (2000) Regulatory reform in the British Columbia petroleum industry: the oil and gas commission. Alta L Rev 38(1):143–169

Reviewable Projects Regulation, B.C. Reg. 370/2002

Ritchie K (2013) Issues associated with the implementation of the duty to consult and accommodate aboriginal peoples: threatening the goals of reconciliation and meaningful consultation. UBC Law Rev 43:397

Rokosh CD, Lyster S, Anderson SDA, Beaton AP, Berhane H, Brazzoni T, Chen D, Cheng Y, Mack T, Pana C, Pawlowicz JG (2012) Summary of Alberta's shale- and siltstone-hosted hydrocarbon resource potential. Alberta Energy Regulator, Calgary

Saunders OJ (1988) Interjurisdictional issues in Canadian water management. The Canadian Institute of Resources Law, Calgary

Shackelford N, Starzomski BM, Banning NC, Battaglia LL, Becker A, Bellingham PJ, Bestelmeyer B, Catford JA, Dwyer JM, Dynesius M, Gilmour J, Hallett LM, Hobbs RJ, Price J, Sasaki T, Tanner EVJ, Standish RJ, Institutionen för ekologi, miljö och geovetenskap, Umeå universitet, Teknisk-naturvetenskapliga fakulteten (2017) Isolation predicts compositional change after discrete disturbances in a global meta-study. Ecography 40:1256–1266. https://doi.org/10.1111/ecog.02383

The Water Security Agency Act, SS 2005, C. W-8.1

Tsilhqot'in Nation v. British Columbia, [2014] SCC 44

Turner N, Berkes F, Stephenson J, Dick J (2013) Blundering intruders: extraneous impacts on two indigenous food systems. Hum Ecol 41:463–574. https://doi.org/10.1007/s10745-013-9591-y

Vermette D (2011) Dizzying dialogue: Canadian courts and the continuing justification of the dispossession of aboriginal peoples. Windsor Yearb Access Justice 29(1):55

Water Sustainability Act, SBC 2014, c 15

Deborah Curran is an Associate Professor at the University of Victoria (Canada) in the Faculty of Law and School of Environmental Studies (Faculty of Social Sciences), and the Executive Director of the Environmental Law Centre where she works with students on environmental law projects for community organizations and First Nations across the British Columbia. Deborah's work is in the areas of land and water law, with a particular focus on environmental protection, collaborative management with Indigenous communities in water law, and municipal sustainability, which includes healthy foodscapes.

Chapter 16
Hydraulic Fracturing in Latin America: Prospects and Possibilities?

Andrés Felipe Sánchez Peña

Abstract During the last three decades, unconventional gas and oil development has substantially transformed the energy section. One of main developments is hydraulic fracturing, or "fracking", a practice which has had substantial impact on people's economic, social and political lives in those areas where unconventional energy reserves are present and are being exploited or could be exploited.

Latin America is one of the world regions with the highest potential for unconventional gas and oil development. Thus, the objective of this chapter is to look at unconventional oil and gas from a geopolitical perspective in the Latin American region, and in particular, the evolution of the industry and related energy policies in key countries of the Western hemisphere (namely Mexico, Argentina and Colombia) since the U.S. shale revolution of the early 21st century.

This chapter begins with a brief review of how hydraulic fracturing works and where in the world it is currently operational. The chapter then presents a review highlighting subsequent developments in the Americas over the last 10 years, giving special attention to Mexico, Argentina and Colombia, countries with large assessed reserves. The chapter thus shows the availability of oil and gas shale resources in Latin America, the developments that have been emerging to regulate the sector, and the enabling regimes/policies. The chapter concludes by considering the extent to which resolving environmental – and particularly water – issues related to hydraulic fracturing may be key for the economic growth for these Latin American countries (i.e. Mexico, Argentina and Colombia). These hurdles must be considered and addressed, in order to better shape the future of Latin American fracking in the coming years.

Keywords Latin America · Columbia · Mexico · Argentina · History of fracking

A. F. S. Peña (✉)
Los Andes University of Colombia, Bogotá, Colombia
e-mail: asanchez@oas.org

© Springer International Publishing AG 2020
R. M. Buono et al. (eds.), *Regulating Water Security in Unconventional Oil and Gas*, Water Security in a New World,
https://doi.org/10.1007/978-3-030-18342-4_16

16.1   Introduction

Currently global population is on the rise and with it the consumption of food, water and – most relevant to the current chapter – the demand for energy. As stated by the U.S. Energy Information Administration (EIA) (2017) Energy Outlook, world energy consumption is projected to increase 28% by 2040. Meanwhile, conventional fossil fuels (still around 70% of the world energy supply) are gradually being exhausted, triggering a search for alternatives, including unconventional hydrocarbons. As we have seen in other chapters in this volume, technological developments together with changing energy sector economics have helped grow the unconventional hydrocarbon sector, especially in the US.

Latin America is one of the world regions with the highest potential for unconventional gas and oil development. Thus, the objective of this chapter is to look at unconventional oil and gas from a geopolitical perspective in the Latin American region, and in particular, the evolution of the industry and related energy policies in key countries of the Western Hemisphere (namely Mexico, Argentina and Colombia) since the U.S. shale revolution.

The chapter is structured in the following way: first, it briefly describes fracking as a technique and its potential impacts, both positive and negative, as are also discussed in other chapters in this book, but here giving particular attention to the region. Second, the chapter then presents a review highlighting subsequent developments in the Americas over the last 10 years, giving special attention to Mexico, Argentina and Colombia, countries with large reserves. The chapter discusses the availability of oil and gas shale resources in the Western Hemisphere, the developments that have been emerging to regulate the sector, and the enabling regimes/policies. Third, the chapter considers whether it is still premature to determine whether fracking is key for the economic growth for these Latin American countries (i.e. Mexico, Argentina and Colombia). The chapter concludes with some recommendations to help address the several challenges that still remain for the further sustainable development of the unconventional hydrocarbon sector, and its potential at the national and regional levels in Latin America. These hurdles must be considered and addressed, in order to better shape the future of Latin American fracking in the coming years.

16.2   Innovation in the U.S. and its Diffusion into Latin America and around the World

Building on earlier work, during the 1940s Standard Oil and Gas Corporation refined the technique and a decade later, the first "official" application of fracking for energy extraction was undertaken in North America. Private energy consultancies such as Halliburton further developed and championed the technique in the 1960s and 1970s. From that point on, the technology gradually started to take off, usually as a collaboration between the government, the private sector and local

Table 16.1 U.S. Gas and oil productions (Modified from U.S. Energy Information Agency (EIA) (2013a, b)

ARI Report Coverage	2011 Report	2013 Report
Number of countries	32	41
Number of basins	48	95
Number of formations	69	137
Shale gas (trillion cubic feet)	6622	7299
Shale/tight oil (billion barrels)	32	345

communities (Manfrenda 2015; Montgomery and Smith 2010). Thus, the 1990s 'explosion' of hydraulic fracturing activities in the US actually built on decades of experimentation and technical innovation.

Shale gas basins are located all around the globe, with at least one formation in each continent (see Figure 1.1). The most recent assessments from the US EIA world assessment report on shale gas have identified 137 shale formations in the world, which account for a total volume of technically recoverable shale gas of an estimated 7299 trillion cubic feet (tcf) and 345 billion barrels (bbc) of oil (Table 16.1).

16.3   Ten Years of Fracking in Latin America

Roughly 80% of Latin America's energy is supplied by fossil fuels (coal, gas and oil), according to the International Energy Agency. As population grows, conventional sources of energy are gradually being exhausted, and the extraction of shale gas through fracking is seen as one of the solutions to ensure access to reliable energy resources in the coming decades to maintain the current lifestyle and consumption rates. Likewise, it could be an opportunity for countries that have stocks of unconventional sources (e.g. shale formations), to start exploring and developing unconventional energy resources. It also represents a geopolitical chance for companies (mainly U.S. ones) to implement this technique in Latin American countries and territories, which at the moment have assessed resource reserves, but relatively little domestic capacity to develop, thus offering a win-win deal for all the involved parties. In this context U.S. Department of Energy launched in 2010 the Unconventional Gas Technical Engagement Program (UGTEP) as a tool to promote knowledge sharing between the U.S. and other countries, to help them to successfully develop their shale resources (oil and gas), including the sharing of technical, regulatory, administrative and diplomatic expertise with all interested countries (Tincher 2015).

One particularity of the Latin American region, as compared to the U.S., is that oil and gas companies—some of the most profitable enterprises in these countries—are usually partially or totally owned by the State. This is the case for Argentina (YPF), Brazil (Petrobras), Colombia (Ecopetrol), Mexico (Pemex), Venezuela (PDVSA) and Ecuador (Petroecuador). This, along with some other factors, will be critical to the expansion of fracking in Latin America, since any increase in profits from this sector could also be beneficial for national budgets. This is important because due to the global economic crisis of 2008 and the 2014 drop in oil prices, there is a huge incentive to develop domestic reserves to address national trade imbalances.

The 2005–2014 period probably represented some of the best years in the history of the energy industry since oil prices reached on average US$100 per barrel of oil, higher than most break-even prices for producers around the world. The story in Latin America was similar, since the oil producing countries (with state-owned companies) were enjoying the biggest oil profits in their histories, and therefore were in a better position to be able to support national budgets at the time. In addition, fracking in the U.S. was seen and promoted as the key for energy self-sufficiency. Aware that there were important shale formations in the Latin American region, soon these national oil companies started exploring the possibilities for extracting shale resources using fracking. This was the case for Colombia, which in 2013 started to search for joint ventures with companies to start developing domestic fracking. However, these early moves towards fracking in the region had to be put on hold due to an external factor—the global drop in oil prices after 2014, which forced companies to focus only on the most profitable oil and gas resources, and therefore excluded fracking (non-conventional oil and gas fields) due to relatively high extraction costs at the time.

Today the economic scenario seems to be different, and there is a renewed willingness to start fracking in Latin America. Nevertheless, it would still take some years, since there is still uncertainty both on its technical feasibility and over its environmental and social impacts. In the following sections, the aim is to put a spotlight on the enabling environment for fracking in Latin America, comparing how favourable is the region for the development of the unconventional energy industry. The fact that the U.S. government has continued to strongly promote fracking on its national territory (though not without some opposition as other chapters in this volume demonstrate), together with the historical relationship between the US and Latin America has allowed them to position themselves as "ambassadors" of the technology in the region.

In order to reduce the costs, if the future energy outlook in terms of demand and supply remain the same, energy companies need to explore countries that have abundant shale gas reserves, and whose enabling environment (e.g. productivity and regulation costs) are favourable to the expansion of this technique. As an example of enabling environments, the cases of three Latin American countries will be briefly analysed (Argentina, Colombia and Mexico), as examples of the evolution of fracking in the region since the turn of the twenty-first century.

16.4 Argentina

Currently fossil fuels account for roughly 90% of the energy needs of Argentina (40% oil & 50% gas), making it extremely dependent on this type of resource. At the same time, Argentina has not been self-sufficient in hydrocarbon production, making the country vulnerable to the market fluctuations and instability in oil producing regions.

As also noted in Chap. 8 of this volume, Argentina has a potential for energy production from shale resources that is nearly unrivalled in Latin America. Four major basins have so far been identified: Neuquén, San Jorge, Austral-Magallanes (shared with Chile) and Paraná (shared with Uruguay, Brazil and Paraguay, see Map 8.1). The country has an estimated shale energy stock of around 27 billion barrels of oil and 802 trillion cubic feet of gas and is ranked among the top ten worldwide potential producers of shale gas (#2) and oil (#4), along with Mexico, China, Russian and the U.S. (EIA 2015).

The main shale play in Argentina to undergo exploration and development so far is located in the centre of country in Neuquén province. It is known as the Vaca Muerta play, and, according to Argentine President Mauricio Macri, it represents a new stage for the energy future of the country, since Vaca Muerta by itself could provide 38% of the natural gas and 60% of the oil in the country (Chequeado 2017). Like the U.S., Argentina is a country rich in shale resources, with the potential to go beyond self-sufficiency and become a net exporter of (unconventional) fossil fuels. As tantalising as this prospect may be, achieving anything like this promise will require a very large investment (as much as US$ 120 billion) from the joint venture created by Yacimientos Petrolíferos Fiscales (YPF) and Chevron. As of early 2019, at least this amount of inward investment had been pledged by major global energy companies such as Chevron, BP and Total.

The current enabling environment with Law 26.741–2012 sets the stage for Argentina's fuel independence, declaring the achievement of the Argentinian energy self-sufficiency to be of "national public interest" (Oxford Institute for Energy Studies 2016; Mares 2015). Moreover, the law also mandated the creation of the Federal Council for Hydrocarbons (Ministry of Justice and Human Rights of Argentina 2012) and required the state to renationalise YPF, the most important oil and gas company of the country and a big energy player in Latin America. As the analysis in Chap. 8 clearly shows, Argentinian leaders (if not all Argentinians!) sought to follow the U.S. in aggressively pursuing unconventional hydrocarbon development (Bertinat et al. 2014). By 2014, around 500 fracking wells to extract tight gas were reported in Argentina, a number that has likely increased since 2014. Given the advances in the country with this extractive technology, some organisations have demanded that more research on this area must be gathered (AIDA 2016).

In terms of regulation of the impacts of fracking, social perception and environmental protection, there are a number of mechanisms, laws, decrees at the national, regional and local level in Argentina aimed at developing a comprehensive legal framework to regulate the energy sector and promote the safe and environmentally-sensitive development of the fracking industry in the country. The overarching framework mechanism for legal environmental protection in Argentina is the General Environmental Law (Law 25,675 of 2012). The Ministry of the Environment and Sustainable Development (MAyDS in Spanish) is responsible of ensuring compliance at the level of national policy, and in specific operations anywhere in the national territory. The 25,675 Law "seeks to supply the minimum financial resources *to accomplish the sustainable and adequate management of the environment, preservation and protection of biological diversity and implementation of sustainable development"*. While Argentina is already performing fracking, one important weakness is that, even though Argentina has in place a general national law for environmental protection, it does not have a targeted fracking-oriented regulatory policy towards the prevention of its environmental and social impacts. As elsewhere around the world, hydraulic fracturing raises challenges that are not always well-accommodated in existing law, policy and regulatory practice.

Like in the U.S., the social perception of fracking in Argentina indicates a significant level of social concern, as reflected in several conflicts and protests that have happened in different parts of the country in recent years (one of these, in Nequen province, is discussed in Chap. 8). Over 60 municipalities, including Río Negro, Choele Choel, Beltrán, Chimpay, Cipolletti, Mendoza, Tunuyan and General Alvear and San Carlos, have passed local legislation or declarations banning fracking on precautionary grounds, It is uncertain if these local actions are to be considered legally binding, since the jurisdiction and competence over hydrocarbons is at the provincial or national (rather than local) level (El Sol 2013; Observatorio Petrolero Sur 2016).

The Argentinian government promotes unconventional oil and gas as "bridge (or "transition") fuels", capable of helping the country to become energy self-sufficient and to grow high wage employment throughout the energy supply chain. To date development has been hampered by a lack of transport and pipeline infrastructure, appropriately rigorous regulatory standards and – most importantly – social legitimacy amongst the population at large.

16.5   Mexico

Natural gas is fundamental to the energy matrix of Mexico. Based on data from the Mexican Commission on Hydrocarbons (CNH in Spanish), in 2017 Mexico was importing 81% of the natural gas used the country, implying a high level of vulnerability to the politics and economic vagaries of trade. Yet Mexico is also known as a world-class exporter of oil over recent decades, producing around 2.1 million barrels per day in 2016, accounting for roughly 30% of the total government income. However, production by the national oil company, Pemex (Mexican Petroleum), has shrunk since 2016 to 1.9 million barrels in 2017 and 1.7 million barrels in 2018. The government needs to find a way to re-establish its domestic energy production in order to maintain its exports, increase government income rates, and achieve energy independence (Export.gov 2017). In this sense, and given the current geopolitical scenario, fracking and the latest findings on shale resources in the country present themselves as a possible way forward.

In the last updated report from the U.S Energy Information Administration (EIA), Mexico is described as a country with "excellent potential for developing its shale gas and oil resources" (US EIA 2015). The EIA ranked the country among the top ten potential producers of shale gas (#6) and oil (#8), along with Argentina, China, Russian and the U.S., with a shale energy reserve of 13 billion barrels of oil and 545 trillion cubic feet of gas. The main shale formations are mostly located along the Gulf of Mexico, an in particular the Veracruz, Sabinas, Tampico, Tamaulipas basins. Also, the Eagle Ford Transboundary Shale (shared by the U.S. and Mexico) is located in the Burgos basin, and it is (as we have seen in other chapters in this volume) a major shale gas play. Its estimated potential yield, estimated at 343 trillion cubic feet of gas and 6.3 billion barrels of oil, dwarfs other Mexican energy basin estimates (EIA 2013a, b).

Oil and gas exploration in Mexico dates back to the early 1900s, and since the ratification of the Constitution of 1917, the state and/or its companies have been the owners and the only entities authorized to explore and extract underground resources such as fossil fuels. This was the scenario in 2013–2014 when, due to multiple factors including a decrease in fossil fuel production and the limited capacity of the Mexican state to explore and increase production rates of shale resources, the Mexican Congress decided to reform the energy sector to provide the enabling environment (investment, lower barriers and private sector participation) for multinational companies to work jointly together with Pemex to extract existing energy resources (Vinson and Elkins 2016). In 2014, within the framework of the Energy Reform in Mexico, the National Agency for Industrial Safety and Environmental Protection (Ansipa in Spanish) was created, to complement the capacity of the country to oversee and regulate industries and production standards.

One critical factor to be highlighted in the Mexican case is water availability. In the U.S., fracking consumes 9–29 million litres per well per year, equivalent to the annual water consumed by approximately 90,000 to 290,000 people (at a rate of 100 l daily per person). The Mexican Secretariat of Energy has estimated that each shale gas well developed in Mexico will require around 21 million litres of water (Posadas and Buono 2016). Mexico is rated to have high average water stress over the shale play area (Reig et al. 2014). In this regard, there is concern from civil society and the community that, with the massive implementation of fracking in the northern states of the country near the Gulf of Mexico, which are arid and semiarid region and thus naturally dry areas, this practice would increase water scarcity in a region which is already water scarce, causing shortages and irreversible consequences to the peoples and ecosystems of this sub-region (Mares 2015).

Water resource allocation for hydraulic fracturing also presents a potential tension with the adoption by Mexico, shortly before the energy reform in 2013, of a human right to access to water (Posadas and Buono 2015). The law adopting the new human right included a mandate to enact a new General Law on Water, offering an opportunity to shift water management paradigms in the country from one based primarily in hydraulic engineering to one that embraces a more sustainable development agenda). As of July 2019, the Congress has not enacted the new water law (Agua.org 2019).

As elsewhere, the negative popular perception of the social and environmental risks linked to fracking remain. Civil society and the government's concerns on the issue are similar to the ones of the other countries of the Latin America region, including the impacts from the increase on GHG emissions, to human health, and to the water quality of rivers and aquifers in and near fracking basins. In addition, some of the most important shale plays in Mexico (e.g. Burgos and Tampico) are located near indigenous and peasant agricultural communities, whose livelihoods could be greatly affected by fracking (Fundar 2017).

Relative to other parts of Latin America, Mexican civil society has probably had the most organized network of stakeholders against fracking, known as the *Mexican Alliance against Fracking*. This is a group of about 40 civil society organizations from across the country that seek to forbid fracking, with the ultimate goal of defending land and water resources. It is a very active group that is constantly organising technical seminars and papers, publishing magazines and press releases, and releasing campaigns and videos with celebrities to raise awareness on the potential threat of fracking to the people of Mexico and especially to its water sources.

Today, Mexican government policy depends on fracking to recover the fossil fuel production and to boost its economy. This has been reflected in the willingness of its government to start making reforms (2013–2014) and to create the enabling environment for engineering and scientific investments necessary to exploit Mexican shale plays. Nevertheless, the current geopolitics (especially current and future fossils fuel prices), social resistance, and the lack of capacity to fully implement fracking in Mexico still remain as big challenges, where further measures need to be taken in several sectors and the final success is yet to be determined.

16.6   Colombia

The Colombian case does not differ much from the cases of Mexico and Argentina presented above. The overall production and stock of oil and gas in the country has been declining in recent years whilst demand has been rising, creating a clear challenge to achieving energy security. Despite the efforts to increase extraction rates—especially oil—by the country and its main oil company (Ecopetrol), only limited inroads have been made into the production-consumption deficit. This is due to the decrease in oil prices over the last years (which made it difficult to invest in exploration and extraction), the exhaustion of current oil stock, and the lack of discovery of new reserves. Based on data from the World Bank's World Development Indicators, the oil rent in Colombia as a percentage of the GDP for the country was 8.5% in 2011, which declined to 6.46% in 2014 and 2.87% in 2016 (KAS 2016). At this point neither the government nor the sector wants to allow further decline in the energy industry, since this may bring consequences for the economy and its macro components, affecting everyone in the country. In this context, and as elsewhere in Latin America, fracking presents itself as the most likely vehicle for rescuing the energy sector and wider economy (National Agency of Hydrocarbons, or "ANH", 2016).

Based on data from the National Agency for Hydrocarbons, Colombia had identified 43 shale plays in the country by 2016. These major shale basins are located in the mid-upper Magdalena, Catatumbo (within the binational Colombia-Venezuela Maracaibo shale basin), Caguan and the Oriental mountain regions. The 2013 U.S. EIA report estimated total technically recoverable unconventional resources of 55 trillion cubic feet of shale gas and 6.8 billion barrels of tight oil for Colombia, figures that triples the amount of known oil reserves and increase by ten the known reserves of natural gas.

Colombia is one of the Latin American countries with the best-established framework for the implementation of non-conventional extractive practices. As the U.S. shale gas boom was taking off, Colombia started to make sure that all necessary arrangements for enabling shale exploration were in place. For example, the 2010–2014 National Development Plan, drafted by the executive branch and issued by the Congress, contemplated as a priority the development of the energy and mining sectors (including investing on exploration and extraction of shale resources). Moreover, in 2014, the Ministry of Energy and Mining and the ANH, after undertaking due diligence and consultations, established the "Technical requirements and procedures for the exploration and exploitation of hydrocarbons in unconventional wells" (Ministry of Energy and Mining, Colombia 2014). This defines all the parameters and standards to develop fracking projects in the country, including social and economic issues.

One differentiating element in Colombia (as compared to Mexico, Argentina, and others) is the role of the Attorney General's Office, the entity charged with ensuring that government actions are lawful. With regards to fracking, in 2012 the Colombian Attorney General released a special report based on the United Nations Precautionary Principle, requiring the Ministry of Environment and Sustainable Development, the Ministry of Mining and Energy, the ANH, and the National Authority for Environmental Licensing (ANLA in Spanish) to reassess and update any mechanisms and norms that foster fracking in the country, due to its potential risk for human and environmental health (water, animals, earthquakes, etc.). It also provided references to the government to consider or recall actions of ban and moratorium done by other countries on this issue.

In December 2015 the Colombian government signed a contract with two multinational companies (ConocoPhillips and Canacol Energy) to explore and exploit hydrocarbons from unconventional deposits through fracking in the northeastern provinces of Santander and Cesar. The first exploration well, Picoplata 1 in the upper Magdalena Basin, was completed in summer 2017 and showed promising results. After significant public opposition, in spring 2019 the Colombian government declined operator permits to the ConocoPhillips and Canacol joint venture, effectively shutting down hydraulic fracturing operations in the country.

Though Colombia is a relatively water-rich nation (average per capita availability of 45,000 m^3/year), there are areas of relative water scarcity, as well as areas where climate change impacts are being particularly keenly felt. One of these regions is the upper Magdalena Basin (southwest of Bogota), precisely the area where there is much current interest from shale gas and oil producers. Resource surveying and extraction (including deforestation) over the last generation have radically altered the regional hydrosphere, altering groundwater-surface water inter-actions and changing seasonal rainfall and humidity As a result communities have been impacted by greater vulnerability to flood and drought, with consequent impact on livelihoods and well-being. It is by no means clear that areas currently being considered for hydraulic fracturing actually have the necessary volumes of water. As of mid-2019 the High Court of Colombia was maintaining its moratorium frack-ing on environmental grounds.

The social perception of fracking in Colombia is similar to other countries in the region, even though social mobilization against fracking in Colombia is still incipi-ent. It was only in 2017 that civil society groups got together and discussed the creation of the group "Alliance: Colombia Fracking-Free", which was expected to be formally established in 2019, composed by nearly 70 organizations representing different concerns around fracking. Additionally, and in contrast to other countries in the region, the guarantees and participation mechanisms for civil society in Colombia are very strong with respect to the extractive industries (including frack-ing). This is, for example, the case of the popular consultations done with the support of citizens in the municipalities of Piedras de la Cruz, which ended with a prohibition of mining activities.

However, with regard to oil production and energy independence, Colombian politicians, like their Mexican and Argentinian counterparts, are increasingly seeing unconventional hydrocarbon exploitation in terms of domestic energy security as well as job creation. The national oil company of the country Ecopetrol expects that by 2019 fossil energy reserves could be tripled by means of fracking. This has been reflected in the willingness of the government in setting the regulatory framework for enabling shale plays in the country. Nevertheless, the current geopolitics (espe-cially current and future fossils fuel prices), further research on impacts and resources availability for Colombia, and future development of social positioning against current fracking activities remain as big challenges, and the long-term implementation and results of the technique are yet to be known.

16.7   Enabling and Regulating the Environment for Fracking in Latin America: Challenges and Opportunities

Latin America is a region that has historically been highly dependent on natural resources exploited by nationalised energy companies. Historically this model has worked well, and even with the strong economic shocks from the 2014 drop in fossil fuel prices, countries such as Mexico, Argentina, Colombia, Ecuador and Brazil are still firmly committed to energy independence and energy-led economic develop-ment. As in other shale gas and oil rich countries however, technical possibility and government enthusiasm does not necessarily translate into public acceptance. On the contrary, it seems that the greater the level of political enthusiasm for fracking there is in Latin American countries, the greater the level of opposition in civil society. Thus, there are several challenges for development and implementation of fracking at the national and regional level in Latin America that must be considered and addressed in the future.

One of the main arguments in favour of fracking is that the technology will help accelerate the energy sector across the world to transition from higher carbon (coal and conventional oil) to lower carbon energy sources, with GHG reductions of 50% touted even in the short term. This claim of course depends on total fossil fuel-related energy consumption remaining flat or even declining – an argument that goes out the window if the prospect of shale gas is actually used to accelerate fossil fuel consumption. It is also the case that the potential for environmental harms to local land and water resources (and therefore other forms of local livelihoods) in Latin American shale basins is as yet poorly understood. Therefore, what follows is a series of recommendations.

- First, it is very important to promote more research and public debate on the economic, environmental and other costs and benefits of fracking in the Latin American region, as this will help to build both the knowledge base and public understanding. Currently, the lack of publicly-available data about energy operations combined with the top-down nature of decision-making in most Latin American countries seems a recipe for popular protest.
- Second, and linked to the above, assessment of water needs in the areas surrounding fracking wells is limited. Latin American countries should adhere to IWRM principles, particularly around integrated assessment of development applications and (meaningful) public consultation prior to initiating or licencing potentially impactful hydraulic fracturing activities.
- Third, regulation is very limited in the region. Financing and development of mechanisms, institutions, laws (national and local), personnel and know-how (especially on environmental criminal law) should be a priority in the countries implementing fracking in the Latin American region. Key to strong regulation will be adherence to the "precautionary principle" (UNESCO COMEST 2005).
- Fourth, exchange of experiences (common approaches to common challenges) is important amongst energy and environmental regulators and activists throughout the region. Countries from the region doing or considering fracking should promote the creation of a hemispheric task force for sharing and knowledge and to collect practical experiences on fracking, their experiences and approaches, policies, management challenges, and community engagement.
- Fifth, the distribution of benefits of hydraulic fracturing activities between the private and public sectors needs to be much more transparent and accountable. Countries should develop policies and systems (tax revenues, programs or others) which seek a proper and equitable distribution of income from fracking activities, designed to compensate stakeholders and governments accounts accordingly.
- Sixth, because fracking is expensive, it is therefore crucial that countries work on the promotion of integrated economic investment opportunities such that local

employment is appropriately developed, creating the enabling environment and guarantees for investors to finance these sorts of projects.

- Seventh, transboundary Shale Basins need a transboundary approach to exploitation and regulation. In the Latin American region, several shale basins are shared among two or more countries (i.e. Paraná and Austral Basins). While basins as geographical units do not recognize borders, conjunctive assessments (multinational), management and use of its resources, needs to be agreed on these areas.
- Eighth, particular care needs to be taken where energy exploitation activities may impact on indigenous communities, given the fact that these communities have historically been entirely excluded from the political process.

As other chapters in this volume have shown, appropriately and effectively regulating hydraulic fracturing is a challenge for all countries with potentially-exploitable resources, whatever their pre-existing level of political and economic development.

References

AIDA (2016) Principio de Precaución: Herramienta Jurídica ante los Impactos del Fracking. Greenprint, Ciudad de México. Available at: https://aida-americas.org/sites/default/files/publication/publicacion_fracking_aida_boell_0.pdf

Agua.org (2019) "La Nueva Ley General de Aguas" (19 July) Available at: https://agua.org.mx/analisis_integral/la-nueva-ley-general-de-aguas/

Bertinat P, D'Elia E, Observatorio Petrolero Sur, Ochandio R, Svampa M, Viale E (2014) 20 Mitos y Realidades del Fracking. Editorial El Colectivo. Available at: www.opsur.org.ar/blog/wp-content/uploads/2015/06/2014-20-Mitos-Final.pdf

Chequeado (2017) Macri: "Vaca Muerta representa la 2° reserva mundial de gas y la 4° de petróleo no convencional". Available at: https://chequeado.com/ultimas-noticias/macri-vaca-muerta-representa-la-2o-reserva-mundial-de-gas-y-la-4o-de-petroleo-no-convencional

EIA (2013a) North America leads the world in production of Shale gas. Available at: www.eia.gov/todayinenergy/detail.php?id=13491

EIA (2013b) Technically recoverable shale oil and shale gas resources: an assessment of 137 shale formations in 41 countries outside the United States (Updated 2015). Available at: www.eia.gov/analysis/studies/worldshalegas

EIA (2015) World shale resource assessments. Available at: https://www.eia.gov/analysis/studies/worldshalegas/

EIA (2017) Energy explained. Available at: https://www.eia.gov/energyexplained/index.cfm?page=natural_gas_where

El Sol (2013) General Alvear: la presión social revierte un veto que anulaba la ordenanza contra el fracking. Available at: www.elsol.com.ar/general-alvear-la-presion-social-revierte-un-veto-que-anulaba-la-ordenanza-contra-el-fracking.html

Export.gov (2017) Mexico – oil and gas. Available at: www.export.gov/article?id=Mexico-Upstream-Oil-and-Gas

Fundar, Centro de Análisis e Investigación, AC (2017) Las Actividades Extractivas en Mexico Estados Actual. Available at: http://fundar.org.mx/mexico/pdf/Anuario2016corr.pdf

KAS (2016) The geopolitics of oil and gas: the role of Latin America. Available at: www.kas.de/c/document_library/get_file?uuid=dec202ba-72f8-4793-2bc7-f65236b972fd&groupId=252038

Manfrenda J (2015) The origin of fracking actually dates back to the Civil War. Business Insider. Available at: www.businessinsider.com/the-history-of-fracking-2015-4?r=US&IR=T

Mares DR (2015) Shale gas in Latin America: opportunities and challenges. Energy Policy Group of the Inter-American Dialogue. Available at: www.thedialogue.org/wp-content/uploads/2015/03/MaresShaleGasforWebposting.pdf

Ministry of Energy and Mining, Colombia (2014) Resolucion 90341 DE – Por la cual se establecen requerimientos técnicos y procedimientos para la exploración y explotación de hidrocarburos en yacimientos no convencionales. Available at: www.minminas.gov.co/documents/10180/23517/22632-11325.pdf

Ministry of Justice and Human Rights of Argentina (2012) Law 26741. Available at: www.sec.gov/Archives/edgar/data/904851/000095010312002488/dp30521_6k.htm

Montgomery C, Smith M (2010) Hydraulic fracturing: history of an enduring technology. J Pet Technol 62(12). https://doi.org/10.2118/1210-0026-JPT

Observatorio Petrolero Sur (2016) Un año de victorias en la lucha contra el fracking. Available at: www.opsur.org.ar/blog/2016/12/26/un-ano-de-victorias-en-la-lucha-contra-del-fracking

Oxford Institute for Energy Studies (2016) Unconventional gas in Argentina: will it become a game changer? Available at: www.oxfordenergy.org/wpcms/wp-content/uploads/2016/10/Unconventional-Gas-in-Argentina-Will-it-become-a-Game-Changer-NG-113.pdf

Posadas A, Buono R (2016) Looming conflicts? Energy reform priorities and the human right of access to water in Mexico. In: Payan T, Zamora SP, Ramón Cossío Díaz J (eds) Estado de Derecho y Reforma Energética en México. Mexico, Tirant lo Blanch

Reig P, Luo T, Proctor J (2014) Global shale gas development: water availability and business risks. World Resources Institute, Washington, DC

Tincher G (2015) The unconventional gas technical engagement program how to ensure the U.S. shares its experience in a socially and environmentally responsible manner. Energy Law J 36:1

UNESCO COMEST (2005) The precautionary principle. Available at: https://unesdoc.unesco.org/ark:/48223/pf0000139578

Vinson & Elkins (2016) Shale & fracking tracker: Mexico. Available at: www.velaw.com/Shale%2D%2D-Fracking-Tracker/Global-Fracking-Resources/Mexico

Andrés Sánchez is an Environmental Economist from Los Andes University of Colombia. He currently works in the Department of Sustainable Development (DSD) of the Organization of American States (OAS), where he serves as the coordinator of the Global Environmental Facility (GEF) projects portfolio executed by the OAS in coordination with beneficiary countries and entities such as UN Environment, United Nations Development Programme (UNDP), Latin American Development Bank (CAF), World Wildlife Fund (WWF), and the Inter-American Development Bank (IDB), among others. He works with the 34 member countries from Latin America and the Caribbean on matters related to integrated water resources management and water diplomacy, regional development and integration, sustainable cities, climate change, energy, biodiversity and land degradation. Within his interests and experience are environmental and economic management, energy production, climate change, social development, policy analysis and the development of multidisciplinary frameworks towards environmental management. Sánchez holds a master's degree in environmental sciences and policy from Johns Hopkins University, where he developed research projects to analyze the impacts of hydraulic fracturing in the U.S. and Latin America.

Chapter 17
The Disposal of Water from Hydraulic Fracturing: A South African Perspective

Loretta Feris and W. R. (Bill) Harding

Abstract Shale gas extraction poses significant risks to scarce groundwater resources in the semi-desert Central Karoo region of South Africa. Review of hastily-compiled Environmental Management Plans, prepared in support of exploration bids, has been scathing. A subsequent Strategic Environmental Assessment did little to assuage concerns about environmental harm or that a regime of regulatory governance, equal to the task, existed at all. In the face of the clear, evident and largely unpredictable challenges, the legal and regulatory tools and experience available for the development of shale gas extraction in South Africa are neophytic at best. There are no existing norms and standards that would transition comfortably into this environmentally-challenging arena. What is needed is a considerable body of further scientific investigation, possibly in parallel with closely controlled, open and transparent pilot-scale drilling and fracturing.

Keywords South Africa · Karoo · Groundwater · Shale gas extraction · Governance

17.1 Introduction

This chapter focuses on the regulatory framework for the disposal of water from hydraulic fracturing in the context of the proposal to undertake shale gas development (SGD) in the Karoo region of South Africa.[1] The South African Karoo is an

---

[1] Between 2008 and 2010 a number of oil companies (Royal Dutch Shell, Falcon, Sasol and Bundu—90% owned by Australian company Challenger Energy- and Sungu Sungu Exploration and Development, a South African company) lodged applications for Technical Cooperation

---

L. Feris (✉)
Office of the Vice Chancellor and Faculty of Law, University of Cape Town,
Cape Town, South Africa
e-mail: Loretta.feris@uct.ac.za

W. R. Harding
DH Environmental Consulting (Pty) Ltd, Somerset West, South Africa

© Springer International Publishing AG 2020   345
R. M. Buono et al. (eds.), *Regulating Water Security in Unconventional Oil and Gas*, Water Security in a New World,
https://doi.org/10.1007/978-3-030-18342-4_17

almost mystical place, a vast expanse of pastoral solitude and spectacular sunsets over nature's sculptures, counterpoised against extremes of semi-desert aridity and enervating heat. Small towns dot the landscape here and there and, for decades, resolute farmers have eked out groundwater-dependent livelihoods based, for the most part, on farming sheep and Angora goats. Droughts are common. In June of 2016 the Central Karoo was declared a drought disaster area, a situation that is likely to continue well into 2017 (Western Cape 2016a, b). It is an ancient landscape, in geological history once a vast inland sea. Sense of place is almost tangible and deeply invested in the well-being of the regions inhabitants. It is a unique area, a globally-recognized hotspot of biodiversity importance—the worlds' only arid hotspot—and part of the Succulent Karoo ecosystem that extends from the Eastern Province of South Africa, well into Namibia to the northwest (Critical Ecosystem Partnership Fund 2003). The environment is largely untransformed and fragile to anthropogenic perturbations. Environmental protection is, therefore, challenging and of paramount importance. The feasibility of shale gas development (SGD), therefore, an activity in which South Africa has no experience whatsoever, presents a novel challenge—both in terms of the technological challenges and the environment in which it may take place. As such it is pertinent here to consider the corresponding regulatory framework.

South Africa's environmental regulatory regime is premised on a number of important principles as expressed in the country's framework legislation, the National Environmental Management Act (NEMA) (South Africa 1998b), as well as its water management legislation, the National Water Act (NWA) (South Africa 1998a). Of particular important is the sustainable development principle, captured in sections 2(3) and 2(4) of NEMA (South Africa 1998b: ss2(3)-(4)). This principle is reiterated in the Mineral and Petroleum Resources Development Act (MPRDA) which defines this principle as "the integration of social, economic, and environmental factors into planning, implementation and decision making so as to ensure that mineral and petroleum resources development serves present and future generations" (South Africa 2002: chap1(1)). The sustainable development principle provides the backdrop against which SGD must be considered, in particular in the context of impacts on water resources, it requires a consideration of the ways in which water use and potential pollution of water resources will impact on the Karoo, its people and on the economic activities that are currently pursued in that area.

Water is indisputably the main environmental concern associated with SGD (Esposito 2013). The NWA makes it clear that the national government is the public trustee of water resources and, as such, the Minister must ensure "that water is protected, used, developed, conserved, managed and controlled in a sustainable and equitable manner for the benefit of all persons and in accordance with its constitutional mandate" (South Africa 1998a: s3(1)) It is trite that in order for water resource protection to succeed, law should follow science. Water pollution, for example,

---

Permits with the Department of Mineral Resources (DMR) under the Minerals and Petroleum Resources Development Act (MPRDA) and in 2010 Royal Dutch Shell, Falcon and Bundu applied for shale gas exploration licences in the Karoo under the MPRDA.

cannot be effectively regulated unless there exists scientifically-substantiated understanding regarding the behaviour and persistence of a specific pollutant in a particular water resource. A Strategic Environmental Assessment (SEA) concluded late in 2016 reveals a number of uncertainties with respect to SDG in the Karoo, including the distribution and magnitude of the gas resource and whether it can be extracted at economically viable rates, as such leading to uncertainties with respect to the impact on the environment (Scholes et al. 2016). It is also trite in environmental law that, where the supportive science is lacking, a risk-averse, precautionary approach should prevail. NEMA thus encapsulates in its description of sustainable development important international environmental law norms such as prevention, (South Africa 1998b: ss2(4)(a)(i) and (iii)) polluter pays (South Africa 1998b: s2(4)(p)) and precaution (South Africa 1998b: s2(4)(a)(vii)) which will need to guide decision makers with respect to SDG in the Karoo.

The disposal of water used in SGD presents, in broad terms, two pollution scenarios, spills or leakages of contaminated water at the surface, either at the wellheads or in transit to or from, or underground as a result of hydraulic fractures or well-casing failures intersecting aquifers (Vidic et al. 2013). The proposed SGD in South Africa could extend over a vast spatial footprint, equivalent to 20% of the surface area of South Africa (244,000 km^2), much of which is arid, sparsely-populated and poorly-resourced with infrastructure that could support an SGD industry. Such a large area indicates that a vast and fragmentary network of drilling pads, roads and pipelines will need to be created in an arid area that is biologically-unique and in which the actual presence and quantity of extractable shale gas reserves remains an unknown factor (Cole 2014; Scholes et al. 2016). Such a vast network compounds the risk of contamination of scarce surface and/or groundwater resources. This chapter considers the disposal and treatment of water used in the process of SGD and assesses whether the current regulatory framework is able to respond thereto.

17.2   Disposal of SGD Wastewater

South Africa is very familiar with a legacy of mining harms, ranging from myriad social injustice issues to acid mine drainage (AMD) (Feris 2012: 1), the latter producing an estimated 350 megaliters of effluent per day (Costello 2012), extending spatially over a large area of the country, albeit significantly smaller than the area in which SGD is currently being considered. There are some similarities with respect to mining effluent treatment challenges, in particular the contamination of AMD flows with naturally-occurring radioactive materials (NORMS). Fracking fluids, however, pose a relatively more complex pollution threat and effluent treatment challenge.

17.2.1 How Much Water Are We Referring to?

The volume of contaminated wastewater arising from a single SGD well is of itself substantial, even before the aggregate use from hundreds, and potentially thousands, of individual wells is considered. Average water use per well in the USA has been determined as 9.2 megaliters (Jackson et al. 2015: A). In more digestible terms, Fig quotes a volume of 20–25 megaliters per extraction event, this requiring some 1667 tanker vehicles to access the site (Fig 2012: 27). An almost equivalent number of vehicles would be required to truck the polluted wastewater to treatment facilities. Neilan and Dooley make an even clearer analogy, equating the volume needed for a single fracturing to the combined displacement of 834 of the world's largest cruise-liners (Neilan and Dooley 2014: 252). The multipliers in an equation of many wells are thus substantial, a scenario that suggests, *inter alia*, the need for a dedicated 'tanker' lane to be added to existing major arterial routes. The precise numbers are irrelevant other than that they are very substantial. More relevant is that "the returned fluid is of poor quality, cannot be reused without substantial treatment and is returned in volumes that can be substantial over time" (Sullivan et al. 2015: 5, citing Gregory et al. 2011; Maloney and Yoxtheimer 2012; Wilson and Van Briesen 2012). A large portion of residual, untreatable wastewaters will, effectively, be removed permanently from the watercycle—which, in effect, is what happens to wastewaters 'managed' by deepwell injection. Hardly a sustainable solution. Regulations already in place in South Africa preclude the option of disposal via deepwell injection, the most common form of disposal in the USA (Nicot et al. 2014). Existing regulations have also disallowed the use of effluent ponds at wellhead sites, requiring that all forms of process liquids, including those in the 'waste' stream, be held in above ground storage tanks, the latter situated in bunded containment areas (Oelofse et al. 2016). While this would substantially offset the leakage commonly associated with lined ponds, the economic feasibility of such a constraint remains to be determined.

17.2.2 Treatment of SGD Effluents

The Karoo has no surplus water available to support SGD (Hobbs et al. 2016). The proposed extraction of shale gas (SGD) will, therefore, require vast amounts of water to be imported and deliberately polluted with a combination of introduced chemicals, combined with those naturally-occurring chemicals eluted during the hydraulic fracturing process.[2] The recovered ('flowback & produced')[3] water will

---

[2] The USEPA has published a list of 1173 chemicals used in the SGD industry, of which 1076 are used in fracturing fluids and a further 134 detected in the flowback water (U.S. Environmental Protection Agency 2016).

[3] Flowback and produced water encompass, respectively, the return of injected fluid and water produced from the formation to the surface, and subsequent transport for reuse, treatment, or disposal.

require treatment before reuse is possible, if at all. Short-term treatment options could be deployed at the wellheads, while a potentially more efficient, centralized longer-term option would require an additional set of extensive pipelines connecting wellheads to dedicated treatment works—which could provide a source of water for new wells.

Precedents for flowback water treatment abound in countries such as the USA, yet their local applicability will require comprehensive investigation and substantiation. The ability to gauge the efficacy of treatment processes will require the establishment of an associated network of dedicated laboratories and appropriately-skilled personnel.[4] Treatment of SGD flowback water will be necessary irrespective of the original source thereof, be it fresh or seawater (imported from outside the region) or brackish (from boreholes within the region). The ultimate level of treatment will likely devolve to an uneasy playoff between environmental and economic interests. Treatment of polluted water arriving back at the surface is theoretically possible to any desired quality, the latter being a function of technology and cost. Any treatment will, however, leave behind residual quantities of noxious filtrates and retentates that will require disposal at an appropriate waste disposal site, as this is likely to be classified as Type 1 Hazardous Waste in terms of Section 7(3) of the National Norms and Standards for the Assessment of Waste for Landfill Disposal (South Africa 2013a: s7(3)). An additional, yet infinitely easier to manage source of water pollution will be that from sewage and greywater effluents originating from the SGD workforce.[5] While again presenting an additional cost item, very efficient 'package' wastewater treatment units are readily available for installation at individual work camps and drilling platforms.

Treatment of flowback/produced water from SGD wells will, of necessity, need to be thorough—if for no reason other than that the chronic toxicity (human health) values for 87% (Yost et al. 2016b) of the 1173 chemicals (including some banned substances) identified in the makeup of fracking fluid have not been determined (Yost et al. 2016a). While a median of some fourteen chemicals are typically employed at individual wells (Yost et al. 2016a: 4791) the absence of exposure risk data is a major knowledge gap in scientific understanding and the extent to which it can underpin regulation of the treatment and reuse of flowback water. While the percentage by volume of chemicals added to fracturing fluids may appear low, these fluids may contain chemicals that pose reproductive and developmental impacts, can be neurotoxic or carcinogenic and, in the case of trace metals and volatile organics, toxic (Coram et al. 2014). The risks are compounded by unknowns associated with the behaviour of mixtures of chemicals and the creation of transformation products (Elsner and Hoelzer 2016). Additionally, fracturing fluids contain biocides to offset microbial growth that could clog the fractures, corrode pipelines or

---

[4] Extant South African water quality regulations do not incorporate any of the potential health-threatening additives used in fracturing fluids. Appropriate analytical skills do not exist outside of the existing corporate petroleum industry.

[5] Here it should be noted that extant wastewater regulations do not provide an adequate measure of protection against eutrophication of surface waters.

produce toxic hydrogen sulphide (Kahrilas et al. 2015). Some of these may biodegrade rapidly, others may become more toxic or have a variable response in waters of different quality, for example salinity. As such, precaution would suggest that SGD flowback waters be subjected to the most rigorous (and costly) form of reverse osmosis purification to provide water safe for potable, stock-watering or crop production use (McLaughlin et al. 2016). Monitoring the removal of so many chemicals will require, at minimum, gas- and/or liquid-chromatographic mass spectrometry (GCMS/LCMS) and the derivation and use of proxy indicators and tracer chemicals, the attenuation of which will be equivalent or better for all other chemicals of concern. In order to comply with South African regulations, operators will be required to disclose the composition of their fracturing fluids—something that is likely to be contentious in the eyes of the public, yet utterly necessary if human and animal health and the environment are to be protected (Scholes et al. 2016: Sect. 6.1.1; Cramer 2016).

An additional and negative consequence of fracking will probably arise from spills and residual accumulation of contaminated effluents at wellhead sites—a reportedly common occurrence in the USA. The waste management section of the Strategic Environmental Assessment for SGD in South Africa notes in its introduction that "[t]he release to the environment of mining waste can therefore result in profound, generally irreversible destruction of ecosystems" (Scholes et al. 2016: Sect 6.1), a consequence that would mostly likely be aggravated in the fragile, arid Karoo environment.

17.2.3   Threats to Groundwater

Treating flowback water is, arguably, the easier part of the threat to water resources posed by the use and disposal of water in SGD. The more insidious and potentially devastating consequence is the pollution of groundwater resources by that portion of fracturing fluids that do not return to the surface and which may somehow migrate through various geological strata in an unpredictable fashion.[6] This is an especially significant concern in arid areas where water availability for human and animal use is sourced from boreholes (Lautz et al. 2014). It is here more than anywhere that developed scientific guidance is lacking. It is here also where a potentially disastrous and unnecessary legacy of mining harm could well be created in the absence of a thorough, pre-SGD appreciation of the structural geology and its undoubted, spatially-variable nuances. For example, induced fractures can extend 600 m upwards—whereas deeper wells in the kilometer range, with similar horizontal distances, could arguably induce a highly variable suite of geological disturbance and potential flowpaths for fracturing fluids to escape through. Assuming that there is an adequate separation between fractured zones and overlying aquifers, the evidence

---

[6]The risk of pollution of groundwater resources is compounded by the extreme depths that SGD will require in the Karoo.

to-date suggests that the contamination risk is very low (Vengosh et al. 2014). However, as reports indicate that as much as 90% of the introduced fluids may not be returned to the surface, it is important to have a clear understanding of where such volumes might end up (Vidic et al. 2013).[7]

By contrast, failure of the well-casing, where it passes through aquifers, presents a recognized direct connection through pollutants may pass. Against this is the observation that "… no groundwater contamination by hydraulic fracturing additives has been irrefutably document in the peer-reviewed literature" (Kahrilas et al. 2015: 26). This statement remains, however, highly contentious in the face of numerous anecdotal, documentary and other reports indicating quite the opposite (Vengosh et al. 2014: 8336 et seq). Many instances of litigation have been and continue to be instituted around this aspect in the USA indicating the existence of real concerns (Nicholson and Dillard 2013). The findings of the long-awaited USEPA report on SGD impacts on drinking water resources, which concluded that the SGD process had not led to "…widespread, systemic impacts on drinking water resources…" (U.S. Environmental Protection Agency 2016) have subsequently been drawn into question by accusations of report manipulation (Tong and Scheck 2016; Chow 2016). Bottom line is that much more investigation is required, benchmarked by extensive pre-drilling investigations of groundwater presence, linkages and quality. From a regulatory perspective, South Africa has yet to develop statutory protections for groundwater resources. Aquifers are not considered as watercourses (which are specifically protected in terms of the National Water Act), even in the vast alluvial fan aquicludes that occur in many river basins. This has obvious implications for the migration of SGD pollutants in underground waters.

From the foregoing it will be evident that the scientific demands necessary to protect water resources against SGD-derived contamination are far from trivial and, as yet, entirely unknown or understood in South Africa. The likely substantial costs associated with pre-SGD development research, the establishment of analytical resources and associated skills or the costs of routine operational and post-operational monitoring have not been included in the economic assessments compiled to-date for SGD in South Africa. None of the economic assessments to-date have addressed the environmental costs of SGD (Wait and Rossouw 2014; Econometrix Pty (Ltd) 2012). A review of an industry-funded economics assessment declared it to be inadequate support for decision making (De Wit 2013).

17.3   Legal and Regulatory Instruments

The disposal of water used in the process of SGD is regulated by four sets of legislation and regulations. First, the National Environmental Management Act (NEMA) (South Africa 1998b) is the overarching framework regulations that regulates the management of the environment broadly and provides *inter alia* for the integrated

---

[7] The SEA reports a range of between 10% and 80% (Hobbs et al. 2016).

environmental management (EIA) process, as well as compliance and enforcement. Second, the MPRDA (South Africa 2002) regulates the minerals and petroleum industry. While the Act was written with off-shore exploration of gas as its focal point (Du Plessis 2015: 1441), the definition of 'petroleum' includes a reference to gas and as such the Act applies to SGD. Third, water resources are regulated under the NWA (South Africa 1998a) whilst some aspects of waste water regulation falls under the National Environmental Management: Waste Act (South Africa 2008). In December 2014 a so-called "One Environmental System" for mining came into effect.[8] In terms of this system all decision-making with regard to mining-related matters, including environmental matters resides with the Minister of Mineral Resources (air quality and water is excluded, however) with an appeal to the Minister of Environmental Affairs on environmental decisions taken by the Minister of Mineral Resources.

17.3.1   Environmental Impact Assessment

As noted above, the NEMA provides for Environmental Impact Assessment (EIA) and as part of obtaining an exploration or production right under the MPRDA, an environmental authorisation must be obtained (South Africa 2002: s5A). The EIA regime provides for a process that assesses 'the potential consequences for or impacts on the environment of listed activities or specified activities' (South Africa 1998b: s24(1)). There are currently two lists of activities that may not commence without an environmental authorisation. The first lists of activities require 'basic assessment' (South Africa 2014a) and is reserved for activities where impact on the environment may be less severe while the second lists activities requires 'scoping and environmental impact reporting' (South Africa 2014b) and is reserved for those activities that impose more serious, unpredictable impacts on the environment and contains listed activities that would be triggered by SGD. It is patently clear that SGD will require the most stringent form of licensing, with the licensing requirements informed by comprehensive regulations. The key role of the EIA process will be to evaluate whether or not SGD may be undertaken in areas of especial environmental sensitivity—a consideration that may apply to most of the deemed exploration area.

The EIA Regulations also require that the environmental reports related to exploration, production and primary processing of a mineral and petroleum resource must address the requirements set out in the Financial Provision Regulations (South Africa 2015c). These regulations provides for the determination and establishing of financial provision for rehabilitation and post-closure management of SGD. The financial provision is designed to address rehabilitation, decommissioning and

---

[8]This was in terms of a process that started in 2008 which required the amendments of several pieces of legislation, most importantly the NEMA and MPRDA. In essence sections dealing with environmental issues in the MPRDA were repealed and reinserted into the NEMA (Department of Environmental Affairs 2014).

closure activities, and remediation and management of latent or residual environmental impacts (South Africa 2015c: reg 5). The applicant must ensure that the financial provision is, at any given time, equal to the actual costs of implementing three distinct, yet inter-related plans: the annual rehabilitation plan; the final rehabilitation, decommissioning and mine closure plan; and the environmental risk report (for latent and residual impacts) and must be available for a period of at least 10 years (South Africa 2015c: regs 6–7). This financial provision is particularly important with respect to the latent impacts of flowback water on groundwater.

Of further importance to the EIA process are the Regulations for Petroleum Exploration and Production issued in June 2015 (South Africa 2015a).[9] These regulations reiterate that an applicant must obtain an environmental authorisation in terms of NEMA as part of the process of applying for an exploration or production right (South Africa 2015a: regs 86(2) and 110(1)). The Regulations for Petroleum Exploration and Production go beyond the NEMA provisions and prescribe additional assessments that have to be undertaken as part of the EIA process. These assessments include an assessment of the geology and geohydrology of the area that would include in-depth analysis of exploration boreholes, as well as groundwater monitoring and deep groundwater investigation (South Africa 2015a: reg 87). In addition, it requires a hydrocensus of the "potentially affected water resources within a 3 kilometer radius from the furthest point of potential horizontal drilling," as well as a full water monitoring report (South Africa 2015a: reg 88). With respect to groundwater, it requires that 'groundwater monitoring (area covered, duration of monitoring, watercourse) and deep groundwater investigation' be specified and that groundwater be monitored on a continuous basis (South Africa 2015a: regs 88(2)(g) and 88(10)). These assessments addresses to some extent concerns raised with respect to the need for prior benchmarking. The broader concern lies, however, with respect to the so-called One Environmental System outlined above that makes the DMR as opposed to the Department of Environmental Affairs responsible for issuing an Environmental Authorisation and ensuring monitoring and compliance thereof.

The Regulations for Petroleum Exploration and Production also attempt to address the concerns raised with respect to fracking fluids. First, it lists in an annexure substances that will not be allowed as additives to fracturing fluids (South Africa 2015a: annexure 1). Second, it requires the submission of a long list of information with regard to the fluids that the applicant intends to use (South Africa 2015a: reg 109), including:

Fluids and their status as hazardous /non -hazardous substances; material safety data sheet information; volumes of fracturing fluid, including proppant, base carrier fluid and each chemical additive; the trade name of each additive and its general purposed in the fracturing process; each chemical intentionally added to the base fluid, including each chemical, the chemical abstracts service number, if applicable and the actual concentration, in percent by mass; possible alternatives; possible risk of the above on the environment and water resources and remediation required if a pollution incident were to occur.

---

[9]A review of draft regulations found them to be deficient in a number of key aspects (Treasure Karoo Action Group 2015).

Third, it requires a risk assessment and risk assessment report to the designated agency and to the competent authority as part of the application for Environmental Authorisation (South Africa 2015a: reg 114), as well as a risk management plan for each well to be fractured (South Africa 2015a: regs 155(1) and (2)). Fourth, the regulations require that the applicant must as far as possible and "to the extent it is technically feasible, maximise the use of environmentally friendly additives and minimise the amount and number of additives" (South Africa 2015a: reg 115(3)(a)). The regulations furthermore require motoring and reporting of type and volumes of water sourced for stimulation operations; volumes and rates of fracking fluid pumped into the target zone; and volumes and rates of flowback received during and after each stimulation operation (South Africa 2015a: reg 112(8)(j)). Management, transportation and storage of flowback water are also addressed (South Africa 2015a: regs 116–119).

17.3.2 The Regulatory Framework for Water Use and Protection of Water Resources

The NWA regulates water resources and the use of water both of which is relevant in the context of SGD. Despite the One Environment System for mining the Minister of Water and Sanitation remains the key authority under the NWA for purposes of water use authorisations and water resource protection measures associated with SGD. With respect to water use section 21 of the Act is key as it defines in a descriptive manner what constitutes a 'water use' under the NWA (South Africa 1998a: s21). They are:

(a) taking water from a water resource[10];
(b) storing water;
(c) impeding or diverting the flow of water in a watercourse;
(d) engaging in a stream flow reduction activity contemplated in section 36;
(e) engaging in a controlled activity identified as such in section 37(1) or declared under section 38(1);
(f) discharging waste or water containing waste into a water resource through a pipe, canal, sewer, sea outfall or other conduit;
(g) disposing of waste in a manner which may detrimentally impact on a water resource;
(h) disposing in any manner of water which contains waste from, or which has been heated in, any industrial or power generation process;
(i) altering the bed, banks, course or characteristics of a watercourse; (j) removing, discharging or disposing of water found underground if it is necessary for the efficient continuation of an activity or for the safety of people; and (k) using water for recreational purposes.

---

[10] 'Water resource' is defined comprehensively in section 1 to include a watercourse, surface water, estuary or aquifer (South Africa 1998a: s1).

A cursory reading of the list above indicates that the activities triggered by SDG include a number of those listed in the section, *i.e.* taking and storing water, discharging waste or water containing waste, disposing of waste that may detrimentally impact on a water resource, disposing of water containing waste and disposing of water found underground if required for the efficient continuation of an activity. The Minister of Water and Sanitation has however declared SDG a 'controlled activity' (South Africa 1998a: s38, 2015b). This will ensure that water use is regulated. In terms of the NWA, a water use can be regulated in four different ways: so-called Schedule 1 use which is reserved for *de minimus* water uses such as domestic use; in terms of a general authorisation (South Africa 1998a: s39); as a continuation of a lawful use (South Africa 1998a: ss32–35); or in terms of a water use licence, which as indicated above will be the likely requirement in the SDG context. The issuing of licences requires the authority to have regard to specified factors and conditions contained, respectively, in Sections 27–29 of the NWA, as well as Section 30 which requires the provision of financial security by the applicant.

In issuing a water use licence, a responsible authority *may*, to the extent that it is reasonable to do so, require the applicant to engage a 'competent person' to undertake an assessment of the likely effect of the proposed licence on the resource quality (South Africa 1998a: s41(2)(a)(ii)). The responsible authority may also direct that such assessment comply with the EIA Regulations issued under the NEMA (South Africa 1998a: ss41(2)(*a*)(ii) and 41(3)). However, while the environmental authorisation needs to have been issued in order for the Minister of Mineral Resources to grant the exploration or production right, in the case of the water use licence, all that is required is the *proof of application* for the Minister to exercise his or her discretion. In practice this may mean that project proponents may be granted exploration and production rights without first having been issued a water use licence by the responsible authority, and may then be compelled to commence operations without such licence in order not to lose their extractive authorisation (Humby 2016).[11]

The Regulations for Petroleum Exploration and Production also requires that the applicant apply for a water use license in terms of the NWA and indicate 'the supply source, quality and location for the base fluid of each stage of the operation and the water usage volume.' In addition the applicant must submit, with its water use application, environmental authorisation application and as part of its EMPr, a water resource monitoring plan and an integrated water management plan (South Africa 2015a: regs 88(2), (3) and (6), and reg 123(2)).[12] The regulations furthermore require motoring and reporting of type and volumes of water sourced for stimulation operations; volumes and rates of fracking fluid pumped into the target zone; and volumes and rates of flowback received during and after each stimulation operation (South Africa 2015a: reg 112(8)(j)).

---

[11] The author points out, however, that operating without a water use licence remains a criminal offence.

[12] The water resource monitoring plan must specify, amongst other things, the sampling methodology, the monitoring points, parameters, frequency and reporting frequency.

The regime for water use regulation is set up to be robust and includes pollution prevention and liability for water pollution. Important in the context of the potential risk of flowback, section 19 of the NWA places responsibility on the person who owns, controls, occupies or uses the land in question to take measures to prevent pollution of water resources. If these measures are not taken, the catchment management agency concerned may itself do whatever is necessary to prevent the pollution or to remedy its effects, and to recover all reasonable costs from the persons responsible for the pollution (South Africa 1998a: s19(3)). It also requires that emergency incident, such as an accident involving the spilling of a harmful substance that finds or may find its way into a water resource be reported and that the responsibility for remedying the situation rests with the person responsible for the incident or the substance involved (South Africa 1998a: s20).

The Regulations for Petroleum Exploration and Production additionally requires that where an anomalous pressure or flow condition or other anticipated pressure or flow condition is occurring in a way that indicates that the mechanical integrity of a well has been compromised and where continued operations pose a risk to the environment, the hydraulic fracturing operations must be immediately suspended, the incident must be reported to the designated agency and immediate remedial action must be undertaken (South Africa 2015a: regs 112(10)-(12)).

17.3.3   The Regulation of Waste

The generation, storage, treatment, disposal and transportation of SDG generated waste water is regulated primarily through the National Environmental Management Waste Act (NEMWA) (South Africa 2008). The latter defines waste in section 1 as inclusive of "waste generated by the mining, medical or other sector." It distinguishes between general and hazardous waste, with the latter referring to "any waste that contains organic or inorganic elements or compounds that may, owing to the inherent physical, chemical or toxicological characteristics of that waste, have a detrimental effect on health or the environment" (South Africa 2008: s1). Flowback is in essence a form of hazardous waste. As mentioned above, in addition to flowback, hazardous waste may be generated through spills and residual accumulation of contaminated effluents at wellhead sites. The NEMWA requires a license for listed waste activities such as the generation, collection, handling, storage, treatment and disposal of waste (South Africa 2013b).

The Regulations for Petroleum Exploration and Production in addition requires that waste be managed in accordance with a waste management plan (South Africa 2015a: reg 116) and that waste must be disposed of waste as set out in the authorisations. The flowback and fluids must therefore be managed in terms of the waste management plan and must be stored in tanks that comply with SANS standards (South Africa 2015a: regs 116(1) and 118(10).[13] As mentioned earlier in

---

[13] SANS standards are developed by the South African Bureau of Standards.

contradiction to the practice in the USA, waste may not be re-injected into the disposal wells nor may it be discharged into a surface watercourse (South Africa 2015a: regs 123(4)–(5)). Drill cuttings and fluids must be stored temporarily in tanks above ground (South Africa 2015a: regs 123(6)–(7)). Solid waste must be disposed of at a licenced waste and treatment facility (South Africa 2015a: reg 124(8)). Finally, radioactive waste must be disposed of in terms of the National Radioactive Waste Disposal Institute Act 53 of 2008 (South Africa 2015a: reg 124(2)).

17.4   Conclusion

In 2011 South Africa instituted a moratorium on shale gas exploration, l argely in response to criticism of Shell Exploration Company BV's draft Environmental Management Plan (EMP). A critical review of the EMP, (the so-called 'Havemann Report') (Havemann et al. 2011) was scathing in its assessment of both the EMP and the associated regulatory regime and concluded that "government [should] put an immediate end not only to [Shell's] application but also to decline any future fracking exploration in the Karoo by Shell or other consortium…" (Havemann et al. 2011: 13). The moratorium was lifted in 2012 and followed by the completion in 2016 of a Strategic Environmental Assessment (Scholes et al. 2016) which, regrettably, did little to assuage the concerns raised in the Havemann document. Notably absent in the SEA was the relation of identified risks to specific sustainability goals or requirements. Also missing were considerations of how the vast quantities of salts (brine) that will be returned to the surface will be dealt with, nor where or how the silica sand commonly used as a proppant will be mined—a process that typically requires lots of water.

The SEA's derivation of surprisingly low overall risks suggests the existence of an industry-specific, robust regulatory regime that is, as yet, non-existent. While the Regulations for Petroleum Exploration and Production are seemingly focused on ensuring that impacts on the environment are assessed and water and waste is regulated, the absence of any compliance and enforcement mechanisms such as the creation of offences suggests that they are written as guidelines only (Du Plessis 2015: 1465).

All forms of mining impact negatively on water resources. In most cases the science has followed mining practice during the last four decades of increasing environmental awareness and the advent of environmental law. Solutions, such as those for the treatment of AMD effluents, have had to be hastily derived in knee-jerk reaction to the increasingly evident environmental threat. Additionally, while the response to AMD pollution has been entirely reactive, planning for SGD in South Africa can benefit from a proactive, knowledge-based assessment of the potentially-irreducible threats. It is, however, unlikely that the SGD industry would provide the necessary funding until such time as the full potential of the Karoo resources has been quantified. For South Africa, SGD presents a whole slew of new scientific challenges necessary to understand and provide guidance for the possible development of this new mining industry. With respect to SGD, various publications have concluded that there is insufficient information to assess the extent and depth of the risk. Others have remarked that the long-term environmental and health costs outweigh the benefits of shale gas as an energy source (Kerner 2012). This would mean that it is not possible to assess whether a regulatory framework is sufficiently responsive (Academy of Science of South Africa 2016; Glazewski and Esterhuyse 2016). Additionally, the effectiveness of a regulatory scheme must be underpinned by the skills and technology necessary to implement monitoring and compliance enforcement—an aspect for which South Africa is ill-prepared. In the USA the State of Illinois has been heralded as having the most comprehensive SGD regulations, yet an inability to enforce them (Callies and Stone 2014: 38).

SGD has been practiced for decades in the USA, largely regulated by state and local-level instruments. More recently countries such as the UK are grappling with a similar challenge to that facing South Africa. The antecedent USA experience provides a well-populated source of 'lessons learnt' examples. At a local level, the physical response to hydraulic fracturing of the many and varied Karoo geologies need to be determined, how the chemistry of 'typical' fracturing fluids may alter in contact with local geologies and groundwaters and how best to treat the residual effluents for reuse need to be substantively addressed, *inter alia*. As such, South African statutory provisions might be guided by foreign experience and the findings of a surrogate exploration phase, amended 'on the fly' as and when concomitant scientific investigation so dictates. South Africa already grapples with substantial water quality challenges arising from AMD, salinisation and/or eutrophication. It simply cannot take the risk of adding thereto as a result of hasty or inadequate consideration of the water pollution risks associated with SGD.

In the face of the clear, evident and largely unpredictable challenges, the legal and regulatory tools and experience available for SGD in South Africa are neophytic at best. There are no existing norms and standards that would transition comfortably into the SGD arena. A considerable body of further scientific investigation is needed, possibly in parallel with closely controlled, open and transparent pilot-scale drilling and fracturing. Future impacts, for example the occurrence of methane in groundwater (Moritz et al. 2015), can only be evaluated against a comprehensive baseline set of data derived prior to SGD exploration. What is ideally needed is a single, concise and tested set of regulations governing all aspects of the SGD process.

References

Academy of Science of South Africa (2016) South Africa's technical readiness to support the shale gas industry. Available: http://research.assaf.org.za/bitstream/handle/20.500.11911/14/final_report.pdf?sequence=14&isAllowed=y. Accessed 10 Nov 2018

Callies DL, Stone C (2014) Regulation of hydraulic fracturing. J Int Compa Law 1:1–38

Chow L (2016) EPA watered down major fracking study to downplay water contamination risks. Available: https://www.ecowatch.com/epa-fracking-study-2121912963.html. Accessed 9 Nov 2018

Cole DJ (2014) Geology of Karoo shale gas and how this can influence recovery. In: Proceedings of the fossil fuel foundation gas conference. 21 May 2014. Available: http://www.fossilfuel. co.za/conferences/2014/GAS-SA-2014/Session-3/03Doug-Cole.pdf. Accessed 7 Nov 2018

Coram A, Moss J, Blashki GA (2014) Harms unknown: health uncertainties cast doubt on the role of unconventional gas in Australia's energy future. Med J Aust 200(4):210–213

Costello G (2012) Greater intervention needed to tackle acid mine drainage. Available: http://www. fse.org.za/index.php/component/k2/item/162-greater-intervention-needed-to-tackle-acid-mine-drainage. Accessed 7 Nov 2018

Cramer BW (2016) What the frack? How weak industrial disclosure rules prevent public understanding of chemical practices and toxic politics. South Calif Interdiscip Law J 25(1):67–105

Critical Ecosystem Partnership Fund (2003) The succulent Karoo hotspot: Namibia and South Africa. Available at: https://www.cepf.net/sites/default/files/final.succulentkaroo.ep_.pdf. Accessed 6 Nov 2018. Figure 1

De Wit MP (2013) Modelling the impacts of natural resource interventions on the economy and on sustainability with specific reference to shale gas mining in the Karoo (Report of 30 August 2013). De Wit Sustainable Options (Pty) Ltd., Brackenfell, South Africa

Department of Environmental Affairs (2014) One environmental system for mining industry to commence on 8 December 2014. 4 September 2014. Available: https://www.environment.gov. za/mediarelease/oneenvironmentalsystem_miningindustry. Accessed 9 Nov 2018

Du Plessis W (2015) Regulation of hydraulic fracturing in South Africa: a project life-cycle approach? Potchefstroom Electron Law J 18(5):1441–1478

Econometrix Pty (Ltd) (2012) Karoo shale gas report: special report on economic considerations surrounding potential shale gas resources in the Southern Karoo of South Africa. Available: https://cer.org.za/wp-content/uploads/2012/06/Econometrix-KSG-Report-February-2012.pdf Accessed 9 Nov 2018

Elsner M, Hoelzer K (2016) Quantitative survey and structural classification of hydraulic fracturing chemicals reported in unconventional gas production. Environ Sci Technol 50(7):3290–3314

Esposito M (2013) Water issues set the pace for fracking regulations and global shale gas extraction. Tulane J Int Comp Law 22(1):167–190

Feris L (2012) The public trust doctrine and liability for historic water pollution in South Africa. Law Environ Dev J 8(1):1–18

Fig D (2012) Fracking and the democratic deficit in South Africa. Available: https://www.tni.org/files/dfig64.pdf. Accessed 7 Nov 2018

Glazewski J, Esterhuyse S (eds) (2016) Fracturing in the Karoo: critical legal & environmental perspectives. Cape Town: Juta.

Havemann L, Glazewski J, Brownlie S (2011) A critical review of the application for a Karoo gas exploration right by Shell Exploration Company B.V. Available: https://royaldutchshellplc. com/wp-content/uploads/2011/04/Karoo4.pdf. Accessed 10 Nov 2018

Hobbs P, Day E, Rosewarne P, Esterhuyse S, Schulze R, Day J, Ewart-Smith J, Kemp M, et al (2016) Chapter 5: water resources. In: Schole, R, Lochner P, Schreiner G, Snyman-Van der Walt L, de Jager M (eds) Shale gas development in the Central Karoo: a scientific assessment of the opportunities and risks. CSIR, Pretoria. Available: http://seasgd.csir.co.za/wp-content/uploads/2016/12/SGD-Scientific-Assessment-Binder1_LOW-RES_INCL-ADDENDA_21Nov2016.pdf. Accessed 7 Nov 2018

Humby T (2016) Environmental assessment of shale gas development in South Africa. In: Glazewski J, Esterhuyse S (eds) Hydraulic fracturing in the Karoo: critical legal & environmental perspectives. Juta, Cape Town, pp 86–109

Jackson RB, Lowry ER, Pickle A, Kang M, DiGiulio D, Zhao K (2015) The depths of hydraulic fracturing and accompanying water use across the United States. Environ Sci Technol 49(15):8969–8976

Kahrilas GA, Blotevogel J, Stewart PS, Borch T (2015) Biocides in hydraulic fracturing fluids: a critical review of their usage, mobility, degradation, and toxicity. Environ Sci Technol 49(1):16–32

Kerner K (2012) Fracturing the environment: exploring potential problems posed by horizontal drilling methods. Univ Baltime J Land Dev 1(2):235–245

Lautz LK, Hoke GD, Lu Z, Siegel DI, Christian K, Kessler JD, Teale NG (2014) Using discriminant analysis to determine sources of salinity in shallow groundwater prior to hydraulic fracturing. Environ Sci Technol 48(16):9061–9069

McLaughlin MC, Borch T, Blotevogel J (2016) Spills of hydraulic fracturing chemicals on agricultural topsoil: biodegradation, sorption, and co-contaminant interactions. Environ Sci Technol 50(11):6071–6078

Moritz A, Hélie J-F, Pinti DL, Larocque M, Barnetche D, Retailleau S, Lefebvre R, Gélinas Y (2015) Methane baseline concentrations and sources in shallow aquifers from the shale gas-prone region of the St. Lawrence Lowlands (Quebec, Canada). Environ Sci Technol 49(7):4765–4771

Neilan PG, Docley FL (2014) Salt of the earth: salt, water and damage to land in the Bakken and the Williston Basin. Int Energy Law Rev 7:252–260

Nicholson BR, Dillard SC (2013) Analysis of litigation involving shale and hydraulic fracturing: part 1. Int Energy Law Rev 2:50–66

Nicot J-P, Scanlon BR, Reedy RC, Costley RA (2014) Source and fate of hydraulic fracturing water in the Barnett Shale: a historical perspective. Environ Sci Technol 48(4):2464–2471

Oelofse S, Schoonraad J, Baldwin D (2016) Chapter 6: Impacts on waste planning and management. In: Scholes R, Lochner P, Schreiner G, Snyman-Van der Walt L, de Jager M. (eds) Shale gas development in the Central Karoo: a scientific assessment of the opportunities and risks. CSIR, Pretoria. Available: http://seasgd.csir.co.za/wp-content/uploads/2016/12/SGD-Scientific-Assessment-Binder1_LOW-RES_INCL-ADDENDA_21Nov2016.pdf. Accessed 7 Nov 2018

Scholes R, Lochner P, Schreiner G, Snyman-Van der Walt L, de Jager M (eds) (2016) Shale gas development in the Central Karoo: a scientific assessment of the opportunities and risks. CSIR, Pretoria. Available: http://seasgd.csir.co.za/wp-content/uploads/2016/12/SGD-Scientific-Assessment-Binder1_LOW-RES_INCL-ADDENDA_21Nov2016.pdf. Accessed 7 Nov 2018

South Africa (1998a) National Water Act, No. 36 of 1998, as amended. Available: https://cer.org.za/wp-content/uploads/2010/05/36-OF-1998-NATIONAL-WATER-ACT_2-Sep-2014-to-date.pdf. Accessed 6 November 2018

South Africa (1998b) National Environmental Management Act, No. 107 of 1998, as amended. Available: http://www.wylie.co.za/wp-content/uploads/NATIONAL-ENVIRONMENTAL-MANAGEMENT-ACT-NO.-107-OF-1998.pdf. Accessed 6 Nov 2018

South Africa (2002) Mineral and Petroleum Resources Development Act, No 28 of 2002, as amended. Available: http://www.wylie.co.za/wp-content/uploads/MINERAL-AND-PETROLEUM-RESOURCES-DEVELOPMENT-ACT-NO.-28-OF-2002.pdf. Accessed 6 Nov 2018

South Africa (2008) National Environmental Management: Waste Act, No. 59 of 2008, as amended. Available: https://cer.org.za/wp-content/uploads/2010/03/59-OF-2008-NATIONAL-ENVIRONMENTAL-MANAGEMENT-WASTE-ACT_2-Sep-2014-to-date.pdf. Accessed 9 Nov 2018

South Africa (2013a) National Environmental Management: Waste Act, No. 59 of 2008. National norms and standards for the assessment of waste for landfill disposal. Government Gazette. 578(36784). 23 August. Government notice no. R10008. Government Printer, Pretoria

South Africa (2013b) National Environmental Management: Waste Act, No. 59 of 2008. List of waste management activities that have, or are likely to have, a detrimental effect on the environment, as amended. Available: Jutastat Legal. Accessed 10 Nov 2018

South Africa (2014a) National Environmental Management Act, No. 107 of 1998. Environmental impact assessment regulations. Government Gazette. 594(38282). 4 December. Government notice no. R983. Government Printer, Pretoria

South Africa (2014b) National Environmental Management Act, No. 107 of 1998. Environmental impact assessment regulations. Government Gazette. 594(38282). 4 December. Government notice no. R984. Government Printer, Pretoria

South Africa (2015a) Mineral and Petroleum Resources Development Act, No. 28 of 2002. Regulations for petroleum exploration and production. Government Gazette. 38855. 3 June. Government notice no. R466. Government Printer, Pretoria

South Africa (2015b) National Water Act, No. 36 of 1998. Declaration of the exploration and/or production of onshore naturally occurring hydrocarbons that require stimulation, including but not limited to hydraulic fracturing and/or underground gasification, to extract, and any activity incidental thereto that may impact detrimentally on the water resource as a controlled activity in terms of Section 38(1) and publication of schedule of all controlled activities in terms of Section 38(4) of the National Water Act, 1998 (Act No. 36 of 1998). Government Gazette. 39299. 16 October. Government notice no. R999

South Africa (2015c) National Environmental Management Act, No. 107 of 1998. Regulations pertaining to the financial provision for prospecting, exploration, mining or production operations. Government Gazette. 39425. 20 November. Government notice no. R1147. Government Printer, Pretoria

Sullivan K, Cyterski M, Kraemer SR, Knightes C, Price K, Kim K, Prieto L, Gabriel M et al (2015) Case study analysis of the impacts of water acquisition for hydraulic fracturing on local water availability (EPA/600/R-14/179). Environmental Protection Agency, Washington, DC, USA

Tong S, Scheck T (2016) EPA's late changes to fracking study downplay risk of drinking water pollution. Available: https://www.marketplace.org/2016/11/29/world/epa-s-late-changes-fracking-study-portray-lower-pollution-risk. Accessed 9 Nov 2018

Treasure Karoo Action Group (2015) Initial review of the Regulations for Petroleum Exploration and Production. Available: http://www.treasurethekaroo.co.za/pdf/TKAG%27s%20preliminary%20review%20of%20the%20fracking%20regulations.pdf. Accessed 9 Nov 2018

U.S. Environmental Protection Agency (2016) Hydraulic fracturing for oil and gas: impacts from the hydraulic fracturing water cycle on drinking water resources in the United States (EPA-600-R-16-236Fa). Office of Research and Development, Washington, DC, USA

Vengosh A, Jackson RB, Warner N, Darrah TH, Kondash A (2014) A critical review of the risks to water resources from unconventional shale gas development and hydraulic fracturing in the United States. Environ Sci Technol 48(15):8334–8348

Vidic RD, Brantley SL, Vandenbossche JM, Yoxtheimer D, Abad JD (2013) Impact of shale gas development on regional water quality. *Science* 340(6134). https://doi.org/10.1126/science.1235009

Wait R, Rossouw R (2014) A comparative assessment of the economic benefits from shale gas extraction in the Karoo, South Africa. South Afr Bus Rev 18(2):1–34

Western Cape (2016a) Declaration of a local disaster. Western Cape Provincial Gazette. 834(61366). 3 June. Government notice no. 7623. Government Printer, Cape Town

Western Cape (2016b) Extension of declaration of a local disaster. Western Cape Provincial Gazette. 1487(54254). 25 November. Government notice no. 7706. Government Printer, Cape Town

Yost EE, Stanek J, DeWoskin RS, Burgoon LD (2016a) Overview of chronic oral toxicity values for chemicals present in hydraulic fracturing fluids, flowback and produced waters. Environ Sci Technol 50(9):4788–4797

Yost EE, Stanek J, DeWoskin RS, Burgoon LD (2016b) Estimating the potential toxicity of chemicals associated with hydraulic fracturing operations using quantitative structure – activity relationship modeling. Environ Sci Technol 50(14):7732–7742

Loretta Feris was until recently Professor of Law in the Institute of Marine and Environmental Law at the University of Cape Town (UCT) where she taught natural resources law, pollution law and international environmental law. Since January 2017, she was Deputy Vice-Chancellor (Transformation) at UCT. Feris is an NRF rated researcher and has published widely in the area of environmental law, including environmental rights, liability for environmental damage and compliance and enforcement of environmental law. She is a board member of Biowatch and Chair of the Board of Natural Justice and, until 2013, served on the board of the South African Maritime Safety Authority. She is a Law Commissioner of the World Conservation Union (IUCN) and a member of the IUCN Academy of Environmental Law, where she served on the teaching and capacity building committee for 3 years. Feris holds a B.A. (law), LL.B. and LL.D. from the University of Stellenbosch in South Africa and an LL.M. from Georgetown University.

William (Bill) Harding is an aquatic ecologist in private practice. His practical experience includes periods in private industry and local government. He recently completed a M(Phil) in Environmental Law at UCT. Harding specializes in aquatic pollution and the ecology of anthropogenically-impacted water resources. He is currently developing his environmental law experience by translating aspects of his ecological understanding of water resources in the form of legal advisory support skills. In addition to augmenting legal argument, he undertakes interrogative reviews of project proposals, administrative decisions and reports. Harding holds a BSc in chemistry and microbiology and BSc(Hons) in microbiology from Rhodes University and a Master and PhD in zoology (limnology) from the University of Cape Town (UCT).

Chapter 18
Hydraulic Fracturing, Shale Development and Water Issues in Poland

Anna B. Mikulska

Abstract Poland is the largest and possibly most promising European country for future shale development. But attempts to recreate a "shale revolution" in Poland for now have failed due to geological, cost and regulatory factors. Even so, the potential for shale exploration is not necessarily lost since technology and the market can at some point deliver more hospitable conditions. Water is probably the single most important element of the environment that hydraulic fracturing utilizes and has the potential to deplete and/or contaminate, resulting in host of adverse impacts across the land and population. This paper is a first take on describing the intricacies that exist between water and potential shale development in Poland. It provides an overview of the legal environment on both shale exploration and water law, and describes and contrasts the geological and water-related features of the regions where shale exploration takes place. On the basis of this initial assessment, the paper provides recommendations with respect to developing water policy and potential shale development.

Keywords Shale · Natural gas · Shale gas · Hydraulic fracturing · Fracking · Water scarcity · Produced water · Oil and gas exploration

After hydraulic fracturing proved successful in extracting hydrocarbons from shale formations in the U.S., many countries around the globe have looked into the possibility of activating their shale hydrocarbon resources. Poland has been one of those most determined to make the new technology work within its borders, including strong governmental and societal support for shale development (CBOS 2013). The interest is rooted in the country's dependence on Russia for natural gas delivery coupled with deep distrust and fear that natural gas may one day be used by Russia as a geopolitical weapon or, at the very least, as a bargaining chip (Collins 2017).

A. B. Mikulska (✉)
Center for Energy Studies, Baker Institute for Public Policy, Rice University, Houston, TX, USA
e-mail: am75@rice.edu

© Springer International Publishing AG 2020   363
R. M. Buono et al. (eds.), *Regulating Water Security in Unconventional Oil and Gas*, Water Security in a New World,
https://doi.org/10.1007/978-3-030-18342-4_18

EIA's early revelation that Poland's shale gas resources total in the range of 5.3 trillion cubic meters (Tcm) (EIA 2011) provided impetus for exploration with many domestic and international companies entering the market, but with time the momentum has diminished. The geologic, economic, legal, and bureaucratic environments have failed to meet the initial expectations and facilitate production. But hydraulic fracturing technology has not left Poland together with shale investors. Instead new applications have been sought to provide for enhanced gas recovery in old gas reservoirs or as a technology to support coal seam gas development (Malinowski 2018). Also, it is not inconceivable that widespread technological innovation may at some point resuscitate Polish shale (PAP 2017).

Thus, it may be that the current hiatus in Polish shale constitutes a good time to assess the current state of affairs and the legal, economic, and environmental needs that potential future increases in hydraulic fracturing may foster. For the Polish government, this may also be the best time to provide for more comprehensive regulation that the country is still lacking, particularly in the area of water use and water protection.

This chapter looks at the most recent attempts by the Polish government to address shale development via legislation and regulation. The focus is specifically on issues of water, given the centrality of this resource to the process of hydraulic fracturing under conditions of Poland's limited water resources.

After providing a brief overview of shale development in Poland and the reasons for its current decline, the chapter will consider the country's water resources, how hydraulic fracturing has affected or may affect those resources and the recent attempts to regulate use of water that have the potential to affect development of shale today and in the future. The chapter concludes with policy recommendations.

18.1 Poland's Unsuccessful "Shale Revolution"

The Polish Ministry of Environment awarded the first government concession for exploratory shale drilling in Poland in 2007. But actual drilling activity did not ensue until 2010, when interest in shale activity began to grow facilitated by promising reports from many reputable sources. The reports estimated the volume of technically recoverable shale resources in Poland to be anywhere between 1.4 Tcm (according to Advanced Resources International (ARI) in 2009), to 3 Tcm (according to Wood McKenzie in 2009), to 5.3 Tcm (according to EIA in 2011). Per the latter estimate, shale resources is Poland were comparable only to the 5.1 Tcm estimated to be located in France, where hydraulic fracturing was banned that same year. In 2011, Poland seemed to hold the biggest potential for a shale revolution in the heart of gas-dependent Europe (EIA 2013).

Investors took note. The number of concessions for exploration of shale resources grew from a modest 11 at the end of 2007 to a peak of 111 in July 2012 (Fig. 18.1). The companies investing in Polish shale ranged from domestic firms

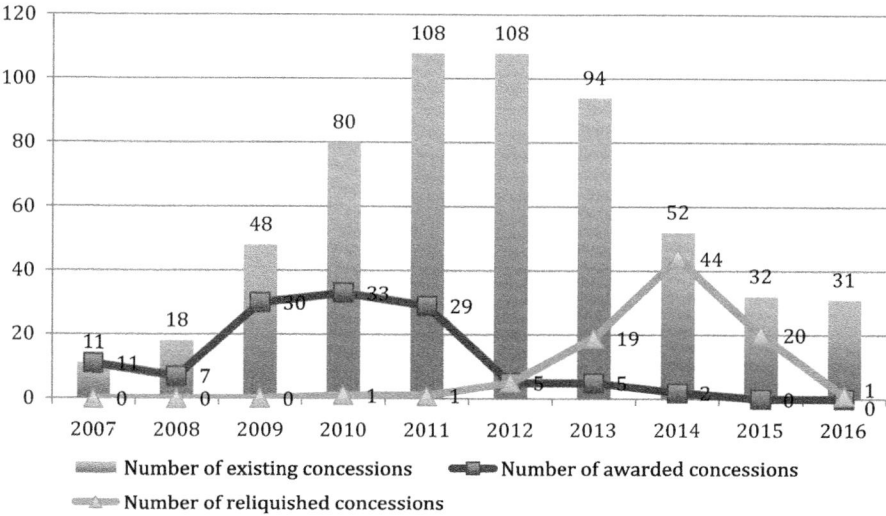

Fig. 18.1 Number of concessions for exploration of shale in Poland, 2007–2016. (Source: Polish Ministry of Environment (MOS) 2016a, b)

such as Lane Energy Poland, Polskie Gornictwo Naftowe i Gazownictwo (PGNiN), Orlean Upstream, and Talisman Energy Polska, to international giants like ExxonMobil, Marathon Oil, and Chevron.

But by the end of 2012 the enthusiasm for Polish shale rapidly lost momentum. The number of concessions began to decrease. Nineteen concessions were relinquished in 2013 and 44 concessions given up in 2014. On July 31, 2017, only 20 concessions were left (Fig. 18.1). The drop in the number of concessions was marked by a hurried exit by the international players. In December 2017, San Leon Energy religuished two final concesions (Trusewicz 2018). This demise has been attributed to a number of interrelated factors.

First, Poland's geology has been much less hospitable to shale development than that in the U.S. (Arthur et al. 2008). Polish shale deposits are located much deeper underground—between 2.5 and 4 km in depth. In comparison, most of producing formations in the U.S. are within 0.6 to 1.8 km range. Shale plays in Poland are also much thinner, maxing out at 50 meters in width, whereas thickness of shale in the Marcellus reaches up to 290 meters. Moreover, the percentage of total organic carbon (TOC), a measure used to assess viability of drilling success in shale has been determined in Poland at an average of 3–5%. The average in the U.S. is approximately 11%. Adding to the difficulties in extraction of hydrocarbons from the Polish shale is its low permeability.

In 2012, the Polish National Geological Institute published its first assessment of recoverable shale resources. Instead of extrapolating from the American experience—as did previous reports—this report was based on historic Polish geological data from 1950 to 1990. The report drastically cut the optimistic figures suggested by EIA, ARI, and Wood MacKenzie. It estimated shale resources in Poland at no

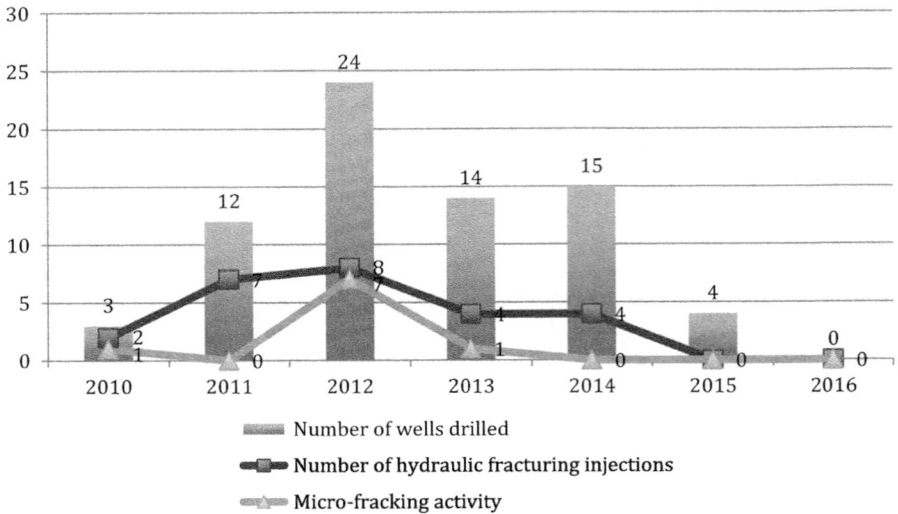

Fig. 18.2 Number of exploratory wells drilled in Poland between 2010 and 2016. (Source: Polish Ministry of Environment (MOS) 2016a)

more than 1.92 Tcm, but most likely in the range between 0.346 and 0.768 Tcm. An updated assessment by the Institute in 2015 confirmed these findings based on data from actual drilling sites undertaken since 2007. Thus, Polish shale may have some potential, but one that is a far cry from the initial promises. This brings us to the second aspect of the shale exodus from Poland: the economic issues.[1]

Given unfavorable geological conditions, most prominently the low depth at which shale resources are located, the average cost of an exploratory well in Poland has been approx. $15 million. This is more than double the cost of the most expensive wells drilled in the U.S. where, depending on shale play, costs range from the low of $5.2 million per well in the Delaware Basin to a high of $7.2 million per well in the Midland Basin. Costs in the Marcellus are approximately $6.1 per well (EIA 2016). The high cost of shale gas extraction in Poland cannot be justified by outputs that have been, on average, only 10,000 m³, half of what could support commercial drilling (Sawicki 2017).

The conflux of these factors visibly discouraged exploratory drilling. Only 72 exploratory wells were drilled by end of 2015 and no wells have since been drilled (Fig. 18.2). For any future success of shale development in Poland, this situation constitutes a vicious circle. Wells are not drilled because they are expensive and do not yield much success, but success cannot be achieved unless more of those expensive wells are drilled in hopes of hitting a "sweet spot," or particularly rich

---

[1] The percentage of Total Organic Carbon (TOC) that is used to assess viability of drilling success in the shale has also been determined in Poland at an average of 3 to 5%, while this average in the U.S. is approximately 11% (Karcz n.d.).

source rock or developing a new technology that reflects different and more challenging geological conditions. It is believed that no fewer than 200 wells must be drilled in order to gain an accurate assessment of actual resources in place and estimate the viability of drilling. The American success has been based on massive activity that, in the Marcellus, for example, ranged between approximately 500 vertical wells drilled in 2009 and 1350 at the peak of drilling efforts in 2013. Over 18,000 vertical wells are currently operating in this one shale play (MarcellusGas. Org 2018).

Besides geological and cost-related challenges, the meager rate of drilling and general lack of success in shale development in Poland can be associated with technical and organizational challenges. The success of U.S. shale was built on the back of thousands of small independent producers, behaving in an entrepreneurial fashion within set of small-scale ventures. Such setup encourages high intensity drilling over smaller territorial entities, providing higher probability of striking the sweet spots. Almost the opposite situation can be found in Poland where the companies involved in shale drilling have been mostly large and vertically integrated entities such as the majors Exxon, Marathon, or Chevron as well as large Polish conglomerates like PGNiG, Orlen or LOTOS. Also, in total, only 39 companies were ever involved in exploration and drilling process. They were all awarded large territorial concessions by the government.

Indeed, the legal environment has not been conducive to shale activity undertaken within smaller territorial units and by smaller entities. Unlike the U.S., in Poland the state owns mineral rights, including on private properties. Thus, to engage in shale exploration an entity needs to apply for a concession from the state. This process has been particularly cumbersome and costly, involving bureaucratic slowdowns and, until 2015, lack of any guarantee that a company finding commercially viable shale resources would also be given right to exploit those resources. The government decided to award large concession territories to attract investors. But the existing law would allow to award concessions only for five years with a possibility of a 2-year extension. Many investors feared that the period would be too short for the lengthy process of exploration to yield successful results. The 2015 Geological and Mining Law attempted to solve this issue (Journal of Laws 2014, item 1133).[2] The new law introduced a comprehensive concession for exploration, development and exploitation of hydrocarbons valid between 10 and 30 years. This new concession replaced several types of concessions an exploration and development (E&D) company needed to apply for in the past. Concessions will no longer be required for geophysical assessment of the shale plays. Also companies will need to keep only one type of records – a geological-investment documentation– in place of the former two. However, the new law was too late for many companies, which by then decided to abandon their investments.

Global market conditions facilitated the exodus as oil and natural gas prices fell drastically at the end of 2014. International oil and gas companies, hit hard by the sudden oil and gas price slump, axed investment in Polish shale, which showed little

---

[2] Act of 11 July 2014 Amending the Geological and Mining Act, Journal of Laws 2014, item 1133.

to no promise but carried high costs of exploration, arguing that they could not justify the lost opportunity cost elsewhere.

18.2 Poland's Water Resources vs. Shale Exploration

Compared to other European countries, Poland is deficient in water supplies. With only 1600 m³ of water available per capita annually, the country is far below Europe's average of 4500 m³ (Polska Fundacja Ochrony Zasobow Wodnych n.d.) and comparable to Egypt in terms of water resource availability (MOS 2016b; PFOZW n.d). Because Polish agricultural water needs are relatively small, this low level of available water resources is not readily detectable. However, coupled with low capacity of artificial retention pools, scarce water resources become a problem in times of drought. A projected rise in climate-related water shortages and increased water use going forward will put additional stress on Poland's water and its users (Gutry-Korycka et al. 2014).

There are two major ways in which shale activity may impact water supplies in Poland: (1) by increasing demand for water due to its use in drilling and fracturing the wells; (2) by negatively impacting water quality; including issues of "produced water."

18.2.1 Water Availability

Regional differences in water availability must be taken into account when considering shale development's impact on water resources, given centrality of water to hydraulic fracturing technology.

Development of shale resources in Poland would require certain amounts of readily available water to be used in hydraulic fracturing. On average, one well requires 17,000 m³ (4.5 million gallons) of water. This is what an average, 4-person, Polish family living in a city uses over a period of 3.5 years.[3] Since only 76 wells have been completed in Poland since 2010, this is admittedly only a 'drop in a bucket' of total water consumption in Poland, which in 2016 equaled 10.5 billion m³. Arguably, even if shale production were to increase, the share of water used by shale activity would be small. If 200 wells were fractured annually in Poland, the amount of water needed for the operations would amount to 3.4 million m³ per year (900 million gallons). This is approximately 1% of total annual water consumption in Poland, similar to the water amounts used for irrigation in agriculture.[4]

---

[3] According to he Polish Statistical Review 2016, an average household water consumption per capita in Poland is 34.3 m³ (Central Statistical Office (GUS) 2016).
[4] The very small amount of water used for irrigation in Poland reflects reliance on rain. (Central Statistical Office (GUS) 2016).

But in a country that is already strapped for water resources, any additional water demand should be evaluated carefully. This is especially important since Poland has been experiencing increases in episodes of drought that are expected to become even more frequent in the future. The situation can be exacerbated if shale is being developed in the regions of the country that have been experiencing water shortages.

Shale resources in Poland are concentrated around the Baltic Basin and Warsaw Through, Podlasie, Lublin Basin and around Fore Sudetic Monocline. This corresponds with the water regions of Lower Vistula and Upper Oder. As Map 18.1 shows, Per Gutry-Korycka et al. (2014) the areas of Warsaw Through, Lublin, and Podlasie, as well as considerable part of the Sudety Mountains, are generally more deficient in water supplies then other Polish regions. The only region where potential shale development aligns with relatively abundant water resources is the Baltic Basin.

Variability in water resources in a region depends on availability of exploitable groundwater as well as surface water, both of which rely on rainfall to sustain their levels. Meanwhile, the droughts that have plagued Poland in the relatively recent past have hit especially hard in the Lower Vistula and Upper Oder water regions,

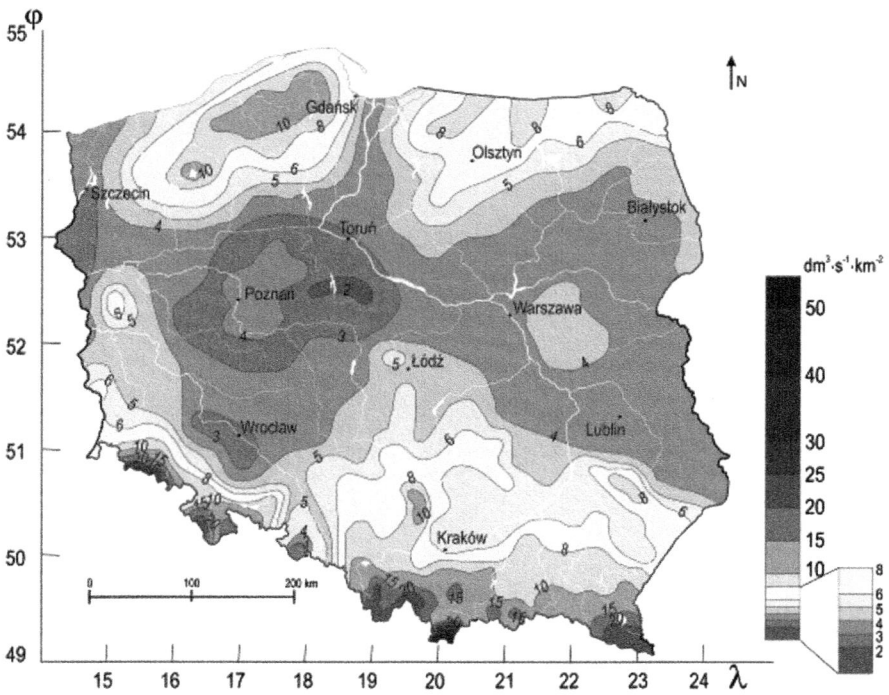

Map 18.1 Average drainage 1951–2000 indicating size and differences in water resources in Poland. (Source: Gutry-Korycka et al. 2014)

where shale resources have been found (See Map 18.2). Hence, regional differences
will be an issue.

While a study coordinated by the Polish General Directorate for Environmental
Protection found no impact of shale on groundwater and surface water (Kantor
et al. 2015; Polish Ministry of Environment (MOS) n.d.-a), the results are hardly
generalizable given the very low level of shale activity experienced in Poland
until now. Consequently, decisions about hydraulic fracturing activity should consider not only possible gas resources in place but also whether enough water will be
readily available at any given time and place to extract those resources. This is a
question that both the government (central and local) and potential investors need to
ask before deciding on concessions and drilling locations.

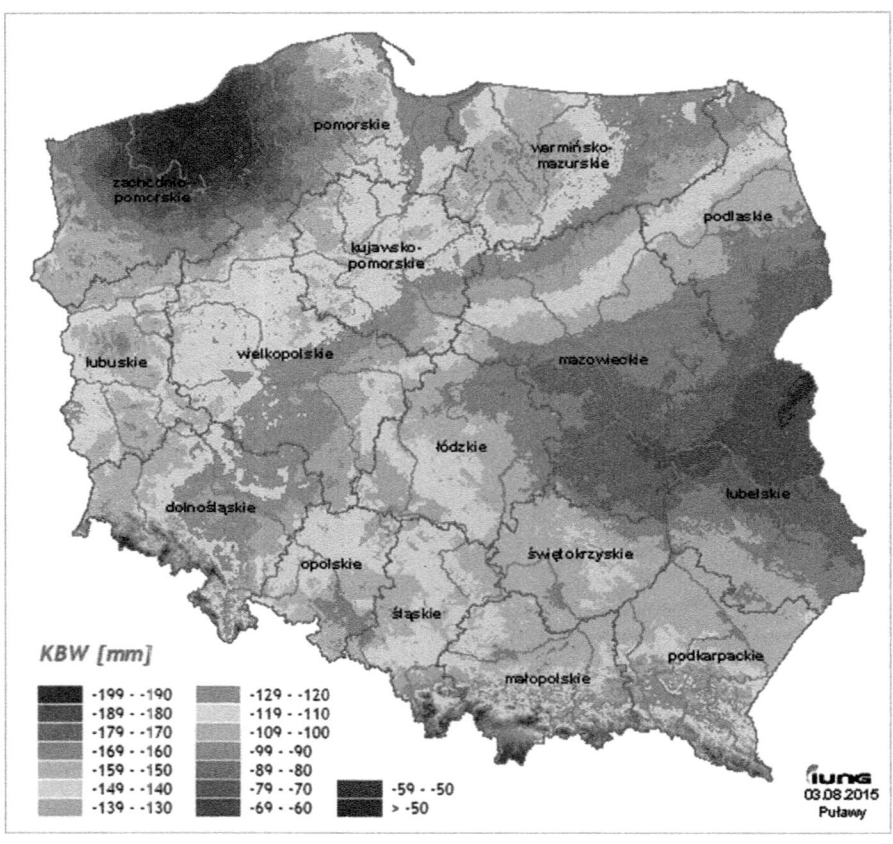

Map 18.2 Rainfall below long-term average (in millimeters), August 2015 Source: Institute of
Soil Science and Plant Cultivation – State Research Institute in Puławy. IUNG. (As reprinted by
Twoja Pogoda 2015)

18.2.2 *Water quality issues and produced water concerns*

Concerns about the impact of hydraulic fracturing activity on water quality in part motivated the aforementioned comprehensive study coordinated by the Polish General Directorate for Environmental Protection (Kantor et al. 2015). The resulting report includes field studies and analyses carried by more than 60 interdisciplinary researchers. The study specifically looked into any possible effects of hydraulic fracturing activities on groundwater and surface water quality. The research found no negative influence of shale activity on water resources in the area, including no changes to chemical composition of the water resources and no deterioration of condition for agricultural use of lands in the area where hydraulic fracturing is performed. Lack of impact on quality of water resources has been associated with the considerable depth at which shale resources are located and lack of permeability of geological material located between the shale and the water resources. The study included all shale regions with varied geological settings. Also, methane emissions detected during testing have not been linked to shale. The report signaled higher level of concern with respect to waste associated with the drilling process, which includes produced water. The report did not, however, make any recommendation beyond referring those who undertake the activity to existing laws, rules, and procedures that govern the processes of transport, treatment and reclamation of the waste product.

18.3 Regulating the Nexus between Shale Development and Water

It is important to highlight that Polish water resources, both on the surface and under the ground, are owned by the state. The new Water Law of 2017 entrusts management and enforcement of water laws and regulations to the newly created State Water Enterprise "Polish Water", which is the state-owned entity, with administrative competences vested upon it. Those competences, among others, include regulatory power over Poland's water and sewage market (under The Act on Collective Water and Sewage Supply of 2001, Journal of Laws of 2006 no. 123, item 858 with amendments).

Poland does not have a law specifically to regulate shale production activity and/or hydraulic fracturing technology. Instead, a plethora of laws and regulations relate to different aspects of the activity (MOS n.d.-b). Some of the most prominent laws include the Mining and Geological Law (Journal of Laws 2011, no.163, item 981) and Water Law (Journal of Laws 2017, item 1566). Other laws that apply to the regulation of water issues include, among others, the 2011 Waste Law, 2008 Mining Waste Law, 2001 Chemical Substances Law, and the 2001 Law on Collective Water Supply and Sewerage. Additionally, it is worth to point out the 2008 Act on the Dissemination of Information on the Environment and Its Protection, the

Participation of the Society in Environmental Protection and Environmental Impact Assessments (Journal of Laws no.199, item 1227), which regulates so-called environmental decision relating to projects which significantly affect the natural environment due to the shale drilling. These laws and several other laws are accompanied by dozens of governmental ordinances.

Lack of specific and detailed law on shale exploration means that the practices of shale companies are governed by a set of general laws and regulations that apply also to other types of hydrocarbon exploration, including the predominant one in Poland: coal mining. Though the Mining and Geological Law was updated in 2014 to accommodate some issues related to shale activity, a new Water Law providing higher levels of protection to water resources came into force on January 1, 2018. The law was passed to address the growing interest in the exploration of hydrocarbons, Poland's limited water resources, and the European Union's Directives, including the Water Framework Directive (Directive 2000/60/C), and the Groundwater Directive (2006/118/EC).

One of the primary ways in which the amendments to the Mining and Geological Law of 2015 relate to shale activity is specification of rights and obligations associated with shale concessions. The new law introduced a comprehensive concession for exploration, development and exploitation of hydrocarbons (art. 49u of Mining and Geological Law "the concession for recognition and exploration of the hydrocarbon field and the extraction of hydrocarbons from the deposit") valid between 10 and 30 years. As discussed above the new concession replaced several types of concessions an E&D company needed to apply for in the past.

The law also increased powers of the monitoring and permitting institutions. The amended law does not include specific provisions about treatment of water resources that are used in the process of hydraulic fracturing or are its byproduct. The one exception is that the law requires a company that applies for investment permit to specify in the application the geological and hydrogeological conditions under which development of hydrocarbons would take place. This could include the depth and characteristics of groundwater, as well as any type of hydrologic conditions that could be impacted by drainage, waste storage, or water injection. The application should contain a description of a balanced approach to exploration and production and should ensure that potential negative environmental effects of the activity are minimized.

Until then, it is likely that the largest impact in the area of shale development will result from the newly overhauled Water Law,[5] which entered into effect on January 1, 2018 (with amendments that came into effect in January 2019). The law incorporates the guidelines provided by the EU's Water Framework Directive (2000/60/EC). The goal of the new law is the protection of groundwater resources, including minimizing the risk of pollution by limiting local activity that could result in contamination of groundwater and maintaining groundwater resources at a balanced level. The new law creates a centralized state agency, Polish Waters, to monitor, manage, and enforce the new law. The agency replaces the prior decentralized

---

[5] Water Law (2017).

system of state agencies. On a local level, the law assigns more responsibilities to local governments related to protection of local water sources, especially during times of natural disasters.

The law introduces water permits, which will be required when an investment decision is made regarding whether to follow exploration activity for hydrocarbons with actual production on a basis of a concession as specified by the Mining and Geological Law. The permit application will involve the formal evaluation of water resources at hand, the impact that proposed activity would have on those resources, and whether this impact would have negative influence on the goals as specified by the new Water Law. The Water Law regulates water management in accordance with the principle of sustainable development, in particular shaping and protection of water resources, use of waters and management of water resources (art. 1 of Water Law). If negative impacts occur, the applicant must submit additional documentation that specifies why the negative impacts cannot be avoided and why they are justified.

One of the most significant changes is the imposition of mandatory fees on the agricultural and industrial sectors and on power generation. Previously, fees were lower or even waived for agricultural and power generation purposes. Higher fees will now be assessed for groundwater and surface water consumption, and sewage. With respect to hydraulic fracturing, a special fee will also be assessed for sand extraction from surface waters, including inland and marine waters (Dyka 2017). The maximum prices for so-called "water services" (viz. water supply), depending on the amount of consumed water under a water permit or an integrated permit, as for the purposes of other mining and extraction (which applies to water consumption for drilling shale gas) are: PLN 0.70 for 1 m^3 of underground water collected, and PLN 0.35 for 1 m^3 of surface water collected (art. 274 sec. 2 subs. b of The 2017 Water Law).[6] The law also creates a system that will contain, among other data, information relevant to hydraulic fracturing activity, such as the amount and locale of water consumption and sewage/waste. This is particularly important as Poland was identified by the European Commission as one of the countries where relevant information on issues related to hydraulic fracturing, including information on chemical substances used in fracturing of the wells, is not published on a regular, systematic basis (COM/2016/794).

The new Water Law creates new requirements for any company that wants to engage in shale activity. The law also, in a small way, mitigates the possible negative implications that can be associated with lack of compliance of the Polish law with the European Commission recommendation of January 22, 2014 (2014/17/EU), regarding the minimum principles for the exploration and production of hydrocarbons (such as shale gas) using high-volume hydraulic fracturing.

The EU recommendation recognizes that environmental legislation written before shale development became a viable way for exploration of hydrocarbons does not sufficiently address all issues related to hydraulic fracturing and horizontal drilling. Thus, it seeks to set minimum principles that should be taken into

---

[6] 1 PLN equals approximately 0.3 USD (04/17/2018).

consideration by EU Member States when applying their regulations. The recommendation neither compels member states to allow shale development and hydraulic fracturing in their territories nor deters them from it. However, if a state decides to pursue this route, the recommendation provides the minimum standards to which the state is obliged to adhere.

Additionally the recommendation requires a risk assessment of potential sites where hydraulic fracturing would be used to prevent any potential leakage or migration of drilling fluid and produced water that could contaminate groundwater and water wells. A baseline study, among other elements, should also determine the quality and flow characteristics of groundwater and surface water and drinking water quality. The recommendation also holds EU Member States responsible for preventing leaks and spills to soil and water that could be associated with the way wells are constructed.

Once wells are operational, the recommendation calls on EU Member States to ensure that good industry practices are used to avoid any potential risks associated with shale development. With respect to water issues, states are expected to promote responsible use of water resources, including development of project-specific water management plans to ensure efficiency in water use and traceability of water flows. Seasonal variations in water availability and water-stress issues should also be included in operators' considerations. The recommendation also encourages member states not to use hazardous chemical substances wherever technically possible and requires minimal use of chemicals in hydraulic fracturing fluid. In addition, operators should make public information on chemical substances and the volumes of water intended to be used and actually used for each well. Water monitoring includes monitoring of composition of the fracturing fluid, the volume of water used, the volume returned and the rate of return, and characteristics of produced water.

As such, the EU Recommendation (2014/17/EU) determines and catalogues important elements of policy related to the use of hydraulic fracturing in shale development. At the same time, it is ambiguous enough for country- and shale-specific application. For Poland, where over 80 different laws and regulations apply to different aspects of shale development and exploration, the EU recommendation potentially offers an organizing and directional tool that allows for greater transparency, efficiency, and predictability.

However, since the EU recommendation neither supersedes national law nor applies automatically, countries may—and often do—chose to shirk their commitments. This is precisely the route that the Polish government chose to take with respect to some of the requirements in the EU recommendation on hydraulic fracturing activities. According to the European Commission's report to the European Parliament and the Council on the effectiveness of the Recommendation 2014/70/EU, Poland continued to apply its domestic rules, according to which a license to drill a well that does not exceed 5000 meters in depth does not require a strategic environment assessment or environmental impact assessment (Nelsen 2014). In April 2016, the European Commission referred the Polish case to the CJEU (MEMO/16/1452). So far the case has not been reviewed by the Court, which is reflective of general understating that shale exploration in Poland would not go forward any time soon.

18.4   Going Forward - Policy Direction

Despite recent setbacks in shale exploration, the Polish government has continued to signal its desire to promote the development of natural gas and its belief that domestically produced gas is crucial to ensure Poland's energy security (PAP 2018). For now, it seems that this gas will not produced by Poland's shale deposits, though going forward, advances in technology that result in lower costs and higher recovery rates, coupled with tighter supply (which generates higher prices), may bring once again shale drilling activity to the forefront. However, lack or limited shale exploration does not necessarily mean that hydraulic fracturing technology also goes away. There have already been successful attempts to use this technology to produce higher recovery rates in conventional natural gas developments and enhanced gas recovery in old natural gas reservoirs. Also, hydraulic fracturing technology has been initially tested in exploration of coal seam gas with encouraging results (Malinowski 2018).

But to facilitate use of hydraulic fracturing in any applications, technological advances and economic incentives need to be accompanied by other elements. This includes transparent and predictable set of laws and regulations that determine conditions under which an activity is undertaken as well as possibly limits and constraints to which such activity can be subjected. Given the unique relationship between hydraulic fracturing and water resources, the government should pay particular attention to issues of water availability, source point pollution, and produced water disposal.

The government took the first step in this direction by adopting the new Water Law. However, the law is general in nature and lacks guidelines that the specificity of shale development and hydraulic fracturing requires. Other laws, including but not limited to the Mining and Geological Law and Law on Chemical Waste or even Atomic Law also apply but - just as in the case of Water Law - are written for general purposes or for the purposes of coal mining or power generation. Additionally, Poland decided against adopting hydraulic fracturing specific regulation as outlined by the European Commission recommendation on minimum principles for the exploration and production of hydrocarbons.

Any new regulations will likely increase the financial burden associated with natural gas development. However, it is also crucial to understand that investors often value transparency and predictability of laws as much as a liberal regulatory environment. The dozens of laws and regulations that currently apply are difficult to navigate and open to interpretation, and many of them have been tailored to address issues and activities that do not necessarily reflect the challenges that lie ahead. Establishing specific standards for environmental protection, including the protection of water resources, may be a difficult and onerous task, but it is one that will pay dividends in the long term.

Because water resources in Poland are relatively scarce, advancing hydraulic fracturing (and even potentially shale) in Poland will require the government to assess whether the activity would impose too much of a burden on already-stressed water resources and whether possible water shortages would negatively affect shale activity in the region. In determining the territorial reach of concessions, the government must carefully consider water availability vs. overall water consumption. This includes extraction-related consumption as well as consumption from already existing demand points, i.e. industry, agriculture, and household use. The discussion on the location of water and shale resources presented at the beginning of this chapter suggests that possibly the best place for hydraulic fracturing activities would be around the Baltic basin, where water resources are generally more abundant and precipitation higher than in other regions.

But regions abundant in groundwater and surface water resources present a different set of challenges for hydraulic fracturing, including water contamination and pollution. While the study coordinated by the Polish General Directorate for Environmental Protection did not register point-source pollution from fracking activity in Poland, opportunities for contamination exist if relevant standards are not set and/or enforced. These issues relate to setbacks; site development and preparation, including baseline water testing and subsequent monitoring; pipeline casing; fluid storage; flowback/produced water monitoring and disposal; as well as terms of well plugging and abandonment. Much can be learned from the U.S. experience, but differences in geology mean that specifics must be worked out at a local level. The geological conditions in Poland differ greatly from the general characteristics of shale in the U.S. Thus, government should invest in research to guide its rulemaking activity in choices of standards to be applied for shale activity in Poland.

Issues of flowback and produced water should also be taken very seriously. In addition to being scarce, water resources in Poland are characterized by low quality (Szekalska 2015). Over the last decade, the Polish government, motivated by its EU member commitments, has invested heavily to offset this issue (Central Statistical Office (GUS) 2017). However, these commitments and undertakings are related mostly to heavily populated urban areas. Rural areas remain often poorly connected to water treatment facilities and surface water is often contaminated with sewage and fertilizers. Lack of enforcement against such practices is problematic but could be catastrophic if practices like these remain unchecked and include produced water disposal. A well functioning system of monitoring and enforcement will be crucial to prevent such occurrences.

Indeed, making sure that standards are set and infrastructure is in place to support environmentally conscious fracking activities in the future can and should be a crucial part of the Polish government's bid for natural gas development. Much of the opposition to hydraulic fracturing activity across Europe has been related to environmental concerns. Protests against and opposition to shale development across France or Germany for example is strongly related to the population's concerns about water contamination as a result of unconventional gas development. Poland's reception, however, has been different. Notwithstanding some protests, the general population is positively inclined towards exploiting the potential economic and geopolitical benefits of a new, domestic source of natural gas. A 2013 report by the Center for Public Opinion Research (CBOS) shows that 78% of the Polish population supports shale development in Poland, while only 7% oppose (CBOS 2013). The figure moves towards more negative assessment when respondents are asked to consider shale development undertaken in proximity of their residence, though even in this case support for shale development reaches 59%. The report notes that an increasing number of Poles believe that hydraulic fracturing is safe for environment. In 2013, 51% of Poles expressed this belief and increase of 8% over those who reported believing this in 2011. Fewer Poles are afraid of negative impact shale extraction could have on their health; the number reporting this fear dropped to 19% in 2013 from 25% in 2011.

Importantly, the numbers appear to be based on actual knowledge of the issue rather than on ignorance. The European Commission's Eurobarometer survey concluded that, among countries where shale exploration is either under way or considered, only Polish citizens felt sufficiently informed about shale gas projects in their regions and possible challenges associated with those projects (COM/2016/0794). Even so, there is room for improvement. As reported by CBOS, a considerable proportion of population expressed doubt about their ability to assess the possible hazards that shale development may pose. Approximately 32% of Poles feel this way with respect to the effect of shale production on the environment, and 37% of Poles consider themselves uninformed about the impact that the effects shale production could have on human health.

Setting direct and specific standards based on a careful consideration of geological and, among others, water-related conditions may be a key element for the Polish government in maintaining the generally positive attitudes of Poles toward shale and other natural gas exploration that may require use of hydraulic fracturing technology. Ensuring that those standards are applied as any activity is undertaken is also crucial.

All in all, if the Polish government wants to propagate hydraulic fracturing technology it needs to consider not only availability of natural gas but also availability of water and water disposal. The first steps to the right directions have been taken, including the updated Mining and Geological Law and the new Water Law, both of which point toward the importance of water resources and set up a more transparent system of supervision and monitoring.

As shale activity has diminished and new applications have not pick up yet considerably, this may be a good time to rethink the policies that govern the activity and embark on a project of comprehensive legislation that, as one of its major elements, would consider use, treatment, and disposal of water resources. If conditions for shale development in Poland improve or hydraulic fracturing is used extensively in other applications, a comprehensive and environmentally conscious policy will be crucial to retaining the generally positive attitude toward the activity that currently is shared by the Polish public. Such legislative and regulatory efforts would also improve Poland's standing with the European Commission on the application of environmental impact assessment directive. And together with predictable laws and easy-to-follow rules, the attitude may be a much bigger draw for investors than a hard-to-navigate and complicated regulatory environment, even if the latter includes fewer environmental considerations. Today more and more companies value an environmentally conscious reputation, and these numbers are likely to keep rising, especially in an increasingly environmentally conscious Europe.

References

Act of 11 July 2014 Amending the Geological and Mining Act (2014) Journal of Laws, item 1133

Arthur DJ, Langhus B, Alleman D (2008), An Overview of Modern Shale Gas Development in the United States. Available at http://www.all-llc.com/publicdownloads/ALLShaleOverviewFINAL.pdf. Accessed 8 Feb 2017

CBOS (2013) Społeczny Stosunek Do Gazu Łupkowego, Available at: http://www.cbos.pl/SPISKOM.POL/2013/K_076_13.PDF. Accessed 14 May 2018

Central Statistical Office (GUS) (2016) Rocznik. Statystyczny Rzeczypospolitej Polskiej 2016. Available at: http://stat.gov.pl/obszary-tematyczne/roczniki-statystyczne/roczniki-statystyczne/rocznik-statystyczny-rzeczypospolitej-polskiej-2016,2,16.html. Accessed 23 May 2017

Central Statistical Office (GUS) (2017) Nakłady na Środki Trwałe Służące Ochronie Środowiska i Gospodarce Wodnej w Polsce w 2016 r. Available at http://stat.gov.pl/obszary-tematyczne/srodowisko-energia/srodowisko/naklady-na-srodki-trwale-sluzace-ochronie-srodowiska-i-gospodarce-wodnej-w-polsce-w-2016-r-,4,6.html. Accessed 2 May 2018

Collins G (2017), Russia's use, of the "Energy Weapon" in Europe. Available at: https://www.bakerinstitute.org/media/files/files/ac785a2b/BI-Brief-071817-CES_Russia1.pdf. Accessed 25 April 2018

Dyka M (2017). Prawo Wodne Po Nowemu- Kluczowe Zmiany Od Lipca 2017. Available at http://www.ekologus.pl/baza-wiedzy/prawo-wodne-po-nowemu-kluczowe-zmiany-od-lipca-2017. Accessed 22 Jan 2018

EIA (2011) World shale gas resources: an initial assessment of 14 regions outside of the United States. Available at: https://www.eia.gov/analysis/studies/worldshalegas/archive/2011/pdf/fullreport_2011.pdf. Accessed 8 Feb 2017

EIA (2013) Technically recoverable shale oil and shale gas resources: an assessment of 137 shale formations in 41 countries outside the United States. Available at http://www.eia.gov/analysis/studies/worldshalegas/pdf/fullreport.pdf. Accessed 23 May 2017

EIA (2016) Trends in U.S. oil and natural gas upstream costs. Available at https://www.eia.gov/analysis/studies/drilling/pdf/upstream.pdf. Accessed 8 Feb 2017

EU Water Framework Directive (2000/60/EC) Available at http://ec.europa.eu/environment/water/water-framework/index_en.html. Accessed 14 May 2018

European Commission (2016) The April Infringements' package: key decisions (MEMO/16/1452). Available at http://europa.eu/rapid/press-release_MEMO-16-1452_en.htm. Accessed 23 Jan 2017

European Commission, Directorate-General for Environment (2016) Report from the Commission to the European Parliament and the Council on the Effectiveness of Recommendation 2014.70/EU on minimum principles for the exploration and production of hydrocarbons (such as shale

gas) using high-volume hydraulic fracturing (COM/2016/794). Available at http://eur-lex.
europa.eu/legal-content/EN/TXT/?uri=CELEX:52016DC0794. Accessed 23 Jan 2018
European Parliament, Council of the European Union (2014) Directive 2014/17/EU of the
European Parliament and of the Council of 4 February 2014 on credit agreements for consumers relating to residential immovable property and amending Directives 2008/48/EC and
2013/36/EU and Regulation (EU) No 1093/2010 Text with EEA relevance. Available at http://
eur-lex.europa.eu/legal-content/EN/ALL/?uri=celex:32014L0017. Accessed 14 May 2018
Gutry-Korycka M, Sadurski A, Kundzewicz ZW, Pociask-Karteczka J, Skrzypczyk L (2014)
Zasoby Wodne a Ich Wykorzystanie. Nauka 1:77–98. Available at: http://www.pan.poznan.pl/
nauki/N_114_07_Gutry.pdf. Accessed 4 Feb 2017
Kantor M, Konieczyńska M, Lipińska O (2015) Prace Rozpoznawcze Dotyczące Gazu z Łupków-
Wyniki Terenowych Badan Środowiskowych. Przegląd Geologiczny (63) Nr. 7. Available at
http://www.pgi.gov.pl/dokumenty-pig-pib-all/publikacje-2/przeglad-geologiczny/2015/lipiec-
4/3163-prace-rozpoznawcze-dotyczace-gazu-z-lupkow/file.html. Accessed 23 Feb 2017
Karcz P (n.d.) Organic matter content in shales, Polish Geological Institute- National Research
Institute.   Available   at   http://infolupki.pgi.gov.pl/en/gas/organic-matter-content-shales.
Accessed 8 Feb 2017
Malinowski D (2018) Gaz Łupkowy Nie Wypalił, ale Technologia Pomaga przy Konwencjonalnych
Złożach. Available at: http://gazownictwo.wnp.pl/gaz-lupkowy-nie-wypalil-ale-technologia-
pomaga-przy-konwencjonalnych-zlozach,317424_1_0_0.html. Accessed 28 Feb 2018
MarcellusGas.Org (2018) Well permit information. Available at: https://www.marcellusgas.org/.
Accessed 5 May 2018
Nelsen A (2014) Poland on road to EU court over shale gas defiance. Available at https://www.
euractiv.com/section/science-policymaking/news/poland-on-road-to-eu-court-over-shale-gas-
defiance/. Accessed 23 May 2017
PAP (2017) Morawiecki: Temat Gazu Łupkowego w Polsce Może Wrócić. Available at: http://
gazownictwo.wnp.pl/gaz_lupkowy/morawiecki-temat-gazu-lupkowego-w-polsce-moze-wro-
cic,295579_1_0_0.html. Accessed 23 May 2017
PAP (2018) Kurtyka:Polska Bedzie Rozwijać Wydobycie Gazu. Available at http://biznesalert.pl/
kurtyka-polska-bedzie-rozwijac-wydobycie-gazu/. Accessed 2 May 2018
Polish Ministry of Environment (MOS) (2016a) Poszukiwania Gazu ze Złóż Łupkowych w Polsce
2007–2016. Available at: http://lupki.mos.gov.pl/pliki/broszura/poszukiwania-gazu-ze-zloz-
lupkowych-w-polsce-2007-2016.pdf. Accessed 23 May 2017
Polish Ministry of Environment (MOS) (2016b) Zasoby Wodne w Polsce. Available at https://
www.mos.gov.pl/kalendarz/szczegoly/news/zasoby-wodne-w-polsce/. Accessed 5 May 2017
Polish Ministry of Environment (MOS) (n.d.-a) Środowisko i Prace Rozpoznawcze Dotyczące
Gazu z Łupków. Available at http://www.gdos.gov.pl/files/aktualnosci/32257/Okreslenie_
zakresu_oddzialywania_procesu_poszukiwania_i_eksploatacji_niekonwencjonalnych_zloz_
weglowodorow_na_srodowisko_raport_koncowy_z_realizacji_badan.pdf. Accessed 23 July
2017
Polish Ministry of Environment (MOS) (n.d.-b) Zestawienie Przepisów Prawnych w Polsce
dot. m.in. Poszukiwania i Wydobywania Gazu z Łupków. Available at http://lupki.mos.gov.
pl/prawo/zestawienie-przepisow-prawnych-w-polsce-dot-min-poszukiwania-i-wydoby-
wania-gazu-z-lupkow.pdf. Accessed on 23 Jan 2017
Polska Fundacja Ochrony Zasobów Wodnych (n.d.) Zasoby Wody. Available at http://www.pfozw.
org.pl/zrodlo-wiedzy/w-budowie-3/. Accessed 24 April 2017
Sawicki B (2017) Az Popłynie Gaz- 5 Lat Poszukiwań Gazu Ziemnego z Łupków. Available at
http://infolupki.pgi.gov.pl/pl/stan-prac-poszukiwawczych/az-poplynie-gaz-5-lat-poszuki-
wan-gazu-ziemnego-z-lupkow. Accessed 11 Feb 2017
Szekalska E (2015), Większość Wód Powierzchniowych Nie Osiągnęła Dobrego Stanu
Ekologicznego.   Available   at   https://www.teraz-srodowisko.pl/aktualnosci/Wody-
powierzchniowe-zly-stan-ekologiczny-841.html. Accessed 23 May 2017

Twoja Pogoda (2015) Polsce Grozi Głęboka Susza, Brak Żywności i Wody Pitnej. Available at   http://www.twojapogoda.pl/wiadomosc/2015-08-11/polsce-grozi-gleboka-susza-brak-zywnosci-i-wody-pitnej_1606398/. Accessed 23 Jan 2017
Water Law (2017) Journal of Laws, item 1566. Available at http://prawo.sejm.gov.pl/isap.nsf/download.xsp/WDU20170001566/T/D20171566L.pdf. Accessed 14 May 2018

Anna B. Mikulska is a nonresident fellow in energy studies at the Baker Institute's Center for Energy Studies, where her research focuses on the geopolitics of natural gas within the EU, former Soviet Bloc and Russia. Mikulska is a senior fellow at University of Pennsylvania's Kleinman Center for Energy Policy, where she teaches graduate-level seminars on energy policy and geopolitics of energy, and a fellow at the Foreign Policy Research Institute. She also serves on the editorial board of the law review at Adam Mickiewicz University in Poland. Mikulska speaks Polish, English, German, Farsi, and Russian. She holds a law degree from Adam Mickiewicz University, a master's degree in international relations from the University of Windsor in Canada, and a Ph.D. in political science from the University of Houston.

Chapter 19
Legal Regulation of Hydraulic Fracturing Activities in Brazil – The Objectives and Achievements to Date

Barbara Bittencourt and David Meiler

Abstract The purpose of this work is to analyze the Brazilian regulations for exploration and production of shale gas through the use of hydraulic fracturing, as well as to discover whether such energy matrix is important for the country. It shall further review the country's requirements for energy sources and provide an overview of the regulations applicable for oil and gas exploration and production activities in Brazil. This chapter outlines the shortcomings of the current applicable regulation for hydraulic fracturing in the country, providing information about the conflicts arising out of the regulatory gaps. Finally, it concludes that there is a long way to go before Brazil develops its shale gas industry and is able to prevent the many problems that may arise out of this controversial method.

Keywords Brazil · Shale gas · Hydraulic fracturing · Fracking · Oil and gas exploration · Energy · Energy security · Regulation · Legal framework

19.1 Introduction

World energy demand is projected to grow, especially as the economies of the Global South grow. Demand for hydrocarbons is on the core of this dependency, driving the need for discovery and development of new oil and gas fields, including unconventional sources such as shale gas, which aims to compensate for the projected decrease of production from conventional sources and source regions.

In this regard, law has a crucial role on regulating the market, giving the necessary background for the development of the oil and gas industry, as a means to guarantee the energy security of the countries (KONOPLYANIK 2008) and, at the same time, safeguard the environment, operations, investors and society, especially

B. Bittencourt (✉) · D. Meiler (✉)
Campos Mello Advogados - in Cooperation with DLA Piper, Rio de Janeiro, RJ, Brasil
e-mail: barbara.bittencourt@cmalaw.com; david.meiler@cmalaw.com

© Springer International Publishing AG 2020   381
R. M. Buono et al. (eds.), *Regulating Water Security in Unconventional Oil and Gas*, Water Security in a New World,
https://doi.org/10.1007/978-3-030-18342-4_19

when it comes to the use of new and controversial technologies, such as hydraulic fracturing.

With the aforementioned proposition in mind, this chapter will analyze the stage of development of hydraulic fracturing in Brazil, demonstrating the country's major energy sources and the role played by natural gas in this context. This chapter will also analyze how the exploration and production of shale gas could positively contribute to the country's energy security. Further, we will describe the Brazilian legal system and legislation applicable to oil and gas exploration and production, providing details about the existing regulation for hydraulic fracturing activities. We will also provide information of the social conflicts and legal cases arising out of the possibility of exploration and exploitation of shale gas in the country with the use of hydraulic fracturing.

Finally, it concludes that although the development of the operations for the production of unconventional gas could conceivably improve the country's energy security, there is a lot yet to be done to develop and implement the applicable regulations for the performance of hydraulic fracturing in Brazil.

19.2   Energy Sources in Brazil

During the last few years Brazil has increased its dependency on non-renewable energy, with oil (together with its by-products) and natural gas the most significant sources used in the country, playing, therefore, an important role in its energy security.

In the year 2016, according to the Ministry of Mines and Energy (BRASIL 2007), the Brazilian domestic market for energy sources was based on 43.5% of renewable energy sources and 56.5% of non-renewable energy sources. Natural gas corresponded to 12.3% of the market, while oil and its by-products corresponded to 36.5%, mineral coal to 5.5% and uranium to 1.5%. The renewable sources were based on sugarcane biomass (17.5%), hydropower (12.6%), wood and charcoal (8%) and other sources (6.1%) (Fig. 19.1).

With specific regard to exploration and production of oil, Brazil ranks amongst the top oil producers in the world, with proven reserves of 12.6 billion barrels of oil equivalent (BOE) and nearly 800 hydrocarbon producing areas under concession. In 2016, the country produced an average of 2730 million BOE per day. The recent discovery of large oil reserves in the so-called pre-salt layers, covering an area of around 800 km across the Brazilian continental shelf, has generated considerable enthusiasm for hydrocarbonbased economic development. The federal government estimates that the country's reserves could amount to nearly 100 billion BOE which, if proven, will indeed place Brazil amongst the most important oil producers in the world (AGÊNCIA NACIONAL DO PETRÓLEO, GÁS NATURAL E BIOCOMBUSTÍVEIS 2017).

The feasibility of profitable exploration and production (E&P) projects in the post-salt and pre-salt layers just offshore but in Brazilian territorial waters has led to investments from major companies worldwide. With this in mind, the oil industry in

Fig. 19.1 Brazilian energy sources. (Source: Ministry of Mines and Energy, "Boletim Mensal de Energia", June 2017)

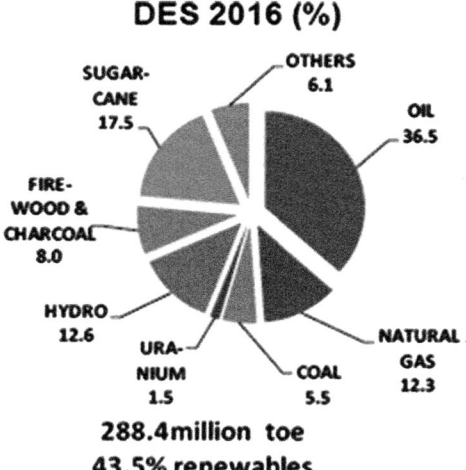

DES 2016 (%)

288.4 million toe
43.5% renewables

Brazil will likely remain a magnet for domestic and overseas investment in the energy sector.

Recent studies estimated that by 2022 the daily production of oil in Brazil will range around 5.5 million BOE. Accordingly, it is expected that Brazil may have a major role in the world's oil market in the next 10 years, acting not only as an exporter of crude oil, but also as an exporter of its by-products, in view of the recent oil discoveries and of the expected enlargement of the countries' refining capacity – currently 18 refineries.

When it comes to natural gas, although demand is not as strong as that for oil and its by-products in the country, it should be noted that currently the Brazilian natural gas consumption is not met entirely through domestic production; thus, there is a need for importation of natural gas to fulfill the country's domestic needs.

For that reason, in order to guarantee the country's energy security, one may conclude that the development of the exploration and production of natural gas should one of the government's top priorities. According to the latest official statistics, from 2004 to 2016, proven reserves of natural gas in Brazil increased 50%, mostly from newly-discovered offshore resources (AGÊNCIA NACIONAL DO PETRÓLEO, GÁS NATURAL E BIOCOMBUSTÍVEIS 2017).

Hence, the production of natural gas in the country is focused on conventional sources, with offshore fields the major contributors – in 2016 offshore production corresponded to 83.3% (AGÊNCIA NACIONAL DO PETRÓLEO, GÁS NATURAL E BIOCOMBUSTÍVEIS 2017) of the total amount of natural gas produced in the country.

In the year 2016, the production of natural gas reached 37.89 billion m³, with conventional gas representing 28.54 billion m³ of such production (AGÊNCIA NACIONAL DO PETRÓLEO, GÁS NATURAL E BIOCOMBUSTÍVEIS 2017).

Even so, demand is such that in order to fulfil domestic needs, the country must import natural gas, mainly from Bolivia.

For that reason, although the exploration and production of shale gas is currently under close scrutiny in respect to environmental and regulatory issues related to its development, the reserves of such unconventional resources may lead Brazil to pursue accelerated development, contributing to the enforcement of the country's energy security, as it would play an important role for the fulfillment of the country's domestic needs.

According to the estimates recently released by the U.S. Energy Information Administration (2015), Brazil may have a shale gas reserve of approximately 226 trillion m^3, the tenth largest reserve of the world and the 2nd largest in South America.

The most promising deposits of shale gas in the country are located in (i) Parecis Basin, in the state of Mato Grosso; (ii) Parnaíba Basin, in the states of Maranhão and Piauí; (iii) Recôncavo Basin, in the state of Bahia; (iv) Paraná Basin, in the states of Paraná and Mato Grosso do Sul; and (v) São Francisco Basin, in the states of Minas Gerais and Bahia. See Map 19.1.

Currently, shale gas operations in the country are concentrated in the southern region of Brazil, more specifically in the city of São Mateus do Sul, located at Paraná State - - the so-called Unit of Operation and Industrialization of Shale ("SIX"), operated by Petroleo Brasileiro SA ("Petrobras"), the Brazilian national oil company. In 2016 the volume of shale processed in the country was 1.6 million of m^3, decreasing on 8.4% when compared with 2015 (AGÊNCIA NACIONAL DO PETRÓLEO, GÁS NATURAL E BIOCOMBUSTÍVEIS 2017). From exploitation of the aforementioned SIX reserves, are obtained the following energy products: shale gas, liquefied petroleum gas ("LPG"), naphtha and fuel oil. Also produced area number of non-energetic by-products. When evaluating the estimate reserves of shale gas and its volume of production in the country, one may conclude that, although Brazil may have important reserves of such unconventional source, its exploration, production and processing are not yet significant.

The production of shale gas in 2016 was of 5.2 tons, being 33.4% less than the production of 2015. The same happened with LPG, which production in 2016 decreased by 14.5%, being produced 20.6 thousand m^3 (AGÊNCIA NACIONAL DO PETRÓLEO, GÁS NATURAL E BIOCOMBUSTÍVEIS 2017) (Table 19.1).

When evaluating the estimate reserves of shale gas and its volume of production in the country, one may conclude that, although Brazil may have important reserves of such unconventional source, its exploration, production and processing are not yet significant.

There are several causes for the shale gas slow development in Brazil, but the most important one seems to be related to the uncertainties brought up by the current legislation/regulations in place. Accordingly, specific laws need to be enacted with regard to the particulars of the operation and environmental impacts linked to application of hydraulic fracturing for energy production.

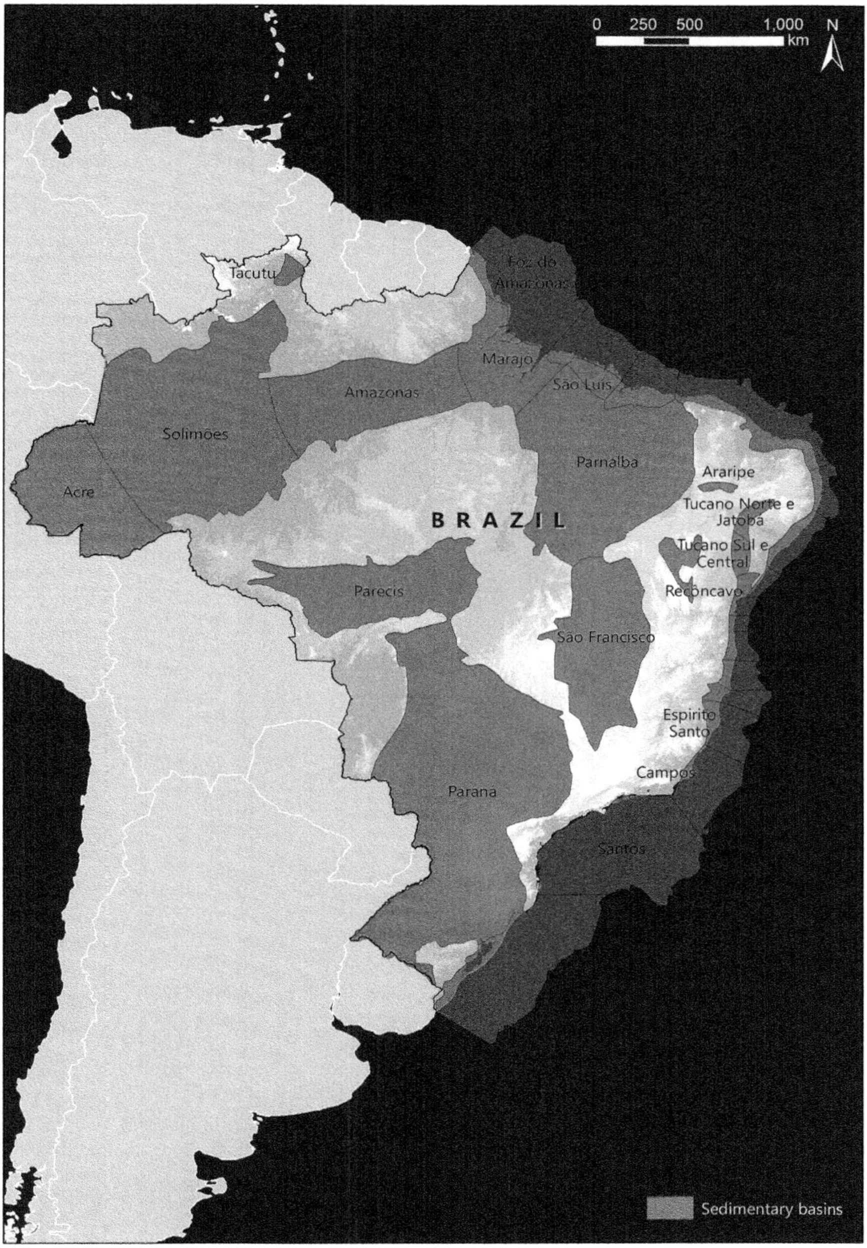

Map 19.1 Onshore sedimentary basins with prospect for shale gas in Brazil. (Callum Foster, UWE Cartography)

Table 19.1 Volume of processed shale oil and production of shale oil products

Specification	Unit	Volume of processed shale oil and production of shale oil products										
		2007	2008	2009	2010	2011	2012	2013	2014	2016	2016	16/15%
Processed shale oil	t	2.343.086	1.925.285	2.117.820	2.069.197	1.579.347	1.732.378	1.458.191	1.655.484	1.696.947	1.564.895	−8,37
Products												
Energy												
Shale gas	t	18.756	13.087	14.314	16.992	13.128	10.619	8.109	8.424	7.752	5.162	−33,41
LPG[a]	m³	23.624	18.529	27.044	26.761	18.766	24.122	21.563	25.419	24.164	20.663	−14,49
Fuel oil	m³	102.544	155.691	270.576	281.779	213.014	244.754	216.689	237.961	219.913	217.955	−0,89
Non-energy												
Naphtha[b]	m³	48.083	37.725	40.809	42.536	33.112	31.689	24.001	28.512	25.824	29.813	15,45
Others[c]	m³	4.012	2.349	1.548	3.145	3.418	2.587	2.374	1.932	296	–	–

Source Petrobras

Source: National Agency of Oil, Natural Gas and Biofuels – ANP, "Oil, Natural Gas and Biofuels Statistical Yearbook 2017", August 31, 2017

[a]Propane and butane included. [b]Naphtha production is sent to Repar, where is used as raw material in the production of oil products. [c]Other non-energy oil products of lesser importance included

19.3   Overview of the Oil and Gas Industry and Regulations in Brazil

Before dealing directly with the regulations related to the production of shale gas in Brazil, we offer an overview of the country's oil and gas industry and its applicable legal framework.

The 1988 Brazilian Federal Constitution grants to the Federal Union ownership over mineral resources within the Brazilian territory, which are deemed an independent asset from the soil wherein they are located.

Article 176. Mineral deposits, under exploration or not, and other mineral resources and the hydraulic energy potential form, for the purpose of exploitation or use, a property separate from that of the soil and belong to the Union, the concessionaire being guaranteed the ownership of the mined product. Paragraph 1 – The prospecting and mining of mineral resources and the utilization of the potentials mentioned in the caption of this article may only take place with authorization or concession by the Union, in the national interest, by Brazilians or company organized under Brazilian laws and having its head office and management in Brazil, in the manner set forth by law, which law shall establish specific conditions when such activities are to be conducted in the boundary zone or on Indian lands. Paragraph 2 – The owner of the soil is assured of participation in the results of the mining operation, in the manner and amount as the law shall establish. Paragraph 3 – Authorization for prospecting shall always be for a set period of time and the authorization and concession set forth in this article may not be assigned or transferred, either in full or in part, without the prior consent of the conceding authority. Paragraph 4 – Exploration of a renewable energy potential of small capacity shall not require an authorization or concession. (BRASIL 1988, free translation).

During the mid-1990s, certain constitutional amendments were introduced in order to create the basis for an open market policy and minimum state intervention in the Brazilian economy. In such scenario, constitutional amendment EC No. 09/1995, which modified article 177 of the 1988 Brazilian Federal Constitution, authorized the Federal Union to hire both state-owned or private companies to perform activities related to oil and gas exploration and production. Up to that moment, only Petrobras, the Brazilian national oil company, was entitled to perform such activities.

Article 177. The Following are the monopoly of the Union: I – Prospecting and exploration of deposits of petroleum and natural gas and of other fluid hydrocarbons; II – refining of domestic or foreign petroleum; III – import and export of the products and basic by-products resulting from the activities set forth in the preceding items; IV – ocean transportation of crude petroleum of domestic origin or of basic petroleum by-products produced in the country, as well as pipeline transportation of crude petroleum, its by-products and natural gas of any origin; V – prospecting, mining, enrichment, reprocessing, industrialization and trading of nuclear mineral ores and minerals and their by-products. Paragraph 1 – The Union may contract with state-owned or with private enterprises for the execution of the activities provided for in items I through IV of this article, with due regard for the conditions set forth by law. Paragraph 2 – The law referred to in paragraph 1 shall provide for: I- a guarantee of supply of petroleum products in the whole national territory; II- the conditions of contracting; III – the structure and duties of the regulatory agency of the monopoly of the Union. Paragraph 3 – The law shall provide with respect to the transportation and use of radioactive materials within the national territory. Paragraph 4 – The law which institutes the contribution of intervention in the economic order levied on activities of importation or commercialization of petroleum and its by-products, natural gas and by-products and alcohol fuel shall observe the following: I- the rate of the contribution may: a) be established in accordance with the product or its use; b) be decreased and re-established by act of the Executive Power, the provisions of article 150, III, b no being applicable; II – the proceeds shall be used for: a) the payment of subsidies to prices or transportation of alcohol fuel, natural gas and by-products and petroleum by-products; b) the funding of environmental projects related to the gas and petroleum industry; c) the funding of infrastructure of transportation. (BRASIL 1988, free translation).

Said EC No. 09/1995 was followed by the enactment of Law No. 9.478/1997, also known as the Petroleum Law, which changed the entire legal framework of the oil and gas sector in Brazil. The Petroleum Law introduced the concession regime for the exploration and production of oil and gas in the country. With this new approach, the government authorized different economic agents to explore and produce oil and gas under par-ticular concession agreements, upon bid proceedings conducted by the National Agency of Petroleum, Natural Gas and Biofuels ("ANP").

The concession regime has been successfully used in Brazil since the opening of the oil and gas market to private companies. Under the Concession Agreement that is entered into by and between ANP and the companies that win its bid proceedings, the exploration and production activities are carried out at the sole risk of the oil companies, who will have the sole ownership of the oil and gas so produced.

Later on, the discovery in 2007 of large oil and gas reserves concentrated in the pre-salt layers encouraged the Brazilian government to enact Law No. 12.351/2010 (the "Pre-Salt Law"), which created a new regime for the pre-salt and strategic areas, by imposing the Production Sharing Agreement ("PSA") in lieu of the concession agreement that is still applicable to all other areas. According to the Pre-Salt Law, Petrobras, the Brazilian national oil company, will have the preference to be the operator and hold a minimum stake of 30% in any consortium formed with private oil companies that win the public ANP bid proceedings. Such preference was regulated by Decree No. 9,041/2017, which provides the rules and definitions that shall be applied for the execution of such preference by Petrobras.

Under the production sharing regime, oil companies shall bear all the risks of the activity, even though the production is the property of the federal government. In case of a commercial discovery, oil companies will then recover the costs and investments made and they will be entitled to a certain percentage of the remainder of the production, according to the provisions of the PSA. The parties to the PSA shall be the Federal Union, represented by the Ministry of Mines and Energy ("MME"); ANP, as the regulatory and supervising agency; PPSA, the state-owned company that was created to manage the PSAs; Petrobras (if it used its preference right) and other relevant oil companies.

Although the execution of the PSA is usually preceded by a bid proceeding, Petrobras may also be directly hired, whenever such measure is deemed relevant for preserving national interests and the attainment of the energy policy's goals.

When it comes to activities related to natural gas, Law No. 11.909/2009 (the "Gas Law") regulates the transportation, treatment, processing, storage, liquefaction, regasification and commercialization of natural gas, as well as the import and export thereof.

The Brazilian natural gas market is also regulated by the ANP along the entire value chain, from exploration and production sites to end user. From there on, regulatory agencies in each state are responsible for the regulation of the distribution of such products to final con-sumers. Both State and private companies, either national or international, are free to participate in all stages of the gas value chain. The Brazilian Federal Constitution authorizes the states to exploit either directly or through concessions the services related thereto. Therefore, such activity is regulated by the states and their respective agencies.

In Brazil, natural gas is mainly used in the domestic market by the industry and transport sectors, since the country's climatic and geographical features make LPG of major use in the residential sector.

19.4   Shale Gas – Regulation and Exploration

Currently, the only piece of regulation dealing exclusively with exploration and production of shale gas in Brazil is Resolution No. 21/2014, issued by the ANP on 10 April 2014, aiming to regulate the operations of hydraulic fracturing in unconventional gas fields in the country. After protracted discussion with the oil and gas industry, the enactment of this Resolution was mainly motivated by the fact that the 12th ANP Bidding Round, which was held in 2013, included under its offered assets areas with potential for exploration and production of unconventional gas through the use of hydraulic fracturing. Among other provisions, Resolution No. 21/2014 imposes the obligation for operators to adopt an environmental management system and to perform previous technical studies, in order to obtain approval for the operations (i.e. fracturing simu-lations and risk analysis). Also, environmental licences, with specific authorization for operations of hydraulic fracturing are required prior to ANP's authorisation for the hydraulic fracturing activities.

Moreover, said Resolution No. 21/2014 also implements standards to be observed by operators during the activities of hydraulic fracturing, as well as requiring determines the preparation of an emergency response plan.

It should be noted that Resolution No. 21/2014 is the subject of numerous complaints, mainly from the Brazilian Public Prosecutor's office and other society representatives, mostly because it may surpass, without the required knowledge and technical basis, the powers and obligation of the environmental authorities. That is mainly because, nowadays, Brazil does not have specific procedures or recommendations from the environmental authorities regarding precisely the activities of exploration and production of shale gas.

This fact brings uncertainties and lead to several discussions in regard to the safety of hydraulic fracturing operations, especially in view of the risks of contamination of water reservoirs and the need for excessive use of potable water.

Accordingly, one of the main issues discussed in Brazil is related to which is the key competent authority to assume responsibility for the licensing procedures and required inspections to environmental licenses for hydraulic fracturing. According to the Brazilian environmental legislation, specifically Resolution CONAMA No. 237 (BRASIL 1997), article 5th, for activities conducted onshore, member states' environmental agencies should be the authorities responsible to perform environmental licensing procedures and the consequent inspection of the operations to be implemented

Considering the location of some fields, specific operations of hydraulic fracturing may affect the territory of more than one member state. In this case, according to Resolution CONAMA No. 237 (BRASIL 1997), article 4, it would be required the involvement of the Brazilian Federal Environmental Agency ("IBAMA") to perform licensing procedures, as well as to assume all of the applicable environmental regulation. However, there is no definition of the specific situations related to hydraulic fracturing where it would be required the involvement and transference of environmental regulation to IBAMA. On other words, it remains unclear which is the competent authority to deal with the environmental licensing for activities of exploration and production of unconventional gas in Brazil.

Another important matter currently discussed is the need for official guidelines for the performance of environmental impact studies required for hydraulic fracturing activities. Such studies would serve as a basis for imposing restrictions to the relevant environmental licenses or even to sustain its denial, if such studies conclude that the impacts to the environment would cause major risks to the affected environment wherever applicable.

Such inconsistencies and regulatory gaps led to judicial disputes related to the legality of the aforementioned 12th Bidding Round, and consequently of the concession agreements subsequently executed (related to areas for exploration and production of unconventional gas) and, thus, caused several discussions regarding the feasibility of the implementation of hydraulic fracturing in the country.

19.5   Conflicts Related to Hydraulic Fracturing Activities

As mentioned under item "IV" above, following the 12th Bidding Round of concessions for the exploration and production of unconventional gas on certain areas located in the Parecis, Recôncavo, Acre, Parnaíba, São Francisco and Paraná basins, several Public Civil Claims were filed in different courts of the country, which areas were included in the scope of the 12th Bidding Round.

Accordingly, the aforementioned Public Civil Claims received different numbers, and were filed before the Federal Courts of the State member which the areas, as per the table below:

Number of the Lawsuit	Court	Area of Fracturing
0005610-46.2013.4.01.4003	Federal Court of Piauí	Parnaíba River Basin
0030652-38.2014.4.01.3300	Federal Court of Bahia	Recôncavo Basin
0800366-79.2016-4.05.8500	Federal Court of Sergipe	Sergipe-Alagoas Basin
0001849-35.2015.4.01.3001	Federal Court of Acre	Acre Basin
5005509-18.2014.4.04.7005	Federal Court of Paraná	Paraná River Basin

The intention of such Public Civil Claims was mainly to block the exploration of shale gas in the aforementioned areas with the use of hydraulic fracturing, con-sidering the lack of proper regulation and technical studies capable to address and avoid environmental damages that could arise out of such activities. In this sense, it has been required under the aforementioned lawsuits (i) to recognize the nullity of the effects arising out of the 12th Bidding Round, as well as of the concession agreements, and (ii) the prevention of ANP to perform new bidding rounds and/or to execute new concession agreements for operations in such areas.

As a result, several injunctions and other judicial decisions were issued (some of the claims do not have a final decision yet), granting the aforementioned Public Civil Claims, so as to prevent the exploration and produc-tion of shale until (i) new regulations are enacted for the use of hydraulic fracturing and (ii) further environmental procedures and studies are performed, so as to avoid risks to public health, economy, and environment.

It shall be noted that the population itself has also reacted against the 12th Bidding Round and organized a protest in the city of Toledo, at Paraná State (where it is located the greater part of the Brazilian shale reserve), when approximately 1000 people participated. Such protest was led by the organization called *COESUS – Coalização Não Fraking Brasil*, which is composed by Brazilian scientists, environmentalists, engineers, among other professionals, and constantly acts, mainly through the participation of hearings with Brazilian authorities and publication of articles, against the exploration and production of shale gas in the country.

19.6   Conclusions

Considering the existing national reserves, the exploration and production of shale gas can play an important role in energy security, since Brazil was listed as one of the most promisor producers.

Although the development of the operations for the production of unconventional gas can lead Brazil to a promising position as a natural gas producer and contribute for the enforcement of the country's energy supply, there is yet a long way to go before the development and implementation of such activities. The main concerns are related to the lack of an appropriate legal framework, which is capable to address the potential risks that hydraulic fracturing may pose on the environment, society and local economy.

As we also see in other countries, mainly in Europe, the activities dealing with unconventional gas in Brazil are in the early stages. It is advisable that the Brazilian government should commission studies and exchange information with other nations (that have experi-ence in the exploration and production of shale gas), in order to create a proper legal framework — especially environmentally oriented — addressing authorities and society's concerns as well as removing uncertainties for companies and potential investors. If these steps are taken, Brazil may finally be in the right track for a sustainable development of its shale gas exploration and production.

Bibliography

AGÊNCIA NACIONAL DO PETRÓLEO, GÁS NATURAL E BIOCOMBUSTÍVEIS (Brasil) (2017) Anuário estatístico brasileiro do petróleo, gás natural e biocombustíveis: 2017. ANP, Brasília

BRASIL. Constituição (1988). Constituição da República Federativa do Brasil. **Diário Oficial [da] República Federativa do Brasil**, Poder Executivo, Brasília, DF, 05 out. 1988. Seção 1, p. 01

BRASIL. Ministério do Meio Ambiente. Resolução CONAMA n° 237, de 19 de dezembro de 1997. Dispõe sobre a revisão e complementação dos procedimentos e critérios utilizados para o licenciamento ambiental. **Diário Oficial [da] República Federativa do Brasil**, Poder Executivo, Brasília, DF, 22 dez. 1997. Seção 1, p. 30841

BRASIL. Ministério de Minas e Energia. **Resenha energética brasileira**: exercício de 2016. Brasília, DF, 2007. 32 p

KONOPLYANIK, Andrey A. Energy security: the role of business, governments, international organisations and international legal framework. **Oil, Gas & Energy Law**, n. 3, 2008. Available at: https://www.ogel.org/article.asp?key=2790. Access 9 Apr 2018

U.S. ENERGY INFORMATION ADMINISTRATION. Technically recoverable shale oil and shale gas resources: Brazil. **Advanced Resources**, sept. 2015. Available at: https://www.eia.gov/analysis/studies/worldshalegas/pdf/Brazil_2013.pdf. Accessed 9 Apr 2018

Bárbara Bittencourt is a senior associate in Campos Advogados' Oil and Gas team, working there since 2013. Previously, she was an associate at Felsberg & Associados and Vieira Rezende Advogados. Bittencourt has experience representing clients in the oil and gas industry in Brazil (mainly drilling and support services companies), providing assistance in a number of transactional areas such as importing, storage and distribution of parts and services, assistance in obtaining permits, licenses or authorizations from the necessary regulatory bodies and agencies, as well as provided consulting support for companies engaged in the oil and gas industry in Brazil. She was recognized as a "Rising Star" in Energy and Infrastructure by the publication IFLR 1000 in

2018, and she holds a B.A. from the Faculdade de Direito Milton Campos and a LL.M from the University of Aberdeen.

David Meiler is a partner at Campos Mello Advogados in the Oil and Gas Area, based in Rio de Janeiro. Before joining Campos Mello in 2013, he was a partner at Felsberg & Associados, and worked as an associate at Steel Hector & Davis, LLP. Meiler advises domestic and international companies involved in the exploration and production of oil and gas in Brazil. He has provided assistance in a number of transactional areas such as importing, storage and distribution of parts and services, assistance in obtaining permits, licenses or authorizations from the necessary regulatory bodies and agencies, drafting and amending contracts specific to the industry, as well as actively providing auditing, consulting and litigation support for environmental, and other civil legal issues within the Oil and Gas industry. Meiler was recognized as a *Leading Individual in Energy* by Chambers Global and Chambers Latin America from 2013 to 2018. He was a recommended lawyer by The Legal 500 from 2015 to 2017, and a recommended leading individual in Energy and Infrastructure by IFLR 1000 from 2014 to 2018. He holds a B.A. in Law from the Universidade Federal do Paraná and a LL.M. from the University of Pittsburgh.

Part V
Conclusions and Recommendations

Chapter 20
Regulating Water Security in Unconventional Oil and Gas: Common Challenges, Trade-Offs, and Best Practices from Around the Globe

Chad Staddon, Regina M. Buono, Elena López Gunn, and Jennifer McKay

Abstract This volume addresses the growing need to improve understanding of effective regulatory and policy regimes in relation to water used to operate unconventional hydrocarbon operations around the world. As the chapters in this book clearly show, legal, policy, regulatory, and political issues surrounding the use of water for hydraulic fracturing are present at every stage of operations. These include direct impacts related to the procurement of water for use in hydraulic fracturing, the collection of flowback and produced water, and the safe disposal via treatment, reuse or sale, or otherwise, of produced or other wastewaters. Also important are more indirect impacts including those related to air quality, induced seismicity and local community support. This book analyses and compares various approaches to these issues from around the globe to glean insights into common difficulties and best practices to develop and advance the interests of all stakeholders, including the natural environment and present and future generations. While it is not possible (or advisable) to simply transfer aspects of law and governing institutions from one place to another (the "cut and paste" approach), there is value in the comparative examination and understanding of legal regimes. International law may also have a role here in terms of helping to create a clear

C. Staddon (✉)
Department of Geography and Environmental Management,
University of the West of England, Bristol, UK
e-mail: chad.staddon@uwe.ac.uk

R. M. Buono
Center for Energy Studies (CES), Baker Institute for Public Policy,
Rice University, Houston, TX, USA

Lyndon B. Johnson School for Public Affairs, University of Texas, Austin, TX, USA

E. López Gunn
ICATALIST, Madrid, Spain

J. McKay
Professor of Business Law, Law School, University of South Australia Business School,
Adelaide, SA, Australia

© Springer International Publishing AG 2020
R. M. Buono et al. (eds.), *Regulating Water Security in Unconventional Oil and Gas*, Water Security in a New World,
https://doi.org/10.1007/978-3-030-18342-4_20

framework for water security in the context of regulating unconventional energy production that individual states can tailor to their local conditions.

Keywords Water · Unconventional hydrocarbons · Hydraulic fracturing · Policy · Law

20.1   Introduction

This volume addresses the growing need to improve understanding of effective regulatory and policy regimes in relation to water used to operate unconventional hydrocarbon operations around the world. As the chapters in this book clearly show, legal, policy, regulatory, and political issues surrounding the use of water for hydraulic fracturing are present at every stage of operations. These include direct impacts related to the procurement of water for use in fracturing, the collection of flowback and produced water, and the safe disposal via treatment, reuse or sale, or otherwise, of produced or other wastewaters. Also important are the more indirect impacts such as those related to air quality, induced seismicity and local community support. This book analyses and compares various approaches to these issues from around the globe to glean insights into common difficulties and best practices to develop and advance the interests of all stakeholders, including the natural environment and present and future generations. While it is not possible (or advisable) to simply transfer aspects of law and governing institutions from one place to another (the "cut and paste" approach), there is value in the comparative examination and understanding of legal regimes. International law may also have a role here in terms of helping to create a clear framework for water security in the context of regulating unconventional energy production that individual states can tailor to their local conditions.

Taken together, the chapters indicate that no country has yet articulated a clear, systematic and effective system for regulating unconventional hydrocarbons, though there are areas of relative success. For all the industry chatter about becoming quicker and more efficient, and therefore cheaper, as more unconventional resource plays come online, there can be no 'one size fits all' approach. Although shale gas has dominated the story so far, there are somewhat different issues attendant on other unconventional hydrocarbons, especially tight oil and coal-bed methane,[1] which have also been explored in this volume. Common themes and key issues encountered in the countries covered in this volume include:

- The need to balance trade-offs between facilitating economic activity related to the production of fossil fuels to meet energy demand and protection of human and environmental health.
- The increasingly intricate linkages between water and energy (the "water-energy nexus"), and the inability of siloed government ministries/departments to achieve the regulatory coordination and integration necessary to appropriately oversee

---

[1] Called "coal seam gas" in some countries.

and regulate unconventional energy operations, including especially the management of necessary water resources.

- The geopolitics of energy and energy security may override environmental or social concerns about pursuing unconventional hydrocarbons.
- The impacts of energy development on the rights of indigenous communities are only partially, and tenuously, recognised the jurisdictions examined in this volume.
- Industrial secrecy regarding use and potential contamination of water is a problem in many jurisdictions, though less so in regions like the EU where there is a well-established "environmental right to know".
- National implementation of super-ordinate regulatory frameworks, such as the EU Water Framework Directive or international operational standards (e.g. ISO), varies across countries. Though there is a better articulated and integrated regulatory system in the EU, problems are still evident as seen in differences between Poland and the UK.

Many of these issues touch on the acquisition and maintenance of the "social licence to operate," a key point raised especially by Bradbury and Cox in their contribution to this volume. A social licence to operate (SLO) refers to the acceptance or approval by local communities and stakeholders of hydraulic fracturing operations in their communities. It is based on the idea that institutions and companies need *not only* formal regulatory permission but also "social permission" from their communities to conduct business. Smith and Richards (2015, p. 89) described the social license to operate as "a tool whereby companies manage sociopolitical risk by conforming to a set of implicit rules imposed by their stakeholders [and ...] an ongoing social contract with society that allows a project to both start and continue operating in a community." In liberal democracies such as the U.S. and U.K., the SLO must be procured by private sector developers who must win community approval including those elements skeptical or opposed to their operations (e.g. the protests against Cuadrilla's operations in the northwest of England). This is an important issue beyond the liberal democracies of Western Europe and North America. Even in countries where energy economics and politics are more state-led, such as China, Russia and the countries of Latin America, the SLO is still necessary. In the chapter by Shao et al we saw that Chinese state authorities are making efforts to ensure that localities are more likely to support development than oppose it.

Given that the United States was the first nation to actively pursue wide-scale hydraulic fracturing of shale reserves, it is not surprising that there is a larger literature on water and shale gas deriving from the American experience than elsewhere. Four of the 20 chapters in this volume directly explore dimensions of the U.S. experience (Collins and Rosen, Webb and Zodrow, Miller, and Ehrman), while three others discuss U.S. issues comparatively (Palmer et al., Bradbury and Cox, and Sanchez). Yet, as the chapters by Collins and Rosen and Webb and Zodrow in particular illustrate, the U.S. experience quickly unpacks into regional or state-by-state case studies expressing considerable diversity of approach. What happens in one state (for example, Texas) may be quite different from what happens in another (say, Pennsylvania or North Dakota). Even more complex is the situation with respect to shale plays and water sources that *cross* state boundaries (Figure 1.1: Major shale gas basins around the world). Depending on how they are enumerated, the chapters in this volume cover as many as 24 separate national, state or regional jurisdictions. In this concluding chapter we summarise the key findings from the preceding chapters and point towards areas for future research.

20.2   Framing Hydraulic Fracturing as an Object of Regulation

The first part of this book focuses on general frameworks and contexts for considering water issues in unconventional oil and gas production. One of the key underlying themes of the book is the need to balance trade-offs between water security and energy security – the "water-energy nexus". Though water is vital to many aspects of energy production and energy is vital to water infrastructure and processes, the "conventional" water and energy sectors often operate in silos, blind to their reciprocal interdependencies as well as those with other sectors such as agriculture. This has the potential to trigger inter-sectoral competition for scarce water resources, for example with public water supply or agriculture or other industrial processes. As Mroue et al. point out in this volume, the array of competing stakeholders involved in decision-making processes includes the food and agriculture, industrial, utility, public health, financial, energy, water, and environmental sectors. More sophisticated knowledge of the interlinkages between water and energy, strong institutional collaboration, data transparency, and effective communication among stakeholders are key to maximizing gain, optimizing trade-offs, and reducing negative impacts. Such awareness is, unfortunately, not manifest in all cases covered in this volume. A clear knowledge gap between the scientific research community and decision-makers persists, greatly hindering realisation of national and global resource management and climate objectives. Tools and methods, including several discussed by Mroue et al. (see also Brown's discussion of the UK), that allow decision makers to understand the scenarios associated with trade-offs inherent in unconventional production will be increasingly critical to effective, systematic regulation of unconventional hydrocarbons.

The political backdrop to water scarcity and limited water resources places an even greater emphasis on the role of public policy and governance to steer choices toward long-term sustainability. In their chapter in this volume Palmer et al. reference the concepts of "peak oil" and "extreme energy" as lenses to describe the range of "relatively new, higher-risk, non-renewable resource extraction processes that have become more attractive to the energy industry as more easily accessible supplies dwindle…". Peak Oil (or Peak Hydrocarbons) is often conceptualised in terms of absolute resource scarcity, though it is more useful to see it in terms of the amount of water, energy, and other resources necessary to produce a unit of energy in a world where hydrocarbons are becoming increasingly hard to get.[2] Conventional hydrocarbons are relatively efficient in these terms, but part of what defines

---

[2] Much work elsewhere suggests that absolute scarcity is not on the horizon, only growing scarcity of relatively easily extracted hydrocarbons.

unconventional hydrocarbons as unconventional is the increasingly high cost of production in terms of water, energy and other necessary inputs.

This is where policy and regulation can play a key role in helping to shape more sustainable energy (and water) futures, or hindering our progress in this necessary direction. Both Palmer et al. (this volume) and Bradbury and Cox (this volume) note that combinations of "weak" and "strong" laws and regulations can create a balance of regulatory structure and fluidity that, perhaps surprisingly, *can* deliver good energy governance. The chapters in Part I also demonstrate the importance of polycentric and multilevel governance of the water/energy nexus. For example, events and actions at the international level, including UN declarations and policy agendas from Mar de Plata (agreed in 1977, see Staddon 2010) through to the SDGS (inaugurated in 2015), have played an important role in advancing norms and policy ideas around the globe. These can either dovetail with national constitutions and legal frameworks, such as the Bolivian Constitution's recognition of a human right to the environment,[3] or clash with them, as in nations like the U.S., which offers a mosaic of state laws that greatly vary as to how hydraulic fracturing is regulated between states and gives little regard to rights and norms under international law. Supranational legislation and normative powers emanating from regional entities, such as the EU's REACH Directive (EC 1907/2006), which incorporates a duty to disclose chemicals used in industrial processes, the Water Framework Directive (EC 2000/60/EC) and the Environmental Impact Assessment Directive (2011/92/EU), have also created a successful combination of hard law (prescriptive and obligatory) and soft law (principles such as the precautionary principle and the environmental right to know).

The chapter by Palmer et al. pays particular attention to the role of international law and, specifically, the role of human rights declarations in helping to legally and discursively frame regulation. They are particularly concerned with the notion of a human right to water and the concept of "vital human needs," wherein drinking water is the most vital of human needs, "for leading a life in human dignity" that "entitles everyone to sufficient, safe, acceptable, physically accessible and affordable water for personal and domestic uses" (p. 53). This introduces the novel concept of "dignity" into the framing of the human right to water, and of what it means to be human and to experience "a lack or a denial to clean freshwater fails to meet this minimum standard of dignity". Though generally welcomed, such declarations have only had limited practical effect, being largely unenforceable, except perhaps by large corporations that can claim such rights as "legal persons" in most legal systems (Staddon et al. 2011).

Here the role of national legal frameworks comes to the fore, as in the case of Bolivia and its explicit recognition of environmental rights and human rights. What is perceived as the inalienable nexus between human and environmental rights— e.g. the Bolivian alternative to the "control paradigm"—is the precursor to sensible

---

[3] Some states are going further and recognising the rights of the natural environment, as in New Zealand's recognition of the Whanganui River as a "legal person" entitled to equal protection under the law and Indian courts' recognition of the Ganga River as a "living ancestor".

resource security: "protection of water from contamination" is a right. Article 124 of the Bolivian Constitution also states that citizens who violate the "constitutional regime of natural resources" are committing an act of treason against the country. This, in reality, will signify potential conflicts between national laws and existing operators, such as the rights of Halliburton in Bolivia to exploit existing reserves and the recognition of a human right to the environment in the constitution and national laws.

As unconventional energy technologies such as hydraulic fracturing develop, they often outstrip the speed of regulatory evolution, meaning that legal/regulatory gaps are un-addressed or that law and policy are borrowed from adjacent or analogous areas (e.g. the application of mine tailings regulations to manage flowback from hydraulic fracturing operations). Whether the regulatory framework is able to appropriately respond to new science and technology depends in part on the ability to develop comprehensive, bespoke guidelines that can regulate these 'pioneer' operations while not strangling innovation. There is much evidence, presented in this volume and elsewhere, which shows that the science-policy interface is anything but smooth and seamless. Palmer et al. point out that even when very good science suggests that the risks are too high, such warnings are often ignored by decision-makers. This is especially likely when regulatory responsibilities and objectives are unclear (best case) or when they are overridden by political expediencies (worst case), as in Russia and Argentina.

The last chapter in Part I looks at the interaction of law and public participation and the importance of sound legislative frameworks that establish fair, transparent, and equitable allocation rules and offer avenues for redress by civil society. Bradbury and Cox offer a pragmatic and rich analysis of conflict and public concern over the regulation of hydraulic fracturing, which is, as we have so often seen, mired in controversy and popular opposition. The authors consider regulations and mechanisms for resolution in a variety of jurisdictions, including key factors for community cooperation and confrontation and insights into cases of social resistance and the loss by the hydrocarbon industry of the social licence to operate. Social conflict regarding unconventional production arises from concerns over water rights and use priorities, potential contamination of groundwater resources, transparency regarding the chemical components of injected fluids and produced water, and seismic activity induced by fracturing or disposal activity. Passage of laws to criminalise protests against hydraulic fracturing has been a feature of some jurisdictions, notably the U.S. state of Oklahoma[4] and in Australia in Tasmania, New South Wales, and Queensland (McKay 2018). Bradbury and Cox emphasize the need for open dialogue between citizens, governments, and industry about effective regulation, which underscores the need for transparency and education, as well as creativity, innovation, and a commitment to finding solutions.

---

[4] House Bill 1123 provides that a person who enters property intending to harm critical infrastructure or impede its operation will face a felony charge and a minimum $10,000 fine or a year in prison or both, and that if successful in affecting the infrastructure will receive a $100,000 fine or 10 years of imprisonment or both. The statute has been earmarked for codification in the Oklahoma Statutes as Section 1792 of Title 21 (Crimes and Punishments).

20.3 Acquiring Water for Hydraulic Fracturing: Conflicts and Regulatory Issues

The chapters in Part II explore issues around how unconventional operators acquire the quantities of water necessary for their activities. As noted throughout this book, production of hydrocarbons via hydraulic fracturing requires huge volumes of water, whilst the reserves themselves are often located in areas of acknowledged water scarcity (e.g. Texas and Oklahoma, eastern Ukraine, western China, the South African Karoo region, etc.). Even in relatively wet environments like the UK or Siberia, it is often the case that reserves are located in areas where pre-existing water demands have prior claims on available resources. There may also be unexpected water-related challenges, as in the sustainable management of permafrost-linked ecosystems in Siberia. In such cases trade-offs are clearly called for, though regulators do not yet have the tools they need to understand the trade-offs they are increasingly required to adjudicate and regulate. Consideration of these issues is the collective purpose of the six chapters that make up this section.

The first chapter in this section looks at water issues in some of the U.S.'s most active shale plays—the Eagle Ford, Barnett and Permian shales—located mostly in Texas. Such is the pressure to further develop shale gas in the United States that it is clear that water demand is likely to continue to grow quickly well into the future. Collins and Rosen tell us that:

> … at its near-term peak in the fall of 2014, daily frac water demand across the U.S. unconventional plays was 8.3 million barrels per day, as calculated from FracFocus disclosure data. To put that number in perspective, it is approximately 3.5 times the daily average water use of Washington D.C., or enough to fill NRG Stadium (home of the Houston Texans) twice over. (Collins and Rosen, Chap. 5, this volume).

Texas accounts for approximately half of total national water abstractions for hydraulic fracturing, with four other states (North Dakota, Pennsylvania, Colorado and Oklahoma) accounting for most of the rest. Collins and Rosen identify a new trend, the rise in the number of "super fracs", where water consumption is significantly higher per well due to either supercharging existing wells or horizontal drilling more densely from existing well pads. Both strategies are designed to boost operational efficiency and, with little regard for the environment, are unlikely to be scaled back without stronger regulation, something that seems unlikely in the current political climate in the United States. Moreover, where shale basins are shared between states there may be a role for either bilateral agreements or for the federal government.

In the United States, the way an operator acquires water for use in operations is conditional on the local/state structure of water rights. As Staddon (2010) points out, east of the Mississippi River the water allocation doctrine of "riparianism" predominates, whilst west of the Mississippi "prior appropriation" is the main allocation principle. Property ownership complicates the picture, particularly with respect to groundwater, as do the legal traditions of specific states. In Texas groundwater (but not surface water) rights are held by the property owner, creating a situation where operators may acquire water rights as part of the same private transaction that gave access to the land necessary for well pad development. As we saw in the UK and other jurisdictions, it is more common for subsurface rights to be vested, as *res communes*, in central or regional levels of government. In certain areas of the U.S., rights acquisition can become even more complex. For example, in Colorado the axis of power between private interest and public regulation is mediated by a system of "water courts," which can and do exercise significant allocative authority, especially in over-allocated catchments and during drought periods (e.g. the last several years). There are cases, reported by Collins and Rosen in their chapter, of water services "middlemen" arising in specific basins, buying rights from rights holders and selling them on to operators as "bundled offers".

State permitting regimes are often required by law or public pressure to include environmental considerations in licence applications, though industry associations and the resurgence of more libertarian viewpoints in state and national governance bodies has rendered more difficult the withholding, or encumbering, of licences on such grounds. As the chapter examining groundwater in this volume points out, "the standard of legal causation required to assign liability in a judicial context may be heightened when compared with the statistical standards underlying scientific correlation" (Miller, Chap. 13). Like Miller, Collins and Rosen worry that the *aggregate* impact on groundwater of numerous independently considered operations seems sometimes to be lost in regulatory considerations of impact. Regulatory rollbacks emanating from the executive branches of the federal and state governments are further weakening the power of state authorities to limit shale gas (and other energy) operations on environmental grounds.

In the UK, where active commercial hydraulic fracturing is only just getting under way, Brown notes that estimations of water demand for hydraulic fracturing are quite variable, but that even high-end estimates potentially aggregate up to total water volumes of less than 1% of total abstraction for industry and agriculture. The national picture, however, may be less relevant than the catchment scale, where it is quite possible for the locally-required volumes to impact negatively on local water balance (Staddon et al. 2016). Moreover, as with Collins and Rosen's comments about Colorado and Texas, Brown recognizes the need for a regulatory approach that is flexible enough to accommodate local and inter-annual variations in water availability. In the U.S., this is greatly complicated by the "rigid" nature of water rights allocation, something the UK is attempting to address through a restructuring of the abstraction management regime. As Brown (Chap. 7, this volume) notes

> … from 2020 all existing licenses will shift to a new system of water abstraction permits. Under the new permit regime, the volume of water that can be abstracted (surface water or ground water) will depend on source availability. A new charging system will mean that water taken from high risk/low resilience sources will cost more, water abstracted from abundant sources will be less expensive.

It may be that a middleman market for third party water suppliers will open up, as it has in the U.S., especially given the commercial water market "opening" effected in England and Wales from April 2017.

China is also addressing the need to meet demand for water for unconventional production of fossil fuels, an issue explored by Shao et al. in their chapter in this volume. Though the shale gas sector in China is still in its infancy, the Chinese government has set aggressive goals for shale gas development, driven by growing energy demand and environmental challenges, including the need to reduce coal use (which China also has in abundance). The Sichuan, Tarim, and Junggar Basins are the largest in the country, but they are also regions with significant water shortage challenges (Figure 6.1). As in the UK, ownership of Chinese water resources vests in the state acting on behalf of the common good. Also like the UK, China is experimenting with more market-based water allocation mechanisms. Though it is still early days, the China Water Exchange (the "CWEX") is the mandated exchange platform where shale gas developers may seek to purchase the right to abstract water from transferors. As of 2018, only ten deals have been completed on the CWEX since its opening in mid-2016, and none of them is related to unconventional hydrocarbon exploitation.

As in the United States, and unlike in the UK or European countries, operators in China exhibit considerable secrecy with respect to quantitative and qualitative aspects of operation including the volume and constituents of hydraulic fracturing fluid and the management of flow-back and produced water. This relatively high level of commercial secrecy inhibits full public participation in operator licensing and water abstraction licensing decisions. Thus, it is perhaps not surprising that of those surveyed about the Weiyuan shale operation, "82% believed that the project would have a big or serious impact on water quality" (Shao et al., Chap. 6). Potentially at risk, even in the relatively authoritarian Chinese context, is the broader social licence to operate upon which smooth development depends.

In July 2011, Jorge Sapag, the governor of the western Argentinian province of Neuquén, inaugurated "the first multi-fractured horizontal well aiming at shale gas in Latin America" (Bernaldez and Rocio, Chap. 8). This was a provocative political act because hydraulic fracturing is quite controversial in Argentina, with more than 50 cities and districts banning it outright. Moreover, energy exploitation in parts of Argentina (as in Canada, Australia, and elsewhere) has an indigenous rights dimension to it, as indigenous communities in Argentina have long claimed territorial rights and denounced the environmental contamination caused by energy extraction and other heavy industries. Now, they also struggle against problems deriving from unconventional exploitation. In Bernaldez and Rocio's view, which may also apply to other case studies in this volume, debates about sustainable water management in unconventional gas exploitation typically fail to challenge the modernist discourse of environmental management that sees resources as put on earth for human use. In other words, there is no inherent problem in draining rivers dry because the alternative, as one Argentinian industry representative put it is that "95% of the water.... flows into the Atlantic Ocean" and is "lost" (Bernaldez and Rocio, Chap. 8). In Argentina, as elsewhere, such "hydromodernist" (Staddon and Langberg 2014) attitudes on the part of government are being challenged by a civil society that increasingly sees beyond the politics of borders and the economics of the short term.

As King's contribution to this volume shows, hydrocarbons other than shale gas can also be exploited by the technique of hydraulic fracturing. King focuses specifically on the implications of the shift toward deep 'tight oil' reservoirs—those located in low-porosity geologic formations at depths up to 4500 metres—for the environment and for the indigenous communities of central Russia. In his view, any promise of better environmental regulation that followed the dissolution of the Soviet system after 1991 was quickly squelched by a combination of state-led corruption and the new global politics of energy. Russia has, since the late 1990s, sought to use its vast energy resources as a source of both economic and political power, including to destabilise potential regional rivals by throttling back exports of energy or, as in Ukraine, by invading outright to compromise new energy supply regions (discussed by Mitryasova et al. in Chap. 10). Western diplomacy appears to have failed to address this dynamic inasmuch as the sanctions imposed by the U.S. and the European Union after the 2013 occupation of Crimea and other parts of eastern Ukraine have not only failed to alter the military-political balance but have also led to an intensification of efforts to accelerate energy production by hydraulic fracturing in other parts of the country.

An additional issue discussed in the Russia chapter involves the fate of indigenous communities, in this case the Khanty people of Siberia, in the face of state-sponsored development of energy resources. As King points out, the fates of the Khanty and other indigenous peoples in Siberia have mirrored those of indigenous people across the world's resource frontiers. In the United States and Canada, the recent campaigns against the Keystone XL oil pipeline and the Dakota Access Pipeline (DAPL) have been framed around issues of indigenous sovereignty over land resources as well as different ways of thinking about human-environment interrelations. Whilst news coverage has centred on police confrontations at sites such as the Standing Rock Sioux's Oceti Sakowin protest camp, these efforts have in fact entailed a multitude of non-violent, co-productive activities, beyond the kinds of civil disobedience that the Khanty have (so far) avoided. These included the creation of educational spaces in which collective cultural identities around water and stewardship are reaffirmed (Roumell 2018). Here, once again, we see that energy operators' social licence to operate is increasingly under challenge from ethno-linguistic communities that do not accept the state's claim to speak on behalf of all inhabitants.

Some of these more political issues are picked up in Mitryasova et al.'s chapter on Ukraine. With technically-recoverable resources that easily dwarf most other European nations, and with increasingly uneasy relations with its northern neighbours (especially Russia) after the 2004 "Orange Revolution", Ukrainian leaders had looked to the energy sector as a way of strengthening the domestic economy and reducing Russia's hold over the country. Even under the most pessimistic projections, Ukraine is ranked fourth in Europe in terms of shale gas reserves—after Poland, France, and Norway—with as much as 7 trillion m^3 of recoverable reserves. Yet two factors have seriously undermined development of these resources to date: concern over environmental consequences and the Russia-backed civil war in the eastern part of the country. As in other regions examined in this volume, the mere presence of significant shale gas reserves does not mean that there are also abundant supplies of other necessary resources, especially water. The central and eastern parts of Ukraine are already quite water stressed, and even in the wetter western Carpathian region there is growing concern over the impacts of shale gas extraction on the environment and water supplies for human consumption. On the geopolitical side, it appears that the Russian-backed rebels fighting in the Donetsk and Luhansk regions of eastern Ukraine (Fig. 10.1) have ensured that at least for the time being, Ukraine's shale gas resources will not be used to challenge Russia's energy dominance.

In the chapters comprising this section we have seen again and again that proving reserves of economically recoverable shale gas or tight oil is perhaps not even half the "battle" of achieving the energy security benefits described in the opening chapters (i.e. energy transition to a lower carbon economy, energy security in an increasingly unstable world, etc.). In many energy-rich regions there may not be sufficient water resources to exploit these reserves, and none of the jurisdictions so far discussed have really come to grips with the new regulatory challenges created. Moreover, civil society in general and indigenous communities in particular are increasingly vociferous and organised in their opposition to the hyper-modernist discourse promulgated by political leaders, who in their turn are becoming increasingly authoritarian. There are glimmers of hope that a new politics and economics of energy can emerge out of this complex situation, perhaps combining benefit sharing approaches (as currently trialled in western Canada, according to Curran's contribution to this volume) with market-based mechanisms for water and energy allocation where appropriate (e.g. China, the UK, the United States). Overall the goal should be to strengthen the social licence to operate through a politics of accommodation, rather than obfuscation and authoritarianism.

20.4   Disposing of Water Used and Produced Through Hydraulic Fracturing

The chapters in Part III of this book discuss issues related to the disposal and management of wastewaters composed of the fluids used for hydraulic fracturing and liquids produced naturally by the geologic formations. As noted elsewhere in this volume, as operators pursue the capture of hydrocarbons, the primary fluid by volume exiting a well—whether production is via conventional or unconventional methods—is, in fact, water. Managing these wastewaters is both difficult and expensive, and volumes have been increasing over time. The primary lessons here centre around the need to balance facilitating economic development and ensuring human and environmental health and safety. The chapters illuminate the complexity of produced water regulation, treatment, and disposal and consider some of the risks that accompany the challenge of dealing with the waste products generated by unconventional production, such as surface and groundwater contamination; potential exposure to hazardous substances, either injected into the well as a component of hydraulic fracturing fluid or naturally occurring in produced water; and inducement of earthquake activity in producing regions.

Webb and Zodrow review the variation in approaches pioneered by six U.S. states in which hydraulic fracturing operations are widespread, with particular focus on Texas and Pennsylvania. In the United States, the primary method of wastewater disposal is to inject it deep underground, either as permanent disposal or as part of enhanced oil recovery (EOR) efforts. The relative economy of disposal in this manner has provided a disincentive for operators to recycle wastewater from hydraulic fracturing operations, but disposal by injection well is highly dependent on geological suitability. Thus, in Pennsylvania, where the geology is ill-suited for deep well injection, recycling and reuse efforts have received more attention and resources. Texas, however, is thinking proactively and—despite leaning more heavily on injection wells for disposal—has loosened regulations and streamlined the permitting process for wastewater recycling. If drought in relevant regions of Texas returns and persists, this regulatory facilitation may bear greater fruit in the future.

Webb and Zodrow review a number of current and developing technologies in use for treating produced water and other wastewaters, noting that numerous factors contribute to the ultimate choice of technology, including the quality of the produced water, the intended use for the treated water, the scale and duration of treatment, and of course cost. Disposal by underground injection or enhanced oil recovery requires less treatment than non-industrial reuse, adding to the economic advantages. Recycling and reuse, however, may act to facilitate the social license to operate, as demand on local water supplies may be lessened. The authors suggest that states that wish to facilitate the uptake of produced water recycling and reuse could do so by streamlining the regulatory requirements on the issue in order to make recycling more economical but warn that protecting the environment and humans in the area must remain a key objective.

Taking a deep dive into coal seam gas (CSG) production in Queensland and shale gas exploration and production in Western Australia and South Australia, Hunter and Campin (2019) provide an accessible overview of the resources ripe for unconventional production in Australia and a broad picture of energy production regulation, before examining how the Australian regulatory system addresses the regulation of produced water and flow-back fluid. Like the U.S., regulation of onshore petroleum activities in Australia is the undertaking of individual states and territories due to the constitutional division of powers in Australia. Regulatory sophistication and development levels vary widely among the states, in accordance with the duration and scope of development historically in each jurisdiction. Though the national government does not regulate production activities, ministers of energy and natural resources have sought to provide guidance relating to development and address local community concerns via the creation of a non-legally binding overview of issues that regulators should consider when developing regulatory tools for coal seam gas (CSG). CSG is currently the sole unconventional resource under commercial development in Australia. Australian states have taken a number of approaches to regulating how operators manage dewatering, a process in which coal seams are depressurised by removing water from the coal measure and allowing the gas to desorb from the coal cleats and flow to the well.

Development of coal seam gas has in some cases encountered difficulty in obtaining and maintaining a social license to operate. CSG has been under development in Queensland since the mid-2000s and has drawn concern from

Queensland communities about the regulation and management of extractive activities, including over the use of water and the disposal of produced water from CSG wells. Spurred on by public interest, the government has taken an adaptive management approach to regulation of CSG production, which has resulted in thousands of regulatory amendments over the last decade, as landowners and communities have adjusted to the impact and consequences of CSG extraction. The Queensland Government has created a hierarchy of options for operators, encouraging beneficial use where possible, and then allowing for treatment and disposal in ways that minimize and mitigate harmful environmental impacts. In other places, such community concerns have led to different results. New South Wales enacted a 5-year moratorium on the development of CSG resources from 2011–2016, though exploration has resumed in the state in recent years. Victoria has permanently banned CSG development via legislation, the first jurisdiction to do so.

The lessons drawn by Miller (2019) also cluster around the need to balance economic use and environmental health and safety. In the context of groundwater, Miller considers the qualitative aspects of water use and the challenges of crafting regulation able to address the complexity and uncertainty of pathways for the potential contamination of groundwater. The threats to groundwater from unconventional operations are many. They include subsurface migration of methane, accidental surface spills, leakage of fracturing fluids, well-casing integrity, and water table interactions with produced water. Miller examines the disconnect between legal and scientific standards of causation and the uncertainty involved in assessing liability where groundwater has been contaminated. He considers the overlapping system of regulations and legal liability regimes used to protect groundwater resources, emphasizing that "regulatory oversight must be backstopped by liability and enforcement regimes, or the regulations will be ineffective". The ultimate take-away from Miller's review of the literature is that hydraulic fracturing operations, conducted in accordance with best operating practices are unlikely to lead to groundwater contamination, but he voices concern that (1) the aggregate impact of thousands of operations conducted by a plethora of operators (some of which will not follow best practices) may ultimately impact groundwater supplies, and (2) existing regulatory frameworks may not adequately account for or consider the implications of that aggregate impact. He notes that where risks are novel—perhaps because new technologies are still evolving or due to underlying scientific and legal uncertainty—best-practice regulations may be insufficient to protect humans and the environment. Exemptions in the law for unconventional operations exacerbate this concern. More recent studies have begun to acknowledge the potential threat of groundwater contamination (Hildenbrand et al. 2015; DiGiulio and Jackson 2016).

Miller also considers the potential groundwater contamination threat presented by unknown chemical components, the composition of which may be protected under U.S. law as commercial secrets. Here, the right of an industry operator to protect a trade secret is prioritized above the health and well-being of nearby humans or the environment. This protective mechanism in the law means that regulators and would-be protectors of groundwater such as environmental or community groups are forced to react to incidents of contamination after they occur, incurring the costs of forensic examination after-the-fact, rather than being able to understand and advocate for anticipatory regulation and management of potentially toxic chemicals.

Miller urges the implementation of "common sense," science-based solutions to regulate interactions between groundwater and unconventional oil and gas production, including aquifer-specific restrictions that limit the depth at which a horizontal well may be fractured and requiring groundwater sampling surveys prior to commencement of drilling. Solutions like these seek to balance the interests so often at odds, between economic advancement and the health of humans and the physical environment. As Miller notes, "[s]cience is on the cusp of understanding the dynamics of groundwater contamination in localized scenarios … but law and regulation maintain the tendency to lag behind." He attributes this lag largely to divisive rhetoric and agendas pushed by special interests, which have the effect of "compounding" the uncertainty that already characterizes the relationship between groundwater and hydraulic fracturing.

Ehrman (2019) also offers insights into how jurisdictions experiencing hydraulic fracturing are balancing the economic benefit of unconventional production with the risk of induced seismicity, often caused by deep-well injection of waste fluids. Induced seismicity is earthquake activity caused by human activities, including waste disposal via fluid injection, secondary recovery of oil, geothermal energy production, oil and gas extraction, reservoir impoundment, mining and quarrying; it is often identified by increased seismic activity relative to historical levels. Ehrman notes that induced seismicity has largely been an issue only in the U.S. and that efforts to understand and address it have focussed on Oklahoma and Texas.

Oklahoma and Texas have both experienced increases in earthquake activity that have forced policymakers to consider how to balance the concerns of property owners shaken by earthquakes and the needs of a booming oil and gas industry. Ehrman highlights efforts by regulatory agencies and scientists in Texas and Oklahoma to understand the geologic context of the states and to predict the effect of volumetric and pressure differentials on seismic stability. The two states' policy makers, however, have reacted differently to the findings of those researchers, with Oklahoma implementing industry-palatable but serious limitations on injection operations in certain regions of the state while Texas has made much more tentative efforts to address the issue. Authorities in other nations have been quicker to identify and accept a relationship between oil and gas wastewater injection and induced seismicity than were their American counterparts. Induced seismic events in the western Canadian provinces, however, appear to be caused primarily by the hydraulic fracturing process itself, rather than by injection of wastewater. Similarly, the hydraulic fracturing process has been determined to have prompted seismic activity in the U.K., and subsurface reservoir pressure and pressure differentials have prompted seismic activity in the Netherlands. Jurisdictions outside of the U.S. have shown themselves more inclined to act decisively on the issue, perhaps a feature of social priorities and national cultures around risk. The U.K. has suspended operations near to seismic events, and the Dutch government announced its intent to halt gas field production by 2030 to limit seismic hazard. Ehrman closes by noting that scientists and regulatory entities are working to understand the relevant geology and the potential effects of volumetric and pressure differentials caused by human activity and calls for operators to help reduce risk by using wastewater treatment technologies and recycling and, where viable, to minimize wastewater injection.

20.5 Regulatory Regimes and Issues: Regional Perspectives

The focus of Part IV is on macro-scale regulatory planning issues across a variety of countries. Key issues discussed include how countries are attempting to address the pursuit of national energy independence, growing civil society opposition to hydraulic fracturing and the need to consult affected communities, cumulative impact assessments, the creation of regional water plans, the operationalisation of sustainability in unconventional energy, and the need to fund effective regulation and enforcement. Indigenous rights and the social license to operate are also key concepts considered here.

The chapter by Curran on Canada deals with the right of First Nations to be consulted in relation to hydraulic fracturing. The 1982 *Constitution Act* and key Supreme Court decisions, such as *Delgamuukw* (1997), acknowledge and affirm certain a priori aboriginal rights to land and resources as well as treaty rights. First Nations are increasingly using these rights to force broader watershed planning and assessment of cumulative impacts, as well as to enforce their right to a share of revenues. In the 1980s and 1990s, battles over logging the remaining areas of old (first) growth forests in British Columbia were particular flashpoints as First Nations began to assert their hard-won rights to be heard and included in natural resource management. Contemporaneously, the Canadian environmental movement was forcing the federal government and some provinces (especially Ontario and British Columbia) to reconfigure resource management systems to incorporate a greater degree of local participation (dove-tailing with the demands of First Nations) and a heightened sensitivity to sustainability issues. Curran describes a case in the western province of British Columbia where plaintiffs, the Fort Nelson First Nations (FNFN), successfully sued to overturn a 2012 water management decision because the Assistant Regional Water Manager relied on incomplete and inadequate information when making the original decision. This placed the Tsea River watershed and FNFN's treaty rights at risk by failing to assess the potential cumulative impact on those rights of such a huge diversion. The decision-making process was harshly criticised as having ignored key ecological and hydrometeorological data, failing to attempt an evaluation of cumulative environmental impact, and failing to engage in meaningful consultation with local communities, especially the FNFN. These faults were critically important for communities in northeastern British Columbia, which is currently seeing a rapid expansion of shale gas exploitation. This case, however, seems exceptional, and Curran (Chap. 15) notes that:

Courts will rarely direct specific consultation and accommodation procedures, nor will they give substantive direction on acceptable impacts to a First Nation's traditional territory. With few substantive remedies or limitations on Crown approvals in traditional territories resulting from this procedural requirement, overarching provincial jurisdiction for lands and water continues, except in a few pockets, and development of natural resources continues apace.

Indeed, legal protections seem limited in practice, particularly in the western provinces where concentration of authority in non-elected statutory regulators (such as British Columbia's Oil and Gas Commission), and limited rights of appeal or challenge mean that the resulting regime is unfit for purpose and, in fact, further erodes public trust and the social licence to operate province-wide.

The chapter by Sanchez highlights the political hegemony of the United States in the Western Hemisphere and the pressure it has placed on Latin American states to adopt U.S.-style laws and regulatory approaches. However, this has not operated successfully, as the respective political systems differ between states as well as between the United States and the Latin American region more broadly. In 2010, the U.S. Department of Energy launched the Unconventional Gas Technical Engagement Program, conceived as a tool to share knowledge, including technical, regulatory, administrative and diplomatic expertise, between the U.S. and other countries to help them recover their shale resources (oil and gas). However, unlike in the U.S., natural resources in Latin America are often owned by the State and exploited through state-owned monopolies, which changes the constellation of actors significantly.

Argentina (also discussed in Chap. 8 by Bernaldez and Rocio) seeks self-sufficiency in energy and so passed the Law 24.741 to declare of "national public interest" the achievement of self-supply of hydrocarbons. In the region of Neuquén, the State has opted to pursue development with only limited environmental regulation. Sanchez, in his chapter, describes it thus:

While Argentina is performing hydraulic fracturing already, one weak but important factor to stress is that even though they have in place a national law for environmental protection in general, it seems not to have a hydraulic fracturing-oriented regulatory policy towards the prevention of environmental and social impacts.

Mexico is similarly keen to achieve energy independence and has, since 2013 and with U.S. encouragement, sought direct private sector investment. Counterbalancing environmental protection regulations were set up in 2014 within the framework of overall energy sector reform in Mexico. The National Agency for Industrial Safety and Environmental Protection (ANISPA in Spanish) was created to complement the capacity of the country to oversee and regulate industries and production standards. As in other jurisdictions discussed in this volume, the new regulatory structures have not been well-funded, nor have they been sufficiently open to public participation and scrutiny to build public trust. Therefore, civil society—including Mexico's indigenous communities—has, unsurprisingly, been especially active in campaigning against hydraulic fracturing.

Feris and Harding examine South Africa and note that as much as 20% of the country's total land area, especially in the drier southern parts including the Karoo, could be subject to shale gas development. South Africa's environmental regulatory regime is premised on a number of important principles as expressed in the country's framework legislation, the 1998 National Environmental Management Act (NEMA), as well as its water management legislation, the 1998 National Water Act (NWA). Of particular importance is the sustainable development principle, captured in sections 2(3) and 2(4) of NEMA. This principle is reiterated in the 2002 Minerals and Petroleum Resources Development Act (MPRDA) which defines sustainability as "the integration of social, economic, and environmental factors into planning, implementation and decision making so as to ensure that mineral and petroleum resources development serves present and future generations" (Feris and Harding, Chap. 17). The sustainable development principle provides the backdrop against which energy development applications must be considered, in particular in the context of impacts on water resources.

Poland has legislation in place to allow hydraulic fracturing, a primary objective of which is to serve the geopolitical aim of energy independence from Russia (the Ukrainian case, discussed in Chap. 10, shares a similar objective). The number of concessions for exploration of shale resources in Poland grew from a modest 11 at the end of 2007 to a peak of 111 in July 2012, and were sought by investors ranging from Polish domestic firms to international giants like Exxon Mobil and Chevron. As of 2019 that number stands at only 20 concessions, all held by Polish entities. Polish environmental protection laws and regulations were recently revised, especially through the overhauled *Water Law*, which entered into effect on January 1, 2018. The law fully domesticates into Polish statute the EC's Water Framework Directive (2000/60/EC) and Marine Strategy Framework Directive (2008/56/EC), which affect regulation of all surface and subsurface waters as well as marine waters out to territorial limits. A centralized state agency, Polish Waters, was created to monitor, manage, and enforce the new law, replacing the prior decentralized system of regional agencies. On a local level, the law assigns more responsibilities to local governments related to protection of local water sources, especially during times of natural disasters, but also strengthens local community rights to meaningful participation in environmental decision-making and re-states the environmental right to know.

However, the Polish case highlights issues related to uniform enforcement. Mikulska points out that whilst EC Directives must be integrated into member state national legal frameworks, application of the implementation principles of "proportionality" and "subsidiarity" mean that national transposition can significantly affect the shape and impact of intended environmental protections. According to the European Commission's report to the European Parliament and the Council on the effectiveness of the Recommendation 2014/70/EU, Poland has continued to apply its domestic rules, according to which a license to drill a well that does not exceed 5000 m in depth does not require a strategic environment assessment or environmental impact assessment (both which are the subject of a separate EC Directive (2001/42/EC, the Directive on Strategic Environmental Assessment).

In the final regional case study, Bittencourt and Meiler discuss the shortcomings of the Brazilian regulatory system. Brazil is self-sufficient in oil but not in gas and, like other states covered in this volume, views energy security in gas as a national priority. Legal reforms in the 1990s authorized the national government to license or contract with state-owned and private companies to perform activities related to oil and gas exploration and production and created a concession regime, generously underwritten by the possibility of state subsidies and indemnifications, to facilitate exploration and production of oil and gas in the country. The concession regime has been successful in opening the Brazilian energy sector to private or joint public-private enterprises. Although revisions in 2014 imposed new environmental management and compliance obligations on operators, there is widespread concern in Brazil regarding who has the authority to regulate production of unconventional gas in Brazil. While national legislation clearly applies when the resource is located in more than one sub-state unit, it is not at all clear which devolved state agencies are responsible in cases of purely intra-state operations. Such technical and geographical uncertainties, as well as a high level of public distrust in state institutions, has meant that attempts by operators to secure concessions and licences are now regularly met by civil claims, often successful, filed by local authorities and/or environmental pressure groups. Moreover, courts have vigorously protected the rights of Brazilian Indian communities, for example annulling state government concessions in Amazonia's Juruá Valley in March 2016. It is unclear what impact on this status quo may come from the new Bolsonaro administration elected in early 2019.

20.6   Future Research in the Regulation of Hydraulic Fracturing

Geographical, legal and cultural variations in the jurisdictions examined in this volume add significant complexity to understanding the regulation of an already complex topic like water security in the context of unconventional production of fossil fuels. Because the geographies, cultures and legal structures of each country differ, there are different regulatory challenges to be addressed—and different social and cultural environments within which to do so. No country has created a perfectly clear, efficient, and effective system for regulating unconventional hydrocarbons, though there are areas of relative success. Even if a perfect system did exist somewhere, it could be simply "cut and pasted" into other less fortunate jurisdictions.

In all jurisdictions, water allocation continues to be practised through a complex and evolving combination of state abstraction licencing, private appropriation (e.g., in the US, where in many regions riparianism applies), prior appropriation, and market trading. Which allocative mechanisms may be best or most efficient will depend as much on local geographical conditions as on prevailing ideologies about market versus state allocation. Also, climate change is rendering regional and national water balances increasingly precarious and volatile meaning that all locally-created allocation mechanisms need to be sufficiently flexible to deal with significant fluctuations in inter-annual water availability. How operators can or should access the large volumes of water necessary needs to be carefully re-thought, as does the management of wastewaters, including 'produced water' volumes.

Once an unconventional operator is up and running, enforcement of regulations regarding acquisition and disposal of the water and liabilities for regulatory breaches needs to be robust and transparent. It is notable, and regrettable, that as yet most of our case studies express rather weak and politically-compromised enforcement mechanisms. There is also an unfortunately high level of ignorance and/or secrecy around key issues like water abstraction management under drought conditions, impact on drinking water supplies, the regulation of hydraulic fracturing fluids, and the disposition of wastewaters. The determination of ways to limit environmental damage must be based on sound science and technical processes that conceive of safety broadly[5] and robustly, not only to respond to incidents when they occur but to prevent them by "failing gracefully". This will require better trained and supported regulators and probably a better appreciation of the distinction between "Safety 1" and "Safety 2".

In unconventional energy operations, "Safety 2" can be linked with adoption of now-common standards of environment management including sustainability, the precautionary principle, and the polluter-pays principle. It is encouraging to see these concepts embedded in more recent laws in the U.S., Australia, South Africa, and Canada. Making clear operators' responsibilities to human populations can also help, as in the use of the human rights-based approach discussed in Chap. 2. The human right to water and sanitation is promulgated under several (largely unenforceable) international instruments, including the International Covenant on Economic, Social and Cultural Rights, the Convention on the Elimination of All Forms of Discrimination Against Women, and the Convention on the Rights of the Child. Such "soft law" (see also Bradbury and Cox, this volume) needs to be translated into "hard law" either by individual nation states or, where the tranboundary nature of the resources concerned makes it easier, through international treaty-making. As with recent moves to standardize water footprinting methods, new ISO type standards can help (e.g. ISO 14046), creating as they do ways for third parties to audit and certify operations or governments. When adopted into national law these mark another step in the globalization of water (and energy) management discussed by Varady et al. (forthcoming). Very useful would be further research and advocacy around supra-national forms of governance and regulation. In the EU the system of centrally-promulgated "Directives" that are then domesticated into member states' national legal frameworks via implementing principles of "subsidiarity" and "proportionality" has proven quite useful and one wonders how well such a

---

[5] It is becoming commonplace in engineering to conceive of at least two types of safety, "Safety 1" and "Safety 2". The traditional view of safety, called Safety 1, has been defined by the absence of accidents and therefore promotes research into how things can go wrong and how things going wrong can be prevented through better systems design and operator training. A different perspective involves thinking about how systems can be designed with "Safety 2" in mind; that is to say the construction of socio-technical systems that not only achieve technical requirements but which, when stressed past tolerance, "fail safe".

system may apply to other "federal" type political systems. As noted above, bilateral or multilateral treaties may be useful, though these are as yet untried as mechanisms for regulating energy sector development. In the absence of such "hard law" approaches, scholars and regulators may elect to look at how far "soft law" can be further developed towards useful outcomes.

A number of lessons regarding the social licence to operate in relation to regulatory structures for hydraulic fracturing may also be drawn. The first lesson concerns the style of community consultation about development proposals. Our case studies, from contexts as disparate as Ukraine, South Africa, Argentina, the U.S. and China, suggest that where there are multiple epistemic communities involved (e.g. pre-existing energy interests, agricultural communities, municipal water supplies, indigenous groups, environmental groups, etc.), the approach to consultation needs to be especially robust. Consultation processes must address these communities in ways that are meaningful to the communities themselves, and not merely try to shortcut or force closure on public discussion (as has happened in the UK, according to the chapter by Brown). If this is not achieved, then the chapters in this book demonstrate that significant political protests may arise.

As illuminated by numerous cases in this volume, siloed government ministries and departments have struggled to achieve the regulatory coordination necessary to appropriately oversee and regulate unconventional energy operations. This is most obvious with respect to managing the role of water through the life cycle of hydraulic fracturing operations, as agency and industry jurisdiction over resources is fragmented in myriad ways. Tensions abound between the need to promote economic development to support energy demand and population growth and the imperative to protect and preserve the environment and human quality of life. These pressures are intensified when the geopolitical context creates additional pressure to achieve energy independence or otherwise threatens national security, and can override environmental or social concerns about pursuing unconventional hydrocarbons. This is especially the case in many Latin American and ex-Soviet Bloc countries.

Energy development can also have a dramatic and adverse impact on the water security and other rights of indigenous communities—impacts that are only partially and tenuously recognised by decisionmakers, even in jurisdictions, like Canada, where the rule of law is perhaps strongest amongst the countries and regions considered here.

Where then ought future scholarship on the social and regulatory dimensions of unconventional energy resources go next? We suggest three research themes that are especially important:

- Continued research, especially comparative research, on the management of water resources throughout the life cycle of unconventional operations. This work will necessarily be technical and dominated by scientists and engineers, but it is critical that these researchers operate in a transparent way and that they effectively communicate their research to broader publics who are often uneducated and confused about the pros and cons of unconventional energy and mistrustful of both the private sector and government.

- Innovation in the structure and functioning of modern regulatory systems is needed. The current and common siloing of regulation around stand-alone functions and sectors—such as agriculture, energy, and environment—is manifestly unsuited to an increasingly interconnected world. Lawmakers and decisionmakers should consider how best for regulators to collaborate and share information across sectors and bureaucratic divisions in order to facilitate better evaluation, monitoring and enforcement of required and best practices. What sorts of policy processes can be crafted that are at once multi-disciplinary and flexible, but also definitive?
- Research and advocacy around public consultation and the social licence to operate. Even the best science and engineering will be insufficient for generating social trust if affected communities continue to feel abused, ignored, or condescended to. Meaningful public participation is a staple of international standards for environmental impact assessment, but it is honoured more in the breach than the observance.

The chapters in this book offer a useful starting point for the above, but much more needs to be done. The comparative perspectives offered here will hopefully contribute to increased understanding and new ideas about how operators and regulators can innovate to achieve energy security while ensuring that water security and the human environment are not harmed.

References

DiGiulio DC, Jackson RB (2016) Impact to underground sources of drinking water and domestic Wells from production well stimulation and completion practices in the Pavillion, Wyoming, field. Environ Sci Technol 50(8):4524–4536

Hildenbrand ZL, Carlton DD Jr, Fontenot BE, Meik JM, Walton JL, Taylor JT, Thacker JB, Korlie S, Shelor CP, Henderson D, Kadjo AF, Roelke CE, Hudak PF, Burton T, Rifai HS, Schug KA (2015) Comprehensive analysis of groundwater quality in the Barnett shale region. Environ Sci Technol 49:8254–8262

Roumell EA (2018) Experience and community grassroots education: social learning at standing rock. New Dir Adult Contin Educ 2018(158):47–56

Smith DC, Richards JM (2015) Social license to operate: hydraulic fracturing-related challenges facing the oil and gas industry. Oil Gas Nat Resour Energy J 1(2):81–163

Staddon C (2010) Managing Europe's water resources: 20th century challenges. Ashgate Press, Farnham

Staddon C, Langberg S (2014) Urban water security as a function of the 'urban hydrosocial transition'. Environ Sci Water Secur 23(3):13–17. Available from: http://eprints.uwe.ac.uk/25128

Staddon C, Appleby T, Grant E (2011) A right to water – a geographico-legal perspective. In: Sultana F, Loftus A (eds) The right to water: politics, governance and social struggles, 1st edn. Earthscan, Abingdon, pp 61–77. ISBN 9781849713597 Available from: http://eprints.uwe.ac.uk/18709

Staddon C, Hayes ET, Brown J (2016) Potential environmental impacts of 'fracking' in the UK. Geography 101(2):60–69. ISSN 0016-7487 Available from: http://eprints.uwe.ac.uk/28892

Varady RG, Albrecht TR, Staddon, C, Gerlak AK, Zuniga-Teran AA (forthcoming) The water security discourse and its actors. In: Springer handbook of water resources management. Springer

Chad Staddon is Professor of Resource Economics and Policy in the Department for Geography & Environmental Management at the University of the West of England, Bristol. Chad's research revolves around the social, political and economic issues related to sustainable resource management. Current projects focus on the historical geography of urban water systems around the world, water-energy trade-offs in unconventional oil and gas operations, and economic policy for resilient urban water services. He received his PhD in geography from the University of Kentucky in 1996 for research on the political economy of water (mis)management in post-communist Bulgaria.

Regina M. Buono is a nonresident scholar at the Center for Energy Studies (CES) at Rice University's Baker Institute for Public Policy and a doctoral student in public policy at the Lyndon B. Johnson School of Public Affairs at The University of Texas at Austin. She previously served as the Baker Botts Fellow in Energy and Environmental Regulatory Affairs at CES. Prior to that, she was an associate with McGinnis, Lochridge & Kilgore, LLP, in Austin, Texas, focusing her practice in the areas of water, administrative, and endangered species law. Buono has also worked in various roles with the Texas legislature and as a consultant to oil and gas companies, designing a habitat credit exchange to achieve compliance with the U.S. Endangered Species Act. She holds bachelor's degrees in international relations, political science, and Spanish from the University of Arkansas, a J.D. from The University of Texas School of Law, and an M.Sc. in water science and governance from King's College London.

Elena López Gunn is Founder and Manager of ICATALIST, an associate researcher in the water group at the University of Leeds, UK, and an associate professor at IE Business School. She has more than 15 years' experience in projects and publications on a wide range of subjects, mainly related to innovation, water governance, agriculture, adaptation to climate change, hydrological planning, partnership models, public policies, sociological analysis on environmental issues and knowledge management. López Gunn has led the elaboration, management, and coordination of various applied research projects in the Horizon 2020 and LIFE programs of the European Commission. She holds degrees in economics and social studies from the University of Wales, a Masters in environment and development from the University of Cambridge, and a PhD in geography from the University of London.

Professor Jennifer McKay is Professor of Business Law at the Law School at the University of South Australia Business School. She has conducted research supported by government and the private sector on sustainable development of freshwater and the development of governance models. This work has been undertaken in Australia, India and the US. The governance models have been developed for urban and rural water and she has made law reform suggestions, appeared before State and Commonwealth Committees and toured and given presentations to state agencies in California and Utah whilst on a Fulbright senior Scholarship to Boalt School of Law at the University of California, Berkeley. Prof McKay has a BA Hons and PhD from the University of Melbourne where she used social science methods to research water and environmental management, an LLB from Adelaide, GDLP UniSA and Diploma in Human Right law from American University in Washington DC.

She has received local recognition for her work and is proud to serve as a sessional Commissioner on the Environment, Resources and Development court of South Australia. Her TedX talk in 2019 was entitled "Duty to Cooperate: Making Messy Mosaic Laws into Jigsaw Laws to manage the environment in a sustainable way". https://www.youtube.com/watch?v=qeGsWmu0RPY

Printed by Printforce, the Netherlands